The Ever-Changing View

A History of the National Forests in California, 1891-1987

By
Anthony Godfrey, Ph.D.

July 2005
R5-FR-004
USDA Forest Service
California

Copyright: When a "source" is indicated beside photographs or graphics in this book, a copyright exists, and therefore, that item may not be used or reproduced in any manner apart from this book unless a separate approval is acquired. Any photographs or graphics that do not show a source are from the USDA Forest Service archives and are part of the public domain.

Cover photo-illustration was created from the Forest Service archive, image # 46977.

First published by the USDA Forest Service, Pacific Southwest Region on July 1, 2005

ISBN 1-59351-428-X

Also printed by the U. S. Government Printing Office as a sales product. The Superintendent of Documents requests that any reprinted edition clearly be labeled as a copy of the authentic work and assigned a new ISBN.

Agency logos are protected by the Code of Federal Regulations (36 CFR Ch. II, Part 264, Subpart A) and may not be used without written permission.

The U.S. Department of Agriculture (USDA) prohibits discrimination in all its programs and activities on the basis of race, color, national origin, gender, religion, age, disability, political beliefs, sexual orientation, and marital or family status. (Not all prohibited bases apply to all programs.) Persons with disabilities who require alternative means for communication of program information (Braille, large print, audiotape, etc.) should contact USDA's TARGET Center at 202-720-2600 (voice and TDD).

To file a complaint of discrimination, write USDA, Director, Office of Civil Rights, Room 326-W, Whitten Building, 14th and Independence Avenue, SW, Washington, DC 20250-9410 or call (202) 720-5964 (voice or TDD). USDA is an equal opportunity provider and employer.

Foreword

*T*o view California on a map is to see a state that is both long and wide. It is therefore appropriate that this book takes both the long view and the wide view of the Forest Service in California. The story begins long ago, before European settlement, and covers the wide range of natural and human events that have shaped both the landscape and its people. One quality that characterizes both of these views is a great capacity for change. Although this volume examines a thin slice of California's history, the national forests, it captures an essential aspect of the Golden State – the ever-changing view.

This ever-changing view defines the Pacific Southwest Region of the Forest Service, which manages 20 percent of California's land and supports cooperative programs in Hawaii and the U. S. Affiliated Pacific Islands. As Regional Forester, I am privileged to work with more than 8,000 dedicated employees who are heirs to the rich tradition described in this book. We owe a huge debt of gratitude to those who came before us, for their hard work, sacrifice and commitment to caring for the land. Today, we continue that tradition of stewardship on behalf of the people who own the public lands. To be effective stewards, we are becoming a workforce that increasingly represents the diversity of views and backgrounds that make California unique.

We manage a landscape in transition. Earthquakes and volcanoes, wildfires and landslides, along with water projects and roads, agriculture and urban development continually transform the physical and social outlook of the region, each on its own time scale. It is a landscape of change.

Its long north-south axis makes California the most geographically diverse among the fifty states. Traveling through the national forests here, one may view desert, chaparral, oak woodland, mixed conifers, redwoods, alpine meadows, and granite peaks. Across its width, California's public lands stretch from sea level to the crest of the Sierra Nevada Range, contributing even more diversity, including the lowest elevation (Death Valley) in the nation and highest elevation (Mt. Whitney) in the lower forty-eight states. Public lands support the largest trees (giant sequoia) and the oldest trees (bristlecone pine) on earth. The terrain is steep and flat, barren and luxuriant, crowded and remote. The climate is hot and dry, wet and cold, and perfect year-round.

This landscape, both welcoming and forbidding, has shaped the near-mythic image of California. Visitors from other countries are astounded to discover how many magnificent natural features are easily available in public ownership, a benefit that Americans sometimes take for granted. It remains, in Dr. Godfrey's term, an Eden-like garden, although many residents find this more a dream than a reality.

Change in California is also defined by motion – as in cars and movies. It is the place where all road trips seem to begin or end. Motorized access to the national forests has shaped the land while it has formed our experience of nature. The ever-changing view of California's national forests can be found along its scenic byways or around Lake Tahoe or in the backdrop for countless motion pictures.

The ever-changing view is found in the evolving attitudes and ideas that Californians have about their land. California's perspective has been shaped by restless adventurers and daring immigrants, by Russians, Spaniards, Chinese, Mexicans, and, in more recent times, by people from countries all over the world who have become Americans.

From the Gold Rush to the real estate boom, people have staked their claim to fortunes from California land. This, as Dr. Godfrey explains, has often caused damage to the natural landscape. On the other hand, California's magnificent scenery has inspired many strong supporters of preservation and environmental protection to become major figures in the history of conservation. California's national forests are a result of these ever-changing and often competing views. Today, the rapidly expanding urban areas are crowding the relatively untrammeled national forests and, for some, the view of paradise is from the parking lot.

The Pacific Southwest Region has been on the leading edge of many changes in the Forest Service, sometimes on its own initiative and sometimes pushed along by outside forces. Dr. Godfrey provides ample documentation for both cases.

As managers and users of the national forests, it is important to learn from this process of transformation. We need to see that what holds true today will probably not be so in the future. While we are guided by basic principles, we must be open to the changes brought to light by the ever-changing view. I hope that this history will serve as a useful guidebook for the road ahead.

JACK BLACKWELL
Regional Forester (2001-2005)
Pacific Southwest Region
USDA Forest Service

Acknowledgements

As with any project of this scale, there are countless people and institutions to thank for their contributions. First and foremost, I would like to extend my deep appreciation to Steven Dunsky of the Public Affairs and Communication Staff of the Pacific Southwest Region, who served as the contracting officer's representative for this endeavor. He diligently and thoughtfully guided me through this project by offering collegial advice during various stages of the research and writing, by opening many doors within the Pacific Southwest Region for my research, by providing significant resource materials at critical times, including oral histories conducted for the documentary film *The Greatest Good: A Forest Service Centennial Film* and for the Pacific Southwest regional retiree oral history project, and by reviewing successive drafts of the manuscript, all the while providing excellent observations and remarks along the way.

Additionally, I would like to acknowledge Jack Blackwell, the regional forester, who initiated and strongly supported this project from beginning to end; Dave Allasia, the contracting officer, who shepherded the contract through the process of expectations, schedules and final products; Donna Dell'Ario, regional publications manager, who thoughtfully managed the design and printing of the manuscript; audiovisual assistant Mario Chocooj, who tirelessly prepared the photographs for the publication;, and Daniel Spring and Richard Spradling for their accurate work on the maps for this book.

While in its draft stages, this manuscript was also read and commented upon by many thoughtful individuals at the request of the Forest Service. Direction came from a variety of Forest Service personnel and retirees during various phases of the study. Forest Service readers and commentators included Gerald W. Williams (Forest Service national historian), Linda Lux (regional historian), Judy Rose (heritage program leader), Mike Chapel (Sacramento representative) and Robert Swinford (special assistant to the chief). Bob Pasquill (Region 8) made valuable contributions regarding Forest Service New Deal history, and Pam Conners (Stanislaus National Forest historian) made lengthy, diligent and discerning annotations pertaining to all of the draft chapters. Additionally, many retired regional Forest Service personnel supported this project. Forest Service retiree Robert Cermak generously contributed his time and support for this study by supplying important resources on the fire history of the region and other subjects and by extensively reviewing the outline and draft chapters. Furthermore, Douglas R. "Doug" Leisz (retired regional forester 1970-1978) provided a lengthy and perceptive review of many of the draft chapters, while Joe Flynn, another Region 5 retiree, contributed a review of the initial chapters. Comments from the above individuals helped the author achieve historical accuracy in presenting the facts and events, and in making interpretations pertaining to the history of the Pacific Southwest Region based on clear evidence. All comments on content and style from the above persons were critical, but good-natured, and the author graciously adopted them whenever and wherever they were appropriate.

The author also solicited several peer reviews for the study, which need to be mentioned as well. Hal Rothman, chair of the history department, University of Nevada, Las Vegas, and former editor of Environmental History, and James Lewis of the Forest History Society, Durham, North Carolina, and author of *The Greatest Good: A Forest Service Centennial Film*, provided rewarding analysis and encouragement for the study. Finally, material for the epilogue came from a number of individuals, including Rick Alexander (deputy director of public affairs), John Fiske (retired regional program manager, reforestation and timber stand improvement), Bob Harris (retired forest supervisor), Vicki A. Jackson (associate regional forester), Matt Mathes (regional press officer), Larry Ruth (Center for Forestry, University of California, Berkeley), David Reider (Public Affairs retiree), G. Lynn Sprague (retired regional forester—1995-1998), Ronald E. Stewart (retired regional forester—1991-1995), Paul F. Barker (retired regional forester—1987-1990) and Steve Dunsky (Public Affairs and Communication Staff).

There are also my debts to various institutions and repositories for their help. They include Juli Hinz, assistant director for public services of the Marriott Library, University of Utah, Salt Lake City, who provided research support facilities to the author to conduct secondary publication research. I also wish to thank Sara Garetz of Pacific Southwest Library and Information Center, and Linda Lux, regional historian in charge of the Pacific Southwest Archival Center for their assistance in locating and photocopying much of the original research for this project in their repositories. Many thanks must also go to National Archives and Records Administration (NARA) Pacific Region archivists John Hedger and Patti Bailey (San Bruno) and Lisa Gezelter (Laguna Niguel) for their commitment and knowledge regarding the holdings of Record Group 95 (Records of the Forest Service) within their repositories and their kind patience in working alongside me during my many weeks of research at these repositories. Their assistance certainly saved many hours of research labor.

Others to whom I owe particular appreciation for aid in preparing this book are my gifted editor Roy Webb, a long-time colleague and friend, who provided significant counsel and timely but necessary stylistic admonitions; the president of WordCraft, Inc., and fellow native New Yorker, Mim Eisenberg, an astute and very competent copy editor; and the staff of U.S. West Research, Inc. (USWR) for the long hours they put into this project, especially Jake "the dog" Mogelthorpe. Needless to say, I accept full responsibility for the final product, including any biases, errors and omissions.

ANTHONY GODFREY, Ph.D.
U.S. West Research, Inc.
Salt Lake City, Utah
April 2005

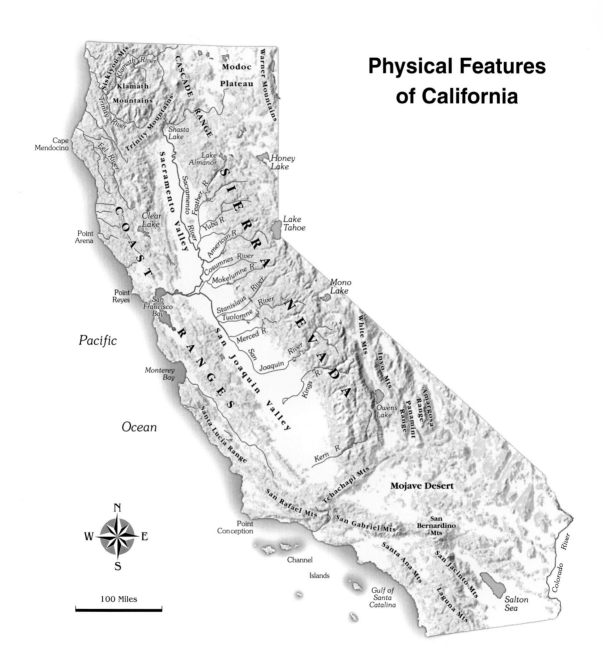

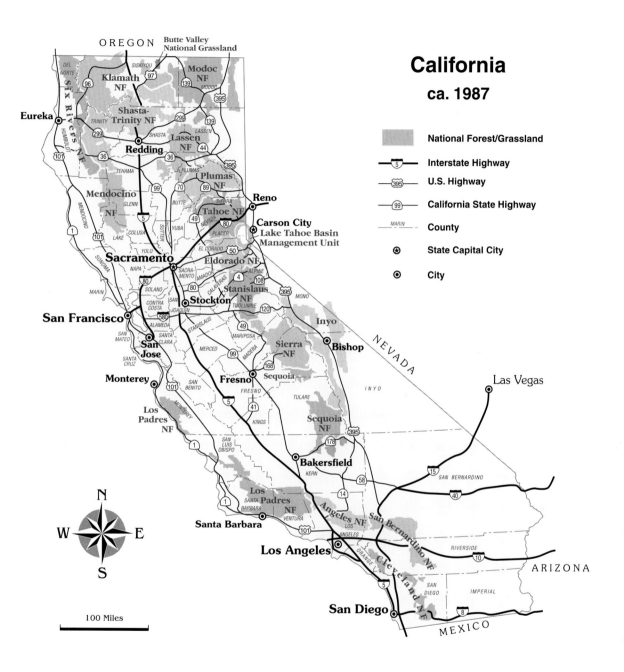

ix ■ Introduction

Introduction

This study is a history of the management of the USDA Forest Service Pacific Southwest Region from the late nineteenth 19th Century to 1987. Based in the State of California and the Pacific Islands, the Pacific Southwest (PSW) Region – historically referred to as District, and later, Region 5 – includes seventeen national forests: Angeles, Cleveland, Eldorado, Inyo, Klamath, Lassen, Los Padres, Mendocino, Modoc, Plumas, San Bernardino, Sequoia, Shasta-Trinity, Sierra, Six Rivers, Stanislaus and, Tahoe; the Lake Tahoe Basin Management Unit, and the Placerville and Humboldt nNurseries and the Chico Tree Improvement Center.

This publication's goals are fourfold. First, this study provides accurate information about Forest Service resource management practices and policies over the last century. Forest Service efforts in managing the diverse natural and cultural resources during the last hundred years are very relevant to today's management problems because the specific events of that history, as well as the Forest Service's reaction to those events, have shaped the region's past and will shape its future.

Second, today's managers need to be aware of the evolution of the conservation history of the region – especially how conservation was defined at the turn of the twentieth century – and what role the PSW Region played in that evolution. Improved conservation awareness thus becomes the second goal of this project.

A third goal is to demonstrate how history provides long-term data that is needed for current crucial decisions. It is to be hoped that this study encourages Forest Service employees to reflect on the development of past PSW Regional policies in order to better understand today's policies. The events that took place in the past clearly relate to many current Forest Service issues, and past resource management decisions have affected both the people of California and the management of the state's national forests.

Fourth, and perhaps foremost, this study was sponsored by the Forest Service to celebrate the 2005 "New" Century of Service – the 100-year anniversary of the creation of the Forest Service. Nationally, this event is being acknowledged and celebrated by a number of events. On January 3-6, 2005, the Forest Service held a Centennial Congress in Washington, D.C., reminiscent of a similar such congress held in 1905. The 1905 American Forest Congress (AFC) was held in Washington, D.C., as well, where almost 400 executives from the timber, railroad, grazing, irrigation, and mining companies, as well as educators, government officials and members of Congress, and foresters from across the country, attended the historic session. Like the 1905 Congress, the 2005 Centennial Congress sought to identify current and future challenges and opportunities for the future of the agency. The PSW Region and the Pacific Southwest Research Station participated in this event by forwarding recommendations from a regional Centennial Forum on California, Hawaii and the Pacific Islands held in Sacramento on November 5-6, 2004. Nationally, the Forest Service produced and released a major film on Forest Service history entitled *The Greatest Good: A Forest Service Centennial Film*, along with a companion volume on the subject. Regionally, this history was written to observe and celebrate the 2005 "New" Century of Service, and its intended audiences include Forest Service employees, other federal agencies, historians, researchers, and the interested public.

Research for this history was conducted primarily in a number of important Forest Service-related government and archival collections. Important primary materials for this study were found at the Pacific Southwest regional library and archives, which holds a tremendous amount of valuable materials. These range from secondary and primary documents related to major management themes in this study – timber, range, minerals, water, fish and wildlife, recreation and fire – to regional-related materials such as annual reports, investigations, and various newsletters, including the *California Log*, to a good oral history collection. A vital and comprehensive historic photograph collection is located there as well. Additional research was also conducted at the National Archives and Records Administrations (NARA) facilities at San Bruno and Laguna Niguel. Located near San Francisco, the San Bruno NARA facilities contain the records of the PSW Region Office (Record Group 95), San Francisco, from 1870 to 1985. This enormous collection (515 cubic feet) contains research material, including publications, reports, memoranda, letters, and clippings, relating to administrative matters and regional planning, the management themes of this study, and the affairs of eleven national forests – Klamath, Lassen, Mendocino, Modoc, Plumas, Sequoia, Shasta-Trinity, Sierra, Six Rivers, Stanislaus, Tahoe, and the Lake Tahoe Basin Management Unit – that comprise the northern portion of the PSW Region. Similar materials documenting the administration of the southern portion of the PSW Region, the Angeles, Cleveland, Los Padres, and San Bernardino National Forests, can be found at NARA's facilities located at Laguna Niguel,

California. Another important research opportunity that presented itself at this time was a series of oral histories collected for the filming of *The Greatest Good*, and a series of oral histories conducted with regional retirees by other retirees with the support of the regional office.

In the initial stages of writing, the author and the Forest Service determined that a chronological, issue oriented, narrative approach should be taken in presenting the history of the region, along with a conservation theme to unify the work. The book is divided into eleven chapters, an epilogue, and appendices. The first two chapters briefly explain the early history of California's forests and how they came under federal control. The next nine chapters provide a descriptive and analytical look at the rise and development of the PSW region, concentrating on the national forests of California from 1905 to 1987. These chapters follow natural historical time periods in American history, such as the Progressive Era, the New Deal, World Wars I and II, and they closely chronicle the history of the PSW Region's first ten regional foresters and the issues the region faced. Each chapter begins with an introductory section on California history to set the stage. The closing date of 1987 was selected because in a span of next eighteen years the PSW Region had five different regional foresters, making the history of the PSW Region less clear. The study ends with a number of features to help the reader. They include a short epilogue, which summarizes some of the significant events that occurred in the Pacific Southwest Region from 1987 to the present, and three appendices, which provide a timeline (1540-1987) of important dates in California and Forest Service history for the reader, a list of regional foresters (1905 to present), and a list of California district/regional office locations (1905-present).

While researching and writing this study, the manuscript underwent an extensive and thorough Forest Service and outside peer-review process to ensure that professional standards were adhered to in achieving historical accuracy in the presentation of facts and events. This review process ensured that interpretations were based upon evidence, and speculations were advanced as such. The Forest Service had final editorial control, in order to correct errors of fact or omission that might have escaped the review process, but the abundance of material analyzed by the author came from Forest Service records and oral histories. The reader should therefore consider the fact that depending on a single source has the potential for inherent bias.

Table of Contents

Chapter I: Prehistory to 1889
California's Forests Prior to Federal Controls 1
Picturing California Prior to American Control 1
American Conquest of California's Land and Resources, 1848-1891 9
California Government and Forestry Conservation 19
Conservation and the Lost Eden (Part I) 22

Chapter II: 1890-1904
California's Forests Conserved Under Federal Control 27
Turn-of-the-Century California 27
California Conservation and Preservation Movements Meet 31
Genesis: 1891 Forest Reserve Act and California's First Federal
  Forest Reserves 36
Administrative Beginnings and Growth of California Federal Forestry,
  1891-1896 41
Department of Interior Management of California's Forest Reserves,
  1897 to 1904 52
Conservation and the Lost Eden (Part II) 57

Chapter III: 1905-1911
Rise and Early Development of National Forests System 63
Progressive California and the National Forest 63
Legacy Year of California Conservation – 1905 68
Initial Organization of Forest Service 74
Further Expansion/Reorganization of California Forests – 1907 84
Creation and Initial Organization of District 5, 1908 88
Early California National Forest Issues and Problems 91
Cooperation and the Changing of the Progressive Guard 109
Fractionalization of the Conservation Movement 112

Chapter IV: 1911-1918
California National Forest System Grows And Goes To War 115
California's Romance with Progressivism and Wheels 115
New Leadership: Coert DuBois Administration (1911-1919): 118
Growing Multiple Use on California District 5 130
World War I and Its Impact on District 5 152
End of War and Reconstruction Questions 157
War and the California Conservation Movement – A Prelude 159

Chapter V: 1919-1932
Maturation of District 5 to Region 5 and the Great Depression, ... 161
The Red Scare of 1919 161

The Big Money: The Rise of Southern California 163
The Great Depression 166
Salaries and Working Conditions – 1919 168
Active Custodial Care of California Forests, 1920-1925 180
Active Administration and Custodial Care of
    California Forests, 1926-1929 201
District 5 Becomes Region 5 220
Great Depression and Region 5, 1929-1932 227
Custodial Care of California's National Forests and Conservation 231

Chapter VI: 1933-1941
A New Deal for Region 5, ... 233
A New Deal for California 233
A New Deal for Region 5 237
Region 5 and Roosevelt's Peacetime Army – The CCC 239
Region 5 and Other New Deal Agencies 246
Rise of the California Forestry New Deal, 1933-1938 248
Multiple-Purpose and Multiple-Use
    Management of Region 5, 1933-1938 257
Waning and Demise of the California Forestry New Deal, 1939-1941 289
On the Eve of World War 294
Conservation Anniversary 295

Chapter VII: 1941-1945
Region 5 At War 297
California At War 297
Region 5 Publicizes the War Effort 301
Defense and War Actitivies of California Region 5 303
Other Actitivies of California Region 5, 1941-1945 320
The Demise of Pinchot Conservationism 325

Chapter VIII: 1946-1954
Golden State of Managing Growth and Multiple Use 327
Postwar California 327
The Late Forties and Region 5 Postwar Inspections 329
The Early Fifties and Multiple-Resource Area Planning 337
Conservation and Multiple Use of Region 5, 1946-1955 344
California: Bellwether for Conservation 372

Chapter IX: 1955-1967
Programmed Multiple Use *Maximus* 375
California: Conformity to Conflict 375

Programmed Multiple Use – The Eisenhower Years, 1955-1959	380
Programmed Management of Multiple Use, 1955-1960	387
Maturation of Multiple Use – The Kennedy Years	401
Multiple-Use Management Guidance, 1960-1963	408
Multiple Use Maximus – The Johnson Years	412
Conservation Broadens to Environmentalism	426

Chapter X: 1967-1978

Region 5 Conservation Contested	**429**
California: A Not So Golden State	429
Conservation Meets Environmentalism	432
Environmental Management – The Nixon Years	440
Conservation Works with Environmentalism	451
Environmental Management Interruptous, 1970-1978	465
California's National Forests: Today and Tomorrow	484

Chapter XI: 1978-1987

Recommitment And Roots Of Ecosystem Management	**487**
California: Eden Or Wasteland?	487
A Problematic Decade Ahead	492
Recommitment and Roots of Ecosystem Management, 1978-1987	507
Roots of Ecosystem Management	523

Epilogue	**525**
Appendix A	**533**
Appendix B	**548**
Appendix C	**549**
References Cited	**551**
Index	**605**

Maps and Charts

Map 1. Physical Features of California	viii
Map 2. California ca. 1987	ix
Map 3. Region in California 1904	26
Map 4. Region in California 1907	64
Map 5. Region in California 1911	114
Map 6. Region in California 1919	162
Map 7. Region in California 1955	374b
Chart 1. Multiple Use and Sustained Yield Management Act	415
Chart 2. Timber Cut 1908-2004	493

Chapter I: Prehistory to 1889

California's Forests Prior to Federal Controls

Picturing California Prior to American Control

"Eden" is a metaphor sometimes used by historians to describe the pre-European California environmental setting. For instance, "Elusive Eden" and the "Contested Eden" have recently appeared in titles of articles and books to describe pre-contact California (Gutierrez and Orsi 1998; Rice, Bullough and Orsi 1996). Alluding to the biblical Garden of Eden, they explain pre-colonial California in terms of a state of perfect happiness, a delightful garden-like region, and an environmental paradise before the "Fall." Much like the story of Adam and Eve in the Book of Genesis, the decline is attributed to a "Serpent in the Garden." The despoiling of California, according to them, came with the westward movement of non-Native Americans into the setting and the "Legacy of Conquest" we left behind (Limerick 1987). Today this idyllic, romantic and static image of early California is more or less a value-laden reflection and yearning for a simpler pre-industrial time, rather than an accurate depiction of the past (Rolle 1973: 111).

Certainly California has environmental qualities that many find agreeable. Arriving in California from almost any other region of the country, one does seem to be entering into the land of Paradise. Its diverse regions, rich vegetation, wonderful wildlife and mostly Mediterranean climate (cool, wet winters and hot, dry summers) draw millions of visitors each year – many of whom return thereafter to reside. Home of many natural wonders, from the High Sierra to the coastline and redwood forests, California has a beauty that has enticed many to this state "west" of the American West. "Nowhere on the continent," said Wallace Stegner, "did Americans find a more diverse nature, a land of more impressive forms and more powerful contrasts, than in California" (Rowell 2002: 52).

When Europeans arrived, they found not only an inhabited land but also one that had already been altered by the hands of humankind. Archaeologists believe that humans moved into California some 15,000 years ago. Starting around then, diverse groups of prehistoric California peoples and their descendents thrived in this land of natural wonders. In time, they adapted, used, and impacted this "wild" Eden from surf to mountains prior to European arrival and the so-called "Fall." In fact, they most likely occupied, and used, every square mile of land that would eventually become part of the California National Forest system. Pre-European California has been described as a "world of balance and plenty" compared to today's post-industrial impacted environment, but it was hardly a world untouched by human hands. What we culturally value today as the remains of past wilderness once "harbored human gatherings and hunting sites, burial grounds, work sites, sacred areas, trails, and village sites" (Gutierrez and Orsi 1998: 12-14). At the time of contact, Native Californians spoke more than 60 mutually unintelligible languages, giving later scholars a convenient means to distinguish the various tribal groups. Penutian speaking groups, who apparently arrived late and pushed earlier Hokan speakers to the surrounding mountainous regions of the coastal ranges and northern mountains, occupied the Central Valley and Sierra Nevada. Both groups practiced intensive acorn harvesting. On the other hand, a mix of language families was found on the northwest coast of California, but with a fairly similar culture strongly focused on salmon harvesting. Finally, Uto-Aztecan speakers occupied the desert areas east of the Sierra and in southern California. They practiced what anthropologists call diverse hunting and gathering subsistence patterns (Rose, 2004; Heizer 1978).

Whereas it was once thought that Indians left "few marks on the land," "lived in harmony with the environment and each other" and "passed the land on to their successors unsullied and undamaged" (Caughey 1973: 3), these

Hive-like Indian acorn caches, Sierra National Forest

naïve views can no longer be supported. Native Americans clearly harvested and gathered California's flora and fauna bounty and fought to defend tribal hunting and gathering territories from intruders. Life was sustained by living off the land, and Native Americans flourished in this environment. But whether we speak of the coastal Indian peoples of California, who were dependent on a variety of marine resources, or the desert hunting and gathering groups of the interior regions of Southern California, they significantly restructured the natural environment to their needs and seasonal land patterns.

The clearest example of Native Californians restructuring "Eden" to their needs is their cultural pattern of repetitive land burning. Burning was the most effective environmental manipulation technique employed by California Indians, although some early scholars argued that native peoples did not deliberately and systematically set fires to mold their environment to their needs. One writer speculated that areas around Indian villages were periodically burned of vegetative cover only to guard against the approach of enemies (Brown 1945a: 15; Clar 1959: 7-11; Cermak n.d.: 16-17). But more recently, several anthropologists and ethnohistorians have irrefutably argued that Native Americans did use fires as a land management tool. They set fires to influence vegetative patterns in order to benefit their proto-agricultural economy, and for many other reasons. Recent scholarly opinion affirms that indigenous burning practices on grasslands and forests, not only in California but also worldwide, were integral to the maintenance of aboriginal economies (see, for instance, Lewis 1973; Pyne 1982; Blackburn and Anderson 1993; Stewart 2002).

In California, Indian groups certainly used burning as a means to enhance plant and animal resources. For instance, some scholars suggest that Indians applied burning techniques to encourage the growth of annual grasses at the expense of perennial plants to provide "supplementary foods as well as browse to keep deer, antelope, and rabbit populations at a high level." They also suggest that fires purposely set by Indians constituted a form of game management or incipient herding of game such as deer and rabbits (Lewis 1973: xxi and xxix). Native Californians' burning practices created forests that they could exploit more easily. Unlike today, fires were not seen as a destructive force threatening valuable tree resources, protective watershed, wildlife and homes, but as a way of modifying the land to benefit their aboriginal lifeways. Whether this practice of "light burning" upset the "pristine balance of nature," making Native Americans the first to disrupt the "Garden of Eden," or whether or not it left permanent marks on the environment is still unanswered (Stewart 2002: 38-39).

According to the Spanish novelist Montalvo, California lay "on the right hand of the Indies" and "very near to the terrestrial paradise" (Watkins 1973: 234). In 1540, thirty years after Montalvo's description, the "Serpent" officially entered in the "Garden" when Melchior Diaz crossed the Colorado River near Yuma on his trek across Arizona. Diaz became the first European to visit what was thereafter named *Alta California*. Several years later, Joao Rodriquez Cabrillo sailed up the Baja peninsula to explore *Alta California's* coastline, reaching San Diego harbor and northward, and landing long enough on the Santa Barbara coast to plant the Spanish flag as a token of conquest (Brown 1945a: 15-16). Needless to say, coastal and central Southern California natives, divided into small remote bands and unaccustomed to organized warfare, were no match for the Spanish invasion, and for the most part they submitted peacefully to Spanish dominion. Eventually, the Spanish established a mission system to harvest the souls of Native Californians of *Alta California* and resettled them to provide a labor force to make the colony self-sufficient. In 1769, the crown assigned missionary work to Franciscan fathers headed by Junipero Serra, who selected mission sites based upon location, soil quality, water availability, and pasturage. The numbers of natives relocated to each mission, California's first urban concentrations, varied from a few hundred to two or three thousand persons. Through the years, native labor developed irrigated fields for grain and other crops, and pasturage for supporting large herds of imported cattle and sheep. Dominated by the mission system, they soon lost much of their indigenous identity and came to be called "Mission Indians." By Father Serra's death in 1784, nine links of the ultimate chain of twenty-one missions were founded and functioning (Gibson 1980: 109-111). Of those twenty-one missions, nine were located adjacent to the general area of the Los Padres National Forest (Brown 1945a: 18).

It took several hundred years from Diaz's initial visit in 1540 before the Spanish were able to install a string of *presidios*, or forts, to protect their California coastal empire from other colonial powers – namely Russia. In 1760, *Visitador-general* Jose de Galvez, disturbed to learn that the Russians were pushing down the Pacific Coast from Alaska, decided that it was time that the Spanish organize and establish *presidios* as a bulwark against further

Santa Barbara Mission, Santa Barbara, California. 1938

penetration. This process was not completed until 1776, when Juan Bautista de Anza led settlers from northern Mexico overland to Monterey and founded a presidio at San Francisco. Other *presidios* were built at Monterey, Santa Barbara, and San Diego. By 1780, the total Spanish population of *Alta California* stood at approximately 600 settlers, an under-whelming number considering 240 years of Spanish investment in *Alta California*!

Twenty years later, *Alta California* still remained a backwater borderland frontier of the Spanish Empire; as late as 1800, the total Spanish population of *Alta California* was only about 1,200 persons. This handful of clerics and soldiers, as well as a few settlers, were spread out along the coast from San Diego to San Francisco. By this date, some Spaniards turned their attention to the material side of the mission system and tried to exploit the resources of the countryside. But other than a few mining endeavors such as the San Francisquito placer deposits (Angeles National Forest), and Antimony Peak and La Panza gold districts (Los Padres National Forest) (Palmer 1992: 136), their activities had little impact on California's forests other than some minor lumbering in the southern mountains. Nonetheless, sparse surviving records, indicate that the Spanish were concerned with several aspects of forestry.

Due to Spanish cultural proclivity for substituting other materials for wood (e.g., adobe bricks and tile roofs), wood use in colonial California was limited. For the most part, the Spanish employed wood in two ways other than fuel wood gathering. First, the Spanish on a limited basis used axe and adze to construct small sailboats, rowboats and seagoing vessels. Second, and more important, Spanish settlers worked with crude whipsaws to fashion rafters and ceiling beams, lintels and doorframes for their churches, mission buildings, and residences. Oak, pine and redwood, and occasionally sycamore and cottonwood, were felled along a strip of seacoast some 40 miles inland from Sonoma to San Diego to meet their needs. Spanish law regulated the procurement of these forest products. Under Spanish rule, a military officer oversaw forest use on crown land, or land outside the boundaries of a town, *rancho* grant or mission holding. On the other hand, the *ayuntamiento*, or town council, oversaw the cutting of wood and grazing of stock in nearby forests. The *alcalde*, or executive officer, was assisted in enforcing local laws and regulations regarding wood procurement by the *juez de campo*, a field judge, or sometimes by a *guardabosque*, or forest warden. By 1813, forest conservation evidently became a concern for the colonists living the missions, because in that year a Spanish government decree ordered *Alta California* reforestation measures (Clar 1959: 11-14; 19-20). This decree indicated that these mission urban centers may have exhausted nearby wood resources and sought to replenish them for the continued prosperity of the community.

One item of special attention for Spanish government officials was fire protection and control in *Alta California*. "Light burning" of fields by mission and non-mission Native Americans became an infraction punishable by a rigorous penalty under Spanish law. Complaints resulting in regulation of this practice date to 1793 and thereafter (Ibid.: 7-11). Nonetheless, elimination of this native forest-management technique under the Spanish regime was unsuccessful.

In 1822, Spain yielded power over *Alta California* to the rebellious Mexican Republic, and three years later, California became a Mexican territory. There was no substantial change in the method of government in California territory as a result of the revolution, but there was a noticeable deterioration in control of the territory with the arrival of *Anglos* to California. Furthermore, the change in government brought numerous incremental environmental consequences.

In Southern California, *Anglo* settlers arrived in increasing numbers after the Mexican revolt from Spain. They brought with them a preference for using wood to build their residences and other structures. Instead of the traditional Spanish materials of adobe and tile, *Anglo* settlers used wood wall construction and wood shingle roofing. Nibbling away at the edges of interior coastal forests, they developed a nascent lumbering and commercial sawmill business, which included exploitation of forests for foreign markets. Mexican and local government officials attempted to regulate these *Anglos*' cutting practices and prevent the exportation of lumber from the territory, but were unsuccessful. *Anglos* continued to chop deeply into the resources and control of the province. Much of this lumbering took place among the Santa Cruz redwoods, as well as among the San Gabriel and San Bernardino Mountain ponderosa pines (Ibid.: 21-38). The harvesting of these living forests had little long-term consequence, but it was a definite shift from the benign indigenous and Spanish forest usage of the past.

Along with the early exploitation of California's timber came a proliferation of livestock into southern California. During the Mexican years, numbers of cattle, through increased importation and reproduction, expanded greatly, especially after 1833, when the Indian mission system ended abruptly and secularization became the law of the land. This event signaled the first land rush to California and a shift in population. In the next decade, more than 300 *ranchos* were granted to Mexican citizens, largely carved out of mission held land. Eventually the *ranchos* numbered more than 800. The granting of these *ranchos* in the 1840s fostered a surge in livestock numbers in order to take advantage of the "burgeoning international trade in hides and tallow" (Preston 1998: 275-276). The era of Mexican cattle ranching lasted only a short time, but the relatively small population produced an immense number of cattle and horses (Brown 1945a: 22). As readily available grazing resources diminished, new forage sources were sought in the mountainous-forested regions of the coast, which would eventually become the Cleveland, San Bernardino, Angeles, and Los Padres National Forests.

In the Spanish-Mexican period, non-Hispanic peoples also invaded northern and central California. First came the invasion of Russians and Aleut poachers, who virtually eliminated the California coastal sea otter and other marine species (Dasmann 1999: 110). British fur trappers soon followed, penetrating central California as well. In 1820s, the Hudson's Bay Company fur brigades, led by trappers such as Peter Skene Ogden and John Work, spread across California's rich beaver and pelt region from the north. In a short amount of time, they decimated the fur-bearing animal population of lands that would eventually become the Six Rivers, Mendocino and Trinity National Forests (Savage 1991: passim; Gates n.d.: passim). Hudson's Bay Company fur trappers also worked tributaries and streams in today's Klamath, Modoc and Shasta National Forests (Brown 1945b: 10-12; Gates n.d.: passim; and McDonald 1979: passim), and to the south, the Lassen National Forest (Johnston and Budy: 1982: 108-111; Strong 1973: 7-10). Historic remnants of some of these targeted marine species and fur trapping activities, which are important in California's and the nation's history, may still be found on these national forests in archaeological form.

Americans were the next group to physically venture into California. Pathfinders and trappers such as Jedediah Smith led the way. In the fall of 1826, Smith set out to discover new trapping grounds. His route ran to the Great Salt Lake and then southward onto the Colorado Plateau to the Colorado River. Smith and his party entered California by crossing the Colorado River and traveling westward over the Mojave Desert to Mission San Gabriel, becoming the first American party to go overland into California through the Southwest. From Mission San Gabriel, the Smith party marched northward through the San Joaquin Valley and over-wintered on the

Stanislaus River. Early in the spring of 1828, Smith and a small party of men crossed the Sierra, probably by way of Ebbetts Pass (now Alpine County's Highway 4), becoming the first recorded explorers to go east from California across the Sierra and the first recorded non-Native American party to cross the Great Salt Lake desert. Later that year, Smith returned to California by retracing much of his previous route to San Gabriel. From there he traveled by ship to San Francisco, met up with the men he had left in the San Joaquin Valley, and wintered in the Bay Area. The Smith party left California by way of the Sacramento Valley to the Trinity River, and exited California by way of the Pacific Coast along today's Smith River on the Six Rivers National Forest.

American exploration of and emigration to California soon followed upon the fur-trapping era. The first American settlers to come to California were the Bidwell-Bartelson party in 1841 who came over the Sierra near Carson Pass after abandoning their wagons. Others followed. In 1843, Lansford W. Hastings led a party of settlers into California by traveling south along the Willamette, Rogue and Sacramento rivers. Encouraged by Hastings' publication, *The Emigrants' Guide to Oregon and California*, pioneer families two years later ventured along the "Hastings Cutoff" westward from Fort Bridger across the deserts of Utah and Nevada, instead of taking the usual northern route to Fort Hall in Idaho. Subsequent emigrants thereafter took this shorter route to California, including the ill-fated Donner-Reed party (Burlingame 1977: 489-490). In 1844 – despite the fact that the region was still under Mexican rule – United States explorer, politician and soldier John C. Fremont forced a passage over the High Sierra south of Tahoe Lake. He was ostensibly searching for emigrant routes into *Alta California*, but after a tense encounter with Mexican authorities, he and his party eventually marched north and reached Oregon (Spence 1977: 406-408). In 1844, Fremont's party left California via Walker Pass in the headwaters of Kern River (Sequoia National Forest). As time went on, more and more American settlers would follow the lead of these explorers into the Mexican province, crossing the High Sierra at several points along the California-Nevada border, including lands encompassed in the Lassen, Plumas, Tahoe, and Stanislaus National Forests. Other emigrant trails passed through the Klamath, Modoc and Shasta-Trinity National Forests. Historic archaeological evidence of these important explorations and emigrant national historic trails can also be found on today's national forests.

One consequence of *Anglo* entry, exploration and emigration to California was its impact on Native Americans throughout the territory. Although many

Native Americans cooperated with the intruders for one reason or other, these early nineteenth-century penetrations of *Alta California* by *Anglos* inevitably resulted in conflicts and even outright warfare. The native Californian population suffered immensely from interference with their accustomed food supply and by outright killings, eventually "pushing them into the rocks" (Shipek 1987). Suddenly turned loose from a well-ordered, disciplined existence which assured ample food and shelter, and unable to return to the native ways of their forebears, the mostly-assimilated Mission Indians faced a future of degradation as non-Indian settlers usurped their lands. Helen Hunt Jackson's powerful novel *Ramona* describes this shameful chapter in California history (Brown 1945a: 20-21). Finally, trappers, explorers, and emigrants inadvertently introduced epidemic diseases of unknown varieties into the region (Supernowicz 1983: 51). By the time war broke out between the United States and Mexico in 1846, at least half of the Indian populations of northern and central California had been killed by disease. With the demise of the Native Americans, the "natural" Eden of northern and central California faded before the thoughtless plundering of California's resources by modern Americans.

The outnumbered *Californio* forces, both military and civilian, resisted the American invasion and the declaration of the Bear Flag Republic by the *Americanos* on July 7, 1846, but after several skirmishes, the *Californios* eventually surrendered to the overwhelming American force. The Treaty of Cahuenga (1847) with the Bear Flaggers and later the Treaty of Guadalupe Hidalgo with the United States (1848) guaranteed *Californio* property rights and civil liberties until California was admitted into the Union in 1850. With the defeat of the *Californio* forces, sleepy Spanish *Alta California* also changed forever.

American Conquest of California's Land and Resources, 1848-1891

On January 24, 1848, James Wilson Marshall, who had a Mexican land grant and a settlement along the mid-section of the American River, discovered gold in the tail race below a mill being built near Coloma by John Sutter. Word leaked out of the American River discovery, and the subsequent Gold Rush unlocked the gates of Eden to the world. A virulent gold fever soon spread worldwide. By February of the next year, gold seekers arrived at San Francisco aboard the ship *California*. Thousands more followed by land and sea from all parts of United States, Europe, and Asia. As a result of the Gold Rush, California was admitted to the Union in 1850, and by 1852 – even though many fortune seekers had already returned home – the population had swelled to 225,000 (Clar 1959: 63). The California Gold Rush was a defining event in the economic history and development of the "Golden" state (Rawls and Orsi 1999), spawning four decades of unparalleled exploration and exploitation of natural resources and resulting in a legacy of over-use and abuse. Much of this mining took place on forested mountain public domain that would later become forest reserve and national forest lands, and much of the remaining gold mining took place in the adjacent foothill lands.

Following the initial Gold Rush, heavy production of placer gold took place. From 1848 to the 1880s, mining was the largest industry in many California counties. During the 1850s, practically every stream in the state was worked or panned. Later, bench gravels and ancient channels drew the attention of miners, who inaugurated the use of hydraulic mining to reach the deeply buried mineral. During the 1850s, quartz mining came into the picture as well. Each new discovery and each new phase of mining expanded settlement of any given area, along with the requisite reservoirs, ditches, and flumes, wagon roads, and mining camps. Exhausting one claim area, most of the people moved on to the next discovery in California or elsewhere in the Rocky Mountain West. Those "dreamers" who remained continued to work claimed areas. In the 1860s, "old timers" reworked surface placers, and in the 1870s and 1880s Chinese miners diligently worked over their leavings. With the depression of the 1890s, large numbers of unemployed migrated to former mining regions in California, and the streams were once more gone over. Nineteenth century records indicate that gold production peaked in 1852 at $81,000,000, and thereafter declined steadily, reaching $15,000,000 by 1900 (Friedhoff 1944: 1).

Mining ventures in California and the West were encouraged by the General Mining Act of 1866, which authorized the exploration and occupation of mineral lands in the public domain, both surveyed and unsurveyed. Essentially, the General Mining Act of 1866 opened all public lands to mineral exploration and patent. A few years later, Congress passed the General Mining Law of 1872, a law intended to settle western lands. This law allowed free entry into public domain land to explore, develop and produce locatable minerals. It also declared that mineral exploration and development would have priority over all uses of the land. The legacy of these two mining acts in California continues to plague the Forest Service, a subject that will be discussed later (Palmer 1992; 138, 146-147).

"Vulcan's footprints," as one historian noted, are found on many California national forests. Though some mining activity took place on eastern California forests, such as the Inyo, the bulk of mining activity contemporaneous with the Sierra Nevada gold rush took place on the western slope of the Sierra Nevada in northwestern forests such as on the Plumas, Tahoe, Eldorado, Stanislaus and, to a lesser extent, the Sierra National Forests. Names of mining towns tended to be expressive: north of Coloma there were Poker Flat, Downieville and Grass Valley, while south of the American River the important mining towns included Placerville, Whiskey Flat and Angels Camp, to name a few. Mining on the Northern California Klamath, Shasta-Trinity, and Six Rivers National Forests centered on the gold-bearing rivers that flowed through those lands with mining towns having names such as Happy Camp, Sawyers Bar, and Weaverville (Ibid.: 137-138; and Watkins 1973: 82).

In their mineral search, gold seekers caused severe and widespread damage to forests, watersheds, wildlife and grasslands of the vast Sierra range and elsewhere. First there was the depletion of timber by lumbermen, who cut into every convenient stand of timber near a mining camp or town. Initially California's timber provided fuel wood for mining camp and local mining needs. Soon the California's forests supplied lumber to build waterworks for California placer mines, lined deep mine tunnels with endless timber structuring, fueled hungry steam engines hauling raw ores from quartz mines to smelters and contributed construction materials for bridges and rail lines associated with mining. The American heritage of "cut and run" and the lack of any conservation ethic led to a vast disfigurement of California's forests. Miners all but used up the yellow pine and sugar pine in the foothills of the Mother Lode in these endeavors (Palmer 1992; 139).

For instance, an early Lake Tahoe history indicated that miners gave no thought to the future supply of lumber needed by generations to follow. When pioneers first came to the area they found a "park-like stand of big timber so heavy that young growth and brush could not make headway." After the California "excitement," mining development took its toll on the Lake Tahoe region, but the real devastation came in 1859, with the discovery of the Comstock Lode over the border in Nevada. California miners, escaping from the decaying placer mines of earlier California discoveries, migrated eastward in droves. The demand for lumber for the Comstock Lode mines was so great that it quickly denuded portions of the Lake Tahoe region of timber. At the time, it was said that for "every ton of ore taken out of the mines an equivalent of one cord of wood went in." After logging operations passed through this area and similar areas in California, intense fires, fueled by the slash and waste left behind, came to these cut-over areas, leaving behind thousands of acres of unproductive brush fields (Bigelow 1926: 8-12).

Perhaps far more damaging to the forests at the time than logging were the ditches, dams and miles and miles of flumes of hydraulic mining systems. Using monitors – powerful jets of pressurized water – miners altered whole mountainous landscapes and watersheds in just weeks. In their wake, they left behind yawning craters of debris and soil eroded hillsides. One consequence of the staggering amounts of water used in hydraulic mining was the massive quantities of silt that poured downstream. Concomitant results of this "fouling of the waters" included raised river beds and subsequent flooding, buried farmland downstream in the broad Sacramento and San Joaquin valleys, and, for a time, the end of salmon runs on the Sacramento and San Joaquin rivers. Eventually this destructive mining practice was shut down by farmers who formed the Anti-Debris Association (Palmer 1992; 139). But "if hydraulic mining had not been stopped," according to one historian, "the gold hunters might well have washed all the soil and loose rock from the Sierra into the Central Valley" (Dasmann 1999: 118). The scars of hydraulic mining, which became widespread after 1854-1855, are still evident on several national forests, such as the Tahoe (Jackson n.d.: 31-34) and Eldorado National Forests (Supernowicz 1983: 75-82). The dams and reservoirs of these

Hydraulic gold mining in operation at Scotts Bar, Klamath National Forest. 1930

large hydraulic systems, which benefited miners, eventually became beneficial in another way. When hydraulic mines were severely regulated by the 1884 *Woodruff vs. North Bloomfield Gravel Mining Company* case, water companies formed around them and used these systems and facilities to develop hydroelectric power projects along the same watersheds, such as on the Tahoe, Eldorado and Stanislaus National Forests (Bigelow 1926: 14; Palmer 1992: 139; Conners 1989: 9-42; Conners 1992: 155-156).

Besides destroying fish runs, mining also impinged on the abundant numbers of large land animals, waterfowl and other fauna living in California's forests. For instance, the Gold Rush touched off the wholesale slaughter of big game such as elk, deer and pronghorn. Each burgeoning mining camp and town required supplies of red meat to feed the many hungry miners and townspeople. Furthermore, in some places, killing deer and elk for skin to be used for gloves, shoes, and similar items outstripped their use as food. In response to these demands, market-oriented hunters filled the need by slaughtering herds by the thousands. They nearly depleted these big game species, and did decimate other species, such as the California grizzly bear, which stockmen and miners killed because they feared them. This carnage stopped only when domestic cattle, sheep and other livestock herded to the camps replaced wild game as a primary food source (Dasmann 1999: 111-112).

The introduction of livestock into the mining mix also caused significant environmental changes and damages to California's natural resources. By the end of the Mexican period, California's grasslands sustained some 400,000 cattle and an additional 300,000 sheep – a figure in keeping with the population of the region (Dasmann 1999: 113-115). But demand by the increasing gold-rush population, and the decline in big game, created a need for additional livestock. One stringy cow was said to be worth five hundred dollars or more to red-meat-starved miners. Soon large herds of cattle were sent northward to meet this profitable demand. Mexican *rancheros* became cattle barons, extracting fabulous prices for their cattle, which were either shipped by boat to San Francisco and the mining fields, or driven in large herds from range directly to market through the Sacramento Valley (Brown 1945a: 27). To replenish livestock, thousands more cattle were driven into Southern California – some from as far away as Texas and the Mississippi Valley (Watkins 1973: 107). By the end of the 1850s, more than three million head grazed on the grasslands. The unrelenting waters of the "Great Flood" of

1862, followed by the droughts of 1862-1864, perhaps the worst in California history, drastically diminished these numbers, but they soon rebounded. In addition to cattle, sheep, which were more adaptive to harsh conditions, multiplied to 5.5 million by 1875 (Dasmann 1999: 113-115).

Overstocking naturally caused overgrazing, and ranchers began to search for viable land in the Sierra Nevada sufficiently gentle in slope to allow livestock to graze with a degree of safety. After the 1870s, few ranches situated along the foothills of the Sacramento and San Joaquin Valleys were able to furnish sufficient yearlong forage for stock. Thereafter, they depended on these mountain ranges for summer forage until such times as the foothills greened from fall rains. Future national forest lands, such as the Plumas, Tahoe, Eldorado, Stanislaus, Sierra, Inyo and Sequoia National Forests, provided cattle and sheep range for these dependent ranches. These forests also served cattle ranches east of the Sierra from north of Reno, Nevada, southward to the Owens Valley. After the Civil War and the corresponding shortage of cotton fiber, the demand for wool increased. Cattlemen soon found themselves competing for these public domain ranges with sheepherders. Competition for range sharpened with increased demand for sheep and wool – along with increased settlement and constriction of open range, and range troubles between cattlemen and sheepherders erupted as a result (Watkins 1973; 243; Cermak 1986: 21).

By 1870, more than three million sheep were seasonally grazing on the forests of central California. For many years thereafter, sheepherders customarily drove their large bands of sheep from the San Joaquin Valley onto the western foothills of the Sierra Nevada, or around the southern end of the Sierra into the rich Inyo and White Mountain grazing areas and onto the steep eastern slopes of the Sierra Nevada. Elsewhere, such as on the east side of what is now the Los Padres National Forest, upwards of 100,000 sheep were ranged on public domain by a number of big landowners in the 1870s (Brown 1945a: 29). By 1890, Californian sheep numbers increased to seven million. With no regulation of the grazing, mountain meadows were severely overgrazed, with ranges denuded of vegetation and trampled out (Robinson 1943: 4).

Furthermore, both cattlemen and sheepherders practiced a type of "light burning" to "improve" forage of this public land. Both parties burned foothill undergrowth to reduce brush and small trees so as to improve grass conditions, eliminate obstacles to livestock, and enlarge the amount of feed available for their animals. Sheepherders, many of Basque descent, were notorious for practicing this technique of forest-forage manipulation. Each fall, dense clouds of smoke billowed over the Sierra Nevada, as sheepherders seemed to be burning everything in sight on their departure from the foothills (Cermak 1986: 21-23). During the period 1875 to 1890, local newspapers were filled with accounts of uncontrolled fires of this nature. Lowland farmers in Los Angeles, Ventura and Santa Barbara counties, and as far north as Monterey County, held an indulgent attitude regarding these fires. At the time, they did not realize that the protective vegetative cover bound the soil in place on the steep hillsides. Loss of that vegetation ensured watershed devastation. A decade or more later, the populated lowlands would experience severe floods that washed countless tons of silt down over their orchards and farmlands (Brown 1945a: 33).

Besides timber, watersheds, wildlife and grasslands, the California Gold Rush also directly shaped and/or impacted settlement patterns of California's national forests. A major ramification of the effect of pick, pan and shovel was the unprecedented influx of people following the cry of "Gold!" Prior to the remarkable initial rush to the Mother Lode, non-Native Americans had utilized and settled only the coastal lowlands and southern montane areas of California under Spanish and Mexican rule. As noted earlier, in 1800, the total population of *Alta California* stood at approximately 1,200 non-native persons. But with the discovery of gold, several hundred thousand additional people entered the state and penetrated deep into California's interior. To link these people spread out in far-flung mining camps of California, several transportation components evolved and settlement took place throughout the state almost at once.

First, inland waterways were used to penetrate the Californian wilderness. Paddle-wheel steamers, packed with gold seekers, churned the waters of the Sacramento River from San Francisco to Sacramento in search of fortune. Naturally, the construction and fueling of this mode of transportation consumed modest amounts of timber. Where paddle wheels left off, wagon wheels took over. Miners, pioneers and others cut wagon roads through largely unmodified wooded regions. Wagon ruts soon laced the hillsides of the Mother Lode as an army of mule and oxen-drawn supply wagons bumped along these dusty, rocky, arterials (Watkins 1973: 150-151). Express companies such as the Pony Express and later Wells Fargo sprung up to carry transcontinental mail, newspapers and packages to Placerville, Sacramento and San Francisco, granting many mining camps their only access to the outside world (Godfrey 1994).

But it was the construction of the Central Pacific Railroad (1869), the western half of the first transcontinental railroad and other railroad lines that truly impacted land uses in California's forests and transformed settlement patterns in California. Though the mining industry exploited the Sierra Nevada for timber for head-frames, flumes, dams, sluices and town construction including houses, saloons and brothels, the arrival of the Central Pacific Railroad elevated this exploitation exponentially. Besides the enormous amounts of lumber required for railroad tracks, bridges, housing, stations and other wooden structures, and the fuel needed by railroads, the iron horse enabled a nascent mining-related lumber industry to expand to a full-blown lumber industry which could market building materials to California's growing urban centers (Watkins 1973: 149-155 passim, 194).

Exploiting the Timber and Stone Act and the Free Timber Act of 1878, valuable timber tracts on public lands were amassed by lumber interests. The first piece of legislation gave public land in a timbered or mineral area unfit for cultivation to a person if he was willing to pay $2.50 an acre for it and "promised" not to accrue more than 160 acres. The second allowed the homesteader to cut lumber on this mineral land, which heretofore had been used exclusively for mining. The purpose of these acts was ostensibly to aid the small landowner, but instead they enriched lumber companies. which easily circumvented the law by having their employees file claims with false promises. They in turn deeded the land over to their employers, who were then at liberty to despoil the best timber tracts for immense profit. By the 1880s, the appalling number of fraudulent timber claims due to these pieces of legislation was clearly recognized by many, but went unpunished (Lockmann 1981: 67-69) because the Sierra Nevada forests and the lumber industry logging them met the needs and demands of the growing communities of the Sacramento and San Joaquin Valleys, and did so for most of the nineteenth century. These fraudulent practices led to logging practices whereby an immediate profit was sought with no regard for stewardship.

Besides spawning a growing and rapacious lumber industry, railroads themselves directly influenced the growth patterns of California. In order to sustain their economic well being, railroads needed people – people to buy the land along their routes, people to settle thereon and develop the countryside, and then people to buy and sell the goods shipped on its tracks in these newly settled areas. For example, the Southern Pacific Railroad, which became the second transcontinental railroad in 1883, turned sleepy southern

Spanish California into choice real estate. Employing an intensive advertising campaign, the Southern Pacific lured people onto its rail cars by its handbills, posters and brochures, depicting southern California as a mythic isle. By the middle of the 1880s, an agricultural and town-site boom emerged, thanks to these solicitations (Watkins 1973: 248-252).

For many years thereafter, railroads such as the Southern Pacific and later the Atchison, Topeka and Santa Fe were the central economic force in California's southern economy. Along with the State of California, the railroads enthusiastically advertised California as an agricultural paradise, encouraging eastern migration to this "new" American Eden. During this time, agriculture in the "land of sunset" made great strides. The citrus industry, dairying, as well as row crops such as grain farming and especially alfalfa, grew rapidly as a result, and even spread into marginal farming areas. Urban centers such as San Diego and Los Angeles grew as well to supply these agricultural communities, producing a speculative land boom. In 1888, the urban land boom ended abruptly. Nevertheless, the growth of Southern California continued. "There were cities now," according to one author, "where there had been towns, towns where there had been villages, and villages where there had been chaparral and creosote bush; farms were planted and irrigation companies formed; at least 130,000 of the boom's population remained to make the land their home…"(Ibid.).

Owners of citrus orchards in California used rapid-growing eucalyptus as a break to the strong winds and a protection during cold weather. 1903

Source: Forest History Society

An important result of California's impressive urban and agricultural growth, which eventually led to the formation of the first California federal forest reserves, was the ever-increasing need for water from California's major drainages. Demand for "liquid gold" came especially from growing urban centers such as San Francisco and Los Angeles. Stimulated by the Desert Lands Act of 1877 and the Wright Act of 1887, large parcels with water rights were acquired in the wake of this legislation, first by timber and cattle interests and later by farmers and water-starved municipalities. The Desert Lands Act authorized the sale of 640-acre tracts of arid public domain lands at $1.25 per acre upon proof of reclamation of lands by irrigation. But like other legislation of its day, faulty design and language allowed an era of exploitation by lumbermen, who bought large parcels of land under the Desert Lands Act ostensibly for the associated water rights, but in reality to strip any forests nearby. Stockmen, on the other hand, gained public domain livestock pasturage under the Desert Lands Act, locking out sheepherders from key pasturages (Lockmann 1981: 67). The Wright Act of 1887 permitted regions to form and bond irrigation districts, allowing small farmers to band together, pool their resources, and bring water to where it was needed. Thanks to the Wright Act, agriculturalists, for instance, were able to tap and divert portions of the Merced, San Joaquin and Kings rivers to their fields in the Central Valley (Watkins 1973: 307). In Southern California, water districts were eventually organized to deliver the feeble water sources of that region to thirsty cities such as San Diego and Los Angeles.

Meanwhile, by the 1880s, it was abundantly clear that major changes in vegetation, uncontrollable erosion and decreased water flow caused by hydraulic and other types of mining, ecologically unfriendly logging practices of the lumber industry and overgrazing by cattlemen and sheepherders alike were adversely affecting farming interests and threatening potable supplies to adjacent metropolitan regions. As noted before, most damage to Central Valley farmland caused by hydraulic mining was halted by the *Woodruff vs. North Bloomfield Gravel Mining Company* case (1884). Much of the flooding of the Central Valley was normal – it was largely an inland sea during high run-off periods. What was different were the immobile population centers in the valley and along the rivers and streams. This fact did not minimize the downstream effects from slickens (a gooey sludge of liquid mud), excessive grazing and logging upstream. Major flooding caused by deforestation in the Sierra Nevada and elsewhere still brought periodic and devastating floods each year, which inundated farming communities along tributaries. By 1890, even people living adjacent to the minimally forested watersheds of Southern California, such as the San Gabriel Mountains, realized that their forests needed protection as well, especially from the scourge of fire and overgrazing caused by livestock. They looked for help to the federal government, instead of the State of California, for many reasons as will be seen.

California Government and Forestry Conservation

The State of California came into existence on September 9, 1850, when it was admitted into the Union with a population standing at approximately 92,500 people. In no time, a surveying district office of California was organized, land offices established in Los Angeles and Benicia, and the survey of public lands in California set into motion. In two years' time, California's population doubled to approximately 225,000.

The shaping of California state policies regarding conservation and/or preservation measures was inextricably involved with and often ran parallel to those on a national level. The primary tenet of congressional legislation in the post-1860 period was certainly to encourage settlement of the public domain and provided for only minimal management of resources located thereon. For example, the 1862 Homestead Act allowed unrestricted settlement of public lands, requiring only residence, cultivation and some improvements on a tract of 160 acres. Under the auspices of this act, more than one million pioneers settled the West during the next seventy years. Other federal land distribution laws such as the Desert Land Act (1877), the Timber and Stone Act (1878) and the Free Timber Act (1878) all favored this *laissez faire* settlement philosophy. At this time, there was no preponderant national question regarding what should be done with forested public domain in the West. Western attitudes prevailed. Miners, lumbermen and stockmen, unconcerned about conservation of natural resources and/or management and protection of forested public lands, wanted the bounties of nature open to them. Without any federal government control to check this western way of thinking, the plundering of California's public domain resources was a natural outcome.

For a better part of the late nineteenth century, the California legislature differed little in its viewpoint regarding settlement and the management of resources. Initially, the State of California did make fledgling attempts to stop the misuse and abuse of California's resources and destructive forces such as fire. But during its formative years, these legislative efforts largely failed. For instance, in 1850, California's legislature passed the state's first forest fire control laws; in 1852, early attempts were made to save the Calaveras and Tuolumne redwoods; and in 1864, California created its first state park, when President Abraham Lincoln approved a congressional act granting land that embraced the Yosemite Valley and the Mariposa Big Trees Grove to the State of California for this purpose (Clar 1959: 60-70). State officials also began to recognize that the timber within their state was not unlimited. As early as

1868-1869, the California Board of Agricultural Transactions addressed the problem. Their annual report regarding "Tree and Forest Culture" stated:

> …we have thoughtlessly come to regard our supply of these materials…as inexhaustible. The facts are quite different…California is far from being a well timbered country. Nearly all the timber of any value for ship and general building purposes, or for lumber for general use, is embraced within small portions of the Coast Range or the Sierra Nevada districts…It is now but about twenty years since the consumption of timber and lumber commenced in California, and yet we have the opinion of good judges, the best lumber dealers in the State, that at least one-third of all our accessible timber of value is already consumed and destroyed! (Clar 1973: 71).

After condemning the reckless destruction of timber through settlement, as well as the consumption of fuel wood by railroads, the California Board of Agricultural Transactions recommended two toothless remedies: (1) prevent the "existing and impending evils" from continuing to threaten the prosperity of the commonwealth of California and (2) restore the lost forests by offering a bounty for the "cultivation of forests and woodlands on every farm and homestead throughout the agricultural portion of the State." The California legislature responded with the passage of a "mild" state law to encourage the production of trees, but there was no indication that this law or later ones succeeded in any major reforestation (Ibid.: 72-86).

The rapid, reckless and wasteful cutting of trees, the ravages of goats and sheep on California's young saplings and brush, and the constantly increasing severity of sudden and devastating floods continued, but the message of prevention and conservation of resources fell upon deaf ears until 1883. In that year, a "concerned" California legislature authorized the Lake Bigler [Tahoe] Forestry Commission to report on a preservation plan for the California side of Lake Bigler. The authorizing resolution decried the "rapidly proceeding denudation of the forest" along the shores of Lake Bigler, and declared it a state duty to preserve from destruction "the most noted, attractive, and available features of its natural scenery…for the health, pleasure and recreation of its citizens and tourists." The final commission report urged that the Lake Tahoe basin be set aside as a national or state park and reserved from private entry. The Lake Bigler Forestry Commission also pushed the governor and the legislature to expand official state interest in other forests as well. Nonetheless, California government failed to take any action in either direction, perhaps because of a "lack of will" to fight special interests, and because of alleged mismanagement of Yosemite Valley State Park (Ibid.: 87-88, 96).

Despite failure to achieve the goal of establishing a park at Lake Tahoe, one breakthrough related to the Lake Bigler Forestry Commission report came out of it – the creation by statute of the California State Board of Forestry (1885). Why the California legislature approved this measure and not the Tahoe Lake Park is speculative. But it appears that several influential political interest groups – farmers, urbanites, and reformers – broadly supported the latter idea. Farmers along the San Joaquin and Central valleys, who often complained about the inertia of the federal government in managing the mountain lands, thought the State of California should intervene. Urban dwellers, on the other hand, supported the statute because they realized that to protect the watersheds that supplied their cities, better forestry practices needed. Finally, early reformers also recognized the abuse and corruption of the federal land distribution laws by lumbermen and livestock owners needed correction (Ibid.: 95-97).

With a board appointed in 1885 of prominent residents from the San Francisco Bay region by the governor, including three members of the former Lake Bigler Forestry Commission, the California State Board of Forestry set out to inventory the state's forestry resources and conditions. "In view of the genuine concern for broad scale watershed protection in Southern California and the relative indifference to the timber lands of the north," according to one scholar, it seemed "unusual" that the membership of the first Board of Forestry be comprised of these citizens. However, the following year, Abbot Kinney, a competent representative of Southern California views, was made chairman (Ibid.: 98).

From 1885 to 1890, the Board of Forestry made a number of special reports, including reports on the redwood reserves counties (Yosemite Valley and the Mariposa) in Sonoma and San Mateo, on the issue of the disposition of forested school lands, on the excessive mining and lumbering activity along the Sierra foothills, on the establishment of a California Arbor Day, on the development of nurseries and experimental farms for reforestation and on the pine species of California. One of the Board of Forestry's most important early actions was the preparation and publication of a forest map of the entire state, "showing the amount and kind of timber standing in the different counties, and its commercial uses and value" (Ibid.: 99-117 passim).

The leadership of the California State Board of Forestry also espoused the cessation of the sale of all government timberlands. The Board of Forestry advocated selling off only the timber and not the land – an important step toward protecting the land, and a program already practiced in the Canadian provinces. Besides this point of advocacy, the Board of Forestry lobbied Congress, proposing that public lands in California be surveyed and those not of agricultural quality be reserved from private use and "permanently maintained in forest for the welfare and best interests of the people in the Commonwealth." They memorialized Congress to amend or suspend the federal statutes under which the many land abuses and frauds occurred. In fact, the Board of Forestry came very close to openly advocating that the federal government cede all forested public domain to the State so that the land and its resources could be placed under the complete control of the Board (Ibid.: 118-129 passim).

Besides the above actions, the State Board of Forestry also filled other public functions. Board member lectures educated the public regarding the links between preserving mountain forests and watershed conservation. In doing so, Chairman Kinney sought to ensure continued water and prosperity for the agricultural and domestic establishments of southern California. The Board of Forestry, under Kinney's leadership, also sought to enforce state forest law regarding the prevention of fires, urging at the same time the "legitimate" harvesting of California's resources by lumbermen, mill men and county boards of trade (Ibid.: 118-129 passim; 152-160 passim). The first Board of Forestry also campaigned against the indiscriminate use of fire in the woods.

Conservation and the Lost Eden (Part I)

In 1848, when Americans first came to California in great numbers, they surely were in awe of California's natural beauty. "There was an outlandishly Eden-like quality to the state's landscape," according to one author. "Who would have believed in the existence of trees thirty feet thick, three hundred feet high, and four thousand years old?....Who would have believed that such wonders as Yosemite Valley or the canyon of the Kings River were anything more than the fancies of some romantic's overheated imagination?" From John Charles Fremont's essay "Geographical Memoir upon Upper California" (1848) to John Wesley Powell's *Report on the Lands of the Arid Region* (1877), exploring and survey party reports all captured and extolled some aspect of California's stark natural beauty in word and image. Nineteenth-century painters such as Albert Bierstadt, Thomas Moran and William Keith also portrayed California's majestic mountains, valleys and streams. Keith especially portrayed panoramic views of Kings River Canyon. As early as 1855, California's natural wonders, such as the mammoth Calaveras County

Artist's depiction of King's Canyon in California that appeared in *Century Magazine*. 1892

Source: Forest History Society

redwoods, were a marketable attraction. Likewise, the State of California often advertised the "glories of the Golden State" as a legendary land of milk and honey. Finally, fanciful images of California abounded as well. For instance, a Currier & Ives lithograph of a "typical scene" along the California coast depicted a Mount Rainier mountain-type overlooking a mythic coastal jungle (Watkins 1973: 234-237). With all these images to choose from, Californians could not help but be aware of the natural treasures of their state. But as adverse impacts generated by the mining, lumber and livestock industries degraded the natural beauty of state, and as a thirsty population grew and spread throughout the land, did Californians develop a sense of conservation in response?

In the past, many have argued that conservation practices in America originated in the East, stressing how conservation was a European idea, imported into the colonies and thereafter picked up by the young republic (USDA Forest Service 1952: 1-3). According to them, thereafter conservation in America developed as an idealistic and scientific interpretation. First they note how America's concern for conservation next derived inspiration from George Perkins Marsh's book *Man and Nature: Or Physical Geography as Modified by Human Action* (1864), which has been called America's first environmental history (Cook 1991; 1). In the conservation movement lineage, the work of Dr. Franklin B. Hough marks the next major milestone. His scientific paper, "On the Duty of Governments in the Preservation of Forests," presented at the annual meeting of the American Association for the Advancement of Science (1873), spearheaded a petition to Congress "on the importance of promoting the cultivation and preservation of forests." Three years later, Congress funded Hough to undertake a "study encompassing forest consumption, importation, exportation, natural wants, probable supply for the future, the means of preservation and renewal, the influence of forests on climates, and forestry methods used in other countries." Ultimately his report led to the creation of the Department of Agriculture Division of Forestry (1881), the precursor to the Forest Service, and his appointment as its first chief. His replacement, Nathaniel Egleston, served from 1883 only until 1886, when Bernhard Fernow, a German-born "professional" forester, replaced him. Hough, Egleston and Fernow concentrated on researching a wide range of forestry topics, but could not put their idealistic scientific knowledge to practical use because they had no forests under their administration. The division would not have any forests to manage until 1905 (Dupree 1957: 239-241; Williams 2000: 3-6; Cook 1991: 2).

While Hough and Fernow dreamed of conservation, a pragmatic need for conservation of resources, especially water resources, developed in the West. This strain of conservation developed in spite of western pioneer attitudes about freedom to exploit the public domain – a conviction that ran counter to any restraint for future benefit embodied in the term conservation (High 1951: 291). But when faced with a clear need, westerners chose conservation, albeit a utilitarian version, unlike the eastern brand rooted in idealism and science.

The desire to conserve California's resources is clearly rooted in its later history. Early Californians, between 1848 and the 1870s, over-used forests, polluted streams and slaughtered game. Though aware of the state's natural beauty, gold dust, sawdust and/or trail dust blinded them. Aggrandizement of riches overpowered restraint. But beginning in the late 1860s, some Californians became aware that their Eden was eroding away. The first warning came in the California Board of Agricultural Transactions report (1868-1869), which stated that one-third of the accessible timber of value had already been consumed and destroyed. The next warning came from the Lake Bigler [Tahoe] Forestry Commission report (1883), which decried the denudation of the Tahoe Lake region. The Commission also declared it a state duty to preserve from destruction California's most noted, attractive and available natural features for the health, pleasure and recreation of its citizens and tourists. The work of the California State Board of Forestry (1885-1890) added to this viewpoint. By 1890, after decades of over-use and abuse of California's natural resources, Californians were ready to embrace conservation principles.

As will be seen in the next chapter, the turning point involved the issue of watersheds and the preservation of water supplies for farming and urban centers. Like their counterparts in other western states, such as Idaho (Godfrey 2003: 17), when industry threatened vital water sources, Californians supported conservation. Californian conservation was purely pragmatic and predicated on actual experience with the devastating effects of a lack of conservation. It was the basic need to protect the purity of California's watersheds for prosperity and posterity that drove early California conservation efforts. The next chapter tells that story beginning with the formation of California's first forest reserve – the San Gabriel Timberland Reserve.

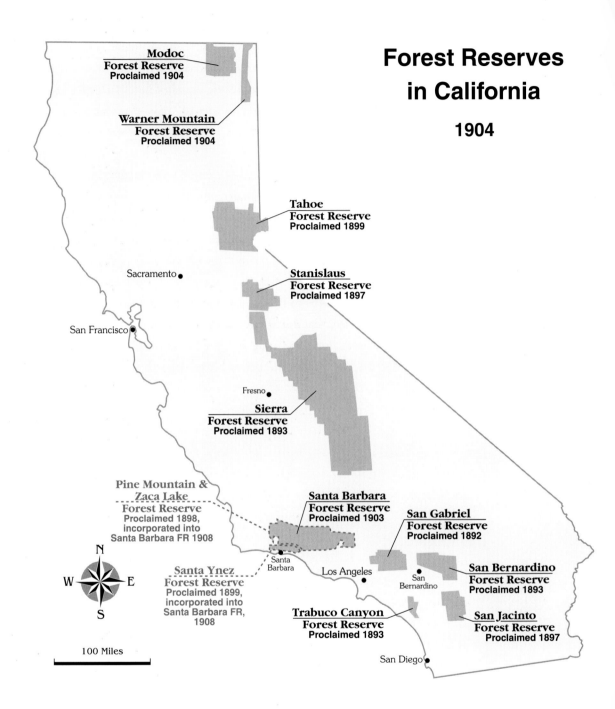

Chapter II: 1890-1904

California's Forests Conserved Under Federal Control

Turn of-the-Century California

In 1890, California was no longer a string of coastal pueblos and *ranchos* but a well-populated state with broad basic rural industries – ranching, agriculture and lumber – and a growing urbanism based on commerce, shipping and manufacturing. The beginnings of modern California were at hand.

Cattle and sheep ranching had long been California's main business, with ranchers fattening their stock on public domain far and wide. American successors to the *rancheros* changed the livestock industry. With an eye for new markets, and in close cooperation with railroads, cattlemen improved beef production. Meatier breeds imported from elsewhere in the United States and as far away as England replaced scrawny, mangy, and bony Spanish cattle. Backed by Eastern and sometimes foreign capital, large cattle outfits were organized. The devastating drought in 1863-64, which left behind the bleached bones of thousands of animals, brought a temporary slump to this development. However, by the 1880s, the cattle industry recuperated, and cattlemen thereafter began to dominate regions. These powerful land barons took over choice pasturelands, which they jealously guarded from competing and ever-increasing and improved bands of Merino sheep. Sheep owners, who lived in towns and tended to be small operators, hired experienced Mexican and Basque sheepherders to tend their stock. In the end, sheepmen were less politically powerful than ranchers in most regions of California. But in either case, roaming sheep or cattle clearly overstocked California's grasslands, causing progressively more damage (Rolle 1963: 342-345).

Meanwhile, by the late nineteenth century, ranching, as the mainstay of California's economy, had given way to a growing California agricultural economic pattern based on wheat production, the citrus industry, and a variety of other crops, such as grapes. The epic story of California wheat production began in the 1870s, when California's unusually hard, dry Durham grain became an important export crop to countries such as England. Other cereals – barley, corn, oats and rye – were produced in large quantities as well. Thanks in part to large-scale, exploitive commercial farming ventures by men like Isaac Friedlander, California's "Grain King," and to financial aid provided by men like San Francisco financier William Ralston, California ranked second in wheat production among all the states by 1890 (Rolle 1963: 346-347; Buck 1974: 21). The epicenter of this wheat boom lay in the San Joaquin Valley, where a struggle developed between farmers and the railroads, replete with speculation, intrigue, ruthless competition, squatters' land rights and settler vigilantism that eventually led to the historic 1880 bloody gun battle at Mussel Slough in which six or more people lay dead or dying (Rice, Bullough and Orsi 1996: 233-254).

The horticulture of perishable citrus fruits was less confrontational than wheat farming. California's citrus industry started with the planting of orange groves in early mission gardens. However, this nascent industry did not become commercially viable because the native fruit was not very palatable and because of California's remoteness from large markets. Apple, peach and apricot orchards were far more important than orange groves. Almost overnight, this situation changed when in the early 1870s the Washington Navel seedless variety of orange was introduced into Southern California along with industrial-scaled grapevine planting. More meaty, juicy and flavorful than the small, pithy, thick-skinned native variety, the Washington Navel was extremely well-suited to California's climate and soil. A decade later, grove after grove of these juicy oranges were planted throughout Los Angeles, Riverside, Orange, San Bernardino, San Diego, Ventura and Santa Barbara counties. By 1892, 60,000 acres were devoted to orange groves. These orange groves started the California citrus culture, which was later augmented by improved varieties of lemons, tangerines, and grapefruit. These citrus varieties competed with the acreage dedicated to other fruits, such as apples, peaches, prunes and pears (Rolle 1963: 350-356; Buck 1974; 22).

Besides grain and citrus fruit, California farmers in the pre-1890 period grew a variety of other agricultural crops in the state's fertile valleys. For instance, "vines in the sun" first bore grapes for wines to celebrate Catholic Mass during Hispanic days. But truly large-scale production of wine in California did not come until after the Civil War, when good vine cuttings arrived from Europe. Thereafter, grapevine planting spread throughout California from Napa and Sonoma valleys eastward to the San Joaquin-Sacramento Valley, and then to places like Cucamonga in Southern California. By 1890, more than 70,000 acres of vineyards were in production, and raisin farming had become a major land use as well. By 1900, Fresno County had become a main raisin-growing center, reaching annual production of 47,000 tons in that year alone (Rolle 1963: 358-362; Buck 1974: 22).

As the beef, wheat and, later, citrus and wine industries grew, the exploitation of California's forests reached its pinnacle as well. California's forest resources contained large stands of ponderosa pine, Douglas-fir, spruce, redwood, pinyon-juniper woodlands, as well as widespread chaparral. These forest lands lay chiefly in the Sierra Nevada and the northern coast counties, where an insatiable demand for lumber led to the development of the forest-processing industry and the associated planing mills and lumber yards. According to Census Bureau statistics, in 1899 there were 276 lumber companies throughout the state, employing more than 10,000 wage earners. In that year, they produced 736,496,000 board feet. Almost half of this amount was redwood, which came from the immense redwood belt of the northern coast. Eureka fast became one of the leading lumber ports of the world. The other half of the timber cut came mostly from the Sierra Nevada region, whose timbered regions supplied scores of farming communities and urban centers up and down the great valley below the timbered mountain range. Railroads and flumes hauled the timber from this rugged terrain. But despite this output, California was not lumber self-sufficient and drew heavily upon nearby Oregon and Washington for lumber as well (Rolle 1963: 341; Buck 1974: 22).

California's economic growth was not tied solely to food and/or lumber production. The counterpoint to these industries located in sparsely populated rural regions of the state was the metropolitan growth of the state. According to the 1890 census, the total population of California was 1,213,398 people, with only 250,189 rural residents. By 1900, official figures listed the state's population at 1,485,053, of which 309,042 resided in rural communities. At the turn of the century, less than one out of four Californians were rural residents (Buck 1974: 21).

Not surprisingly, by 1890 urban San Francisco, with a population of close to 350,000, became the political, economic, and cultural center of the state. Despite a history of civic government graft and corruption, along with ugly labor disputes, this world-renowned cosmopolitan city, which ranked ninth in size in the United States, was on the eve of becoming a truly important commercial gateway to Japan, China and the vast reaches of Siberia. In the south, Los Angeles, "half city, half sprawling, overgrown town" with a population of just over 100,000, was also on the eve of a great commercial awakening. Founded largely by national railroad transportation links, which created a land boom in the latter half of the century, its chief resources were climate, oranges and tourists. In conjunction with the agricultural development of the Imperial Valley in the 1870s and 1880s, a new land boom in Los Angeles awaited only the development of oil resources in San Luis Obispo,

Santa Barbara, Kern and Los Angeles counties, and unprecedented opportunities in California shipping and Pacific trade expansion. Other factors, such as construction of the deep-water harbor at San Pedro, the annexation of Hawaii, the acquisition of the Philippines in the Spanish-American War and the construction of the Panama Canal promised to transform Los Angeles as well. In the wings, the motion picture, automobile and airplane industries would soon take root and help to catapult Los Angeles into the future (Cleland 1962: 223-225).

The great transformations in California's lifestyle and growth from the Gold Rush to 1890, and especially the growth of the capitalistic order, caused a variety of emotions in people, whose responses ranged from outright optimism to critical apprehension of the effects of industrial capitalism on rural people and resources. Others grew nostalgic for the past as a means to cope with the changing society and modernism. The works of several outstanding California writers reflect and span the range of emotions evoked by California modernization at the turn of the century.

Optimism, a California cultural trait from the days of panning gold, could be found almost anywhere in the state. As one person in 1873 plainly put it, "I am willing to take the country barefoot and wait for the shoes and stockings." A 1880s Southern California land boom real estate pamphlet that read, "buy land in Los Angeles and wear diamonds," mirrored this confidence and buoyancy. Furthermore, in 1888, a self-confident citizen forthrightly declared, "I do not mean to say that everybody in Southern California is rich – but everybody expects to be rich tomorrow." As a growing number of individuals gained wealth from commerce, shipping, ranching, agriculture and lumber, a growing number moved to cities and became patrons of the arts, which led to a flourishing cultural renaissance. They supported the arts and encouraged prose, poetry, art and learning by constructing new museums, art galleries, theatres and opera houses that served as paeans to California's cultural growth. In 1883, one writer soon dubbed San Francisco the Paris of the Golden State; Los Angeles, the Lyon; and San Diego, the Marseilles (Watkins 1973: 257, 260).

Not everyone shared this optimism or approved of the means that many employed to acquire it. Writers like Henry George, Josiah Royce, Frank Norris and Helen Hunt Jackson strongly attacked the new capitalistic order and published works that influenced related social, political and economic problems of the day. For instance, Henry George's *Progress and Poverty* (1880) championed the "landless laboring masses" and criticized the wealthy who were further enriching themselves at the expense of the downtrodden. Idealist and non-conformist Josiah Royce's novel, *The Feud of Oakfield Creek* (1887), effectively presented the consequences of railroad dominance in state and national politics. The violence at Mussel Slough inspired California writer Frank Norris to write his inflammatory novel, *The Octopus* (1901). Norris adopted the settlers' point of view of the tragedy, and his book intensified public outrage against the railroads, and later helped pave the way for Progressives to demand stricter controls on corporations. Finally, Helen Hunt Jackson combined a concern for the mistreatment of native Californians and evinced some nostalgia for dying Spanish traditions. The works of other California writers such as William Heath Davies, Charles Fletcher Lummis and Mary Austin were either simple reminiscences that looked to the past with uncritical pride, or they celebrated the glories of the past. Davis's *Sixty Years in California* (1889) gave a personal view of San Francisco society as he knew it. Charles Fletcher Lummis' *The Land of Poco Tiempo* (1893), The Spanish Pioneer (1893), and his magazine *The Land of Sunshine*, which later became Out West, glorified California's Hispanic heritage and extolled the beauties of California life. Mary Austin's incisive articles and books, like *The Land of Little Rain* (1903), were more of a fictional nature and provided penetrating sketches of people and how the physical environment of the country south of Yosemite and north of Death Valley affected them (Rolle 1963: 416-419, 422-424).

In California, optimism, apprehension and nostalgia crossed paths when the California conservation and preservation movements met at the turn of the century.

California Conservation and Preservation Movements Meet

The California preservation movement was rooted in the exuberance of men like the naturalist John Muir. Born in Scotland but educated in the United States in geology and botany at the University of Wisconsin, Muir arrived in San Francisco in 1868 after spending the year before on a thousand-mile walk from Wisconsin to the Gulf of Mexico. Thereafter, he walked across the Central Valley and spent his first summer in the Sierra, which he thought should be called the "Range of Light." In his journals, Muir wrote in glowing terms of the natural beauty of the mountains, flora and fauna – such as his beloved Yosemite Valley. For instance, in his first book, *The Mountains of*

California (1894), he wrote, "Nature's peace will flow into you as sunshine flows into trees. The winds will blow their freshness into you and the storms their energy, while cares will drop off like autumn leaves" (Bean 1968: 342).

By 1873, "John of the Mountains" began to publish articles for the *Overland Monthly* and other magazines on the natural wonders of California. For instance, regarding Tuolumne Canyon, he wrote, "Here was no field, nor camp, nor ruinous cabin, nor hacked trees, nor down-trodden flowers, to disenchant the Godful solitude…." At the same time, he also witnessed the disappearance and destruction of California's wilderness by the lawless forces of mining, lumbering and livestock interests. He hoped that nature would someday heal every raw scar left by "the roads blasted in the solid rock" or by the "wild streams dammed and tamed and turned out of their channels and led along the sides of canons and valleys to work in mines like slaves." In regard to the timber industry, he penned, "God has cared for these trees, saved them from drought, disease, avalanches, and a thousand straining, leveling tempests and floods; but he cannot save them from fools." He continued, "Any fool can destroy trees. They cannot run away." The destruction of grasslands by sheep, which he likened to "hoofed locust," particularly bothered

Sheep grazing near Ellery Lake, Tioga Pass, Inyo National Forest. 1927

him, but he admired Native Californians, who in his romantic eyes "hurt the landscape hardly more than the birds and squirrels…" (Browning 1988: vii-viii, 9, 18).

Love of pure unblemished nature led Muir and other naturalists like him to seek its preservation with religious zeal. In 1890, disturbed by the mismanagement of Yosemite State Park by the State of California, Muir and important allies such as Robert Underwood Johnson, editor of the *Century* magazine, convinced the California legislature to return Yosemite Valley to the federal government, which re-designated the area as Yosemite National Park. At the same time, an aroused national movement to save California's "big trees" (*Sequoia gigantea*) led Congress to set aside Sequoia and General Grant national parks (later incorporated in King's Canyon National Park) (Bean 1968: 343-344). These large tracts, which became the nation's newest national parks since Yellowstone (1872) and Mackinaw Island (1875), were withdrawn owing to their scenic values and not for their utilitarian values. They were clearly dedicated as public parks for "the pleasure and recreation of the people" (Lockmann 1981; 75; Robinson 1975: 5).

It should be noted that at the same time, efforts had been made by Tulare County officials to include the entire western slope of the Sierra in this federal withdrawal, from the present boundary of Yosemite National Park to the southern end of the forest belt in Kern County. Tulare County officials worked from the principle that "it was necessary to protect the watershed of all major streams flowing into the San Joaquin Valley from the Sierra." In the end, however, they, along with Muir, concluded that to ask Congress for the reservation of such a large area would "necessitate an educational campaign on the value of land preservation for which there was not time." Therefore, they chose to work to preserve small areas that contained the giant sequoias that were in immediate danger, rather than to launch a lengthy campaign to preserve the entire western slope of the Sierra (Strong 1967: 5-6).

In the eyes of men like John Muir, society was based on cooperation, not competition. According to one biographer, Muir believed that "brought into his true relationship with wilderness, man would see that he was not a separate entity but an integral part of a harmonious whole." "No man had a right to subdue his fellow creatures or to appropriate and destroy their common heritage; to do so brought unbalance in nature, and loss and poverty for all." To nineteenth-century naturalists, the beauty of a preserved wilderness in its natural state was the most important of natural resources. This school of thought became known as "aesthetic conservationism" or "preservationism." In 1892, John Muir help organized the Sierra Club in order to support preservation and the further exploration of the Sierra Nevada through mountaineering (Bean 1968: 342-344).

Sometimes allied with it, but often in sharp opposition to preservationism was "utilitarian conservationism," which eventually formed around the scientific forestry idea of promoting sustained-yield forest management espoused by men like Gifford Pinchot (Bean 1968: 344). Pinchot, who graduated from Yale University in 1889, was at the forefront of this new movement. After spending time in France and Germany studying advanced forestry practices, which emphasized this philosophy, a youthful Pinchot returned to the United States in 1892. He took charge of the Biltmore Estate in North Carolina and began drawing up and implementing working plans for the estate based on the principles of scientific forest management he had learned in Europe (Hayes 1969: 28-29). Unlike the preservationists, utilitarian conservationists believed that forests should not be saved solely for their beauty and wilderness values, but "should be developed for commercial use rather than preserved from it" (Bean 1968: 344).

In the interim, a "pragmatic" conservation movement was in full swing in California, led by men such as pioneer California conservationists Abbot Kinney, chairman of the California Board of Forestry. For years, Abbot Kinney watched from his ranch the unrestricted use of the San Gabriel Mountains and the effects of uncontrolled fires, floods, unrestrained logging and overgrazing by cattle and sheep. In 1886, as chair of the California Board of Forestry, he wrote Governor George Stoneman, "The necessity of the hour is an intelligent supervision of the forest land and brush lands of California.... The destruction of the forests in the southern counties means the destruction of the streams, and that means the destruction of the country" (Brown 1945a: 19). In 1889, however, Abbot Kinney was dismissed from the chairmanship of the California Board of Forestry because of partisan politics. The California State Board of Forestry nevertheless continued Kinney's efforts. The Board pointed out the evil effects of fire, timber trespass and the delinquency of the federal government's neglect of the public domain in California. In 1890, they petitioned Congress to tend to the proper administration of California's timberlands, pointing out the "danger of a decreasing water supply in California and the Southwest because of the mistreatment of the watersheds." Nonetheless, the Board drifted from this mission into petty political squabbling and political patronage, straying from its larger aims and becoming "more of a Southern California institution." In 1893, it met an ignominious and indifferent end when the California legislature abolished it (Clar 1959: 118, 128, 130, 132-160 passim).

In the meantime, Theodore P. Lukens, also a leader in the crusade for watershed protection, pushed for regeneration of forests by planting efforts.

During the 1880s, Lukens, a Pasadena civic and business leader and citrus grower, had been interested in tree culture. He believed that burned-over mountainsides could again be covered with timber and the watershed protected. To this end, he collected seeds and cones and experimented with them on the mountain slopes above his home in Pasadena. His perseverance with reforestation would eventually earn him the soubriquet "father of California forestry" (Robinson 1991; 101). Reforestation was also a high priority for the California Board of Forestry. Many believed at the time that trees improved climate and assured increased natural water production. With dire predictions of national timber famine by the late 1880s, conventional wisdom suggested that tree planting would also satisfy this need as well. In 1887, the Board established experimental plantations in the north, south, and central parts of California at Santa Monica, Chico Creek, Merced, Hesperia, Livermore and San Jacinto (Clar 1959: 112-113; Robbins 1985: 1-2).

Mark Twain Cabin where he reportedly wrote "The Celebrated Jumping Frog of Calaveras County" near State Highway 49 north of Sonora, Stanislaus National Forest. 1945

To be sure, Kinney and Lukens were not the only people who were conservationists for pragmatic reasons. Aided and abetted by the preaching of these apostles of conservation use, local leaders such as Fred Eaton, mayor of Los Angeles; J. B. Lippincott, water expert and consulting engineer of the same city; Colonel Adolph Wood, manager of the Arrowhead reservoir; and General H. G. Otis, publisher of the *Los Angeles Times*, backed endeavors to have public watersheds properly protected (Brown 1945a: 19-20). In the end, men like Kinney, Lukens and others perceived early on the interrelationship between the conservation of forests and the conservation of water for metropolitan areas' domestic needs and hydroelectric development – the lifeblood of many thirsty and power-hungry cities. Protecting the ability of soils of mountain watersheds to hold water and to slowly release it downstream, particularly in Southern California, where a whole year's rainfall might be limited to one or two winter cloudbursts, was the fundamental key to this "pragmatic" conservation movement. Witnesses to the severe erosion of the surface soil of mountain slopes caused by indiscriminate cutting of timber and overgrazing of livestock and the flash floods, they joined forces to lobby the federal government to take action to protect California's forests from further damage (Bean 1968: 344-345).

Genesis: 1891 Forest Reserve Act and California's First Federal Forest Reserves

On March 3, 1891, while debating the issue of land frauds related to the Timber and Stone Act and other homestead laws, Congress at the eleventh hour attached a one-sentence amendment (Section 24) to an omnibus bill entitled the "General Revision" Act. Section 24 was drafted by the American Forestry Association (AFA), which was organized in 1875 to advocate the "protection of the existing forests of the country from unnecessary waste," and to promote the "propagation and planting of useful trees." This amendment directly granted the President of the United States the power to establish forest reserves from forest and range lands in the public domain. Unfortunately, Section 24 – also known as the Forest Reserve Act or Creative Act – provided no forest management infrastructure for these forest reserves. Nevertheless, less than thirty days later, President Benjamin Harrison used Section 24 to create the first forest reserve in the country – Yellowstone Park Timberland Reserve. This was the first step toward the creation of a national forest system. For the first time in history, the nation had a law that retained public lands for the future use of citizens, replacing the previous practice of rapid disposal of all federal lands (Williams 2000: 8; Robinson 1975: 5; Lockmann 1981: 73-74; Smith 1930: 18-19; Robbins 1985: 4).

The creation of this initial forest reserve was a remarkable event, and an idea that many Californians eagerly supported. In sharp contrast to many other western regions, who opposed any sort of land reservation system that "locked" the public out of the forests, less opposition to the concept of forest reserves developed in California, largely because many in the state recognized the importance of mountain watersheds (Lockmann 1981: 70-73, 82-83; Clar 1959: 163-164). Setting aside federal forest reserves promised not just watershed protection but also fire suppression and judicious cutting – goals that Californians also supported. However, it must be stressed that from the onset, California forest reserves were not considered as preserves for unique landscapes, like Yosemite National Park, which aesthetic conservationists such as John Muir had vigorously fought to set aside; they were justified by men like Abbott Kinney and other California conservationists as necessary and functional entities.

The first book in the genesis of California's federal forest reserve system came shortly after President Harrison's creation of the Yellowstone Park Timberland Reserve, when Colonel Benjamin F. Allen, a former Iowan banker and clearly a patronage appointee, was made a special agent for the General Land Office (GLO). Allen was assigned the task of investigating watersheds and timbered lands in Arizona, California and New Mexico for potential forest reserve status (Rose 1993: 2). Upon hearing the news of Allen's assignment, Southern California residents, associations and individuals immediately pressured Allen to conduct a California field trip as soon as possible.

Los Angeles residents and public officials led the effort in California. On November 2, 1891, eight months after the passage of the Forest Reserve Act, Los Angeles County citizens petitioned the GLO to withdraw the San Gabriel Range watershed as a forest reserve. "Water," according to the petition, "will be preserved in the mountains, the snow water saved as it melted, the waters [would be] protected from pollution by large droves of cattle and sheep, disastrous floods will be prevented in winter, and the valley's [sic] below furnished with water in the irrigation season." A month later, the Los Angeles Chamber of Commerce passed a resolution urging the GLO to permanently withdraw from sale all public domain included in the San Gabriel and Los Angeles river watersheds. Then, in the early weeks of 1892, members of Los Angeles County irrigation districts endorsed the establishment of a "forest reserve," and the County Board of Supervisors passed a resolution in favor of the move as well (Brown 1945a: 21-22).

By early 1892, Southern California public awareness of the relationship between forest conservation and sustained water supply had grown rapidly as the consumption of indigenous waters by irrigation and domestic users swelled. A shift to urban centers, such as Los Angeles, along with the establishment of lowland irrigated citrus farms in the area, very naturally pointed up the urgency of safeguarding watersheds (High 1951: 298). Southern California residents also realized early on that the lowly chaparral – despised elsewhere – should be appreciated along with the forests for its soil-holding capacity (Clar 1959: 104).

Based upon these petitions and requests, as well as pressure by California Congressman W. W. Bowers, Secretary of the Interior John M. Noble took action. Special Agent Allen was sent to investigate the need for establishing the "San Gabriel Forest Reserve" and other reserves throughout California. Following an extended field trip in the San Gabriel region, Allen submitted his report to GLO Commissioner Thomas H. Carter, who in mid-summer transmitted it to President Harrison. Commissioner Carter wisely wrote, "The future prosperity of Southern California depends upon protecting the water supply of the numerous streams which have their source in the mountains embraced in the reservation" (Brown 1945a: 21-22). On December 10, 1892, President Harrison signed the proclamation establishing the San Gabriel Forest Reserve (part of today's Angeles National Forest) – the first such reserve in the State of California. Composed of chaparral-covered land in the Transverse Ranges, it extended from Pacoima to Cajon Pass in both Los Angeles and San Bernardino Counties and embraced 555,520 acres (Clar 1959: 144; Rose 1993: 3; Lockmann 19981; 75).

While the establishment of the San Gabriel Forest Reserve was debated, Special Agent Allen considered the potential for other California forest reserves. On February 14, 1893, President Harrison withdrew California's largest federal forest reserve – the 4,057,470-acre Sierra Forest Reserve. This area at the time was practically as unknown and unconquered a wild mountain region as anything its size in America (Anonymous n.d.: 5-6).

The history of California's second forest reserve began in 1889, when Tulare County officials petitioned Congress and federal officials regarding the despoliation of the Sierra by lumber and livestock interests. In their petition, they explained how private individuals had developed irrigation in the San Joaquin Valley after the 1870s, leading to rapid agricultural growth in Fresno County, connecting it to world raisin markets. "This budding prosperity," according to the Tulare County petition, "depended on protecting the source of water which was endangered by timber speculators in the low mountains and by countless sheep in the high mountains" (Strong 1967: 8-9). This petition was either ignored or lost, until October 1891, when Special Agent

Allen was sent to California to investigate this and other matters. After interviewing "a number of interested and informed citizens," Allen immediately advised GLO Commissioner Carter to set aside *at once* the "Tulare Reserve" an area covering nearly five to six million acres, and embraced parts of eight counties from Tuolumne on the north to Kern on the south, and Mono and Inyo on the east side (Ayers 1938: [11]) In response to Allen's counsel, Carter immediately withdrew 230 townships in the Sierra pending further investigation (Strong 1967: 9). According to Allen, with the exception of a few sheepmen, a few miners around Kernville, and a handful of settlers near Fresno Flats, there was little opposition to creation of the "Tulare Reserve." "On the contrary," he wrote, "the leading citizens there complain of the devastation caused by flocks of sheep from the valley passing over their lands, and think that if included in the reservation, they would escape this annoyance." Allen focused his work on watershed protection for the Sierra, and in his report he made little mention of the timber or other forest resources located thereon. He concluded that a great wrong was being done to the public at large by those parties illegitimately using these public lands for pasturage and lumbering purposes (Rose 1993: 3). Though largely ignored in Allen's report, sheepmen, along with anti-park associations, vigorously protested his action. Allen, however, dismissed their protests and countered that they were "primarily Italians and Frenchmen who did not own land, speak English, or pay many taxes." In Allen's opinion, "foreign sheepmen were growing rich by pasturing sheep on government land which they damaged severely by permitting overgrazing and by setting fires" (Strong 1967: 11).

In the end, the protests of sheepmen were outweighed by the great majority of local valley residents who either supported the Tulare Reserve or were neutral about the vast withdrawal of land. Important California conservationists such as Abbott Kinney and Robert Underwood Johnson also lent their support to the creation of the Tulare Reserve. "This would not be reserved for Park purpose, of course," wrote Underwood to Secretary of the Interior Noble, "but to save water supply for irrigation below and to preserve timber." In late 1892, Allen revisited the area and completed his investigation, which included listening to the suggestions of John Muir and the Sierra Club. Early in January 1893, Allen sent his final report to Commissioner Carter. However, to eliminate a great deal of potential opposition, Allen reduced the size of the Tulare Reserve, especially on the east side of the Sierra. He also changed its name from the Tulare Reserve to the Sierra Forest Reserve (Strong 1967: 11-

13). Named after the snowy Sierra Nevada range of mountains, its northern portion bordered most of Yosemite National Park, and its western boundaries practically surrounded General Grant and most of Sequoia national parks (Rose 1993: 3). The size of this forest reserve increased over time, reaching a maximum of 6,602,353 acres in 1908, including large portions of seven counties (Anonymous n.d.: 2).

Only days after setting aside the vast Sierra Forest Reserve, President Harrison withdrew public domain for two additional California forest reserves. California's third forest reserve was the 109,920-acre Trabuco Canyon Forest Reserve (now part of today's Cleveland National Forest) in the Santa Ana Mountains. Named for a prominent peak and canyon within its boundaries, the Trabuco Canyon Forest Reserve was withdrawn for watershed protection purposes like the previous two reserves (Ayres 1938: [2]; Buck 1974: 27). More importantly, on the same day, President Harrison also set aside the 737,280-acre San Bernardino Forest Reserve.

Forest homes among western yellow and Jeffrey pines on land that eventually became part of the Angeles National Forest. Photograph taken by Theodore P. Lukens considered by many as the "father of California forestry." 1898

As with the pressure to establish the San Gabriel Forest Reserve or the Sierra Forest Reserve, a series of petitions from prominent citizens were sent to the GLO requesting forest reserve status to protect the montane vegetation cover of the San Bernardino watershed. Allen carefully inspected the area, canvassed the people in the region to gain their opinion independently, found the San Bernardino Range suitable as a potential forest reserve and informed Washington officials that "a large territory is depending on the water to be supplied from this watershed – it is therefore very important that the remaining forest belonging to the Government be preserved." Government bodies, such as the San Bernardino County Board of Supervisors, favored the idea, but a few people, acting either singly or in groups such as the San Bernardino Society of California Pioneers, expressed opposition to the idea. They feared that a federal forest reservation would displace local residents, owners of the reservoir companies and the larger public from access to these lands, and that large amounts of timber and minerals would be wasted. This opposition was never vigorous, nor coalesced, and President Harrison established San Bernardino Forest Reserve by proclamation (Lockmann 1981: 75, 83-85). It adjoined the "San Gabriel [Reserve] on the east at Cajon Pass and embraced all of the San Bernardino Mountains, extending to the Mojave Desert on the north and to its edge at Whitewater Canyon in the eastern extremity of the range" (Brown 1945a: 22).

By early 1893, four very large national forest reserves (San Gabriel, Sierra, Trabuco Canyon and San Bernardino) had been withdrawn from the public domain in California under the Forest Reserve Act. However, this action was not the panacea that conservationists expected. As pointed out earlier, the Forest Reserve Act made no provision for the active management of forest reserves and/or legitimate use within them – either for timber cutting, grazing, mining or any other use (Robinson 1975: 6). Forest watershed destruction by mining, lumbering and livestock interests, or by fire, did not halt simply because Washington "designated" an area a forest reserve. In fact, during the interim period between the passage of the Forest Reserve Act and the demarcation of the final boundaries of each forest reserve, private owners filed on millions of acres of California's choicest timberlands under the Timber and Stone Act. These tracts easily passed into the hands of the lumber interests. It was "safe to say that only a fraction of one percent of such lands were ever used for any purpose by the entryman personally" (Lockmann 1981: 74).

Administrative Beginnings and Growth of California Federal Forestry, 1891-1896

GLO officials understood the problem of a lack of administrative authority at once, and tried their best to protect these newly-born forest reserves from further exploitation and damage. Following the passage of the 1891 Forest Reserve Act, GLO officials set forth on a policy of "active and concerned administration," not one of "benign neglect" as some historians have contended. The latter coloring of the GLO as a "custodial manager" during the 1891-1897 period probably stemmed from long-repeated biases and criticisms against the GLO vocalized by Gifford Pinchot, who felt the reserves should be managed by trained professional foresters and scientists and not by "political hacks." During the years 1891 to 1897, the USDA Division of Forestry had no part in the reserves. But once Pinchot became head of the division in 1898, he worked to wrest control of the forest reserves from the United States Department of the Interior (USDI) and place them under the USDA Division of Forestry (later elevated to the Bureau of Forestry) (Muhn 1992: 259; Cermak n.d.: 48; Dupree 1957: 244-246).

In any event, in 1891 and subsequent years, GLO leadership actively sought to prevent further destruction by woodsmen and sheepherding trespass and wished to establish a protective force to patrol the reserves. But with no appropriations, and without any direction from Congress on how to administer or protect the reserves, the GLO did the best with the situation. It took the position that the reserves should be "closed" to any use until further notice, or until Congress provided the answer as to how the GLO should manage these lands, newly placed under their auspices. Neither Commissioner Carter nor Secretary of the Interior Noble was a preservationist. They believed that the reserves should be fully utilized and not be locked up as "pristine parks." Nevertheless, they took a responsible "caretaker" course of action and prohibited timber cutting, grazing and other uses until Congress decided how the reserves should ultimately be managed (Muhn 1992: 262-263). This position naturally created difficulties, and the question was raised as to how to enforce this closure without any appropriations or personnel. The GLO had no hope of effectively patrolling the forest reserves unless they received help from elsewhere.

To solve this problem, the GLO first turned to the War Department for assistance, much as it had done in 1890 following the establishment of Yosemite, Sequoia and General Grant national parks in California. The War Department, while sympathetic, refused their request because it felt it had no legal authority to use soldiers as a "*posse comitatus*." Without their assistance, the GLO next turned to mere bravado. In 1894, GLO formally promulgated regulations prohibiting trespass and/or any depredations within the forest reserves. The Land Office then published them in local California newspapers and also posted hundreds of notices along forest reserve boundaries. Public reaction was immediate. The general public outside of Southern California, ignorant of what a "reserve" meant or might mean in the future, and unfamiliar with the concept of conservation, almost unanimously thought that a "reserve" meant something "selfish, useless, locked up, taken from the community and the people" (Anonymous n.d.: 10-11). Many a California politician protested this action against important constituents, declaring that the prohibition would bring "absolute ruin" to the livestock industry. Others, especially sheepherders, openly defied the "paper tiger." Sheepherders tore down the notices posted along the Sierra Forest Reserve's west boundary, and even worse, there were said to be a "half a million sheep in the reservation, more than anyone had previously remembered seeing." In fact, the condition of the Sierra Forest Reserve had become so bad that it was said that, if not for the pines and tamarack trees, which the sheep could not prey upon, it could be termed a desert (Muhn 1992: 259, 264-266, 268; Strong 1967: 13-14).

Failing to prevent trespass and further depredations of the natural resources on the newly- established forest reserves with this tactic, the GLO finally turned to the federal courts and filed legal actions, obtaining mixed results with this approach. Lawsuits in Southern California against timber trespassers proved successful. Vigorous prosecutions had "scared timber cutters with the real possibility of arraignment and imprisonment." By 1896, the GLO had halted timber depredations on the San Gabriel, San Bernardino and Trabuco Canyon Reserves (Muhn 1992: 266-267). Stopping grazing trespass was another matter.

At first, the GLO sent agents to the reserves to prevent further grazing, but after a few days, they returned to Washington to report complete failure. "After all," according to T. P. Lupkin, special agent for the GLO, "what could one man do in millions of acres against seasoned sheepherders who knew the terrain and remained primarily in remote areas?" (Strong 1967: 13). Therefore, the GLO turned to prosecuting grazing trespass cases in Southern California. This action also proved more difficult and less successful than similar techniques used against fixed-in-place lumber interests. The GLO took several Southern California livestock owners to court, charging them with trespassing on the federal forest reserves. U.S. attorneys for the Southern District of California and GLO special agents worked hard to gather sufficient evidence against stock raisers for using the range – perhaps the first such cases in the nation. However, they won no convictions. On reserves that had been overgrazed for decades, such as the San Bernardino, witnesses could not pin the destruction of the range on any one single band of sheep (Muhn 1992: 266-267). Sheepmen continued to treat forest reserves as unreserved public domain.

The problem of forest reserve management in California and elsewhere in the West continued without resolution, even after the reelection of President Grover Cleveland in 1892. President Cleveland promptly foisted the problem on Congress by refusing after September 1893 to withdraw more land for forest reserves until "Congress made provision for the administration of the already-existing forests" (Robinson1975: 6).

For several years, a tortuous battle was fought in Congress over legislation to resolve the problem, causing a stalemate on the issue. The primary bill under consideration, known as the McRae bill, after Arkansas Congressman Thomas Chipman McRae, "provided for protective administration by the Secretary of the Interior, regulated the sale of lumber *apart* from the land, and restored to entry, lands primarily valuable for agriculture" (Smith 1930: 20). The complexity of national opinion for and against forest reserves revealed itself in the debates, which provided an arena of diatribes that either violently blasted preservationists like John Muir and other persons, who were allegedly misguided by "sentimental emanations from forestry clubs" and/or repudiated and damned "great lumber and water corporations," which many congressmen felt had "engineered the creation of the forest reserves for their selfish exploitation" in the first place (Clar 1959: 164).

Finally, in 1896, at the urging of the AFA and others, the Cleveland administration requested that the National Academy of Sciences appoint a committee of experts to investigate the problem and break the stalemate (Clar 1959: 164-165). A National Forest Commission was set up to probe the problem of managing the forest reserves in order to halt the rapid and wasteful deforestation of Western lands by private individuals in California and elsewhere. Many were quite familiar with California geography and conditions. Bostonian Charles Sprague Sargent, founder and editor of *Garden and Forest* (1887-1897), whose monumental *Report on the Forests of North America* (1884) was the first authoritative report on national forest conditions, chaired the commission. Other prominent members of the commission included William Henry Brewer, a scientist and early AFA member who assisted in the geological survey of California (1860-1864) (Clepper 1971: 40-41; 282-283; Robbins 1985: 3); aging naturalist Alexander Agassiz, who served on the U.S. Coast Survey in California (1859); distinguished geologist and mining expert Arnold Hague, who served on the United States Geological Survey (USGS) 40th Parallel Exploration Expedition (1867-77) and who later became an expert on Yellowstone Park; and Harvard chemist Oliver Wolcott Gibbs (ex-officio member) (Williams 2000: 8; *Who Was Who* 1962: 10, 451, 500). Finally, Gifford Pinchot, the least known of the men at the time (Clar 1959: 165), served as Secretary. Together, the commissioners traveled throughout the country touring existing forest reserves and areas proposed for future reserves.

Before the National Forest Commission tour reached California, the passionate preservationist John Muir joined them. Muir had just spent the previous year fighting off efforts of various interests to halve the size of Yosemite National Park in order to open it for lumbering and grazing. Regarding reservations of land, the fiery Muir proselytized:

> The very first reservation that ever was made in the world had the same fate. That reservation was very moderate in its dimensions and boundaries were run by the Lord himself. That reservation contained only one tree – the smallest reservation that ever was made. Yet, no sooner was it made than it was attacked by everybody in the world – the devil, one woman and one man. This has been the history of every reservation that has been made since time; that is, as soon as a reservation is once created then the thieves and the devil and his relations come forward to attack it
> (Browning 1988: 52).

Muir's and Gifford Pinchot's differences over "preservation" versus "utilization" of forest reserves, and especially sheep grazing, most likely emerged at this time. John Muir, a Scotsman who loathed sheep, did not want any on the reserves; while Gifford Pinchot, trained in scientific management of forests, felt they could be managed and restricted (Williams 2000: 8).

After a three-month nationwide field survey, Chairman Sargent submitted the commission's findings to the Secretary of the Interior. The commission recommended that the President follow the provisions of the unsuccessful McRae Bill. Their recommendations also pushed for the creation, forthwith, of thirteen additional or enlarged reserves and two national parks from 21 million acres of public domain. On February 22, 1897, just days before he left office, the President reneged on his earlier commitment not to create more forest reserves, and did as Chairman Sargent advised; surprisingly, Pinchot protested this action (Clar 1959: 165; Lockmann 1981: 87). On that date, Cleveland created thirteen reserves in California, Utah, Washington, Montana, South Dakota, Idaho and Wyoming, which became known thereafter as "Washington's Birthday Reserves" (Williams 2000: 9). Howls of protest from the western states ensued. From this action, however, California gained two new reserves – the

Stanislaus Forest Reserve and the San Jacinto Forest Reserve (now part of the San Bernardino National Forest).

The 691,200 acre Stanislaus Forest Reserve, named for the river whose headwaters rose within its boundaries, contained what is now the southern part of the Eldorado National Forest, as well as a great deal of the east side of the Sierra, which comprised the former Mono National Forest (today's Humboldt-Toiyabe National Forest) (Ayres 1938: [12]; Ayres 1911: 1; Clar 1959: 177). Like the vast neighboring Sierra Forest Reserve to the south (created in 1893), the lesser Stanislaus Forest Reserve quietly became a reserve with little public opposition other than a few sheepmen. But unlike the other Southern California forest reserves (San Gabriel, Trabuco Canyon and San Bernardino, set aside in 1893), the 737,280-acre San Jacinto Reserve did not come about due to local pressure over threatened watershed (Clar 1959: 177). Civic and agricultural leaders felt there was plenty of water to go around because Hemet Dam had just been completed. Instead, they were concerned about the denudation of the San Jacinto Mountains, which had been seriously logged out and overgrazed. "The worst enemy [of] the forests," according to botanist Harvey Monroe Hall who studied the San Jacinto Range, was "not the forest fire, but the sawmill. Many a pine-clad slope has been stripped of its best trees in order that they might be converted to lumber." Though cattlemen had kept the "woolly lawn mowers" off the high mountain pastures of the San Jacinto Mountains, the area was still heavily overgrazed by cattle and the range needed rehabilitation (Robinson and Risher 1993: 73, 96, 177-178).

Within weeks after President Cleveland's last-minute creation of the Washington's Birthday reserves by executive proclamation, Congress passed a measure to eliminate the new forest reserves, as well as the older ones. Cleveland pocket-vetoed the bill. Thereafter, President William McKinley took office in early March 1897. He called a special session of Congress, during which intense Congressional debate took place over forest reserve management. With McKinley's inauguration, supporters of federal forest reserves, including both conservationists like Gifford Pinchot and preservationists like John Muir, feared that the entire system was in jeopardy and might be abolished (Robbins 1985: 8) by powerful Western interests who worked to eliminate the reserves entirely (Clar 1959: 165). Finally, in mid-May 1897, after much compromise in conference committees, Congress presented to President McKinley the Sundry Civil Appropriations Act, which he signed into law on June 4, 1897.

Known today as the Organic Act, because it provided the main statutory basis for the management of federal forests, the Sundry Civil Appropriations Act was the first legislative step toward the proper care, protection and management of forest reserves, as well as their overall administration. This carefully-crafted piece of legislation had many features. First, the 1897 Organic Act specified the purposes for which forest reserves could be established, as well as their administration and protection. Second, the Secretary of the Interior was granted authority to make rules and regulations for forests. Third, the act allowed the GLO to hire employees to administer the forests and open the reserves for use. Fourth, it provided that "any new reserves would have to meet the criteria of forest protection, watershed protection, and timber production" (Williams 2000: 10; Clar 1959: 165; Strong 1967: 14-15). And fifth, it delegated responsibility for mapping of forest resources in detail to the United States Geological Survey. This latter feature gave two separate branches of the Department of the Interior responsibility for the forest reserves. The GLO was charged with sales, claims and administration of the forest reserves, while the USGS was given the task of producing land cover maps and delineating reserve boundaries (Lockmann 1981: 87-88).

To appease and secure the support of Western delegates for the Organic Act, Cleveland's "Washington's Birthday Reserves" were suspended for nine months (Williams 2000: 10). This action temporarily restored the land in these reserves to the public domain, but with no explanation excluded California's two newest reserves – the Stanislaus and San Jacinto reserves (Smith 1930: 21). Western interests, which feared that a strong reservation policy threatened their livelihood, supported the measure because it made very clear that the purpose of the reserves was "to furnish a continuous supply of timber for the use and necessities of the citizens of the United States." Essentially, the function of the reserves would be economic, guaranteeing that forests would be utilized and not closed to economic uses. Furthermore, the Organic Act stated that forest reservations could also be made "for the purpose of securing favorable conditions of water flow," a provision that pointedly provided for California's needs (Robbins 1985; 6-8; Lockmann 1981: 87). In the end, the 1897 Organic Act paved the way for Gifford Pinchot's future resource utilization policies (Jackson n.d.: 126).

The Organic Act was not perfect. Probably its worst feature was the so-called "forest-lieu" clause that stated that owners of unperfected or patented lands within the reserves could exchange said lands for selected vacant land open to settlement, which equaled the acreage of their abandoned land. The intention of this clause was relieve settlers who found themselves surrounded by forest reserve. However, this clause resulted in many scandals, as large landowners such as railroads and lumber companies, and even small homesteaders traded near valueless lands within the reserves (e.g., worthless railroad grant lands, denuded timber tracts and/or poor agricultural lands) for valuable land elsewhere (Robbins 1975: 7; Smith 1930: 21-22).

In the wake of passage of the Organic Act, President McKinley set aside several important new California reserves. In March 1898, he set aside 1,644,594 acres for Pine Mountain and the Zaca Lake forest reserves in the Sierra Madre Mountains of Ventura and Santa Barbara counties that today comprise the Los Padres National Forest. A year later, McKinley set aside the 145,000-acre Santa Ynez Forest Reserve (October 1899), which lay mainly in Santa Barbara County. In December 1903, the Pine Mountain, Zaca Lake and Santa Ynez forest reserves were combined to become the Santa Barbara Forest Reserve. McKinley's proclamations were in direct response to numerous petitions from local citizen groups asking for federal protection of the vital montane vegetation cover of the watersheds of the Sierra Madre and the Santa Ynez mountains. For several decades prior to this action, local newspapers were filled with account of uncontrolled fires that burned for weeks on end, and the loss of good Santa Clara farmland washed away by floods directly resulting from fires in the higher hills (Clar 1959: 177; Lockmann 1981: 88-89; Brown 1945b: 33-37).

By 1900, huge patches of federal forest dominated Southern California mountains, and in this part of the state, federal forest control was certainly welcomed. Watershed protection was a civic goal, which Southern Californians early on endorsed, and an emerging Southern California conservation movement educated the public on the matter. For instance, in 1899, the Los Angeles Chamber of Commerce and the Southern California Academy of Science formed the Forest and Water Society of Southern California, whose mission was to spread the message that "local mountains store water" and that they were subject to depletion unless stringent precautions were taken. Timber growth and scenic values were highlighted as well, but control and protection of watershed to ensure stream flow and groundwater storage for future metropolitan expansion on the coastal plains was paramount in the minds of these conservation-minded reformers – especially at national expense (Lockmann 1981: 91-93; High 1951: 304).

The California Society for Conserving Waters and Protecting Forests formed at this time as well. In 1899, twenty-four organizations, including horticultural and agricultural societies, boards of trade and chambers of commerce, and groups as divergent as the Sierra Club and the Miners Association met in San Francisco to form this organization and thereafter adopted resolutions requesting that the governor appoint a commission to study and report on the "status of water and forest conditions in California." Abbot Kinney, one of three vice presidents of the organization, a year later published *Forest and Water* (1900), a popular account of general forestry problems that soundly summed up the argument for watershed protection (Lockmann 1981: 93; Clar 1959: 168-169).

The McKinley administration also moved toward conservation of the northern Sierra Nevada Range. In April 1899, the President, by proclamation, created the Tahoe Forest Reserve as a "forestry reserve and public park." Since the 1870s, many Californians expressed concern for the Tahoe-Truckee basin, which had been stripped bare of timber, triggering the 1883 Lake Bigler Commission Report. Unfortunately, by this time, the south shore of the lake had been completely cut over by the Tahoe Lumber and Flume Company, as well as large tracts both in California and Nevada. In the 1890s, interest in resource conservation of the region arose again. The Sierra Club, fresh from the Sierra Forest Reserve fight and with the support of top California and Nevada officials, campaigned for protection of northern Sierra forest range. Their ultimate goal was the creation of a 260,000-acre Tahoe National Park (Jackson n.d.: 127-128; Pisani 1977: 12-14).

In response to this pressure, GLO Special Agent Allen was sent to investigate, and he readily concurred that the area should be withdrawn from the public domain for a national park. When Allen's report became public, unfriendly forces quickly mobilized in strident opposition and petitioned the GLO against any creation of any national park. The opposition included county officials, livestock, mining and lumber interests, who argued that the "proposed reserve would reduce taxable property of the community, be detrimental to the grazing rights of sheepherders and restrict development of fruits and potatoes on land suitable for agriculture." Lumbermen especially felt threatened by the proposed reserve. One local newspaper editorialized that the park would become no more than "a shady resort for Forest Commissioners and nonproducing loafers," while another stated that the San Francisco Sierra "Sporting" Club "had no right to create a game preserve and recreational playground at the expense of the local economy." To placate local interests, the GLO suggested that President McKinley withdraw a smaller tract of land – a suggestion he followed, when in 1896, he proclaimed 136,335 acres (less than half of what the proponents had suggested) as the Lake Tahoe Forest Reserve (Jackson n.d.: 128-129; Strong 1981: 82; Clar 1959: 177).

Immediately following its designation, leaders from the California Water and Forest Association, the State Board of Trade, the Sierra Club and other influential organizations pushed to expand the Tahoe Reserve to close to one million acres. Various economic interests, such as railroads, hydroelectric firms, irrigation companies and a nascent tourist industry, joined them. But the Tahoe Forest Reserve expansion movement failed, largely because some supporters realized the hidden danger presented by the "forest-lieu" clause of the 1897 Organic Act. In a highly influential article, the *San Francisco Examiner* condemned the expansion movement, charging it would result "in the gift of thousands and tens of thousands of acres of the choicest lands – timber, oil, mineral, agricultural and grazing – to private parties." Any enlargement of the Tahoe Reserve would most likely have included thousands of acres of cut-over lumber company land, or rocky, barren and precipitous lands owned by the Central Pacific Railroad (CPR). Under the Organic Act, these barren lands could be exchanged for valuable land of equal size elsewhere on the public domain (Jackson n.d.: 129).

The last forest reserves to be created in California under GLO management were the 306,518-acre Warner Mountains Forest Reserve and the 288,218-acre Modoc Forest Reserve – the most northeasterly of California's

Forest supervisor and ranger on the Trabuco Canyon and/or San Jacinto forest reserves. 1904

forest reserves. They encompassed the headwaters of the Pit River, largest feeder of the Sacramento River, and were both created by proclamation on November 29, 1904, by President Theodore Roosevelt. Four years later, these two units were consolidated into the Modoc National Forest. The background and circumstances behind of the formation of these isolated reserves differed from the creation of most other California forest reserves. Neither watershed protection nor timber depredations drove the impetus to invite the federal government's protection for this part of California. Instead, overstocking and overgrazing the ranges with cattle, horses and sheep, along with open hostilities between local cattlemen and itinerant sheepmen led to their founding (Ayres 1938: [5]; Brown 1945c: 1; Gates n.d.: 217-220)

When the first pioneers came to Modoc country, virtually the entire public domain was covered with stands of waving grass, and public rangelands were used on the basis of first come, first served. Little attention was paid to conservation measures. "Stock were turned out on the wild lands when the first green feed showed in spring and the ground still wet and soft from winter storms," one historian wrote, "the greater part of the growing grass being tramped into the ground before it could establish a sturdy growth. Only when the snows of winter forced the cattle out of the hills into the lower valley areas were livestock – and not all of them – taken off the open range." For the first decades after settlement, most of the livestock were cattle and horses. Then about 1880 came hordes of "outside" sheep from the Sacramento Valley and as far away as the interior of Oregon, monopolizing the ranges. Local cattlemen tried to stop them. First they passed county ordinances to impose taxes on the outsiders. Next they tried to prohibit grazing sheep within a certain distance of any homestead. But in the end, their tactics proved powerless. By the turn of the century, the Modoc ranges were in a "sorry condition, and in places had become mere dust beds from the trampling of myriads of sharp hoofs." In 1903, in a last ditch effort, the Modoc stockmen organized and signed a petition "almost to the last man among the livestock and business interests of Modoc County," and presented it to Washington officials, asking that a forest reserve be created. The GLO agreed with cattlemen's position when the Warner and Modoc Forest Reserves were created, and sheep were soon prohibited from the range. However, the GLO also had in mind other factors, such as the conservation of watersheds and potential timberlands, when these reserves were created (Brown 1945c: 25-26).

At the turn of the century, "reserving" anything in the way of large amounts of public domain was simply revolutionary. Yet between the period 1891 and 1904, a remarkable twelve forest reserves with more than 9.4 million acres of "reserved" forested lands (close to one-tenth of the state) were created in California. Americans woke up to learn that the nation was now in a "peculiar new business of rather amazing proportions. This was a new something called forestry" (Clar 1959: 166).

Department of the Interior Management of California's Forest Reserves, 1897 to 1904

Under the 1897 Organic Act, jurisdiction for forest reserve management rested with the USDI – specifically with an ill-prepared GLO, whose previous major obligations were more related to disposing of public domain rather than administering reserved land. In fact, the old expression "doing a land office business" was derived from the speed at which the GLO disposed of public land. At the end of the nineteenth century, GLO administration of its California forest reserves entered a new era. The Organic Act effectively opened all forest reserves to timber cutting, mining and livestock grazing. Thereafter, the GLO set out to manage its forest reserves accordingly. On the other hand, the United States Department of Agriculture (USDA) Division of Forestry, which had been created twenty-two years earlier, had no jurisdiction over these forest reserves or any other federal land. At this point, the USDA Division of Forestry had only ten people, and Gifford Pinchot made eleven, when he was became "chief" of the division on July 1, 1898 (Buck 1974: 20), although some might say his boundless energy would soon compensate for the small office numbers.

To confuse forestry matters, in 1901, the GLO created a Forestry Division (Division R) of its own, and a trained forester was placed in charge, with three other foresters to assist him. The new administrative machinery was planned along the lines of the existing force of special agents. California's reserves were divided into two districts (southern and northern) with each district under a different superintendent. B. F. Allen administered all four reserves in Southern California, while each district was further divided into reservations with supervisors in charge. Each supervisor had a force of forest rangers to perform detailed work (Robinson 1975: 7; Buck 1974: 17; Smith 1930: 26). To take some of the burden off the GLO, in 1901, the agency also signed a formal agreement with Pinchot's USDA Division of Forestry to provide forestry

Early horse drawn fire cart on the Modoc National Forest. 1920

services, such as writing manuals for fire control, while the GLO patrolled and enforced the law on the reserves (Cermak n.d.: 49).

The GLO had openly advocated for the establishment of a supervisory corps to protect the forest reservations prior to the passage of the Organic Act, and now officials believed that legislation called for the "issuance of rules and regulations that would provide for the use of the forest reserve sources" (Muhn 1992: 260-263). For instance, in 1897, the GLO, under pressure from cattlemen, issued a ruling "that the pasturing of livestock other than sheep would not be interfered with so long as it was neither injurious to the forest nor to the rights of others." This ruling allowing cattle to graze on public domain and prohibiting sheep softened in 1898, when the GLO Commissioner permitted sheep to be grazed on private or leased lands within California forest reserves if they were "guided" to and from the prescribed areas. However, it seems that paid guides cared little for boundaries, were susceptible to supplementing their income with payments from sheepherders to direct them to the finest meadows on the reserve and allowed herds to move to the leased lands at a "near-glacial pace." Overgrazing and trespassing on California forest reserve land continued unabated. In effect, there were "no limitations on stock, no grazing fees, and no established season" (Strong 1967: 14-15; Lockmann 1981: 143). In the area of timber sales, in 1899 the first GLO-approved timber sale on California forest reserve lands (thirty-three acres for little more than a million board feet) occurred on the Sierra Forest Reserve, and cutting began a year later. This was a simple application, probably with few, if any, restrictions and no supervision. By 1901, the GLO developed printed timber sale contract forms that contained twenty-five stipulations, and by 1905, the GLO had approved seventeen timber sales on California forest reserves based on these stipulations. However, there is no indication that GLO staff supervised any of the sales (Buck 1974: 20, 26).

Though the GLO had full authority to administer forest reserves, it clearly lacked sufficient funds and personnel to adequately protect them from overgrazing or timber trespass. At first the duties of California's first GLO forest rangers, who were often political appointees with no outdoors experience, were to build firebreaks and trails, fight fires and keep trespassing sheep off any particular reserve. They had to contend with "poor communications, conflicting orders, lack of tools, slow travel time, and [an] underlying antagonism from local people and forest users" who had no concept of the word "conservation" or understanding of what a "forest reserve" was in the first place (Cermak n.d.: 50-52).

The early administrative officers between 1897 and 1904 had little money to spend on any of the forests, but their jobs differed little from

An early ranger and his collie on the West Walker River within today's Stanislaus National Forest

reservation to reservation. On the Sierra Forest Reserve, as a rule, they were faithful, honest, and hardworking, but also as a rule, "they knew nothing whatever about forests, grazing problems, the human side of everyday life, or mountaineering in general." This summer force of three to eight men were paid $60.00 each to ride the vast slopes of the reserve and do all of the fire control jobs, including arresting fire violators, while maintaining good public relations (Anonymous n.d.: 8-9). According to the GLO *Forest Reserve Manual* (1902) (Cermak n.d.: 55):

> Forest officers should inform transients and others concerning the rules and regulations. This must be done cheerfully and politely. A Forest Officer must be able to handle the public without losing his temper or using improper language.

Oftentimes, however, it was a dangerous job, in which a ranger's politeness fell quickly into rigid, strict, uncompromising courage. This character change happened especially when it came to trespass issues. Rangers had the authority to tie up sheep dogs, separate herders from their bands, mix and scatter the sheep and even drive sheep bands off the reserve. However, it took a very brave ranger, a man like Richard L. P. Bigelow or George Naylor, to face down an irate sheepman, threatening him bodily harm with a stick, or revenge, if he took these actions. Often the encounter resulted in fist blows and even gunfire, as in the case where Ranger Naylor had words with a Basque sheepherder that resulted in gunplay. Naylor was adjudged to have acted in self-defense in the affair (Anonymous n.d.: 9-11; Cermak n.d.: 55, 65). On Southern California reserves such as the San Gabriel, the job of a ranger was just as dangerous, if less confrontational. Most of the San Gabriel rangers were colorful woodsmen, like shotgun-carrying "Barefoot" Tom Lucas, who lived on a claim within the reserve. They patrolled the reserve, built trails, roads and firebreaks, and strung telephone lines, but also had "as many encounters with grizzly bears as they had with fighting fires" (Cermak n.d.: 53-54).

Early fire warning system included both the heliograph and telephone. Heliographs were devices for signaling by means of a movable mirror that reflects beams of sunlight.

The work of fighting fires in the brush, timber and saplings on the newly-established reserves usually fell to a few rangers and a small crew. They would grab their tools, some food and blankets for an overnight stay and ride or hike to the fire. Thereafter, they assessed where to build a fire line, and then worked independently until it was put out. But when these reserve fires threatened nearby communities, ranches and farms, local citizens and businesses aided forest reserve rangers in a coordinated effort (Cermak n.d.: 57-58).

Up until 1903, there was no coordinated statewide fire protection plan or effort. This situation changed after Governor George C. Pardee was elected that year. Pardee, a California Progressive, and a conservationist, covered many forestry issues in his inaugural address. In particular, he spoke about the danger to second-growth timber because of fire hazard in logging slash, as well as the waste of natural resources, and fraudulent land claims on federal lands. Governor Pardee fought off a California legislature antagonistic to the further creation of forest reserves, and signed a bill authorizing the State of California to enter agreements with the proper federal agency to conduct a joint survey regarding six mutual areas of forestry interest: (1) preventing loss by forest fire; (2) improvement of forests following logging; (3) reforesting parts of Southern California; (4) regulating grazing; (5); producing a vegetation type map; and (6) developing a plan to administer forest lands. A contract of agreement was signed between California and the USDA Bureau of Forestry, and Gifford Pinchot initiated a joint survey of the forest situation in California. The survey was led by his young professionals, such as William Churchill Hodge Jr., who received his master of forestry degree from Yale and was assigned to supervise the survey (Clar 1959: 186-197, 213; Cermak n.d.: 59).

The joint survey of California's forest situation took several years and included seventeen comprehensive scientific reports on various subjects, and forty-five special papers. Unfortunately, this large body of scientific literature on California's forests has never been compiled into one study under one cover, and regrettably, much of it is unpublished and the manuscripts are either deeply buried in national and state archives, or are lost. Important titles of reports range from E. T. Allen's "National Forest Reserves for the State of California," to Hodge's "A Report on a Forest Policy for the State of California" and his "Forest Conditions in the Sierras," to R. S. Hosmer's "Forest Conditions in Southern California (Clar 1959: 195-207).

Ultimately, the joint survey recognized that the forest conditions in California were in deplorable shape, and that watersheds were being damaged yearly by fire, widespread overgrazing and land clearing. Timber management was nonexistent, and exploitive, heedless logging had badly deteriorated the forests to the point where some open stands lacked the ability to reproduce. Regarding fire control, the survey pointed out that post-settlement, human-caused fires were ever increasing due to general carelessness and an indifference to fire. Its conclusions led to the passage of a series of laws by the California legislature, the most important being the Forest Protection Act and the State of California entering an era of modern forestry. It provided for a State Board of Forestry, a State Forester and many other features, leading one historian to call it a "milestone in the progress of forest conservation in California" (Cermak n.d.: 59-60).

Though the joint survey may not have directly attacked the GLO, the cumulative effect of all the reports, many prepared by Pinchot's Bureau of Forestry professionals, was a clear indictment of the ineffectuality of the GLO's centralized administration, as well as its tenure managing California's forest reserves since the passage of the 1891 Forest Reserve Act. Gifford Pinchot was convinced that the USDA Bureau of Forestry could better manage California's forest reserves. In 1904, when a landslide Progressive tide elected President Roosevelt, Pinchot was prepared to seek the transfer of the forest reserves to the USDA.

Conservation and the Lost Eden (Part II)

The San Gabriel Forest Reserve, California's first federal forest reserve, revealed the need for watershed protection in order to promote a growing and very thirsty urban modern California. The vast Sierra Forest Reserve, California's second, demonstrated the consequences of unbridled utilization of this huge rural region by various economic interests. A close examination of the condition of John Muir's "Range of Light" illustrates how darkness had befallen what some considered a "pristine paradise" by the turn of the twentieth century.

The diaries, letters, reminiscences and reports written by early visitors to the Sierra Nevada Range, whether they be gold seeker, explorer in search of passage or emigrant crossing over to a new life in California, revealed the basic appearance of the northern and central Sierra when it came under American control. In their writings, they commented and described at length a wide variety of terrain, vegetation, wildlife and natural phenomena. Many accounts were purely poetic, romantic and impressionistic descriptions, while others selectively commented on subjects that interested or impressed them, such as the *Sequoia gigantea*, while ignoring others, such as chaparral and brush cover. Later, scientists filled in the gaps for the northern and central Sierra, as well as describing the entire southern Sierra. For instance, after the Civil

War, the historic records of the California Geological Survey, the University of California excursion party led by Joseph LeConte, and various observations by naturalists such as John Muir, John W. Audubon and others left a basic record of conditions in the southern Sierra (Cermak: 1994: 27, 62, 64-79).

In his pioneering work, *Range of Light – Range of Darkness: The Sierra Nevada, 1841-1905*, Robert "Bob" Cermak argues that from the above early descriptions a picture of the original Sierra Nevada terrain, vegetation and wildlife can be derived. Paraphrasing Cermak's work, the lower foothills were a mixture of grass and oak savannah, changing to foothill pine and brush as elevations rose. As one ascended the ridges and flats, a person came into mixed conifers dominated by huge sugar pines, some more than 300 feet tall. As one climbed higher, this forest was replaced by true firs. Canyons were brushy at lower elevations, but then graded into a mix of brush and trees at higher elevations. Lodgepole pine was found at higher elevations, sometimes in extensive stands, but sometimes killed over large areas by insects, or perhaps fire.

Openings created by fire, insects, disease and the California Native American cultural pattern of repetitive burning to manipulate the environment to their needs occurred throughout the forests. Some areas had little ground debris, but most observers found an understory of downed logs, woody debris, brush and small trees. Birds and wildlife were abundant. Eagles, hawks, condors and other birds of prey were common. Large and small carnivores such as grizzly bears, gray wolves, coyotes, mountain lions and other wildcats were to be found throughout the range. Deer were found in the foothills and lower slopes of the higher mountains, and smaller game animals such as hares, rabbits and squirrels inhabited most of the range. Anadromous fish, such as salmon, were plentiful throughout the Sierra, but any stream or lake whose entrance was blocked by a natural barrier was devoid of ocean-going fish, such as salmon and steelhead. Many Sierra lakes never had fish in them, until they were planted in the late 1880s (Cermak: 1994: 71-82).

Between 1860 and 1890, the original "Edenic" Sierra forest described above was lost to the exploitation of miners, loggers, railroads, livestock and wildfires, whose impact on the soil, water, wildlife and vegetation was unimaginable devastation by today's values. The first resources to suffer were water, soil and wildlife. Placer mining during the Gold Rush destroyed nearly all the riparian areas of the west half of the Feather, Yuba, Bear and American rivers, and the lower reaches of the Consumnes, Mokelumne, Calaveras, Stanislaus, Tuolumne and Merced rivers. Hydraulic mining continued the damage, destroying lesser streams with deposits of mining debris, raising riverbeds and flooding farms downstream with slickens. An estimated 2.375 million cubic yards of all types of waste were put into California's river systems between 1849 and 1914. Scars from placer mining eventually healed, but damage left by hydraulic mining is still present (Cermak: 1994: 89-95).

Logging, a natural associate of mining, shadowed each mining operation. Logging companies supplied the props and shoring for tunnels, drifts and shafts. By 1855, there were about 100 sawmills in the northern Sierra alone, producing about 100 million board feet. In addition to mining materials, miners needed shelter from the elements, and the estimated 3,400 mining camps and towns in the Sierra stripped the woods for miles beyond each community. Logging intensified and reached deep into the mountains for timber as mining companies developed improved transportation techniques, such as the "gravitation chute," the "v" flume and the inclined tramways. The advent of railroads such as the Central Pacific also gave rise to a growing lumber industry, and required enormous amounts of timber. Just to cross the Sierra, the Central Pacific basically leveled a forest eighty miles long and several miles wide. Each mile of track required 2,500 ties; while the snow sheds over Donner Summit alone consumed 65 million board feet. As railroads built north and south along the Central Valley, seemingly limitless regions of timber were culled all along the western Sierra as far as the southern Cascades. Sugar pine was the logger's favorite species, although ponderosa pine was also taken. As pines disappeared, Douglas-fir and other species, especially redwoods, were cut. All in all, in the nineteenth century, loggers nearly stripped the Truckee-Tahoe Basin completely and had cut a swathe over the mountains for the Central Pacific in the northern Sierra. They also made heavy inroads into the forests of the central and southern Sierra (Cermak: 1994: 95-96, 102-114).

Matching the detrimental exploitation of the Sierra by mining and lumber interests was the overall impact, following the Civil War, of the livestock industry on the region. While many livestock fed in the valleys on local ranches, many more made the annual trek to the Sierra, depending on public lands for feed, causing long-term ecological change. "The effects of overgrazing and trampling by millions of cattle and sheep upon the Sierra Nevada," according to Cermak, "cannot be quantified, but a measure of their impacts on the land can be gained from first hand accounts." One citizen wrote, "There can be no doubt that sheepmen are a curse to the State; they penetrate everywhere, destroy the roots and seeds of the grasses; in traveling over the hills they keep the rocks and earth moving, destroying vegetation and denuding the hills of soil." The southern Sierra felt the hardest impact of livestock overgrazing because it was the destination for many transhumance sheep trails. This practice of moving sheep between different grazing lands according to season began in the spring from trails that led northward from San Bernardino, San Gabriel and Bakersfield up and through the Owens Valley. By summertime, they were moved up to mountain pastures. Using one of several mountain passes (e.g., Carson, Ebbetts, Sonora, Virginia and Tioga) they were brought back down into the Central Valley by winter and looped back along the western foothills to their home destinations. Other trails existed in the northern Sierra and the southern Cascades as well (Cermak: 1994: 116-122).

After everything else – the abuse of soil, water, timber and forage resources – much of the Sierra wildlife was decimated either by survival hunting by emigrants and miners who needed food or by market hunters, who made it their trade to supply the camps with fresh game and hides. Market hunters operated throughout the Sierra until at least 1903. All game, such as elk, antelope, deer, rabbits and even the fearful grizzly bears turned overnight into a promise of dollars for meat. Deer especially were slaughtered and taken by the thousands until 1883, when the decline of deer herds caused the state legislature to pass a bucks-only law, and then in 1893, the sale of deer for market was completely prohibited. Game birds were also intensely hunted, especially with the growth of urban communities such as San Francisco and Los Angeles, where there was a restaurant market for ducks, geese, doves, pigeons, quail and other birds. Finally, fish also became a market item. Similar hauls of salmon and trout were taken from the Sierra rivers and streams. At the turn of the century, the status of fish and wildlife reached its lowest point in history (Cermak: 1994: 123-130).

Despite the devastation by miners, loggers and livestock owners, probably the worst destruction to the Sierra forests came from wildfire. These infernos were caused by the indifferent attitude of emigrants, miners and settlers toward wildfire. Growing up in the humid Eastern and Southern forests, where fires spread slowly, they saw no harm in setting them, or did not think to watch over them carefully. But coupled with California's Mediterranean climate of long, hot and dry summers, indifference resulted in conflagrations of immense proportions. For instance, novelist, short-story writer and lecturer

Mark Twain recounts in his book, *Roughing It*, how one night he lit a campfire along the north shore of Lake Tahoe, turned for a moment only to find it "galloping all over the premises" until the flames roared out of sight and up and over the ridge. Wildfires followed people wherever they settled in the forest, and no part of the forest has not been ravaged by them. Despite this type of universal carelessness by the general public, most fires were deliberately set either by miners, seeking to expose mineral veins, by loggers to clear away slash, by hunters to improve visibility and by stockmen, both cattlemen and sheepherders, who sought to clear and remove brush and trees. The latter group's pyromaniacal habits were especially damaging in the central and southern Sierra Nevada (Cermak: 1994: 131-136). However, it was not the presence of fire that was the problem – for Native Americans routinely burned forests for thousands of years – the problem was the size, extent and randomness of burning that was so destructive (Lux 2004).

The culminating effects of years of mining, logging, livestock overgrazing and wildfires were recorded by the USGS in their reports covering the Sierra – a stipulation of the 1897 Organic Act. John B. Leiberg's report on the northern Sierra found that a "third of the forest had been cut over and that most of the forest had a heavy undergrowth of brush." In addition, "half or more of the timber on the area of 715,000 acres from the North Fork Feather River southeast to Lake Tahoe had been destroyed by fire." George B. Sudworth's report for the central Sierra found "widespread damage to soil, water, meadows and forests due to heavy cattle and sheep grazing and decades of repeated fires," and noted that parts of the forest had been "so heavily grazed by sheep that in some areas the surface was bare ground." The USGS reports established without doubt that unrestricted use of the Sierra over the past fifty years had destroyed the "forest values of the pristine Sierra Nevada" (Cermak: 1994: 138-139).

In 1904, despite seven years of attempted management by the GLO, California's twelve forest reserves had reached their nadir in terms of depreciation of forest quality and natural resource values when compared to the undeveloped conditions of the ecosystem at the advent of American incursion. Between 1848 and 1904, these once plentiful and unspoiled lands had been severely over-utilized, altered and damaged. A year later, however, a recovery period began when President Theodore Roosevelt transferred all forest reserves to the USDA. Thereafter, they began a journey of "rational" recuperation based on the conservation principles and leadership of Gifford Pinchot, first chief of the Forest Service (1905-1910) and his new Forest

Service. At this time, many new forest reserves, needing recovery too, were formed in California – completing for the most part, today's Pacific Southwest Region, which encompasses 20 million acres on eighteen national forests located in the North Coast, Cascade and Sierra Nevada ranges and stretches from Big Sur to the Mexican border in the south Coast range. The story of the resurgence of California's forest reserves and their rescue from despoliation begins in the next chapter.

Chapter III: 1905-1911

Rise and Early Development of National Forests in California

Progressive California and the National Forests

On September 12, 1901, Theodore Roosevelt's formal services to the nation as president began when President McKinley died from an assassin's bullets fired six days earlier in Buffalo, New York. This event in some ways ushered in the Progressive Era in America nationally, but by this date many Californians were already on board. In the 1890s, Californian's conflicts with the Southern Pacific Railroad, and an increasing corporate economy, spawned a Populist agrarian revolt in California, convincing many that traditional politics could not represent their interests. Many measures and ideas introduced by the Populists, such as the initiative, referendum and recall, were a prelude to those later embraced by California progressives. By the time of Roosevelt's ascendancy to the presidency, progressivism as a reform ideology was widespread in California and throughout most of the nation (Rice, Bullough and Orsi 1996: 340-349, 354).

All strata of society in California eventually enlisted in this crusade for "social justice," but initially there was a comparatively small group of business and professional men who were motivated to participate in this reform movement. These middle-class urban men found themselves caught between big corporations and labor unions and were either provoked to act by economic discontent or by status-class anxiety (Mowry 1951; Hofstadter 1955). They tried to wrest control of their municipal governments from city machines, which were supported by various interests such as saloons, brothels and a variety of businesses that profited from their close ties to city bosses. Reformers wished to reorganize governments scientifically and use them as instruments of social and economic reform, and to bring efficiency to government and order to economic life. This certainly was the case in San Francisco under "Boss" Abraham Reuf. Before and after the devastating 1906 San Francisco earthquake, reformers ran on municipal reform platforms, seeking to end the political corruption, labor troubles, graft and bribes from firms such as Pacific Gas and Electric, United Street Railway Company and Spring Valley Water Company – the highlights of the Boss Reuf era. The San Francisco reform efforts marked the opening of the Progressive Era in California (Rice, Bullough and Orsi 1996: 354-356). In the statehouse, California progressives tried to circumvent a corrupted legislature by electing crusading governors such as George Pardee and Hiram Johnson, vowing to end the political influence of "interests" such as the Southern Pacific Railroad upon the state.

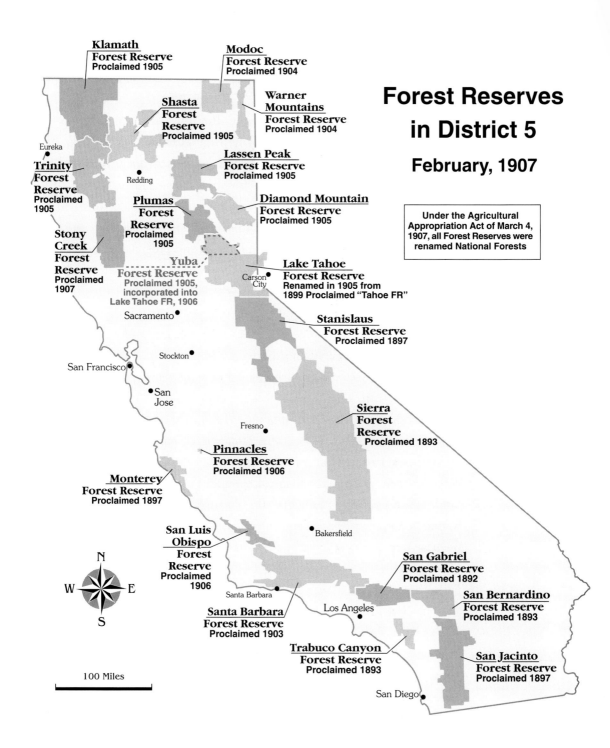

Despite individual differences and motivations, most California progressives shared a common fear regarding the private exploitation of public resources, a value shared by President Roosevelt. The previous chapter indicated how many Californians wanted the federal government to reserve particular forests to protect them from commercial abuse by mining, lumber and livestock interests. These views prevailed in Southern California in particular, where urbanites, along with irrigators, demanded public action to protect watersheds essential to the economic health of municipalities and agriculture (Hays 1969: 23-24).

As an outdoorsman, Roosevelt was no stranger to conservation principles, forestry issues and the American wilderness. In fact, Roosevelt learned of McKinley's death by messenger while hiking on Mount Marcy, the highest peak in the Adirondacks. According to his biographer Edmund Morris, Roosevelt also had a "profound, almost Indian veneration for trees, particularly the giant conifers he encountered in the Rockies." Along these lines, in 1888, he, along with others, established the first Boone & Crockett Club in America. Named after two of Roosevelt's personal heroes, he presided over it until 1894. As president of that organization in 1890, he petitioned Congress to pass key conservation legislation, including protecting Sequoia groves in California. A year later, in concert with the American Forestry Association (AFA), Roosevelt fervently lobbied on Capitol Hill for the Forest Reserve Act (1891). Ironically, a decade later, when he became President of the United States, he inherited the powers to "set aside at will any wooded or partly wooded country whether of commercial value or not" (Morris 1979: 384-385).

During the remainder of the McKinley term, Roosevelt, as an "accidental" occupant of the White House, gradually attracted progressives into the government. By the time of 1904 election, he was in firm control of the Republican Party thanks in part to the personal loyalty of these progressive-minded individuals. During the interim, Roosevelt used the power of the Forest Reserve Act sparingly. On October 3, 1903, he consolidated the Pine Mountain, Zaca Lake and Santa Ynez forest reserves into the Santa Barbara Forest Reserve (1,982,100 acres). And, in late November 1904, shortly after his victorious election, he established the Warner Mountains (306,518 acres) and the Modoc (288,218 acres) forest reserves in northeastern California. Both reserves thereafter came under the aegis of the GLO. Neither of these actions, nor his presidential campaign swing through California, gave any indication of any forthcoming radical change in forestry policy.

Elected by a popular vote of 2.5 million voters on December 9, 1904, Roosevelt made his conservation intentions known in his annual message to Congress. The forty-three-year- old President devoted a major section of his annual message to conservation and provided specifics regarding the management of America's forest reserves. With the work being divided among three bureaus (GLO Division of Forestry, USDA Bureau of Forestry, and the USGS) in two departments (Interior and Agriculture), Roosevelt recommended centralization under the USDA Bureau of Forestry (Smith 1930: 31). This transfer action had been recommended by the GLO, the AFA and the Society of American Foresters (SAF) five years earlier but was held up due to opposition from those western states without any watershed concerns. In 1900, Gifford Pinchot founded the SAF, and Vice President Roosevelt attended many of their meetings in Pinchot's Washington, D.C., home (Pendergrass 1985: 3-4). Within days of Roosevelt's annual message, the House passed a bill embodying this change, and a month later, the Senate reported favorably on a strengthened amended version. Meanwhile, the American Forest Congress convened in Washington in early January 1905 – an event that culminated the long struggle to bring coherent control and use to forest reserves that began in 1891 (High 1951: 306). Agriculture Secretary James Wilson acted as the presiding officer, and Roosevelt addressed the Congress at this important meeting devoted to forestry. The President declared that "the object of forestry was not to 'lock up' forests," but to "consider how best to combine use with preservation" (Smith 1930: 31).

The effect of the convention was to push the pending legislation into law. On February first, Congress, at the insistence of the president, passed the Forest Transfer Act (33 Stat. L., 628), providing for the transfer of the forest reserves and execution of all laws affecting forest reserves from the USDI to the USDA under Secretary Wilson. The Forest Transfer Act, which took effect on July 1, 1905, also provided that "revenues derived from the reserves during a period of five years from passage of the act be expended for the protection, administration, improvement, and extension of such reserves as the Secretary of Agriculture might direct" (Smith 1930: 31-32). For the first time in history, forestry on public lands was under the same executive branch – the USDA. The days of untrammeled exploitation of California's forests were over (High 1951: 306-307).

On the day the Forest Transfer Act passed, Agriculture Secretary Wilson sent out a charter letter, ghost-written by either Pinchot himself or with

Frederick E. Olmsted, District 5's first district forester (1908-1911)

the assistance of Frederick "Fritz" Erskine Olmsted. The latter was a young, discipline-minded engineer who, after attending Yale, entered the USGS in 1894 as a surveyor, where he met Pinchot. Inspired by him and following Pinchot's advice, Olmsted attended the Biltmore Forest School and then trained in the field of forestry in Germany and India. Olmsted returned to America in 1900 to join the USDA Division of Forestry at the age of twenty-eight. Olmsted's first assignment was to direct the important work of locating forest reserve boundaries. He became one of the famed "boundary boys," so-named because they thought they were superior to the Bureau of Forestry working plan groups (Clepper 1971: 242; Show n.d.: 12, 20; Pendergrass 1985: 80-81; *California District News Letter* 1925: 1).

Whether drafted by Pinchot or Olmsted, Agriculture Secretary Wilson's letter affirmed quite clearly the mission of the Bureau of Forestry (shortly to become the USDA Forest Service). Wilson's letter stated:

> In the administration of the forest reserve it must be clearly borne in mind that all land is to be devoted to its most productive use for the permanent good of the whole people and not for the temporary benefit of individuals or companies. All resources of forest reserves are for use, and this use must be brought about in a thoroughly prompt and business-like manner, under such restrictions only as will insure the permanence of those resources . . . .The permanence of the resources of the reserves is therefore indispensable to continued prosperity, and the policy of this Department for their protection and use will invariably be guided by this fact, always bearing in mind that the conservative use of these resources in no way conflicts with their permanent value. . . . The continued prosperity of the agricultural, lumbering, mining, and livestock interests is directly dependent upon a permanent and accessible supply of water, wood, and forage. . . . In the management of each reserve local questions will be decided upon local grounds; the dominant industry will be considered first, but with as little restriction to minor industries as may be possible; . . . and where conflicting interests must be reconciled, the question will

always be decided from the standpoint of the greatest good of the greatest number in the long run (Smith 1930: 33).

A year later, forestry management in America was completely reformed, based on Wilson's letter, which set forth the principles of conservation and multiple use. Most Californians joined the chorus of public opinion in favor of federal forest conservation. In 1905, they supported the Roosevelt administration even more when he nearly doubled the size and number California's federal forest reserves. "Roosevelt's conservation program," according to one historian, "was to be among his most impressive and enduring achievements" (Mowry 1958: 214).

Legacy Year of California Conservation – 1905

Credit for the rapid development of California's forest reserves and to the revolution in forestry was due in part to a young man named Gifford Pinchot and to Theodore Roosevelt, who listened to Pinchot's arguments and was persuaded by them. The "tall, lithe, dreamy-eyed," Pinchot had for years been one of Roosevelt's main sources of ecological information. For instance, during Roosevelt's second term as New York Governor, Pinchot convinced him to put a conservation section in his annual message to Albany. It was a revolutionary plea for a "system of forestry gradually developed and conducted along scientific principles." At that time, Pinchot theorized, "controlled, conservative lumbering of state and national forests would improve not only the economy, but the forests themselves" (Mowry 1958: 214; Morris 1979: 714).

On March 4, 1905, the vital and virile Roosevelt marched in his own inauguration parade. He had finally come into his own right and out of the shadows of McKinley. During the interregnum period from passage of the Forest Transfer Act and the inauguration, Pinchot busied himself with restructuring the agency to meet his long-dreamed ideas for a new agency. Under Pinchot's tutelage, the USDA Bureau of Forestry had grown from eleven employees in 1898 to 821 by 1905. At the same time, appropriations had increased from $28,520 in 1898 to $439,873 in 1905. The Bureau of Forestry in these early days was a technical bureau with a "small upper-story of mature foresters educated in Europe and an under-story of enthusiastic young men from the East, where most of them had newly graduated from a forest school." The Bureau of Forestry's name was changed to the Forest Service in order to reflect the agency's commitment to public service. Pinchot also became the first "Chief" of the Forest Service, although he wished to be known as

"The Forester." On July first, the administration of the former GLO Division of Forestry, comprising some five hundred employees, was merged into the newly designated Forest Service (Smith 1930: 30-31; Ayres 1941: 4).

To establish management principles for the Forest Service, Pinchot assigned Olmsted, one of his key men, to rewrite the GLO's *Forest Reserve Manual* (1902), which had essentially frowned upon forest users. Pinchot wanted Olmsted to write a manual to explain what the forest reserves meant, what they were there for and how to use them – in effect a policy blueprint to clarify forest officer duties concerning general public claims, rights-of-way, charges, duration of permits and other issues likely to arise. Ranger Mainwaring of the Sequoia Forest Reserve (R.L.P. Bigelow's brother-in-law) was sent to Washington to work with Olmsted and others on this task. On July 1st, the Forest Service issued the first "Use Book" of regulations and instructions for personnel, one of the first administrative government manuals. Prepared by Olmsted, the *Use Book* included the essential material from the GLO version, but the tone of the guide was positive instead of negative. It represented the Forest Service as not just another "enforcement agency dedicated to prohibiting use," but as an agency which was "willing to consider use under certain conditions" (Pendergrass 1985: 66, 81). In just sixty pages, the *Use Book* set forth for supervisors and rangers the goals of the organization, regulations governing each form of use and activity, methods to be followed, administrative powers and responsibilities in bringing orderly use, and protection and development to the reserves. Another forty pages listed the applicable laws (Show n.d.: 22).

Ranger's outlook, summit of Bear Valley Range. San Bernardino County, California

In 1905, the enlargement of Forest Service through the absorption of the GLO staff was accentuated by fresh responsibilities for new reserves around the nation. During President Roosevelt's first term, expansion of the nation's forest reserves seemed inevitable. However, a major blockade to expansion was the "forest lieu" clause of the 1897 Organic Act. This clause gave owners of unperfected or patented lands within the reserves the right to exchange these lands for selected vacant land of equal acreage. In 1900 and 1901, Congress constrained this process when it passed two pieces of legislation. The first law restricted selections made under the "forest lieu" clause to only vacant, non-mineral, surveyed public lands, which were subject to homestead entry. The second law gave lieu selectors an extra period of grace in which they might select land from unsurveyed and surveyed lands. Congress' action resulted in major land frauds in Oregon and California, which stopped only when the act of March 3, 1905 (33 Stat. L., 1264) abolished the "forest lieu" privilege as a matter of general application. Up until the passage of this legislation, prominent newspapers, such as the *San Francisco Examiner*, had condemned the expansion of reserves in California. Editorials charged that any expansion would result only in the "gift of thousands and tens of thousands of acres of the choicest public lands – timber, oil, mineral, agricultural and grazing – to private parties." When Congress repealed the lieu land law, a major barrier to expansion was removed (Smith 1930: 30; Strong 1981: 87-89; Jackson, Herbert, Wee n.d.: 129-130). With the blockade broken, expansion of the national forest system materialized rapidly. In California, seven additional California forest reserves were created under the Forest Reserve Act in a ten-month period between late March and mid-November 1905, making 1905 a legacy year for California conservation. During this period, Roosevelt also expanded the Lake Tahoe Forest Reserve and changed its name to the Tahoe Forest Reserve. Local newspapers did not voice opposition to this expansion, but one noted that an enlarged Tahoe Forest Reserve would be beneficial to the mining interest because it would furnish a permanent supply of mining timber (Jackson, Herbert, Wee n.d.: 130).

Created on March 27th, the 787,742-acre Plumas Forest Reserve was the first California reserve created under the auspices of the Forest Service. Situated in the Sierra Nevada, its boundaries roughly encompassed areas surrounding the branches of the Feather River. Forested with Douglas-fir, incense cedar, California black oak, red and white fir, ponderosa pine and sugar pine, the area had been used for at least 8,000 years by Native Californians. In historic times, it was the homeland of the Mountain and Konkow Maidu Indians and was also utilized by the Great Basin groups, such as Washo and Paiute, whose members lived along the eastern boundary. In 1820, Captain Luis Arguello led the first non-Indian exploring expedition into today's Feather River drainage. Thereafter, he named the river *El Rio de las Plumas* because of the great number of feathers of wild fowl floating on the river. Within a short time of the creation of the Plumas Forest Reserve, President Roosevelt also set aside 649,837 acres of the rugged high country to the north of the Plumas Forest Reserve as the Diamond Mountain Forest Reserve. On July 1, 1908, the Diamond Mountain Forest was combined with the Plumas National Forest by executive order (Ayres 1938: [7]).

Roosevelt's work under the Forest Reserve Act continued. On April 26, 1905, he designated the 1,243,042 acres between the interior Coastal Range on the west and the Cascade Range on the east in the central part of Northern California as the Trinity Forest Reserve. Members of the Wintu Tribe still use this native soil for food and other resources, such as native plants for basket making. The forest reserve appellation derives from the Trinity River, which Spanish Captain Bruno Heceta named after he discovered Trinidad Bay on Trinity Sunday, 1775. The Trinity Forest Reserve took in almost the entire headwaters of the Trinity River and included the picturesque Trinity Alps (Ayres 1938: [14]). Prior to its designation as a forest reserve, early settlers had brought cattle, sheep and horses; private loggers had harvested the forests; and, since the early Gold Rush days, miners had developed the minerals in the area.

Ironsides Lookout, Trinity National Forest

Days after setting aside the Trinity, Roosevelt next created the immense Klamath Forest Reserve, north of Mount Shasta, on May 6, 1905. Lying along the border of California and Oregon, with the majority of the forest in California, the 1,896,313-acre Klamath Forest Reserve spanned an area between Siskiyou County, Northern California, and Jackson County, Oregon. Named for the river that flows through the middle of the forest, the Klamath region was home to prehistoric peoples more than 10,000 years ago and to the Karuk, Yurok and Shasta Indian people 2,000 years ago. These native Californians made full and respectful use of the area's abundant natural resources, hunting deer and small animals, fishing for salmon and steelhead, and gathering berries, plants and acorns. At the time of its creation, the major activity on the Klamath Forest Reserve was still mining, including hard-rock or lode mines in the mountains and placers being worked on the rivers. Lumber operations centered on the eastern portion of the reserve, and livestock, mostly beef cattle, ran on the open range, grazing in the mountains in the summertime. In 1903, the Bureau of Forestry sent field agents to determine the suitability of the region for a forest reserve. At that time, local citizens were told that the forest reserve would provide employment opportunities for them. There was little public protest of these federal actions, which may have been due to the isolation of the area and the citizen's lack of knowledge of the situation (Davies and Frank 1992: 1-2).

A month after the creation of the Klamath Forest Reserve, the Lassen Peak Forest Reserve was established. On June 2, 1905, an area called the Crossroads – because it is where the sagebrush of the Great Basin meets the granite of Sierra Nevada, the lava of the Cascades and the Modoc Plateau – was set aside by presidential proclamation. The 897,115-acre Lassen Peak area was the central point of the territories of four tribes – Yahi, Yana, Atsugewi and Maidu. Living in the foothills of the Lassen Peak region was not easy, and native Californian seasonal patterns here were severely interrupted with the arrival of trappers, pioneers, miners and loggers. The tribes resisted as best as they could, but the reaction of some American intruders was too often untempered and unjust brutality, and slaying starving Indians for supposed offenses without distinguishing between innocent and the guilty (Strong 1973: 31-32). Settlers met the stiffest resistance from the Yahi, but time and luck ran out for them, and they vanished into the dense thickets and rocky cliffs of Deer Creek, which offered them refuge. In 1911, the sole survivor, Ishi, came out of the hills of his own accord, but died in 1916 of tuberculosis. Theodora Kroeber beautifully tells his story in her book *Ishi in Two Worlds: A Biography of the Last Wild Indian in North America* (Kroeber 1976).

Although not as majestic as Mount Shasta to the north, the 10,457-foot Lassen Peak region at an early date drew the attention of conservationists seeking to protect the watershed for northern Sacramento Valley communities. The reserve was dominated by 250,000 sheep and 100,000 cattle, which either used the range in the summer months or passed through it on the way to other range. Aside from forage, the Lassen Peak Forest Reserve contained valuable stands of untapped timber. Then again, almost annually, frequent fires caused by stockmen, careless logging operations or lightning destroyed large tracts of this excellent forest cover. Chaparral, which was not as effective for protection of this vital watershed, too often replaced the lost timber stands. When it came time to consider federal forest reserve status, powerful political interests, mostly stockmen, opposed it, fearing that grazing of sheep, cattle and goats would be either prohibited or restricted. Local county officials also saw no virtue in protecting timberland (Strong 1973: 31-32).

The last forest reserve created by President Roosevelt that year was the Yuba Forest Reserve, which lay immediately west of the Tahoe Forest Reserve and included the lands within the watershed of the forks of the Yuba River. Roosevelt set it aside by proclamation dated November 11, 1905, but the following year consolidated the Yuba with the Tahoe Forest Reserve (Jackson, Herber, Wee n.d.: 130).

Over the course of the next two years, Roosevelt and Pinchot turned their attention to creating five additional forest reserves in California. Three reserves were devised along and near to California's beautiful coastline – the San Luis Obispo Reserve (363,350 acres) and the Monterey Reserve (335,195 acres) on June 25, 1906, and then the smaller Pinnacles Forest Reserve (14,108 acres) on July 18, 1906 (Brown 1945a: 36-37). None provoked strong disagreement or public debate.

Several months later, the large inland Shasta Forest Reserve (1,523,770 acres) was created, although not without controversy. As early as 1885, snowy-peaked Mount Shasta (14,162 foot), perhaps the most sacred mountain to California's Native Americans (Ayres 1938: [10]), came into the sights of preservationist and naturalist John Muir. Appalled at the destruction of the forests on Mount Shasta's slopes, Muir and others were amazed that no voices had been raised in protest. For many years, Muir and the Sierra Club unsuccessfully tried to convince the powers that be to make the area into a national park. But a combination of various interest groups vigorously opposed the idea and expressed antagonism to a forest reserve as well. For instance, the

Trinity County Board of Supervisors was uneasy about losing a large amount of tax revenues from railroad lands if the reserve was to be established. Other opponents of the forest reserve were concerned that such an action would result in uncertainty over timber supplies needed by the local mining industry, thereby forcing them to cease operations. Despite this challenge, Roosevelt went ahead and on September 24, 1906, proclaimed the Shasta Forest Reserve (Martin, Hodder, and Whitaker 1981: 120-121).

Resistance to federal forest reserves spread to Mendocino County when the Stony Creek Forest Reserve was under consideration. The first examination of the Stony Creek area for forest reserve status was made in 1902. At that time, the majority of the local residents in and adjacent to the proposed area, with the exception of a few individuals, definitely were against this action. For instance, in 1903, these citizens sent a petition of protest to Washington, signed, after a mass rally, by 300 persons. Hostility to the idea continued. When the Bureau of Forestry sent R. W. Ayres to inspect the area in 1905, he was "almost mobbed" at one meeting he attended (Price 1946: 1-3). Despite the opposition of local residents, President Roosevelt, under the advice of Pinchot, issued the proclamation creating the Stony Creek Reserve on February 6, 1907. It was named after a stream whose source was within the forest reserve (Ayres 1938: [5]).

With the addition of the Stony Creek Forest Reserve, California had the largest number of forest reserves in the country – twenty. Two others were yet to come, but not until the forest reserves were renamed as national forests.

Initial Organization of Forest Service

Subsequent to Agricultural Secretary Wilson's charter letter, supplementary orders followed which began to set up a new order. To manage the "old" California GLO reserves and the "new" Forest Service reserves, Gifford Pinchot sought the help, influence and advice of future District 5 Foresters Olmsted and Coert DuBois. Both men had done much to help Pinchot build up the Bureau of Forestry into an "aggressive, bold, respected, close-knit and fearless band of zealots champing at the bit to take on the undone task of saving the West by public forestry." DuBois, like Olmsted, graduated from the Biltmore Forest School and had a brilliant and imaginative mind "lethal in spotting and dealing with inefficiency, abrupt and daring in devising and applying new methods" (Show n.d.: 12). Pinchot recruited him into the Division of Forestry in 1900.

During their many years of travel throughout parts of the West and California, Pinchot, Olmsted and DuBois witnessed and studied the "gross and widespread evils" affecting the forest reserves – "season long and destructive fires, destructive cutting and stealing of the public timber, incredibly severe overgrazing, deterioration of watersheds already vital or to be vital to irrigation agriculture and communities, wholesale and often fraudulent patenting of the public lands under loosely liberal and vilely administered land laws." They saw the "growth of monopoly in control of timber, range and water, blocking opportunity for the *bona fide* settler, the little man who was the intended or at least purported beneficiary of the lands laws" – which conflicted with their Progressive ideals. They saw the "hopeless ineffectuality, at best, of the General Land Office system of Forest Reserve administration, so highly centralized that rules and decisions were made by clerks in Washington who had never been in the West…" Pinchot recognized an urgent need to redeem the blunders of the GLO which, by "locking up" the reserves, by centralized control and by gross inefficiency had "created such powerful antagonisms that the whole public forest system balanced precariously on the knife edge of life or death" (Ibid.: 12-15, 17). To them it was a great crusade to curb control of monopoly while protecting and preserving the nation's forests for the "little guy" and future generations. The answer to the problems they observed was to organize the Forest Service under a decentralized plan.

For many years, at the Cosmos Club, at luncheons of the Geological Survey and at gatherings attended by academics, editors and other professionals, Pinchot discussed the best and worst aspects of organizations, administration issues and bureaucracy with knowledgeable persons. He also ardently studied, pondered over and discussed the subject with his bright young Civil Service qualified "technical" forest assistants of his staff. Thanks to Olmsted and Sir Dietrich Brandis, Pinchot's mentor in Europe, Pinchot learned of the decentralized organization and personnel methods of the Forest Service of British India. They seemed successful under conditions that broadly resembled those found in the American West – "burning to clear land, overgrazing by the sacred cow, hill agriculture, usage rights based on past use, a native population predominately opposed to change and authority, [and] a vast gulf between the haves and the have nots." Pinchot also decided that there was no American precedent to guide him, and he clearly rejected the field administration approach of the 1896 National Academy of Science Committee studies under the leadership of Charles S. Sargent (Ibid.: 17-17a).

Ultimately, Pinchot synthesized the initial pattern of organization, personnel and administration of the Forest Service from these experiences. Pinchot then gambled the fate of his venture on a decentralized field organization, placing its success or failure in the autonomous and practical hands of forest supervisors and the rangers underneath them. But more importantly, Pinchot established six inspection districts to oversee the fiscal and personal matters in various parts of the country. The idea of decentralization was as different from the previous GLO organization as it could be. As former Regional Forester S. B. Show put it, the doctrine of decentralization said in effect, "We trust you." "In a word, it brought responsible and responsive administration directly to the people concerned; they could deal face-to-face with decision-makers who knew conditions; they could get prompt action; if they tried the tricks of corner-cutting and trespass, such things were no longer ignored" (Ibid.: 23).

In California, placing the fate of his organization in the hands of forest supervisors was quite the risk. Early California forest supervisors were a highly diverse group. Some were politically appointed carryovers from the GLO, such as the notorious head of the GLO Binger Hermann, who represented something of the worst-case scenario. Hermann favored nepotism when selecting personnel for the Sierra Forest Reserve. His appointees included three brothers-in-law, a son-in-law, an elderly friend and a neighbor, who readily admitted they knew "nothing about forest conditions." Hermann's nepotism and the incompetence, malfeasance and inappropriate conduct of other personnel seemed to define many of the early California GLO supervisors. They were "more concerned with paychecks than the rules of the reserves." Failing to exercise any centralized authority over them, they largely did as they pleased in the days of GLO authority (Rose 1993: 14-15). Needless to say, most GLO holdovers did not last long in the new organization under Pinchot, especially after they took the 1906 Civil Service exam for Forest Supervisor.

For those GLO holdovers who were not fired or did not quit, and for those newly qualified Civil Service "professional" forest supervisors recently hired in these initial months, on October 3, 1906, their Herculean responsibilities were described to them at the first meeting of California forest supervisors under the authority of the Forest Service. This meeting was held at North Fork on the Sierra Forest Reserve. They did not know it then, but their job would be nothing like the scientific forestry practiced in hand-grown forests of Europe. They would have to make their own technical rules to fit American conditions (Ayres 1941: 5).

There were many pressing and immediate problems on each supervisor's mind at that meeting. First, and perhaps most important, they had to win the confidence and the goodwill of the forest users and the local communities, made difficult because of the sullied reputation of the GLO. Forest supervisors had to convince the public that they were capable and were honestly trying to make the resources that they protected as a government agency available to those who were dependent on them, whether it be timber, forage, minerals or water (Ibid: 5).

This was a difficult task because there were differences of public opinion regarding the forest reserves between northern and Southern California forests – much as there is today. In Southern California, supervisors met a public that had begged for federal "involvement to help them cope with the problems of fire and erosion, which required substantially improved watershed protection." Early on, the Southern California public accepted restrictions such as burning permits, while Southern California courts also willingly imposed harsh penalties on those who violated forest regulations like those pertaining to grazing permits and trespass. On the other hand, the general public in Northern California concluded that they had been "deprived of their natural right to do whatever they wanted" in the forests when President Roosevelt with a stroke of a pen "unfairly" imposed regulation on timber, forage and water. In addition, recent proclamations of the Stony Creek and Shasta forest reserves against popular local opinion increased public antagonism against the Forest Service (Pendergrass 1985: 54).

The situation was not helped when administrators openly vied over policy decisions. Under Pinchot, supervisors were given considerable latitude in their job and decisions, but they were expected to support the good of the whole organization over the needs or views of the individual administrator. If they did not "toe the line," they were fired. A case in point occurred on the San Bernardino and San Gabriel forest reserves. The Forest Service fired Acting Superintendent Theodore Parker Lukens (noted protector of watersheds and "father of California forestry") when Lukens disagreed with a reduction in force decision. After only one year's service, Lukens was discharged for "inciting local groups such as the Pasadena Chamber of Commerce against the policies of the Forest Service" (Ibid.: 56-59).

Besides dealing with a public that could be by turns hostile or friendly, supervisors were often confronted with difficult policy questions and decisions, such as how to handle agricultural land on forest reserves or how to establish forest grazing on a business-like basis or how to provide fire protection – issues that will be discussed later in this chapter. Finally, supervisors had to craft an infrastructure of much needed administrative improvements from limited budgets. Ranger housing, barns, fencing pastures, telephone lines, trails and firebreaks all needed to be planned for and constructed – not to mention having to conduct inventories of timber and grazing resources. All of this had to be done with very limited resources (Ayres 1941: 5, 7)

The weight of these responsibilities and decisions upon supervisor's shoulders caused some to quit soon thereafter, while others thrived when given the authority to put forest conservation into practice. One such man was long-time Forest Service Supervisor Richard L. P. Bigelow. He started out as the first supervisor of the Klamath Forest Reserve (1906-1908), and then for close to three decades, until his retirement in 1936, managed and made his home on the Tahoe National Forest. Bigelow was a "foothill cowboy – a jack of all trades, and by modern standards a workaholic who was ready to do whatever it took to make the reserves work." He exhibited many of the classic qualities needed to be a standout Forest Service supervisor. Practical experience early in his life helped

Tahoe National Forest Supervisor Richard L.P. Bigelow.

him understand the viewpoint of those he regulated. Dedication to and acceptance of the demands of long hours and days of constant work gave him the tenacity needed to endure his long service. His directness and firm delegation of responsibility to his rangers, his judicious scheduling of his staff and his patience when working with his colleagues and the public all were necessary qualities needed to maintain the rapport and fraternal comradeship of the early Forest Service. Supervisors who did not join in for dinners, story telling and singing usually left the job within a short period of time (Pendergrass 1985: 65-67; Rose 1993: 28-29).

Another California forest supervisor who exhibited the "right stuff" and who helped the sprawling Sierra Forest Reserve through its formative years was Charles H. Shinn. In 1902, Shinn landed a job as head ranger of the Sierra Reserve after a literary career writing newspaper and magazine articles on plants, forestry and natural resource issues, and after a stint as an inspector for California's Bureau of

Forestry. By 1906, Shinn had become Supervisor of the Sierra, a position he held until 1911, and thereafter, he worked as a forest examiner until his retirement in 1924. Shinn and his wife Julia, who served as his forest clerk for many years, went to work on the Sierra in a way that would be legendary. Julie ran the office while her husband handled the field operations. A disciple of Pinchot, Shinn was also an acquaintance of the prolific and entertaining California writer Stewart Edward White. The popular author had a cabin in a meadow near Shinn's field headquarters, where White wrote affectionately of the mountains and forests (Rose 1993: 17-18, 20, 22, 24; Show n.d.: 35; Cook 1991: 4).

Supervisor Shinn's most important administrative quality was his ability to recognize that a well-trained and motivated staff was imperative to success. He sought out "exceptional, dedicated, hard working field rangers who could work on their own and under the most difficult conditions." Early on, he was able to separate the "do-nothing" holdovers from Binger Hermann's patronage system on the Sierra Forest Reserve, and spot talent among the ranks of the "hard-working, self-reliant cowboys and teamsters" in the area, such as pioneer

Angeles Forest Reserve Supervisor's staff. 1905

ranger and cowboy, teamster and general handyman Gene Tully. Over the years, Shinn, with his wife Julia's approval, hired other outstanding personnel, and gradually their joint management brought many new ideas to the Forest

Service, such as organizing and hosting the first Forest Supervisor's meeting, held on October 3, 1906, at his North Fork headquarters on the Sierra Forest Reserve (Rose 1993: 26, 29-30, 33; Show n.d.: 35; Cook 1991: 4). On July 1, 1924, the "Spirit of the Sierra" retired. At that time, many old friends, colleagues and associations gathered to honor Shinn. Chief Forester W. B. Greeley praised him for the effective role he played in spreading the message of forestry, and California District Forester Paul G. Redington paid him tribute for the important part he played in the "upbuilding of forest protection and conservation in this state." By December of that year, Shinn had "passed over the Great Divide" (*California District News Letter* 1924: 2).

Supervisors like Shinn and Bigelow saw their rangers as part of the family and emissaries to the public. Not unlike them, California's earliest forest rangers were an extraordinary lot. While it may be easy to stereotype them, in the long run they seemed to be earthy, rough-hewn individuals, who were short on formal education but who had the ability to work hard and to live and care for themselves in the wilderness. Some were ex-cow punchers and ranchers. Others were miners, lumberjacks and other frontier types. "Even an occasional bartender, tin-horn gambler, and itinerant evangelist made the grade" (Ayres 1941: 21).

In general, the GLO-appointed rangers were for the most part incompetent. There are many stories that attest to this deserved reputation. Some were reproved for not doing the jobs they were assigned. Others were reprimanded for going home every day to work on their farms or businesses. Still others were "unwilling or unable to undergo the rigors of living in the wilderness for long periods of time." Many simply did not have "any knowledge of what they were doing." In some cases, they were "actively involved in land frauds committed by their friends, or in accepting money to 'assist' homesteaders in obtaining forest land that was immediately sold to speculators or timber companies" (Williams 2000: 19).

With the changeover to the USDA, most GLO rangers quit or were drummed out. To clean up the situation, one of Pinchot's earliest actions was the creation of the famous "Forest Ranger" Civil Service exam. Devised in the summer of 1905, this written and field exam tested an applicant's ability in a number of fields, as well as his self-reliance. Applicants were expected to be able to ride, pack, shoot, cook, sleep on the ground, do carpenter work and identify common trees and range plants. Only healthy, husky men who could do work with their hands were encouraged to apply. "This is a strenuous life," read one announcement, "and no weaklings need apply." By design, local men of good character and standing in the community became rangers and often later moved up to the rank of supervisor (Ayres 1941: 20-21; Williams 2000: 20).

USDA forest rangers were expected to work hard, and they were on their own to a degree unheard of today. As the "Government Man," they set forth on long trips – sometimes six to eight weeks at a time – covering huge ranger districts by horse and pack train, in all types of weather conditions. Guided

Preparing a typical pack train to access California's national forests.

only by the *Use Book* and its embedded conservation ethic, they had a very dangerous and solitary job. A typical trip might include evicting trespassing livestock and dealing with recalcitrant owners, handling fraudulent or misguided land claims, halting timber trespass and putting out wildfires. In their "idle moments," they were required to build administrative improvement such as cabins, pasture fences and trails, and to survey timber tracts. In the later case, one hardy ranger wrote the following doggerel (Show n.d.: 26):

> Oh its pace from morn till night thru the snow so soft and white
> Over hill and brush and snag, heavy snowshoes we must drag.
> While our toes and fingers freeze, we must tally countless trees
> Till we cry in desperation – "Oh to hell with conservation."

Outpourings such as the one above indicated the humorous side of ranger disposition in the face of challenging weather conditions and workloads. For years to come, "The Ranger" – the so-called foot soldiers for Pinchot – symbolized the Forest Service to the public. Their conviction, integrity, courage, firmness, fairness and resourcefulness, and their hard and willing work, their knowledge of country and people, and their savvy and common sense were the essential character ingredients that earned them the respect of locals.

Most lived by a set of rules which were often printed and displayed in forest headquarters. The author of the following early list of pragmatic maxims is unknown – but it captures the Forest Service's initial concern for establishing good public relations (Fishlake-Fillmore News (Utah) n.d.).

Ten Commandments for a Forest Ranger

1. **Be Agreeable.** If your voice is disagreeable and your manner of speech indistinct, see specialists. Don't get mad. You should be sunny, but don't get freckles.

2. **Know Your Business.** And when you tell anything talk plainly.

3. **Don't Argue.** When you argue with a man you are trying to push him. Be patient, overnight he may change his mind.

4. **Make It Plain.** Get a grasp on the fellow you are talking with. Do not get out that little book [Use Book] that will only puzzle him. Answer his questions without looking at your books.

5. **Tell The Truth.** By the law of averages, honesty gives the greatest profits. If you are working in the Service where you cannot tell the truth, quit and go elsewhere.

6. **Be Dependable.** If you tell a man you are going to do a thing, do it if it takes off a leg.

7. **Remember Names and Faces.** Don't call me Green when my name is Crane. I am sensitive about my name.

8. **Don't Be Egotistical.** I am. You must not be. Don't show off.

9. **Think Success.** Radiate prosperity. Do not mention calamities. Be a little Pollyanna.

10. **Be Human.** If the Service merely wanted to disseminate information they would use a catalog, not you.

Of course, individual rangers' temperaments and experiences varied, but perhaps the events in Robert Harvey Abbey's first year with the Forest Service illustrates the typical pathway between taking the Forest Ranger Civil Service exam and then becoming a forest ranger.

As a youth, Abbey worked as a tender and foreman for his father's band of sheep in Plumas County. In June 1905, while herding sheep, he encountered a forest guard, who informed him that the Forest Service was holding exams in July for the position of Forest Ranger. Being able-bodied and able to care for himself and animals in "regions remote from civilization," Abbey and fourteen other area residents participated in three days of field exams. This ordeal included marksmanship with rifle and revolver, axemanship and slash piling, saddling, packing, and riding a horse, using a compass, surveying and finally cruising and estimating a timber stand of growing pine and fir. The exam results were sent off to the U.S. Civil Service Commission in Washington, and Abbey would not learn how he had fared until the following spring (Abbey 1968: 1-10).

While waiting for the results, Abbey hired onto the Plumas Forest Reserve as a temporary forest guard for $720.00 a year – an appointment that terminated in December. During his tenure, he spent three weeks learning and seeing the conditions of the newly created Diamond Mountain Forest Reserve (July 14, 1905), which became his future assignment. The young forest guard found out early that the Diamond Mountain Reserve was "overrun with many bands of foreign migrating sheep from the State of Nevada and owned or managed by French sheep men or else a mixture of Spanish and French Basques." Fortunately, his background as a sheepherder and his limited knowledge of the Spanish and Basque languages enabled him to get along with the trespassers. No quarrels or bloodshed resulted from these encounters. During this time, he routinely put out small fires caused mostly by lightning (Ibid: 1-10).

Sometime in the spring of 1906, the Civil Service Commission notified him he had passed the Forest Ranger exam taken nine months earlier. Despite losing money as a forest guard because of unreimbursable expenses and because he had to provide his own transportation and horses, Abbey accepted the position. On April 1, 1906, he reported to Forest Supervisor L. A. Barrett, and received a probational appointment as assistant forest ranger for Diamond Forest Reserve at a salary of $900.00 per year. After receiving preliminary instructions from Barrett and setting up his summer headquarters on Last Chance Creek, Abbey set out to perform his year's assignments, which included making a boundary survey of the forty-mile-long and thirty-mile-wide Diamond Forest Reserve; locating and reporting on suitable administrative sites along the way; investigating and reporting on three homestead entries, one clearly not filed for agricultural purposes but for the stands of timber located on it; building a ranger station house and pasture; and putting out numerous fires with the help of stockmen and other users of the reserve. In late October, he was furloughed for the remainder of the year and returned home (Ibid: 11-16).

Like so many other rangers at the end of the season, Abbey lived through the winter on saved salary or by working at other jobs such as trapping, mining or doing odd jobs. Unlike Abbey, many rangers left the Forest Service after the first year because of the hard life and low pay. Even so, as a sign of their devotion, Pinchot expected his rangers to work for the cause of conservation "rather than the emoluments of office." Nonetheless, over time, the low pay scale for long hours of demanding work often weakened the morale of the best of these good men (Show n .d. 31a-33).

Further Expansion/Reorganization of California Forests – 1907

In early 1907, the initial expansion pattern of the Forest Service, instituted by Pinchot and his staff two years earlier, ended when the Agricultural Appropriations Act of March 4, 1907, forbade further "creation of forest reserves except by act of Congress in the states of Washington, Oregon, Idaho, Montana, Wyoming and Colorado – six states that contained by far the

Fishing camp San Gabriel Mountains. 1907

Source: Forest History Society

heaviest stands of timber in the West" (Dana and Fairfax 1980: 91). In the same piece of legislation, all forest reserves were thereafter renamed "National Forests" to emphasize "that the wood, water, forage, and recreation of the reserves was open to conservation use if compatible with the preservation and perpetuation of the resources" (Cook 1991: 5). This positive action only marginally offset the negative aspect of the 1907 Act.

There is no satisfactory explanation for why California was excluded from the prohibition list for creating additional forest reserves by presidential proclamation. Perhaps California was left off the list because private interests already owned twice as much timber as was protected in California's federal forests. Furthermore, this private timber was more accessible and of far better quality (Ayres 1958: 12). Or perhaps California was not listed due to the political strength of Southern Californian lobbying favorable to the Forest Service. As will be seen, important watersheds encompassing Owens Valley and the San Benito Mountains east of Monterey were yet to be protected. It might have been a combination of these two factors. Nevertheless, any further national forest expansion in California officially ended in 1912, when restrictions were finally extended to California by amendment (Robinson 1975: 9).

Meanwhile, Pinchot and his associates turned away from their initial inclination toward total decentralization and faith in trained supervisors and rangers when they created six "Inspection Districts," with a group of inspectors assigned to each. On the one hand, this action was a further effort to decentralize to give more authority to the inspection districts, where on-the-ground local people could handle problems more efficiently. It was envisioned as a means to facilitate and coordinate decision-making between the general policymakers in Washington, with their "hardboiled derbies and choker collars" and their "fluffy ruffled female clerks," and the autonomous, hardnosed, field-knowledgeable forest supervisors. On the other hand, the effect of the formation of these inspection districts was to plant a kernel of centralized structure into the newly named national forest system. In time, this seed would strengthen and grow in power, changing from California Inspection District 5 to simply California District 5, and thereafter would overshadow the autonomy of the supervisors and rangers (Show n.d.: 39; Williams 2000: 26).

In December 1908 and January 1909, the new inspection districts were staffed with persons from among the cadre of young professionals groomed by Pinchot. Olmsted was placed in charge of California Inspection District 5, with its headquarters located in the First National Bank Building at the corner of Post and Montgomery Streets in San Francisco. Under Inspector Olmsted were men like Coert DuBois, John H. Hatton, George W. Peavey, George B. Lull and William G. Hodge. DuBois and Hatton stayed with the Forest Service for many years, and their Forest Service careers will be covered from time to time in the pages that follow. After a short time of service, Peavey and Lull pursued different career opportunities. George Peavey left the Forest Service to become the first dean of the Forest School at Corvallis, Oregon, and then president of Oregon State University. Eventually George Lull left

First National Bank Building at 65 Post Street, San Francisco, California, second home of District 5. Circa 1906-1914

Source: San Francisco Public Library

the Forest Service and became California's second state forester. After 1912, Hodge left the service, and his later career directions are unknown (Show n.d.: 34-35; Buck 1974: 28; Ayres 1942: 8).

Initially, Inspector Olmsted and his staff had little executive authority other than reviewing the work of California's forest supervisors and pounding out reports to Pinchot on their Olivetti or Underwood typewriters. But during the years prior to the creation of California District 5, several key administrative actions took place in California that can be attributed to their influence and recommendation.

First, in early 1907, two new national forests were created. The Inyo, created on May 25, 1907, by presidential proclamation, became California's first designated national forest. As created, the 221,324-acre Inyo National Forest took in most of the floor of Owens Valley. Ostensibly, it was created to protect valuable public watershed and largely prevented further settlement in the valley and in the Inyo Mountain Range. One might say, however, that its sole purpose was to protect the nascent but critical Owens or Los Angeles aqueduct project, for the creation of the Inyo National Forest prevented further homestead and other claims from interfering with the aqueduct's right-of-way (Robinson 1933: 13).

As early as 1900, Southern California communities from San Diego to Santa Barbara sought to augment their water supplies. Los Angeles chose to go far afield to acquire water rights, and in 1904, to secure water for its future, city officials clandestinely turned to the Owen Valley, some 240 miles away in the Eastern Sierra Nevada (Lockmann 1981: 108-109). In the so-called "Owens Valley Caper," the Los Angeles Board of Water Commissioners, in conjunction with City Engineer William Mulholland, quietly began to secure options on land with associated water rights for the construction of the Owens Aqueduct. These activities remained secret until 1905, when several newspapers exposed it. Despite the scandal, the parched citizens of Los Angeles issued a bond for the construction of the aqueduct to tap the far-off Owens Lake. Preservationists and Owens Valley residents alike opposed the project. They rightly feared losing their own water resources if the project went through. Utilitarian conservationists, like President Roosevelt, supported it. As a major ally of the project, Roosevelt affirmed "it is a hundred or thousand fold more important to the State and more valuable to the people as a whole, if [this water is] used by the city than if used by the people of the Owens Valley." The creation of the Inyo National Forest for the benefit of the thirsty citizens of City of Los Angeles created bitter complaints from Owens Valley residents. Construction of the aqueduct began shortly after the establishment of the Inyo National Forest, and in 1913, the first Owens River waters poured into the San Fernando Valley. Owens Valley residents for years to come fought the aqueduct, and even resorted to sabotage, but in vain. To this day, many see Gifford Pinchot and President Roosevelt as "no good" culprits and scoundrels (Robinson 1933: 13; Rice, Bullough, and Orsi: 1996: 371: Hundley 2001: 141-155).

The second new California national forest was the San Benito, created on October 27, 1907, by presidential proclamation. The San Benito National Forest had a far less dramatic history than that of the Inyo. Roosevelt proclaimed this relatively small, 140,000-acre national forest to protect watershed in the San Benito Mountains east of Monterey. Months later, it, along with the Pinnacles National Forest, was consolidated with the Monterey National Forest (Brown 1945: 35).

The consolidation of several forests into the Monterey National Forest on July 1, 1908, was one of many consolidations, renaming, additions and eliminations of California's piecemeal forestry system. The July 1st reorganization of California's national forests was part of a Pinchot's larger plan to decentralize the Forest Service. On December 1, 1908, by administrative action, Pinchot created a new administrative system composed of six district offices, each under a district forester. On this date many administrative actions were taken, including the following: establishment of the Angeles National Forest from the San Bernardino and parts of the Santa Barbara and San Gabriel National Forests; adding the San Luis Obispo National Forest to the Santa Barbara National Forest; renaming the Lassen Peak National Forest the Lassen National Forest; setting up the Mono National Forest from parts of the Inyo, Sierra, Stanislaus and Tahoe National Forests; bringing into being the California National Forest from parts of Trinity and Stony Creek National Forests; incorporating the Warner National Forest into the Modoc National Forest; and integrating the Diamond Mountain National Forest into the Plumas and Lassen National Forests. A day later, on July 2, 1908, the southern half of the Sierra National Forest was cut off and renamed the Sequoia National Forest.

When the smoke cleared, California's patchwork quilt of twenty-two national forests was reduced to just seventeen. They were the Angeles, California (changed to Mendocino on July 12, 1932), Cleveland, Inyo, Klamath, Lassen, Modoc, Mono (absorbed by Nevada's Toiyabe National Forest in 1945), Monterey (changed to Los Padres on December 3, 1936), Plumas, Santa Barbara, Sequoia, Shasta, Sierra, Stanislaus, Tahoe and Trinity National Forests. Several other changes happened in the course of the next two years. First, the Calaveras Big Tree National Forest was created on February 18, 1909. A year later, two new forests were created. On July 1, 1910, the 1,434,750-acre Kern National Forest was established from part of the Sequoia National Forest. Then a few weeks later, the 841,211-acre Eldorado National Forest was created from parts of other national forests, making it the nineteenth California national forest. Public hostility toward the establishment of the Eldorado National Forest was minimal because most of the land had been previously included in the Tahoe and Stanislaus National Forests (Anonymous n.d.: 16; Supernowicz 1983: 153-155). That action occurred on July 28, 1910, and signaled the end of the expansion of California's national forests begun in 1905 by Roosevelt, Pinchot and Olmsted.

Creation and Initial Organization of District 5, 1908

Because of the number of national forests in California, the state became a district unto itself. With the formation of District 5, forestry policies and procedures geared specifically to California conditions could be developed. Naturally, Inspector Olmsted was made District Forester Olmsted, and San Francisco, which had served as the location for California Inspection District 5, was made the headquarters for Forest District 5. Under Olmsted were several strong and highly individualistic men, many of whom served previously with him in California Inspection District 5. They included Coert DuBois (assistant district forester), E. A. Lane (district law officer), R. L. Frome (chief of operation), C. Wiley (district fiscal agent), G. M. Homans (chief of silviculture), J. H. Hatton (chief of grazing), and C. S. Smith (chief of products – later called research) (Show n.d.: 40;Buck 1974: 29).

At this time, because of his powers, the district forester was likened to being an "autonomous king" and overlord of a domain – controlled only by laws broadly interpreted and by general policy. Olmsted and other district foresters thereafter were subject to orders on special projects such as boundary surveys, the size of timber sales and sending in annual grazing statistics to Washington, which were "solemnly and meaninglessly reviewed and approved." On the other hand, they had "virtually complete authority in hiring and firing; promoting and demoting people; creating new jobs; setting projects and priorities for using money; [and] deciding on methods of work and standards." Under Olmsted's control were the supervisors, deputy supervisors, forest assistants and between 130 and 140 district rangers (Show n.d.: 41-42).

Olmsted's term of office (1908-1911) was a time of "Arcadian simplicity." With the reorganization and decentralization, the creation of District 5 gave Olmsted and his staff a fresh start. Unencumbered with past failures, or limited by massive rulebooks, District 5 personnel were ready to save the "West from itself through public forestry and with the West's consent and support, to break monopoly and favor the little man." They held an unwarranted optimism that "conservation of water, timber, and range would readily follow once a determined effort was made." The year 1908 saw the general conservation movement and the Progressive movement in full swing, and they were ready to "crusade against sin and sinners, evils and devils, and for the little man" (Show n.d.: 42-43a). In time, their ebullient outlook, change-the-world attitude and faith in progress eroded away under the impact and weight of hard facts on hopeful assumptions.

Almost immediately, there were disputes between District 5 leadership and forest supervisors over jobs and responsibilities, salary issues and other administrative concerns. For instance, just locating the District 5 office in San

Francisco and not Los Angeles created problems. Jealousy emerged between the southern and northern forests. Southern California national forests, which had a public sympathetic to their creation and needs, felt like the "poor relatives" of District 5. It may have appeared that the San Francisco office favored the northern forests, but in reality, a recalcitrant northern public demanded more time and resources from staff, and the northern forests simply outnumbered those in the south by a two-to-one margin (Pendergrass 1985: 55).

Optimism also crumbled under the crushing burden of reality associated with timber management, fire protection, the ubiquitous grazing mess, archaic mining laws and watershed protection. Before tackling these inherited problems, Olmsted and his District 5 staff first and foremost had to settle the forest/agricultural land question in California – a problem that grew from the passage of the Forest Homestead Act several years earlier.

When the initial boundaries of California forest reserves and national forests were affixed after their creation, they often encompassed the lower slopes of the mountains, and sometimes part of the foothill country. Naturally, this included some land, mostly in small tracts, which had some measurable value as farmland. In order to appease those who thought any vacant government land was fit for the plough, Congress passed the Forest Homestead Act on June 11, 1906 (Ayres 1941: 8).

This piece of legislation, which was not repealed until 1964, created major administrative policy headaches, which for many years overloaded District 5 resources. The Forest Homestead Act required the Forest Service to examine all of its lands to determine whether any were more suitable for agriculture than for growing timber. If such lands suitable for agriculture were located, then the Forest Service was compelled to open them for homestead entry. As a concession to stockmen, lands found valuable exclusively for forage were excluded from entry. At the time of its passage, Pinchot publicly supported this legislation for one notable political reason – to reverse the intense dislike that most of the West had for his agency (Pendergrass 1985: 5-9).

In California, homestead applications flooded the San Francisco office at a rate of 150 a month, taxing and debilitating the manpower and budget of the embryonic District 5. Common sense indicated that much of the land applied for was not appropriate for agriculture because of water, soil, climate, altitude, and slope conditions. The tracts of submarginal agricultural land under application simply would never become a paying farm. In spite of that fact, land hungry Californians believed that there was no such thing as worthless land and were convinced they could succeed where others had failed. By July 1912, district staff examined approximately 12,000 homestead applications encompassing over 1,144, 359 acres. By this date, District 5 listed roughly 350,000 acres as opened to entry. But a check twenty years later on these entries in California revealed that more than 80 percent of the homesteads taken up under the Forest Homestead Act were no longer used for any agricultural purpose (Ibid.: 22; Ayres 1941: 8). Though the last legal "land rush" in California was over, there was still the problem of illegal squatting and fraudulent homesteading claims which needed resolution.

From earliest days, squatters have settled on California national forests. Some squatters, such as old-time miners and early-day mountain types, were tolerated until they passed away because they knew no other place to call home. Others were not. The latter group included men like John Knox, a reputed killer, who squatted in San Gabriel Canyon for twenty-five years, or like George Findlay, a former preacher, chicken farmer and miner, who went blind, was moved to a County farm and then drowned while trying to find his way back to his home (Brown 1945b: 76-78).

Along with unlawful squatting, homesteading abuses also occurred on California national forests. Unscrupulous "land sharks," known as locators, took advantage of the land-hungry "suckers" and foisted upon the unsuspecting what Assistant Regional Forester Louis A. Barrett once called "skim milk" lands. The Forest Service challenged many homesteader applications, but when they did, the government was subjected to years of protracted legal battles in the courts. Homesteaders even sometimes appealed directly to the chief forester or to the agriculture secretary. However, according to one source, "over ninety percent of the time, the initial judgments made in San Francisco proved correct" (Pendergrass 1985: 10-16).

Early California National Forest Issues and Problems

Handicapped with the problems associated with the agricultural land question and homestead applications, District 5 personnel still managed to address various forest issues and resource management problems. Timber management was by far one of the major responsibilities of District 5. The task of providing regional direction to the management programs on California's forests fell upon Chief of Silviculture G. M. Homans. Silviculture was a concept imported from Europe and taught in American forestry schools. It assumed that with the cooperation of timber operators, a fully productive and conservatively utilized forest estate could be managed for continuous production; in other words, sustained-yield management. This optimistic goal made several suppositions. First, it confidently assumed that the reservation of a substantial residual stand on timber sales (favoring pines) would insure prompt restocking of superior species, which would recapture the ground from invading brush. Second, it adopted the idea that through forest extension or reforestation, either seeding or direct planting could reclaim established brushfields. By pursuing these silviculture goals and achieving them, District 5 hoped to meet the Forest Service's statutory goal – to provide timber through sales for the use and necessities of the citizens of the United States (Show n.d.: 44-45).

The first government timber sale on a California national forest occurred in 1899 on the Sierra Forest. GLO Supervisor J. W. Dobson conducted it with little or no supervision. According to Dobson, timber prices paid at the time were considered higher than those paid to private owners for timber of equal value (Ayres 1958: 13). By 1905, GLO timber sales took place under stricter regulations, one that even stated that all lumber and forest products were to be consumed only in the state of California (Show n.d.: 1941: 22). This provision would have practically prohibited sales of any size to lumber companies if it were not ignored, for in practice, officials treated this requirement in a highly casual manner. Many timber sale contracts made no mention of the export clause, and generally speaking, in GLO days, timber sales were considered a "pain in the neck" to both purchasers and the government, in the opinion of one early District 5 official. The GLO made only seventeen recorded timber sales in California (Ayres 1958: 13-15).

In 1905, the first contracted sale under the authority of the Forest Service in California was made with the La Moine Timber and Trading Company on the Shasta Forest Reserve. Interestingly, this timber sale disregarded the "old" 1901 GLO printed form and was instead written on plain paper. It had twenty-five clauses, twelve of which were new (Ayres 1958: 15-16). Two years later, the largest of the early-day California timber sales (50 million board feet) was made on the Stanislaus National Forest with the Standard Lumber Company. In these early Forest Service contracts, no attempt was made to separate lumber values by species and grades, and stumpage rates were simply set by a "fair" agreement with the purchaser, based on the judgment of the officer in charge. In time, the amounts of timber purchased increased gradually, as well as the length of time of each contract. Forest Service sale contracts were printed up to replace the 1901 GLO form. The new contract form contained fourteen standard stipulations, including conditions regarding "care of unmarked trees, use of oils in locomotives, spark arrestors, and cooperation by the company in preventing and extinguishing fires" (Buck 1974: 27).

Before selling timber in California and elsewhere, Washington officials endeavored to get a baseline inventory of the timber stand on each national forest in the nation. In response, District 5 asked each forest supervisor to estimate the timber on his individual forests. Essentially having no inventory experience or staff to do so, supervisors simply made their best guess. Standing on high points with a pair of field glasses, rangers estimated the timber stand within sight of their district, and then supervisors sent the ocular figure along to Washington. These inventories pleased the head office so much that they wanted to turn all rangers into timber cruisers. Nonetheless, these "extensive" versus later "intensive" cruises were highly inaccurate and had little value when it came to timber sales. Furthermore, for more than two decades, much of the lumber industry ignored or paid little attention to these inventories and, for that matter, even to the details of their timber sale contracts. Because of limited staff and untrained men, early Forest Service timber regulations in California were violated with impunity (Show n.d.: 21, 23; Ayres 1941: 23; USDA Forest Service 1910: 1).

District 5 officials realized that the conduct of timber management in California in the first few Forest Service years left much to be desired and were not conducted to pursue the first goal of silviculture – sustained-yield management. Determination of the proper cutting rotation by sustained-yield studies was essential to good forestry. Therefore, in 1911, they set into motion the creation of silvicultural work plans for California's national forests. These work plans were concocted to determine how much timber could be cut each year continuously without interfering with a forest's productiveness. By 1912, a seventeen-page plan outline, asking for past, present and future information regarding all classes of forest business, as well as timber management on California's national forests, was worked out. Within two years, forest work plans for the Plumas, Monterey, Mono and Inyo National Forests were prepared. Though these plans were supported by a mass of related data, these documents, in one supervisor's opinion, were mostly made to give the files a "scientific aroma." In 1914, the project was suspended (Ayres 1941: 24-25).

The second supposition of silviculture, reclamation of brushfields through forest extension or reforestation, was also a major District 5 activity in its formative years, especially in Southern California. Unfortunately, it too did not succeed very well – but not for lack of effort.

In March 1904, a public meeting in Los Angeles attended by many businessmen of the region was called to discuss the protection of the watersheds. Grave doubts were expressed as to the future of the city unless water could be obtained. The consensus of the meeting was to increase forest covering and protecting the watersheds. Toward this end, they appropriated monies to carry out field plantings (Munns 1913: 1). Many Southern Californians were optimistic that the Forest Service could replace dense native chaparral on their mountainsides with conifers or some other exotic species (Ayres 1941: 20). Early on, GLO and Forest Service rangers were encouraged to plant trees of any species during spare time from their other duties (Brown 1945b: 117). Civilian groups also supported this dream. For instance, after leaving the Forest Service, Theodore Lukens, along with early conservationist Abbot Kinney, devoted a considerable amount of his time and personal money to planting trees, including many eucalyptus trees, on the sharp slopes of the hills above Pasadena – efforts applauded by Pinchot (Pendergrass 1985: 59; Brown 1945a: 92).

To meet the need for reforestation, District 5 established several tree nurseries on national forests in Southern California. By early 1905, the Angeles Forest began to produce trees at the Henninger Flats Nursery on Mount Wilson. The first year, nearly 222,000 seedlings were planted in the field or distributed to individuals who planted them in the mountains and parks at their own expense (Munns 1913: 1). The Los Padres Forest followed the next year with the establishment of the San Marcos Pass Nursery. By the conclusion of 1906, rangers had planted 30,000 trees in the Santa Ynez Range. To encourage and support this District 5 effort, Washington sent out Fred G. Plummer to locate five additional nurseries on what is now the Cleveland and San Bernardino National Forests. Soon everybody seemed to be forest extension minded – even in Northern California, where planting or seeding was thought necessary to assist natural reproduction in the reclamation of areas deforested by fires (Brown 1945b: 116; Brown 1945a: 91; USDA Forest Service 1910: 43).

The high tide in tree planting in California came in 1911, when District 5 contracted with the War Department for the afforestation of Angel Island, Fort Barry, Mare Island and, later on, at Yerba Buena, which was then called Goat Island. Other important nurseries established by this date included the Converse Flat and Lytle Creek Nurseries on the Angeles National Forest and the Pilgrim Creek Nursery on the Shasta National Forest. However, a year later, a "doleful" report written by Assistant Chief of Silviculture T. D. Woodbury and Forest Assistant Edward N. Munns, indicated that forest extension in Southern California was a foolish failure. The Forest Service's determined efforts to convert indigenous chaparral into the forests desired by Southern California residents fell short, through no fault of their own. Lamed-back tree-planting rangers knew before their superiors did that the reforestation work was futile. Each season, they watched the recently planted pine seedlings die in the hot summer sun or be consumed by great numbers of rabbits and small rodents. Despite the project's failure, Woodbury's report suggested that planting become a research matter and recommended that it be assigned to the newly established Feather River Experiment Station on the Plumas National Forest (Brown 1945b: 116; Brown 1945a: 91; Ayres 1941: 20; Buck 1974: 28; Munns 1913: 1-21).

As to fire, District 5's optimistic goal was that fire's destruction could be halted or reduced on California's forests and ranges with only a modest and orderly effort. From 1905 to 1909, District 5 met this goal with a thinly spread net of rangers and forest guards patrolling, as well as protection improvements such as firebreaks (Show n.d.: 44). Additionally, logging companies were required to pile brush and debris to eliminate potential fuel loads, and by 1906, lumber company employees and subcontractors were contractually obligated in timber sale contracts to prevent and assist in extinguishing forest fires (Ayres 1958: 20). Finally, a lookout system was established.

Patrolling the Inyo National Forest by pick-up.

The practice of placing men on commanding points on California's forests began in 1907, when a lookout was put on Claremont Peak on the Plumas Forest. At about the same time, permanent lookout stations were built on the Stanislaus Forest at Pilot Peak and on the Sierra Forest at Signal Peak. Within

Old Mount Ingalls Lookout Station erected 1910-1911, Plumas National Forest. 1932

a short time, other lookout points were established in the Sierra Nevada. A lookout system in Southern California forests soon followed (Ayres 1941: 14). With a broad system of regular patrols by rangers and guards, firebreaks and lookouts, as well as logging operators cooperating in prevention, District 5 felt that they were prepared to handle any fire situation. Then came the fateful year 1910, and several dreadful developments.

First, on January 7, 1910, Gifford Pinchot, the "father of American conservationism," was "fired" by President Howard Taft in a controversy over Pinchot's incendiary public criticism of Interior Secretary Richard Ballinger regarding withdrawn coal claims in Alaska. Long-time colleagues and protégés like Olmsted most likely were stunned at the dismissal of their champion of conservation. They had always assumed that Forester Pinchot would be there to lead them in the fight. Pinchot's replacement, Henry S. Graves, was appointed less than five days later. Though the "strongly puritanical and no-nonsense" Graves was a former dean of the Yale Forestry School and close friend of Pinchot, his "low-key," "strong-willed" personality differed considerably from that of Pinchot's and would be hard for some District 5 staff to get used to (Smith 1930: 39; Williams 1993: [2-4]; Cermak n.d.: 110; Dana and Fairfax 1980: 94-96; Robinson 1975: 10).

Then, to make matters worse, the 1910 fire season wreaked havoc on the western United States, burning two and one half million acres on national forests alone. The "Big Blow Up of 1910" was an unmitigated disaster in Idaho. In that state, the "Milestone Blaze," so dubbed for its effect on the public conscience, killed eighty-five people, seventy-two of whom were firefighters (Cermak n.d.: 108; Williams 2000: 32).

During the 1910 season, California was not struck as hard as such other western states as Idaho, Washington and Oregon. Yet by the end of the summer, 330,000 acres of California national forest land – "half a forest" – had gone up in smoke. The devastating conflagration was spread out over 278 different fires, two-thirds of which were caused by people, either accidentally – prospectors and others seeking to clean up the underbrush for one reason or another – or deliberately, in order to obtain work. In the latter case, District 5 officials seriously discussed the creation of a "secret service of plains-clothes agents" to spy on potential arsonists, hoping to convict and punish anyone who started a fire on purpose (USDA Forest Service 1910: 12, 76-81).

The size and number of fires burning, as well as the excessive losses pointed to unrevealed weaknesses in District 5's fire policy. In a post-season analysis, Forest supervisors readily admitted that they had been "running along in more or less complacent frame of mind, smugly sitting at their desks each fall turning out figures to show what splendid work they had done in preventing fires." Olmsted admonished them for this complacency, declaring that "if one-hundredth of the damage from fire this past summer had occurred in any German state, the whole forest force would have been promptly dismissed." Associate District Forester Coert DuBois scolded the supervisors further, proclaiming that "until we can handle fire on the Forests entrusted to our care, we cannot practice forestry on them…to protect them, is our first duty to our employers – the people of the United States…It's time we got war-like – time we put a fighting force on our frontier to stand over the enemy with a club" (Ibid.: 76-81).

The seriousness of the 1910 fire season, along with Pinchot's dismissal, convinced Olmsted to convene District 5's first supervisor's meeting. At this meeting, the major issues affecting California's national forests, as well as the past fire season, were discussed at great length.

Held in San Francisco in early December 1910, the gathering was attended by sixteen forest supervisors, five deputy supervisors and practically the entire District 5 staff. Olmsted led off the five-day meeting with an oblique reference to Pinchot and his philosophy that the West's timber, range and waters should be developed "for use, profit and enjoyment." [Pinchot's name was never mentioned directly over the span of the conference, probably because Washington staff was also present.] Olmsted soon turned to several broad questions in his introduction. What does the Forest Service mean? What is your everyday work about? Toward what end are we striving, and why? Olmsted tried to convey the idea of total decentralization and the responsibility of the individual forest supervisor for decision-making, pointing out that a district office was "merely a passing phase in the chain of organization" and would "almost disappear from view in the near future." "The Supervisor will be the Forester, and will run his Forest," said Olmsted, "without restriction except necessary to keep his own policy uniform with that of his brother Foresters throughout the West." Olmsted concluded his speech by emphasizing that the Forest Service should be there for the "protection and use of natural resources of the West for all time" (Ibid.: 1-12).

Following Olmsted's introduction, pointed questions were asked about every aspect of District 5's program – including timber sales, reforestation, working plans, fire protection, grazing, wildlife, etc. Regarding fire management, District 5 staff and California's forest supervisors set about trying to resolve the current crisis in fire control. At this time, Coert DuBois introduced a forest protection plan devised that fall for the Stanislaus National Forest by him, Forest Supervisor R. W. Ayres and Ranger Brownlow, which emphasized control of

Side view of Blue Mountain Lookout, Modoc National Forest. 1929

incendiarism, developing a patrol and lookout organization and ways to ensure that rangers would be able to complete their patrols. The introduction of the forest protection plan stimulated additional discussion regarding paying trained "standby" fire crews, giving rangers fire assistants, urging the state to require burning permits during fire season, requiring campfire permits for the 75,00 to 100,000 people going into the forests during the course of a year and developing various public relations prevention messages and postings. Because accumulation of fuels appeared to be the prime reason for the large, high-intensity fires, some forest supervisors advocated "light or controlled burning" to reduce the fuel load. Nonetheless, the majority of the supervisors rejected the idea (Ibid.: 78-113 passim; Cermak n.d.: 107-108).

As a result of the 1910 meeting, all forest supervisors were required the next year to submit forest protection plans similar to the Stanislaus plan. In addition, from 1911 onward, there were yearly increases in the number of forest guards on California's national forests (Ayres 1941: 15).

Range management was the next major question addressed at the meeting. Nationwide, annual products of the livestock industry in the United States were valued at 1.3 billion dollars – greater than the wealth production of iron and steel (second), and lumber and timber (third). Furthermore, 1.75 million cattle, horses, and hogs, and 8 million head of sheep and goats grazed on national forest forage. Early Forest Service administration was confined to regularly approving grazing applications, requesting nominal fees for grazing privileges and adjusting difficulties with stockmen over particular ranges and rights. These grazing issues and economic factors brought the Forest Service "closer to the people than any other activity" and "affected more persons than did the management of timber, water, wildlife or other natural resources." Therefore, cooperation with stockmen was one of the cardinal policies of District 5. Political reality determined that they work along "lines of least resistance," and all regulations were purposely framed to interfere as little as possible with past and present use, while at the same time, to not detract from the practice of pure forestry (USDA Forest Service 1910: 126-129; Ayres 1941: 9).

Regarding range management, the statutory goal was to establish forest grazing on a business basis. District 5's optimistic goal was moderate control of use. Though everyone realized at the meeting that their chief business was forestry, they acknowledged the fact that grazing was interwoven with forestry, and they could not relegate it to a minor position in policy decisions. Led in the discussion by the quiet-spoken District Chief of Grazing, John Hatton,

District 5 policy sought two central objectives. First, Hatton wanted to promptly and permanently restore "beat up" ranges to past pristine productivity. Secondly, he wished to adjudicate and distribute grazing privileges and end the "evil of monopoly" (Show n.d.: 45). Though forage conditions appeared to need some attention, favorable periods of above-normal precipitation, at least before 1917, obscured the effect of deleterious overgrazing on the range. It also helped exaggerate the opinion that forage could sustain continued heavy use and convinced many to believe that close grazing helped eliminate possible fuels for fires (Fox and Walker n.d.: 9).

Questions regarding trespass of unpermitted stock, relationship of grazing to other forest uses, interaction and cooperation with stock associations, and range reconnaissance to find out if the Forest Service was using the forests to the best advantage were discussed at length for the Eldorado, Shasta, Sierra, Stanislaus and Tahoe National Forests. Despite problems, when posed with the question whether or not to sell the range within each forest to the highest bidder, no one rose to support the idea (USDA Forest Service 1910: 126-155 passim). Conclusions regarding grazing issues were few at the 1910 forest supervisors' meeting, and important studies of California grazing conditions would not begin for another few years. These studies would look at the complex biological system of soil, water, vegetation and animal use involved in grazing (Ibid.: 126-155, 171-175 passim).

Timber, fire prevention and grazing were the main subjects covered in the five-day San Francisco supervisors' meeting. Other subjects covered to a lesser extent were mining and watershed management.

From the beginning, mineral deposits on California's national forests were regarded by the mining industry as one of their major sources, one reason why they were withdrawn subject to the operation of the general mining laws (Friedhoff 1944: 5). In 1905, when the forest reserves were transferred to the USDA, Gifford Pinchot was mindful of establishing a good working relationship with the mining industry, and he met with them frequently to iron out their problems. Forester Pinchot believed that mining fit into his concept of utilitarian management of the forests – a fact that no doubt comforted the mining community. What's more, the 1907 *Use Book* stated explicitly that it was the "policy of the Government to favor the development of mines" (Dempsey 1992: 100; Palmer 1992: 142-143). Nonetheless, in 1910, fraudulent or questionable claims on national forests were a problem that needed consideration.

In California and elsewhere, mining claims on national forest lands had to meet certain mining law requirements in order to be valid. These restrictions in many ways paralleled the limits placed on unperfected homestead claims regarding residence, cultivation, improvements and "good faith" efforts. Though rangers could examine the legality of a particular mining claim, normally it was left to an expert miner to determine the propriety of an individual claim. To be valid, a mining claim had to meet three basic criteria during an investigation. First, under various mining laws, the claimant had to prove "discovery," which meant the "reasonable expectation of developing a paying mine." Next, mining laws required an expenditure of $500 worth of improvements or development work on a claim before it could be patented. And third, a claim could not be patented for purposes "foreign" to mining. In other words, if a claim was located for its timber, as a power site, for a saloon, in order to gain control of a trail or for any other purpose other than mineral development, it could be declared an invalid claim (USDA Forest Service 1910: 226, 230-213). Prior to 1910, nearly 83,000 acres of California national forest land met these requirements and were legally patented under mineral laws (Friedhoff 1944: 5).

In the area of watershed management, the statutory goal was to ensure favorable conditions of water flows (Show n.d.: 45), an issue which included the development of hydroelectric power for heat, light, local transportation and industry. In 1901, provision was made for hydroelectric power development on public lands, and in 1905, Forester Pinchot initiated a Forest Service water power policy whose essential features were fourfold: first, to charge for use of Forest Service sites for power development purposes; second, to prevent the speculative acquisition of sites; third, to provide for the orderly and full utilization of power resources; and fourth, to protect the public's interests with provisions regulating rates and services (Smith 1930: 46-47).

By 1910, most national forests on District 5 were at the end of the second era of water development. In the first era (1850-1895), which preceded the formation of most California forests, water development was directly linked to mining activity, where ventures were dependent on local initiative and capital. The topographical, technological and economic conditions associated with active mining districts on the Stanislaus National Forest mimicked such a pattern (Conners 1989: 9-47 passim). The second era of water development (1895-1910) represented a formative period for both the hydroelectric industry and the Forest Service and was a period "characterized by the push-pull between enterprise and regulation." Also by this time, improvements in electrical machinery and in long distance transmission of energy increased the value of the waterpower resources on the forests (Conners 1992: 154-157; Conner 1989: 112-166 passim).

Of course, the watersheds of California's national forests offered hydroelectric companies many rivers and streams with ideal sites for waterpower. Control and development of these power sites was very important to the Progressive cause, and the dialogue regarding this "experiment on public lands" was festooned with regulation ideas associated with Theodore Roosevelt's vision of conservation and public control. Washington Chief Engineer Oscar C. Merrill brought this message home to District 5 staff and forest supervisors. At the 1910 meeting, Merrill explained to California's forest supervisors the "gospel of efficiency." "In its nature," Merrill said, "water power is a natural monopoly," because to be efficient, the generation, transmission and distribution of electrical energy must be under unified control. To Merrill, the only safeguard against the improper use of the great power reposed in such a far-reaching monopoly, and the only way to defend against speculative holding and non-use and to ensure proper development was public control. He advocated federal control of the sites and state control of rates and services (USDA Forest Service 1910: 70-72).

In 1910, through recent consolidations, a handful of corporations controlled perhaps 90 percent of the waterpower in District 5. Two corporations, Southern California Edison and Pacific Light and Power, dominated the market in Southern California. Pacific Gas and Electric controlled Central California. Other sections of the state were managed by smaller groups, such as the Nevada-California Electric Company in Inyo and Mono Counties, the Truckee River General Electric in the Lake Tahoe Region, the Northern California Electric Company in the upper Sacramento Valley, and the Pacific Light and Power Company and the San Joaquin Light and Power Company in that area. Each corporation sought new hydroelectric sites on Forest Service land, but in 1910, the Forest Service had made no attempt to survey valuable water resource sites within District 5 because of a lack of manpower and appropriations (Ibid.: 70-74).

Issues related to hydroelectric power were usually intertwined with the protection of watersheds to meet the thirsty demands of California's growing urban populations. Take for example, the familiar controversy over the development of a reservoir in the Hetch Hetchy Valley, a canyon 170 miles east of

San Francisco and on federal land in the northern part of Yosemite National Park. This project was seen as both a municipal water project and a source of hydroelectric power. A dam placed there not only captured the Sierra runoff to meet San Francisco's drinking water needs but also provided an additional dividend – the "generation of hydroelectricity to supply the Bay Area's growing demand for power, thereby producing revenue to help underwrite the cost of the project" (Hundley 2001: 173-175).

The Hetch Hetchy Valley project presented a dilemma for the Forest Service and probably for District 5 officials. According to Stanislaus National Forest historian Pam Conners, "the Hetch Hetchy O'Shaughnessy Dam," while not on Forest Service land, had many related facilities on the Stanislaus National Forest, such as tunnels, hydroelectric generation plants, penstocks and additional water storage facilities and dams. Moreover, regulation of actual flows on the Tuolumne River were ultimately negotiated with San Francisco officials. Therefore, there was a great deal of hand-wringing and policy development with District 5 regarding the Hetch Hetchy project and the Raker Act" (Conners 2004).

While still in office, Gifford Pinchot supported the project as a prime example of "full utilization" and persuaded President Roosevelt to do so as well. Utilitarian conservationists such as Pinchot emphasized conserving resources through careful management for later use, but they also believed that "water supply comprised a more important public use of an area than

Hetch-Hetchy Reservoir showing corner of O'Shaughnessy Dam—San Francisco's power and water supply.

did recreation." Preservationists disagreed. The seventy-year-old John Muir, along with the Sierra Club, valiantly fought against damming up the Hetch Hetchy Valley. They were "dedicated to preserving forever in their natural state unique and beautiful wild places." In their view, San Francisco should tap other resources, and not destroy this place of rare beauty. Though each side contended that they represented the "public interest" and their opponent the "private interests," in the end the utilitarian conservationists won out. In 1908, a permit was issued by the Interior Department, and in 1913, the Raker Bill, authorizing the development, passed and was signed into law by President Woodrow Wilson. John Muir died broken-hearted a year later (Ibid.: 175-189; Hays 1969: 192-194).

While San Francisco efforts to obtain water and power floundered because of a number of issues that went well beyond the Muir-Pinchot debate, the Los Angeles aqueduct had been in operation for many years. Los Angeles was much more successful in its pursuits for several reasons.

First, residents perceived water supply, as well as hydroelectric development, as a legitimate public enterprise. Sustained public interest in Southern California's water and forest situation was striking from the very beginning. Though the forests were extensive and valuable, municipalities realized that the perpetuation of their water supply depended largely upon absolute conservation use of the forest cover in the mountains. Water in the southern portion of the state was also highly valued for agricultural purposes. Deficient and poorly distributed rainfall made agricultural development dependent on irrigation, unlike northern interior valleys, where the rainfall was greater (Sterling 1906: 1).

Second, unlike San Francisco, citizens were not afraid to finance the project with municipal bonds. Third, city officials knew from the beginning they had to have absolute control of the water rights involved and thereafter set out on a policy of "water imperialism" (Kahrl 1983: 18-19). Finally, they succeeded earlier than San Francisco because the Forest Service's "passive" assistance in helping Northern California cities acquire water turned to open assistance when it came to Southern California cities like Los Angeles. "Local elites and corporations," according to one historian, "co-opted the administration of the public land by the Forest Service for their own requirements," and "their system, in effect became the official policy of the Forest Service as applied to the San Gabriel Mountains (Headley 1992: 226; Hundley 2001: 175-189). The creation of most all of the southern national forests, especially the Inyo National Forest, was part and parcel of a urban watershed protection policy for communities from Santa Barbara to San Diego.

That Pinchot chose water supply over recreation use is not surprising. In his original plan of organization, he was preoccupied with wood, water and forage. He ignored forest recreation altogether, even though the federal government recognized and responded to recreational demand as early as 1899, when Congress passed the Mineral Springs Act, allowing the leasing of sites on forest reserves for "health" or "pleasure" (Bachman 1967: 1). That Roosevelt supported him is surprising, especially given his personal attitudes toward and his experiences in the outdoors. Even so, the value of recreation and the outdoors in California had long been appreciated by its residents as well as tourists to the state.

Though recreation was allowed on forest reserves by statute as early as 1899, long before that date, recreation use of forests was a rather common thing in California's Eden-like lands. For many Californians, outdoor recreation was considered a necessity of life. Each year, the mountains, lakes, streams, woods and natural parks of California drew forth its citizenry for health and enjoyment. The simplest form of recreation was the ubiquitous

Picnicking at the Big Trees within Stanislaus National Forest.

camping party, where California families took a team and wagon to anywhere there was good fishing, hunting or attractive country. Early California recreation also took the form of the seasonal visit to a mountain resort area, such as the Mount Shasta region, accessible by railroad, or Yosemite Park, easily reached by horse stage. Camps and resorts also sprang up at an early date in Southern California. Here, where the mountains were practically in the backyard of towns and cities, and where urban communities pushed out into them, recreation was an important asset. These montane areas offered the citizens of Southern California opportunities for sightseeing, picnicking, camping, fishing, hunting, hiking and eventually winter sports (Show 1963: 139-141, 148; Ayres 1941: 12; Jarvi 1961: 1). During the 1890s, the back-to-nature movement became extremely popular in Southern California, and hiking clubs formed to "take advantage of the rugged trails that dotted the San Gabriel Mountains and Mount Wilson east of Los Angeles." Men and women visited and enjoyed the natural surroundings at important resorts such as Switzer's in the Upper Arroyo Seco, Colby's in Coldwater Canyon, Strain's on Mount Wilson or Sturtevant's in Big Santa Anita Canyon. At the same time, the Sierra Club, based in San Francisco, organized rather elaborate camping parties into the high Sierra every summer, starting as early as 1893 (Lux 2003: 18-19; Show 1963: 140; Headley n.d.: 22-24).

Local rural and urban residents were not the only visitors to California's scenic wonders. Tourists from around the nation came to recreate and refresh themselves physically and mentally in the state. Prior to 1879, the outside public learned of the virtues of recreating in California from romantic stories and through painters such as Albert Bierstadt and photographers such as Carlton Watkins. In the late 1800s, several publications such as Norman W. Griswold's *Beauties of California: Including Big Trees, Yosemite Valley, Geysers, Lake Tahoe, and Donner Lake* (1883) or N. C. Carnall's *California Guide for Tourists and Setters* (1889) emphasized California's tourist attractions and "extolled California's natural wonders and unparalleled beauty, and played an important part in attracting people to the state's rural areas" (Lux 2003: 18-19). Countless others learned of California's natural treasures from the naturalist John Muir and the popular writings of Stewart Edward White.

Despite all of this attention and recreational use in California, prior to 1910, early administrators had no overall goal or plan regarding recreational use. Frankly, many rangers considered campers and tourists simply as "nuisances" who caused fires and additional work for them. This attitude slowly changed, starting with the GLO's 1902 *Forest Reserve Manual*, which was the first time camping and travel for pleasure or recreation was even noted. Rangers were charged with instructing campers not to build large fires and to put their fires out before leaving – all this to be done very politely, without losing their tempers or using abusive language. Under the Forest Service, rangers began visiting among the various camping groups, trying to instruct the users in the ways of the woods, many of whom were amateurish at best. Continuing this tradition, sometime around 1908, William G. Hodge of California Inspection District 5 wrote what was perhaps the first camper handbook, "telling campers what to do, how to do it, and why, when they camped in the mountains" (Show n.d.: 45; Show 1963: 142-144, 150; Tweed n.d.: 1). This handbook was so popular nationwide that in 1916, Washington asked District 5 to reprint it when supplies of the publication were exhausted (*Weekly Bulletin: Forest Service, District 5* 1916: 1-2).

As recreational use of California's national forests grew, the Forest Service could not ignore this form of utilization of California's national forests for very long, especially since mountain roads were being built, allowing fleets of flivvers filled with outdoorsmen and campers into the wilderness. People began to demand, and expect, certain forms of attention and amenities as well, especially in forests near urban centers. Others enjoyed the experience so much that they requested special-use permits to build summer homes – to avoid the dirt and dust and hard work of camping. For instance, in 1906,

Summer home at Huntington Lake showing flower garden, Sierra National Forest. 1937

the Angeles National Forest, which for a time included the San Bernardino National Forest, issued the first special-use permits for summer homes under the 1897 Organic Act – with the first summer permit signed on August 13th for fifty acres in San Gabriel Canyon above Azusa. With no overall District 5 recreation plan or program, users selected their own sites, with minimal guidance or oversight by the Forest Service; they provided their own water and sanitation, much of which were wholly inadequate; and there was no control over the "rustic" architecture of these summer cabins, allowing for some very individual expression. In time, a very common pattern for summer home living evolved, with women and children moving to the cabin while "papa" went back to work in the valley, coming back up on weekends as he could. And soon neighbors and related groups selected spots in the same locality to be together (Ayres 1941: 12; Show 1963: 146; Lux 2003: 68). The most intensely developed areas were those in the Angeles Forest north of Los Angeles, where every canyon that had running water was crammed full of well-furnished camp houses, which were occupied for several months of the year. Many of these summer areas eventually turned into permanent communities and towns (Waugh 1918: 21-22).

As recreational needs boomed, conflicts grew between recreation and other forest uses such as timber and grazing. Prior to 1910, these conflicts were either overlooked or smoothed over largely for economic reasons. Recreation residences and associated permits and fees produced an early source of independent income for the Forest Service's special reserve fund. This fit very well into Pinchot's philosophy of revenue generation – a fact that explains the early Forest Service dependence on the permits and fees (Lux 2003: 21).

At the same time the Forest Service was starting to recognize the value of recreation, it also acquired the responsibility for protecting cultural sites on national forests. On June 8, 1906, Congress passed the American Antiquities Act, authorizing protection of antiquities (prehistoric and historic remains) and features of scientific or historical interest on land owned or controlled by the government. Excavation and gathering of objects by reputable scientific investigations were allowed under this act, and the Smithsonian Institution was charged with issuing permits for them to museums, universities, colleges or other scientific and/or educational institutions. Criminal sanctions for unauthorized destruction or appropriation of antiquities were established as well by the act, and a failure to obey the rules could result in arrest (Boyd 1995: 1-2).

At the time of its passage, enlivened by Helen Hunt Jackson's popular novel *Ramona* (1884), concerning the oppression of California Indians, popular interest in historic preservation in California centered on preserving the state's Spanish mission heritage. A year earlier, the Historical Society of Southern California was formed to rescue crumbling missions, and thereafter other groups formed to preserve important historical sites such as Sacramento's deteriorating Sutter's Fort and the Custom House in Monterey. By 1902, a state Historic Landmarks Committee was established which prioritized the restoration and preservation of old Fort Ross in Sonoma County, Colton Hall at Monterey, where California's first constitutional convention convened, as well as the old missions of California. Thereafter, the Historic Landmarks Committee erected monuments and plaques throughout the state to commemorate other historic places and events important to California history. The State of California proved generally supportive of these early preservation efforts (Hata 1992: 3-6).

Unfortunately, prior to 1910, there was no matching Forest Service component or cultural/historical/archaeological conscience – even after Congress passed the American Antiquities Act. In California, Forest Service administrators ignored the act because of one primary reason – ignorance of the law and its stipulations. This unawareness and consequent widespread non-compliance resulted in the loss of countless prehistoric and historic sites. Land managers and field personnel unknowingly destroyed or disturbed important sites. Awareness of historical sites eventually developed over time. As Forest Service administrators recognized the historical and recreational value of sites, preservation efforts began with the erection of monuments to historical events, and lands were withdrawn for that purpose. But awareness of the significance of historical sites developed in a piecemeal fashion on individual California national forests. Conscientious preservation of prehistoric sites took more time. In Northern California, in places such as the High Sierra, few prehistoric sites beyond the obvious bedrock mortars and scattered lithic debris were preserved. In Southern California, adobe structures and Native American sites were more easily visible due to the drier climate and building materials used and therefore were preserved (Boyd 1995: 2). However, full awareness of historic preservation would not blossom in District 5 until the late 1960s and the establishment of a cultural resources management program which grew out of necessity to meet and comply with regulations and a national policy of preservation stimulated by the passage of key federal legislation, such as the Reservoir Salvage Act (1960), National Historic Preservation Act (1966) and the National Environmental Policy Act (1969).

Cooperation and the Changing of the Progressive Guard

By the time President Taft dismissed Forester Pinchot in January 1910, the general outlines of a government-industry cooperative policy with private landholders had been established as a federal program goal. Pinchot recognized the importance of private forests to the nation's welfare, and only months after his appointment to the Bureau of Forestry, his office issued Circular No. 21, which offered advice "to farmers, lumbermen, and others in managing their forestlands." In addition, he continued the cooperative agreements initiated by his predecessor with scientists and private timber corporations (e.g., Weyerhaeuser Timber Company and Great Northern Paper Company) and large landholders (e.g., William G. Rockefeller and E. H. Harriman). He also embarked on new agreements, especially with other federal agencies such as the GLO, the USGS, Bureau of Reclamation and the Bureau of Entomology, to name a few (Robbins 1985: 11-19). Political capital was gained with these cooperators, but cooperative agreements with state forestry agencies did not come until the changing of the Progressive guard from Roosevelt to President Taft. That accomplishment fell to Pinchot's successor, Forester Henry Graves, and the passage of the Weeks Act in 1911.

In 1905, following a series of uncontrolled and disastrous fires, the California State Board of Forestry, which had been abolished in 1893, reconstituted due to pressures by George Pardee and others. Then the state legislature passed the Forest Protection Act outlining state forest policy. At that time, Forest Supervisor E. T. Allen of the San Gabriel Forest Reserve left the service and became the first California state forester. Following this development, the subject of cooperation with the State of California regarding issues such as forest fire protection, waterpower and regulation of timber production permeated discussions at District 5. However, no cooperative action took place until the disastrous 1910 fire season hit California. It then became apparent to Olmsted and others that the Forest Service could no longer "go it alone." With the size of the forestry tasks at hand in the nation and in California, state cooperation on these matters was necessary and desirable. The solution was the Weeks Act. Passed on March 1, 1911 (36 Stat. L., 961), this law authorized Forest Service cooperative efforts with states in the purchase of lands (mostly in the East) needed to regulate the flow of navigable streams and to provide forest fire protection (Smith 1930: 42-43, 163).

Premised on the "commerce clause" of the U.S. Constitution, the new law was epochal in three ways. First, it established a National Forest Reservation Commission authorized to examine, locate and recommend for purchase lands necessary to the regulation of the flow of navigable streams. Loosely interpreted, this ability included the purchase of timberland in the upper reaches of navigable rivers to provide erosion control. Importantly, the Weeks Act established the principle of purchasing private lands to incorporate into national forests and "gave the federal government, for the first time, the power to expand the national forest system by acquisition." As a result, numerous additions, eliminations and consolidations of national forest lands were made permissible. Second, the Weeks Act introduced into national forest policy the principle of the federal government contributing to state fire-suppression organizations if they complied with Forest Service standards. Section 2 allowed for "better preparation, greater resources, and extensive construction of improvements to equip forests with means of communication and transportation" to fight fires. Specifically, the agriculture secretary was authorized to enter into agreements with states "to cooperate in the organization and maintenance of a system of fire protection on any private or state forest lands within such State or States situated upon the watershed of a navigable river." And third, the Weeks Act provided in Section 13 that 25 percent of each national forest's receipts be returned to the states to fund public schools and public roads in counties where the forests were located. Section 13 may have been the most important provision because it extended a fiscal bond between state cooperation and the Forest Service first established by the Agricultural Appropriation Act of May 23, 1908, for forest reserves (35 Stat. L., 251, 259). Revenue sharing enlisted the support of key western congressmen who would thereafter fight for Forest Service appropriations and policies that would benefit their states (Smith 1930: 41-44, 161, 163-167; Robinson 1975: 10).

In June 1911, following on the heels of the passage of the Weeks Law, District Forester Olmsted resigned from the Forest Service, thereby bringing a change in the Progressive guard. The thirty-nine-year-old Olmsted left his job ostensibly because he desired to be "closer to the woods" and because he wished to promote good forest management by private timber owners. One of the nucleus of young professionals that Pinchot gathered around him while the Division of Forestry expanded into a bureau, and then into the present-day Forest Service, Olmsted may also have missed Pinchot's leadership.

After leaving the Forest Service, Olmsted enjoyed a long career before his death in 1925. After a brief stint in Boston, he returned to California in 1914 to open an office in San Francisco as a consulting forester. Thereafter, the Diamond Match Company employed him until his retirement in 1923, introducing conservative cutting and good management to the company's holdings in California and elsewhere. During his later years, he served as president of the Society of American Foresters (1919) and wrote numerous articles promoting effective fire protection and good forest management in the

United States (Clepper 1971: 242; *California District News Letter* 1925: 1). Olmsted's resignation marked the beginning of a new era in the history of the California national forest system – an era discernible by a search for order.

Fractionalization of the Conservation Movement

The creation of the Forest Service, the ascendancy of Pinchot as its first forester and the years 1905 to 1910 marked the triumph of utilitarian conservation in America. In California, Frederick Olmsted's career as California inspector (1905-1908) and then as California's first district forester (1908-1911) brought this philosophy to the regional level. But attainment of utilitarian conservation in California came at a heavy price – the fractionalization of the conservation movement into those who called themselves conservationists and those who referred to themselves as preservationists.

The roots of this split in philosophy lay very deep in the American political, economic, social and ecological psyche, but in California it first became evident and vocalized over water issues and urban growth – namely the Hetch Hetchy Reservoir and the construction of O'Shaughnessy Dam to serve San Francisco, and the Owens Aqueduct that served metropolitan Los Angeles. The Hetch Hetchy project in Yosemite National Park brought on the bitter struggle and inevitable split between preservationists and conservationists. These two groups splintered when Roosevelt and Pinchot openly and publicly disagreed with John Muir and other preservationists in regard to protecting the scenic values of Hetch Hetchy Valley for the public. Pinchot battled for the protection of all publicly owned natural resources, including waterpower and storage sites, for the public good. But, though Pinchot clearly appreciated great natural wonders and believed they should be in public ownership, he also felt that forest recreation had no place or role in the Forest Service (Show n.d.: 50-51).

Then again, the Owens Aqueduct project also highlighted a crack in the Progressive paradigm of who truly represented the public interest. The creation of the Inyo National Forest to assist Los Angeles' Owens Aqueduct project put Progressivism on the spot regarding protecting the "little guy" against monopoly. Correlative to the problem was the growing population dichotomy in California between urban and rural centers in which the needs of rural Californians were subsumed to those of these growing metropolitan regions.

Nevertheless, preservation opposition to the Forest Service's inherent utilitarianism – trees were to be used wisely, but not protected forever – raised the issue of the role of scenic quality and recreation within the spectrum of

Ranch abandoned because of lack of water, Owens Valley, south of Bishop, California.

multiple-use management (Hundley 2001: 175-189). It was a concern the Forest Service would have to grapple with in the future. Out of Muir's activities came a growing appreciation of places of natural wonder or cultural patrimony, leading to a movement that demanded a national park service. Because of the dominance of utilitarian conservation in the Forest Service, naturalists, preservationists and aesthetic conservationists turned to the Interior Department for the establishment of a single-purpose agency to protect areas with important natural and scenic values. This led directly to the creation of the National Park Service in 1916. Eventually, this shift in the conservation movement led to a formal fragmentation into two legally equal services, each with conflicting aims, laws and policies. Thereafter, preservationists, represented by the Sierra Club after Muir's death, found kindred spirits in the National Park Service and virtually acquiesced to all Forest Service plans and policies in California until World War II. At that time, a new generation of preservationists emerged in California to carry on the legacy of John Muir and bring a new challenge to the Forest Service's philosophy of utilitarian conservation (Show n.d.: 51; Pendergrass 1985: 26-28).

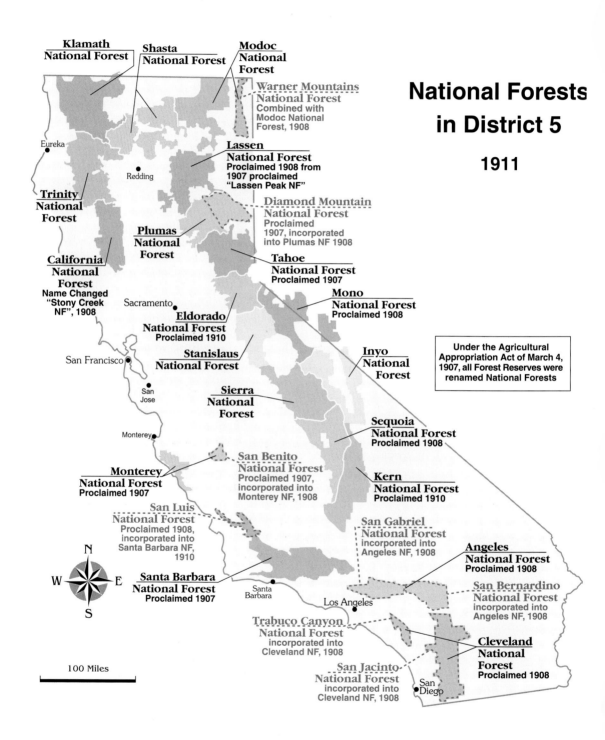

Chapter IV: 1911-1918

California National Forest System Grows and Goes to War

California Romance with Progressivism and Wheels

In the pre-World War I period, transformations in California's politics, economy, society and lifestyle were many and diverse. Many of these changes can be attributed to the triumph of Progressivism in California following the 1910 election, and they reflected directly upon Forest Service policy and District 5 programs in California in the 1910s and in later decades.

Politically, the 1910 election of Governor Hiram W. Johnson, who came to be called the "western Theodore Roosevelt" for his aggressive demands for reform, marked a milestone for progressive political change in the state. Following the election, Governor Johnson and other victorious progressives committed themselves to end corruption in the country, as well as exploitation of public resources – including human resources. Johnson and others in the state also wished to "take the politics out of politics." Toward this end, they created a cross-filing political system and implemented "direct democracy" through a number of measures such as the initiative, referendum and recall. California progressives also sought and won woman's suffrage in 1911 (Rice, Bullough and Orsi 1996: 357-361).

Economically, progressives in California's legislature took swift action on issues related to railroad and utility regulation, as well as unethical business practices. In only a short time, progressives accomplished what a generation of reformers had failed to achieve in the state. Increasing the authority of the Railroad Commission, passage of a Public Utilities Act along with a regulatory commission, and the creation of a state superintendent of banks were among some of their major regulatory achievements. Some people claimed these accomplishments were anti-business in nature, but in reality, these reforms were tempered with much anti-union feeling. Many California progressives were less than enthusiastic about supporting organized labor; they were even more so once the radical Industrial Workers of the World (IWW or "Wobblies") entered the picture, making speeches and singing anti-capitalist songs from San Diego to Sacramento. Local anti-IWW laws were passed, but all were declared unconstitutional. Furthermore, in 1913, deadly riots occurred in Wheatland, north of Sacramento, when the IWW organized farm workers to protest grossly inadequate working conditions. In the melee, shots were fired and five people were killed, including two deputies, two workers and the Yuba County district attorney. An investigation led to the vindication of the IWW's position and passage of the California Labor Camp Act of 1915, but it did not lessen suspicion of all radicalism (Ibid.: 358-360; 365-368).

Socially, California went through a period of discrimination and outright racism. Tens of thousands of Mexicans crossed the border into California at this time, fleeing the political and economic dislocations in their homeland caused by the overthrow of the hated dictator Porfirio Diaz. In the turmoil, many Mexicans moved into *barrios* in Los Angeles, Santa Barbara, San Diego and other communities. Others settled in *colonias* in agricultural areas, where they became the mainstay of the farm labor force. The majority of Californians ignored the Mexican migration, much as Native Americans, Chinese and African-Americans went unnoticed by California society. Such was not the case for Japanese migrants. Around 1900, *issei* (first-generation Japanese-Americans) began migrating to California from Hawaii, and then from Japan. Through hard work, they soon became successful truck farmers, raising tomatoes, vegetables, and a variety of produce. Their success brought them in direct competition with local small farmers, resulting in several bills restricting Japanese land ownership and permitting segregation in city schools and residential neighborhoods. Sponsors of these bills wished to curry favor with the farm and labor vote. Fortunately, these bills failed passage, but the effort did result in the passage of the Alien Land Law of 1913, which stated that aliens ineligible for citizenship (namely *issei*) could not purchase or lease California land. This law led to a major international crisis, bringing with it tense relations and troubles between Japan and America (Ibid.: 368-372; Link 1963: 84-87).

Despite the political, economic and social upheavals and controversies described above, California's economy burgeoned in some sectors. There was the rise of several new industries, most notably motion pictures, the generation of hydroelectric power and the bewildering romance that began between Californians and their automobiles. All of these developments would eventually have an effect on Forest Service resource development, but the latter development would have the greatest impact.

In 1909, the California State Highways Act initiated a program of surveying and mapping out a state highway system running north and south through the state to service the 36,000 automobiles registered in California at the time. A decade later, there were 604,000 automobiles, and California became a state on wheels (Cleland 1962: 280) Before the United States entered World War I, the Automobile Club of Southern California, with its increasing membership, began to take an active interest in the subject of surveying routes in California, but America's entrance into the war set aside the surveys and plans (Hoffman 1968: 311).

"Nothing," according to the noted Californian historian Robert Glass Cleland, "ever influenced California life and society so spontaneously and profoundly as automotive transportation." Said Cleland:

> The automobile broke down isolation, diffused population, encouraged rural and suburban life, relieved urban congestion, greatly increased the inflow of tourists, opened the mountains and deserts to endless throngs of visitors, made the beaches universal playgrounds, acquainted Californians with the beauty and varied resources of their own state, carried them far a field into other states, and radically affected styles, dress, customs, manners, culture, and morals…. The effect of the automobile upon California's economic development was even more revolutionary than on the state's social and cultural life. In addition to ushering in an entirely new era in transportation, it transformed the petroleum industry, greatly stimulated certain branches of manufactures, initiated an entirely new era of highway and bridge construction, and created innumerable subsidiary enterprises from which California profited even more than most other states (Cleland 1962: 281-282).

As time passed, the automobile became not just means of transportation, but one of entertainment, and California's forests became a recreation destination, not just a resource for lumber, forage, minerals and water.

By the end of the decade of the teens, California's romance for wheels turned to the dream of flight. The California airplane industry began in 1909, when Glenn L. Martin began manufacturing airplanes in Santa Ana. A year later, thanks to the sponsorship of William Randolphs Hearst's *Los Angeles Examiner*, the first documented powered flight in California took place near Long Beach in Southern California (Watkins 1973: 342-347). The infant industry grew thereafter, especially once America entered into World War I. The outbreak of war in Europe occurred in 1914, but America did not enter the war until April 1917. World War I especially stimulated both the airplane and automobile industries.

To coordinate America's war effort, a slim, bespectacled President Woodrow Wilson, in an unprecedented exercise of regulation and control, centralized American industry, transportation, business and labor under one agency – the War Industries Board (WIB). California benefited greatly by the outbreak of war and Wilson's centralized wartime economy. Wartime demands stimulated a sluggish California economy by accelerating food and war-related material production. By the time the State Department announced that

Germany and the Central Powers had agreed to an armistice on November 11, 1918, California agricultural production had increased in value to $612 million; lumbering, fishing and manufacturing had reached $714 million; and the state's mineral and petroleum output rose to $202 million. Shipping imports and exports also improved measurably during the war years, until German submarine warfare caused a shortage of ships, leading to a decline in harbor business. Even so, the enormous growth of America's navy in size and efficiency during World War I stirred government officials to locate several important naval bases in California's well-suited harbors in order to protect the nation's western coastline (Cleland 1962: 280, 286). As the economy strengthened, Californians, distracted by this economic development, turned their attention away from reform issues that marked the end of California progressivism. For the next decade, Californians would ride a cresting wave of prosperity until it crashed on rugged economic shores in 1929.

New Leadership: Coert DuBois Administration (1911-1919):

Throughout the pre-World War I period, the young Coert DuBois served as District 5 forester. Chief Forester Henry S. Graves' appointment of DuBois at the youthful age of twenty-nine to oversee California's many national forests surely indicated Graves' confidence in the young man's ability to lead (Show n.d.: 54; Williams 2000: 34). It might also have helped that DuBois was Frederick Olmsted's brother-in-law (Pendergrass 1985: 83)!

Coert DuBois, District 5's second district forester (1911-1919)

Still, Chief Forester Graves could have found no better man to help him meet his goals of stabilizing the Forest Service and strengthening its forestry foundations by putting the agency on a more scientific basis. Born in Hudson, New York, in November 1881, Coert DuBois joined the Forest Service in 1900 shortly after graduating from the Biltmore Forest School. As one of Gifford Pinchot's boys, DuBois rose rapidly through the ranks of the agency. In 1908, he came to California along with Olmsted to establish the California Inspection District. He afterward became assistant district forester for California District 5 (Show n.d.: 54; *Who Was Who* 1962: 239).

In his search for administrative order, Coert DuBois sought standardization and specialization in the district office. His ideas clearly were linked to the scientific management ideas of Frederick Winslow Taylor, which was one response to modern American industrialism. First systemized by Taylor in the 1890s, this way of thinking started as a manufacturing technique based on interchangeable parts. It later evolved into a widely accepted "scientific" management principle that basically sought the greatest output from workers with the least amount of waste and cost. Corporate executives were encouraged to take away the arbitrary powers of foremen and place them in the hands of a specialist (Wiebe 1967: 151, 294; Hays 1957: 10-11; Cochran 1977: 68, 158).

Ever since its founding, the Forest Service wished to put national forests on a more business-like footing, and by 1910 or so, some saw Taylorism as the answer. The application of this theory of scientific management reached an apex on California District 5 in 1912 during a regional conference. At this meeting, Taylorism's principles of scientific management were thoroughly discussed, especially as to how they could be applied to the district and the field in order to put the Forest Service on a paying basis. Principles discussed included (1) development by management, not the workmen, of the science of executing the work, with exact rules and standardization of implements and working conditions; (2) careful selection and training of rangers into first-class men, and elimination of all men who refused to or were unable to adopt the best methods; (3) bringing rangers and the science of doing the work together through management and through paying a bonus for efficiency; and (4) providing an almost equal division of work and responsibility between the rangers and the management – supervisors. In other words, science, harmony, cooperation and maximum output were to be emphasized over rule of thumb, discord, individualism and restricted output (USDA Forest Service 1912: 1-15).

This discussion was quite a change from the previous Olmsted administration. As late as December 1910, District 5 supervisors had passed a resolution at their regional conference stating that it was the "sense of this meeting that specializing of work with the exception of special duties was not advisable" (USDA Forest Service 1910: 234). Following the 1912 regional meeting, DuBois led the district in the direction of specialization on both the district and the field administration level.

In 1912, District 5 was deeply preoccupied with silviculture, and DuBois adopted Taylorism principles primarily in that area. DuBois' first step toward scientific management was the employment of technical specialists in forest pathology, entomology, silvics, timber and forest influence questions, timber sale scaling and reconnaissance. These young "technicals," or "tech asses," as they were referred to by some, hoped to educate and build up the skill and capacity of generalists (both rangers and supervisors) to meet the future challenge of scientific forestry. They were also hired to develop tighter standards in a number of field areas. The establishment of the Feather River Forest Experiment Station in 1912 aided District 5 in this effort and to a degree lifted the burden of making every forester a scientist. However, hardy old-line district supervisors and rangers held the technicals' "purported learnedness" in little regard. The specialists' repute and standing with these types often depended on their abilities and willingness to learn from the mountain man! Interestingly, many young technicals ascended into the esteemed management posts of supervisor and deputy supervisor. For instance, by 1911, seven out of nineteen forest supervisors, and by 1915, ten out of eighteen forest supervisors were considered "technical" supervisors (Show n.d.: 59-63, 87). Additionally, by 1914, there were one to three technical men on each of California's nineteen national forests. These lower-level specialists were used on timber sale work and in special silviculture work (Ayres 1941: 27).

Alongside specialization, Taylorism brought the principle of standardization to District 5, where there was a clarion call among staff, supervisors and rangers for uniformity in claims reports, scales used in field maps, marking rules, methods of timber reconnaissance, spacing on plantations – even uniforms. This appeal for standardization was often discussed in conjunction with ranger work cost and efficiency, for during the early teens, it had become evident to inspectors that rangers were becoming less and less efficient.

There were several causes for ranger inefficiency. One primary cause was low morale due to the statutory roll – a federal government-wide fixed salary system that failed keep up with inflation or with increases in job responsibility. Forest Service personnel realized that they were being underpaid, but most early Forest Service personnel found their career rewarding and enjoyed the comradeship (Pendergrass 1985: 50). At the 1912 regional conference, there was considerable talk and criticism of Pinchot's old system of "initiative and incentive." Pinchot had demanded initiative, but the federal government's statutory roll system provided little financial incentive to secure it. Another cause of ranger inefficiency was ever-increasing new job demands that whittled away the time each ranger could spend on any particular job. Timber sales, fire control, settling grazing and range disputes, boundary and land examination, building ranger stations and lookouts, fencing pastures, stringing communication networks, not to mention attending to the growing recreational use on each national forest as Californians began to explore the state's scenic wonders by automobile, certainly reduced a ranger's efficiency. And unlike old-timers who worked long hours with low pay, newer employees did not have the fire-bellied enthusiasm of the Pinchot era. Instead, they saw their work as a job to be performed and not an avocation. Therefore, it was natural for them to worry about workload and schedules. To make matters worse, in the 1910s, the heyday of ranger autonomy waned. More often than not, they were increasingly assigned as "helpers" to projects run by specialists and experts. The net effect of all of the above seriously lowered ranger morale, eroded their independent image, and deteriorated their prestige. These circumstances left them to tasks to which specialization had not been applied yet – hewing wood and drawing water (Show n.d.: 69-73)!

Therefore, at the 1912 meeting, DuBois and others proposed to reorganize and classify the workload of rangers by relative importance with protective assignments at the top, followed by income producing business, public benefit and investigative work, in that order. To induce incentive, officials at the meeting suggested grading the rangers and increasing salaries for the higher grades. By thus bringing pressure from above, they hoped to force field men to systematize their work and pay the most attention to what they regarded as important issues. Discussion regarding cost, efficiency and supervision issues followed on various national forests such as the Angeles and Monterey forests, along with several proposed cost-cutting methods (e.g., larger districts, a yearlong ranger to do all administrative work). But though they tried to standardize ranger work, they made little progress in that regard at the meeting or, for that matter, during DuBois' tenure (USDA Forest Service 1912: 16-50 passim).

Besides addressing ranger work, they evaluated and defined the position of forest clerk staff and field duties, in an effort to make each forest more efficient. Clerical work on each forest was divided between two forest clerks. In the supervisor's office, the lead forest clerk was to open mail, answer all letters, attend to vouchers, allotments sheets, homestead claims cases, timber sale contracts, property accounts, improvement records, requisitions and the purchase of supplies, compose all letters of transmittal for grazing, special use, timber sales and so forth – and pleasantly attend to visitors. In short, the forest clerk's duty was to be the supervisor's "lieutenant," anticipate his and the ranger's needs and do everything possible to relieve them of office work. The

Plumas National Forest office staff. Back, left to right: Lambert Hiley, Exec. Asst.; John Edwards, Fire Dispatcher; Ray Arr, Timber Frgmn; D.N. Rogers, Supervisor. Front, left to right: Rosie Katz; Gladys Herckinson Robinson. 1911

forest clerk also was responsible for running the office efficiently in the absence of the supervisor. Between the lines, a good forest clerk did not run into the supervisor's office to take up each matter as it occurred, and he or she got "bores away from him when they accidentally get to see him." All of the above office work was shared with the second forest clerk, who was more or less an understudy to the above activities (*California District Newsletter* 1919: 5).

In the field, the lead forest clerk was given the opportunity to visit all parts of the forest for the purpose of learning about operations and the lay of the land. The second forest clerk was also considered an understudy in this regard, but was occasionally allowed to go on trips so that they could better understand their clerical work. During field trips, the primary forest clerk attended to the following lines of work: checking up on property, inspecting ranger files, closing cases and putting files in order, inspecting equipment to determine the need for repairs and replacements, assessing surplus property and arranging for transfers, passing along ideas and methods used by other rangers, and instructing the ranger on how to improve office procedures – in other words, doing everything possible to relieve the ranger of unnecessary office work (*California District Newsletter* 1919: 5).

In California, many forest clerks were women. If it were not for the assistance of these "forgotten foresters," much of the Forest Service's mission would not have been accomplished (Pendergrass 1990: 17). In the early years, rangers' wives played this critical role. In the pre-1910 era, they were unacknowledged and unpaid, such as Julia, the wife of Sierra National Forest Supervisor Charles H. Shinn, author of the classic *Mining Camps: A Study in American Frontier Government* (1884). In addition to her "wifely duties" of cooking, cleaning and sewing, Julia Shinn performed all the general administrative work of a clerk as well. She also traveled with her husband on field trips, and served up "square meals" as the camp cook on these road trips. Beyond these contributions, the wife of a ranger was expected to provide important social contact and sociability at community picnics and other social functions, to boost her mate's morale when needed and to temper her husband's bad character traits such as a "weakness for liquor." Typically, wives of career rangers, such as Julia Shinn, "assumed these extra duties early in their marriage and continued them until their husband retired or died." "Lacking sufficient staff, money, and time to meet both the physical and administrative demands of the job," men like Charles Shinn appreciated their helpmates. He was very proud that she was his equal and stood "shoulder to shoulder" with him, but "she deferred to his judgment whenever policy issues were at stake" (Ibid.: 17-18; Pendergrass 1985: 128-136).

But as America's middle class grew with doctors, lawyers, teachers, journalists and social workers, accredited and specialized organizations developed and began to establish professional standards. Driven by the women's suffrage movement, women entered professional life at this time as well, even though at first they were only allowed to advance in a few professions, namely teaching and social work (Wiebe 1967: 111-123). Such was the case in the Forest Service. The Forest Service had given tacit approval of women helpmates clerking for their supervisor and ranger husbands – a job that did not threaten their male partners. But before long, wives evolved into professional, paid forest clerks. For instance, in 1907, Julia Shinn became a permanently paid clerk when a replacement clerk and two male successors' work proved unsatisfactory and deficient. Nevertheless, nepotism gradually gave way to hiring women in general for the forest clerk position. Women thereafter quietly dominated that occupation and its routine duties of mastering filing systems and accounts and typing correspondence. Though the title of forest clerk started out as a male-dominated job, over time it came to be considered "women's work" and was eventually accepted as such by male colleagues (Pendergrass 1990: 19, 21).

Besides forest clerks, during the 1910s, District 5 also hired a small number of women as telephone operations, and a few women (usually unmarried) even worked as lookouts. For instance, in the summer of 1913, thirty-year-old Hallie Morse Daggett became the first woman to work for the

Hallie Daggett, first woman to work for the Forest Service as a lookout. Klamath National Forest

Source: Siskiyou County Museum

Forest Service in the nation as a forest guard and lookout – a field dominated by men. For the next fourteen years, Daggett worked in this capacity at Eddy's Gulch Lookout Station, atop Klamath Peak overlooking the Salmon River watershed. By 1918, Hallie was no longer the only fearless woman lookout in California. The coming of World War I made it harder to hire and retain men, so the Plumas National Forest turned to twenty-three-year-old Mollie Ingoldsby, and the Tahoe National Forest employed Harriet Kelley, to scan the horizon for fires day and night, with naked eye and telescope. They led the way for a host of later female lookouts (Williams 2000: 51-52; Pendergrass 1990: 21-22; Cermak n.d.: 116).

However, despite the ability of women like Daggett, Ingoldsby, Kelley or even Julia Shinn, women did not rise to or begin to fill the ranks of the Forest Service from the ranger position on up. They did receive adequate and equal pay for the limited positions they held as clerks, but wage and job discrimination for women did not change until the 1960s, when the women's movement brought pressure on the Forest Service in general to hire women for a variety of field and management positions. Even with affirmative action, as later chapters will describe, it was an uphill battle for career-minded women who wished to move into upper management ranks or even to receive equal pay for their work. Alongside addressing critical personnel issues regarding the work of rangers and forest clerks, several administrative issues confronted DuBois and the District 5 office staff. One important concern was administrative improvements. District 5 had outgrown its offices in the First National Bank Building at the corner of Post and Montgomery Streets because professional and clerical staff numbers had increased considerably since they had moved into it in 1908. Therefore, in April 1914, DuBois moved the San Francisco headquarters down the block to the Adams-Grant Building located at 114 Sansome Street. The new headquarters in the heart of San Francisco made staff more efficient and left room for new clerical staff. After the move, DuBois instituted a policy of excluding college men from administrative staff and clerk positions, which marks the time when these positions became "women's work." One unfortunate result of this policy was that the District 5 office lost considerable knowledge that these technically trained men brought to this level of administration (Ayres 1941: 27).

One reason District 5 had outgrown its previous office was because the district's responsibilities were growing. For instance, in 1916, Chief Forester

Lookout platform at Bald Mountain, Klamath National Forest

Graves created in Washington a new branch of public relations, later called information and education (I&E). He formed this branch under the perceptive and scholarly Herbert A. Smith. In order to cooperate with a number of Smith's projects, District 5 added office staff and began a program of public education to improve community relations. By this date, District 5 staff and supervisors were regularly making public addresses and lectures at libraries and high schools regarding forestry matters. District 5 also early on used motion pictures as a public relations tool. For instance, in 1912 the Edison Moving Picture Company, in cooperation with District 5, made a film illustrating the work of the Forest Service in firefighting on the Sierra National Forest, which played at local theaters such as the Market Street Theater in San Francisco and was reproduced and sent around the country as a training film for other forests. Additionally, by 1916, various district forests, such as the Angeles, Lassen, Stanislaus and Sierra National Forests, held displays at local fairs and prepared displays for events such as the National Orange Show in San

Forest Service Exhibit at National Orange Show, San Bernardino, CA. 1917

Bernardino (USDA Forest Service 1912: 24; *Weekly Bulletin: Forest Service, District* 5 1916: 3, 6-7; Rose 1993: 60). Region 5 also had exhibits at the Panama-Pacific California Exposition in 1915 (Conners 2004).

In December 1916, to better communicate with the far-flung nineteen national forests overseen by the San Francisco office, DuBois augmented the district's normal communication of ideas – carried on through letters, orders, dispatches and word of mouth – with a news bulletin, an idea he garnered from house organs published by modern California corporations such as the Standard Oil and Western Electric. Called the *Weekly Bulletin: Forest Service, District 5*, this corporate-styled publication filled demands for news about the district from supervisors, rangers and members of the district office.

The first issue provided a section entitled "Happenings," detailing events around District 5. It covered a variety of topics, ranging from research, fire studies and land classification to roads and insignia for authorized automobiles. DuBois hoped that the five- or six-page mimeographed *Weekly Bulletin* would not degenerate into a "stereotyped dry-as-dust circular letter.... or die a lingering death, but evolve someday into a five-color cover and real type"(*Weekly Bulletin: Forest Service, District 5* 1916: 1-2). It was also common for individual forests to have their own newsletters or bulletins. For instance, the Sierra National Forest had an employee letter as early as 1912 (Cermak 2004). Bulletins such as the *Stanislaus Bulletin*, published by the Stanislaus National Forest, touched on both local matters as well as wide-ranging research on silviculture, grazing and other current topics that applied to their jobs (Conners 2004).

At the same time the San Francisco headquarters moved, several new supervisors' headquarters were built on California's national forests in the 1910s. The supervisor's office was the nerve center of any national forest because it was there that all office work, record keeping and files were concentrated. Physical improvements such as ranger stations, pastures, barns, trails and telephone lines were also steadily built with limited appropriations – Congress allotted only $650.00 for any one improvement. With this limited amount, DuBois wisely implemented the first set of standard building plans for District 5 (Rose 2004) – an action influenced by Taylor's call for standardization. To make rangers themselves more efficient, ranger headquarters on many forests were relocated in order to be nearer to supplies, mail and telephone. They were also moved to more central locations for better assistance to firefighting, to users and to important timber sales. Far more important to California, the stations were concentrated at points where they could be easily accessed by the Forest Service's growing highway system.

Before the creation of the forest reserves, ranchers, stockmen, lumbering and mining industries constructed low-standard roads on them in order to carry on their business. These two-track dirt and gravel, single-lane wagon roads with an occasional turnout became the nucleus of California's forest highways. As time passed and commercial volume increased, many of the old trans-mountain roads were gradually improved until several major highways extended over most of California's mountain ranges in an east-west direction. Of course, many of these highways passed through national forest land in the High Sierra. As Californian's love for wheels mounted, the public, especially tourists and campers seeking to enjoy the scenic values of national forests, began to demand from the Forest Service both primary highway and secondary road improvements. Many Californians with registered automobiles wanted the major highways running from east of the Sierra to the Pacific Coast enhanced to facilitate commerce and accessible travel to the rest of the nation. On the other hand, tourists and campers wanted the somewhat disjointed and illogical secondary road system within the forests themselves improved as well (Burnett 1933: 1-10).

Sunday morning in a Plumas National Forest camp

From the inception of California's forest reserves in 1892 until 1912, no federal money was appropriated for the specific purpose of improving forest roads. Then in 1912, Congress authorized an expenditure of 10 percent of the gross receipts from forest products for the improvement of roads and trails of a given national forest. In the first year, District 5 received $24,821 dollars from this fund, and increasing amounts as receipts increased. With this amount, District 5 immediately began the construction of a forest transportation system, which was greatly needed not just for recreation but for administration, protection and development of California's nearly twenty million acres of rugged, mountainous forests. However, due to the cost of roads compared to trails, District 5 spent the first 10 percent funding mostly on constructing much needed fire trails (Ibid.: 14). In addition to the construction of fire trails, one important trail begun at this time was a segment of the John Muir Trail under construction from Lake Tahoe to Yosemite, which crossed national forest land. Forest Service officials predicted that "when this trail is tied in with the John Muir Trail, the crest of the Sierra will be open to travel for a distance of two hundred miles, [becoming] one of the most famous in the world" (*Weekly Bulletin: Forest Service, District 5* 1916: 2-3).

The 10 percent appropriation proved wholly inadequate, especially in California, where the number of automobiles on the roads seemed to grow exponentially. Widespread demand to speed up forest road construction came from every part of the state. Subsequently, the so-called Section 8 Fund was created, under which Congress appropriated $10 million to be expended in ten annual installments nationwide. California's national forests consumed the greater slice of this federal funding pie. By the end of the installment period, District 5 received almost 15 percent of the fund ($1,464,333), which it spread out over its forest system (Burnett 1933: 15).

In the expenditure of 10 percent and Section 8 funding, District 5 faced many issues and difficult problems that needed resolution. First and foremost, one of the main problems were conflicts between desired highway locations and hydroelectric power withdrawals. Early on, many sites along streams were set aside for potential power purposes by executive order for administration by the Reclamation Service. However, streams and canyons often were also the natural location for highways. Therefore, difficult development decisions had to be made regarding which sites were more valuable for highways as opposed to reservoirs and other appurtenant water power works. Studies weighing the general importance and values of the two elements took time, and the decision-making process often caused great delays (Ibid.: 17).

Then there was the issue of mineral claims. As noted earlier, mining claims were often staked as a subterfuge to acquire title to national forest lands for purposes other than mining. After Congress authorized funding for Forest Service highway and road construction, District 5 had to sort out the legitimacy of various mining claims along proposed routes. They found that many mining claims were made along the only feasible route for a highway and "sometimes for the purpose of forcing their purchase at an exorbitant cost" before highway construction could get under way (Ibid.: 18).

Growing Multiple Use on California District 5

Under Coert DuBois, District 5's attention continued to focus on the major multiple-use management areas promulgated by Pinchot – timber, fire control and protection, grazing and mining. The work demands for each resource were time consuming, but were made easier through administrative work plans. However, during Coert DuBois' tenure, one new concern was formally added to District 5's responsibilities – recreation. The following describes the major accomplishments and problems of both the "old" and "new" forest uses prior to World War I.

In August 1911, Chief Forester Graves, in a letter to all district foresters, encouraged the preparation of administrative work plans for each forest. In accordance with the *Use Book*, forest plans were divided into sections such as silviculture, grazing or protection. To meet this requirement, forest work plans were prepared for the Plumas, Monterey, Mono and Inyo National Forests by 1914, and additional work plans were made for the Sierra and the Tahoe National Forests by the end 1915. However, these work plans were premature, as the foundations of national forest management were still being laid in the 1910s. As one writer put it, it was a "time of strenuous physical construction and not of the leisurely study so necessary for technical accomplishment." After the preparation of these early plans, District 5 for the most part ignored the planning process. Some staff could not see the need for them, and others did not believe they would work if followed. According to Assistant District Forester T. D. Woodbury, in the 1910s, the "policy of national forest management in California was very simple: it consisted in selling mature timber wherever the operator wants it unless some reasons were known why the sale should not be made" (Ayres 1958: 43-45). Woodbury's blunt statement clearly indicated that silviculture and timber sales took precedence over all other forest management areas and responsibilities.

Under the direction of District 5's office of silviculture (1908-1919), the pre-war period was marked by steady and simplified timber sales, improved cruising and marking techniques, better timber reconnaissance or surveys, a turn toward experimental reforestation and the beginnings of insect control work and pathology. Priorities for these management areas slowly changed once war broke out in Europe.

Even though many of California's forests were devastated by the previous century's misuse and abuse, sales of merchantable timber continued. During the period 1911 to 1916, timber sales within District 5 were solid, but spread out over the district. For instance, on the Eldorado National Forest, timber sales were a low priority and in 1912 only amounted to $233.43 (Supernowicz 1983: 162). Timber sales on District 5 were seen as beneficial to the public for several reasons. First, a growing automobile-oriented populace, as well as tourists, appreciated the income derived from the sales because that went to road and trail building on the national forests. Second, in the opinion of some, Forest Service timber sales prevented monopolies of this resource, and therefore, the public was protected from timber being tied up in large holdings and held for speculation purposes. And third, the public was assured of the continuation of forest stands because the public believed that proper regulation and distribution of sales was confined to the most mature timber (USDA Forest Service 1912: 109).

One important aspect of timber sales was administrative cost, and under Assistant District Forester Woodbury, timber sale costs were kept down, largely because of improved marking techniques – one of the most controversial of all subjects between forest officers and timber purchasers. The object of marking was to harvest ripe timber, to secure reproduction after cutting, to accelerate growth of reproduction and to make a timber purchaser's operation profitable. Prior to 1910, California timber operators often complained that the Forest Service misled them in the amount of timber they could get from a given sale. Responding to the situation, Woodbury inaugurated the practice of sample marking in July of that year. Under this scheme, forest officers marked a representative area of sufficient size to give the prospective purchaser, who was taken over the ground, an idea of the value of the timber sale. This proved to be so successful that in 1911, Chief Forester Graves recommended sample marking as a standard practice for the entire Service. By 1916, local and regional marking boards were also organized and instituted for large timber sales. Composed of a representative of the district office of silviculture, the forest supervisor, timber sale men and a representative of the purchaser who understood logging, these boards' main responsibility was to examine and critique at the end of each season important timber sales on a given forest, as well as other aspects of sale administration. California marking boards took the approach of saving about one- third of a total stand for future cut. This recommendation was based on a thirty-year first- cut cycle, and a fifty- to seventy-five-year second-cut cycle. The main effect of this system was to "leave intact groups of thrifty even-aged trees usually under 30 inches in diameter breast high." Marking boards in California operated from 1912 to 1924 (Ayres 1958: 26-28).

In addition to improved marking techniques, District 5's office of silviculture conducted better cruising or timber reconnaissance than the previous visual methods used in Pinchot's day. In 1912, the object of reconnaissance was defined as a way to "learn the amount, location and accessibility of timber, together with a description of the timber and factors affecting its growth." This knowledge was necessary to District 5 in order for forest officers to determine where, what and when to sell timber from any region (USDA Forest Service 1912: 123; Ayres 1958: 55).

Important changes in reconnaissance methodology began in 1911, when District 5 Forester Olmsted sent out general instructions on cruising methods for the first time. A year later, DuBois supplemented Olmsted's instructions with more details and instituted and standardized the two-man strip method of cruising at that time. Under this system, one man ran the compass, computed the distance and took field notes for mapping, while the other tallied the trees on each side of the line. This method resulted in a percent cruise, which was considered for the day an "intensive" survey (Ayres 1958: 56).

Reconnaissance work under the DuBois administration was conducted in the wintertime to keep rangers and guards employed, a chilling idea credited to Supervisor R. F. Hammit of the Shasta National Forest. Hammit's crews began cruising on his forest during the winter of 1910-1911, a practice soon copied by other supervisors. For example, the next season there were winter reconnaissance crews working on the Plumas and Eldorado National Forests, as well as the Shasta. For the next six years, crews conducted timber reconnaissance in horribly brutal and trying weather conditions in California's northern forests, including the Lassen and Modoc National Forests. Each year, crews set out on lengthy winter trips, moving their supplies (about 1,000 pounds of dunnage) along on Yukon River sleds little by little on snow-packed trails, usually on an upgrade. As reconnaissance crews trekked along, they moved from one remote one-room cabin to another. Sometimes they bivouacked in tents in the wilderness, often in well below zero weather. Under average conditions, one man pulled 125 pounds or so on his sled and was able to cover about ten to twelve miles each day. To get around they used skis and snowshoes. Louis Margolin of District 5 was placed in charge winter reconnaissance work until his untimely death in June 1914, when he slipped and fell into Dinkey Creek on the Sierra National Forest. The creek was swollen to a torrent by the melting snows, and Margolin was swept away; his body was never recovered. W. M. Gallagher replaced him. In the end, this "he-man" experience proved to be very ineffective and inefficient because of equipment failures caused by adverse and severe weather conditions. Fortunately, it was discontinued before other employees lost their lives (Ibid.: 57-59).

Meanwhile, District 5 silviculture staff began to doubt the value and accuracy of the 5 percent cruise, which had been the standard for "intensive" reconnaissance for fifteen years. As a result, the silviculture staff decided to make a 10 percent cruise the district standard, especially where detailed figures were required by a sale. District Forester DuBois approved the idea in April 1914, but the 10 percent cruise standard did not go into effect until 1916. In that year, Washington prepared a new set of regulations for timber surveys entitled *National Forest Surveys and Maps-Topographic Surveys*, which ushered in a new era. The new regulations, which served as the benchmark until the late 1950s, changed the name of the activity from reconnaissance to timber surveys and adopted the California 10 percent standard. The first work in California under the new regulations took place on the Stanislaus National Forest. At about the same time, California District 5 established timber marking rules that provided for 200-foot scenic corridors along roads, lakes and riverfronts. Marking in these locations was to be light and aimed at improving the appearance of the forest (Ibid.: 60-63).

In addition to improved timber surveys and marking standards, District 5's office of silviculture also paid closer attention to the pathology of tree stands at this time. One significant enemy of trees in California's national forests was disease. To handle the growing problem, District 5 assigned Dr. E. P. Meinecke, a European-trained forest pathologist, as a consultant on tree diseases to advise them. More than any other person, Meinecke recognized the "vast problem of converting ragged, irregular, defective over mature, stagnant and only partly merchantable forests into productive and managed stands." Dr. Meinecke pointed out that the forest sanitation clause in timber sale contracts, which originated in District 5 and was thereafter adopted for the entire Forest Service, was a valuable management tool for the elimination of cull (snags, diseased and malformed trees) which would produce high-grade timber in new reproduction. Meinecke

Dr. E.P. Meinecke. 1937

also indicated that the lack of accurate data and reliable methods of interpreting such data was the main obstacle in the way of applying the science of pathology to District 5's problems. As a matter of good business practice and to ensure maximum yield, he encouraged District 5 to acquire that cull data (Show n.d.: 57-58; USDA Forest Service 1912: 133-134).

Another key enemy of California's forests was insects. The first authentic record of insect control in the California's forests began in 1907, when ranger Roger Baldwin carried out a small control project in the Santa Barbara (Los Padres) National Forest. In 1911, C. Stowell Smith, then the chief of District 5 branch of products, reported the first mountain pine beetle infestation on the southeastern portion of the Lassen National Forest. To clean up this infestation, District 5 promoted a timber sale – perhaps the first time forest management had been called in to check a bark beetle attack. Ironically, in this case the lumber company sued the Forest Service, claiming that the government misrepresented the amount and value of the timber. The timber operator, however, lost the case when the court ruled in 1921 "buyers had to beware when they bought timber from the Government" (Ayres 1958: 75, 82).

To meet the growing threat of further insect damage, in April 1912, the Forest Service established a cooperative relationship with the USDA Bureau of Entomology in Washington, whose official duty was to control insect damage on forests. The bureau had conducted yearly surveys for insect and disease conditions on the forests as early as 1908. Under this agreement, the Forest Service independently conducted all administrative work relative to insect infestations in national forests – calling on the Bureau of Entomology only for scientific data and advice (Ibid.: 75-76). After this arrangement was recognized, Ralph Hopping, a ranger on the Sequoia National Forest, was appointed forest examiner in charge of insect control work in the office of silviculture. J. M. Miller, a trained Stanford entomologist and ranger on the Sierra National Forest, was placed in charge of blocking out and initiating insect control projects. Under Hopping and Miller's direction, several early insect control plots were located on the Klamath, Shasta and Sierra National Forests (Ibid.: 76-77; Show n.d.: 57-58). In addition, rangers were encouraged to spend time with insect control parties so that they could become better acquainted with the condition of infested trees (USDA Forest Service 1912: 119-120).

In the early years of insect control in California, controversy over the proper method or procedure in control work developed between the Forest Service and Bureau of Entomology. The bureau men based their control methods on work they had previously conducted in other parts of the country, while California forest officers, unhampered by any established "scientific" methods, proceeded by conducting work based on practical experience. Before long, a "thinly veiled" acrimony between the two agencies occurred over their theoretical approach, which was not cleared up until 1920. By the time a cooperative procedure was established, Hopping had left for a better paying position with the Canadian Forest Service and insect control had been severely curtailed because of World War I (Ayres 1958: 77-81)

Unlike insect control and pathology, which were growing in importance in the office of silviculture, experimental and extensive reforestation projects on District 5 continued only on a marginal basis. By the end of the Olmsted administration, direct seeding projects on both northern and southern forests were termed a complete failure, and the program was dropped (Show n.d.: 61).

Not much later, the reforestation program in the south was also found to be a universal failure. That program, with the exception of a small body of experiments, was dropped as well – but not soon enough – for before the termination of this planting program, District 5 put considerable time, effort and expense into the promotion of eucalyptus as the answer to the tree planter's prayer – at least in California (Ibid.). By 1910, Southern Californians became very interested in the tall Australian evergreen. Bolstered by private and sincere scientific studies, as well as by some cooperative efforts between the Forest Service and the California state forester, the general public and speculators came to believe that the aromatic tree, which also produced an oil with medicinal properties, could grow rapidly and yield as much as 100,000 board feet in ten years. These extravagant claims led to a "Eucalyptus Gold Rush" in California. In the rush to plant, an investment boom followed, with exorbitant stock selling as well as outlandish promotion of real estate for planting. The Forest Service jumped on the bandwagon. Based on the initial information, District 5 immediately started several nurseries on southern forests geared to grow eucalyptus, such as at Oak Grove (Cleveland), Los Prietos (Santa Barbara), Merrick Canyon and San Bernardino (Angeles). The proverbial eucalyptus bubble burst when further experimentation and study proved that the yield and return on the eucalyptus species did not meet the exaggerated expectations that earlier studies found regarding growth rate and timber quality. In order to warn the public, Assistant District Forester Woodbury without delay published Circular No. 210, "Yield and Return of Blue Gum." At the same time as its release, the Forest Service quickly abandoned their eucalyptus nurseries. Eucalyptus fever lasted in California for several more years but finally died out, leaving many investors penniless (Ayres 1958: 67-68).

The conifer planting program in the north faced a slightly different fate. Widespread conifer reforestation projects ended during the early years of the DuBois administration, with the exception of those conducted on the Shasta National Forest. Forest Examiner S. B. Show (who in 1926 became district forester) and Eldorado Forest Supervisor Edward I. Kotok were assigned to review the reforestation problem. After some study, they concluded that experimental reforestation work was well worthwhile. Nonetheless, in 1917, the last two early District 5 nurseries, the Converse Nursery (Angeles National Forest) and Pilgrim Creek (Shasta National Forest), were closed, along with the Feather River Experiment Station on the Plumas (Show n.d.: 61; Ayres 1958: 70-71).

The accomplishments and failures of the office of silviculture in the pre-World War I period were many and indicated a preoccupation with the timber industry. However, during the same period, fire control and protection work matched the intensity of timberwork and resulted in several important developments as well as failures.

The disastrous 1910 fire season rudely awakened District Forester DuBois and others to the real fire problem on California national forests. They had never in their careers witnessed fires that burned with such intensity. A shocked DuBois, who was not yet appointed as district forester, quickly realized that the old days of one or two rangers putting out fires in the backcountry with little coordination were indeed over. Never one to sit on his hands, the young DuBois focused his energies on meeting the problem and began to systematically organize a plan of intensive protective measures.

To begin with, DuBois went to the Stanislaus National Forest, studied a fire-ravaged range district, and worked out District 5's first fire protection plan. Afterwards, he reviewed the plan with California's forest supervisors at the winter annual meeting. Based on their reaction, in January of 1911 DuBois outlined his planning ideas in a booklet entitled *Fire Protection Plans*. Within these pages, DuBois likened fighting forest fires to a military operation. He encouraged aggressive, proactive preparation before a fire started, rather than the customary reaction afterwards. He promoted devising forest fire plans for each forest – plans that should include building up a patrol organization and a better communication network. Following this direction

Fire tool boxes used on the Mono National Forest. 1922

and his judgment, District 5 thereafter began a speedy campaign of erecting telephone lines, building lookout stations and distributing tool caches on each forest at strategic points. DuBois' plan also called for individual ranger district fire plans and, importantly, required improved transportation and communication maps (Cermak n.d. 111-112; Benedict 1930: 708; Ayres 1942: 7). In conjunction with this course, the position of inspector of fire plans was added to the district forester's staff. William G. Hodge, a holdover from the Olmsted administration and author of the influential paper "A Report on a Forest Policy for the State of California" (1905), initially filled this position (Ayres 1942: 8; Clar 1959: 195, 223). In 1914, the duties of the job changed and included not just review of fire plans, but also all fire protection work in the district. By the time of the change, Hodge had died and DuBois named David P. Godwin, former supervisor of the California National Forest (renamed the Mendocino National Forest in 1932), as district "fire chief." This newly-created position was placed under operations headed by Assistant Forester Roy Headley. Together, DuBois, Godwin and Headley began a rational, orderly analysis of the fire problem. They methodically gathered a tremendous amount of fire information for each forest. Meanwhile, DuBois zealously drove home the theme of fire control to his forest supervisors and rangers. He browbeat them into believing that this nemesis of the forest was not just an "unfortunate act of God," but a human-caused problem that could be managed and controlled. Starting in 1911, DuBois gathered needed data by requiring forest supervisors to regularly report in great detail all fire activities on their forests. Next, he assigned different problems of fire protection to selected forest officers to study, expecting them to report their results at a special fire protection meeting in the winter of 1913 (Ayres 1942: 9).

Joint meeting held at Nevada City, California during the winter of 1911-1912 between forest officers of Tahoe and Eldorado National Forests. 1912

In early 1914, while at his Sausalito home, DuBois organized fire control solutions based on this collected data. Editing and distilling this contributed material, he published *Systematic Fire Protection in the California Forests*, which postulated a scientific approach to fire control. "Here for the first time was gathered," according to one source, "the practical knowledge upon the subject of forest fire control which had been building up among field and office employees of the Federal Forest Service since the Department of Agriculture had assumed supervision of the forest reserves in 1905" (Clar 1959: 372). This fire manual analyzed District 5's "norms" regarding weather, vegetation and fire history, and then tried to predict abnormal conditions and how to handle them. DuBois' manual ended the "go it alone" policy of the past and instituted a district-wide coordination of effort as the main purpose of District 5 policy. Moreover, the publication provided down-to-earth, detailed information on planning, organizing and supervising fire control activities. Forest protection and emergency mobilization plans were required of all supervisors, who, along with rangers, were made accountable for any failures on their forest or ranger unit. Important sections in *Systematic Fire Protection* covered fire detection, communication and light burning as a means of hazard reduction. DuBois' fire control manual set policy and procedure for District 5 for years to come. This was "planning with a vengeance" and "must have come as a shock to weary supervisors already burdened with timber, grazing and land adjustment plans" (Ayres 1941: 15; Cermak n.d.: 117-122, 154; Rose 1993: 61; Clar 1959: 372-376).

DuBois' fire control manual also called for systematic fire plans such as the one he devised earlier for the Stanislaus National Forest, along with better roads and fire trails, permanent lookouts and improved communications. With it in hand, many were convinced that District 5 had fire control conquered. DuBois' enthusiasm also stimulated new innovations – some not so successful. For instance, the Angeles National Forest, which had recently built hundreds of miles of firebreaks in the San Gabriel and San Bernardino mountains thanks to local funding, could not keep up with cutting new chaparral sprouts. Goat herds were hired to graze the firebreaks, but this solution failed miserably (Show n.d.: 60, 70; Cermak n.d. 114, 152; Ayres 1942: 9).

Ancillary to DuBois' work was the enlistment of the State of California in carrying out its share of the load of fire control. The passage of the Weeks Act (1911) facilitated cooperative forest protection efforts between the Forest Service and the states by granting matching funds for this purpose on forested watersheds of navigable streams. Though the federal government saw the wisdom of these expenditures, it would be eight long years before the California legislature acknowledged a similar responsibility and provided proportionate fire suppression funds (Clar 1959: 311).

Matters involving fire control did not concern the timber industry, which had little regard for protection until the 1920s. Around the year 1900, California's more important lumber companies formed loose business associations, and there were a few attempts made to form cooperative fire protection associations. During the Pinchot years, District 5 timber contracts required the purchaser to clear all railroad rights of way and all donkey engine sites of inflammable material. These contracts also required the piling of all brush and debris left after logging operations in California national forests. In 1911, District 5 tried to encourage cooperative agreements with the lumber industry by offering to extend federal protection to private land adjacent to or intermingled with the national forest land. To foster more cooperation, the California Forest Protective Association (CFPA) was incorporated in 1912, but only a minimal amount of cooperative fire control policy on private lands followed. By 1916, lumber companies were required to have a supply of shovels, axes and barrels of water at each donkey site on government land in case of fire. Even so, during a fire, lumbermen were content to save their own equipment, and contributed little to suppressing fires not in their immediate area. This attitude was not fully overcome until at least 1925, when the Forest Service threatened to "shut down" operators who did not contribute (Ayres 1958: 20-21; Benedict 1930: 707).

Nonetheless, Forest Service policy regarding fire on lands outside national forest boundaries was challenged in 1914, when the historic Sisson fire broke out near the town of that name (now named Mt. Shasta), which was adjacent to the Shasta National Forest. The Sisson fire began as a manageable fire. It was not a very big blaze, only 2,000 acres. But because of the lack of private and federal organization and cooperation, it soon got out of control and took the combined efforts of almost the entire northern part of California to put it out. All available District 5 personnel, some from as far away as the Stanislaus National Forest, were brought in to fight the fire. Fighting the fire alongside them were hired itinerant laborers, mostly "hoboes," who were transported to the blaze on special trains. By the time the Sisson fire was suppressed, DuBois realized that a "massive infusion of manpower without the proper organization to feed, blanket and supervise them was just as inappropriate as providing too few men." When it was all over, thousands of dollars had been spent on what today would be considered a routine brush fire (Cermak n.d. 152-156; Benedict 1930: 708; Pendergrass 1985: 92-94).

Vowing not to let this happen again, DuBois resolved "future fire control activities would include economic objectives." On May 1, 1915, DuBois sent out a circular letter to all the forest supervisors, stating that the "District Forester wished to correct the impression that speed rather than economy in suppression was the sole criterion of efficiency." This circular also contained instructions that led to what became known as the "let burn policy" – a fundamental change in fire policy on District 5. Urged on by Headley, henceforth large areas of non-timbered government land would be allowed to burn if expenditures were disproportionate to the value of the resources under protection. In addition, DuBois gave cautionary conditional permission for cooperative associations to exercise controlled burns on private lands along the edge or near the boundaries (inside and out) of national forests, under certain conditions. District 5 fire Chief Godwin granted approval if the action was to protect homes and property, to facilitate the handling of livestock, to assist systematic prospecting, and to clear agricultural land. This "cooperative burning" policy, along with DuBois' economic fire protection policy of "let burn," was established in only certain areas of California's national forests from 1915 to 1919. To compensate for potential crimes of incendiarism, Chief of Operations Roy Headley prepared District 5's first law enforcement manual, entitled *Instructions Relating to the Apprehension and Prosecuting of Fire Trespassers on National Forests* (1918). Based on this manual, one permanent staff person on each forest was trained in law enforcement. Under the watchful eye of this "arson squad," incendiary-caused fires plummeted from 325 to just 62 during World War I (Ayres 1942: 10-11; Cermak n.d. 152-156; Benedict 1930: 708; Ayres 1941: 15; Pendergrass 1985: 92-94; Show 1963: 155).

Timber management and fire control and protection were not the only areas that received special attention during the pre-World War I DuBois administration. In this era of Taylorism, "scientific" range management began as well.

By 1911, District 5 had already passed through the major struggles over grazing fees, range allotments and other grazing restrictions such as carrying capacity. In 1907, shortly after the Forest Service implemented these changes, a major test case came to federal court involving the pasturing of sheep on the Sierra Forest Reserve without obtaining a permit and paying fees. The grazing fee system charged according to numbers and types of stock entering the forest and for the length of time they stayed on the range. Four years later, the Supreme Court found in favor of the government. The Supreme Court ruled that any use of national forest land for pasturage was subject to rules and regulations promulgated by the secretary of agriculture. The Court also concluded that fees were necessary "to prevent excessive grazing" and that they were needed in order to provide for administrative expenses to manage this resource. On the other hand, grazing allotments were determined by the condition of the range and the numbers of livestock using the range. *Maximum* and *protective limits* were set. The *maximum limit* was the greatest number one outfit was allowed to graze. *Protective limits* represented the number of livestock a settler or company needed to support a family or operate economically. Finally, range carrying capacity was influenced by available water, the climate, nature of the land, value of the different grasses, plants, and brush, how much each head of stock required and seasonal grazing practices (Rowley 1980: 64-72).

Soon after the Supreme Court upheld the very-low-fee system instituted by the Forest Service, DuBois became district forester. By this time, Chief of Grazing John H. Hatton, who served under Olmsted, had transferred to another district. M. B. Elliott replaced him, but for unknown reasons he was replaced by "cowboy" C. E. Rachford, who was "hoisted out the ranks of his peers" to become DuBois' grazing "expert" (Show n.d.54, 72; Ayres 1941: 10).

District 5's stated grazing policy, and that of the Forest Service in general, was to help the poor, small stockman make a living and not fall victim to larger outfits that tended to dominate rangeland. Initially, this noble, unwritten law worked as an instrument of social policy. For instance, a 1908 grazing report from the Stanislaus National Forest openly frowned upon larger outfits, stating "they are the men who, if there were no National Forests, would in a short time monopolize the range without justice to any one." But as time passed, it was clear that small livestock enterprises could not survive without Forest Service aid. While the Forest Service continued to protect the small livestock owner, they showed little sympathy for sheep owners. During the 1910s, District 5 practiced an "unconscious and unwitting discrimination against sheep and goats," even on the Southern California forests such as the Santa Barbara National Forest (Ayres 1941: 10-11; Rowley 1985: 72-74, 82-83). This discrimination against sheep and goat herders was also partly institutionalized. Also, there is some evidence to suggest that there was also a strong ethnic bias associated with these herders (Conners 2004).

For most of the 1910s, District 5 range work revolved around routine duties: settling of range boundaries and allotments; cooperating with stockmen in making trails, building counting corrals, salt licks, and drift fences; and developing permanent water improvements such as springs, irrigation ditches and erosion dams. Other duties included eliminating cattle diseases such as anthrax and blackleg, eradicating poisonous plants such as larkspur, which accounted for 95 percent of cattle loss at the time, and the extermination of predators such as coyotes and bobcats (Ayres 1941: 10; USDA Forest Service 1912: 165; 1916: various; 1917: 14). A conservative estimate of livestock losses in 1916 amounted to more than 6,600 cattle and 15,000 sheep. Livestock diseases accounted for the largest loss of cattle, followed by larkspur. Local stockmen joined with the Forest Service in these efforts, but together they were only partially successful. Unexpectedly, the destruction of natural predators caused a dramatic population increase in prairie dogs and other range-destroying rodents – an infestation problem that plagued the Sequoia and the Modoc National Forests in particular. All in all, the ranges on California's national forests prior to World War I appeared in good shape, with only a few areas of localized overgrazing. This overgrazing was attributed to a number of factors, including grazing too early in the season, too few and improperly located salting boxes, a lack of sufficient watering places and the absence of range riders to prevent congregation in favored areas (USDA Forest Service 1917: 10, 17-20).

California's grazing program succeeded in the early years because supervisors and rangers took into account local traditions of the people on their forest. Forest officers attended local livestock association meetings and asked locals for help in formulating grazing work plans. Chief of Grazing C. E. Rachford and the Modoc forest supervisor, for instance, worked closely with livestock associations on the Modoc National Forest, helping them to attain the direction that the permittees wanted on such issues as salting, number of bulls allowed and roundup restrictions (USDA Forest Service 1912: 164, 171-172). They also encouraged the development of grazing advisory boards composed of prominent citizens. Stockmen were encouraged to bring disputes to these advisory boards for settlement, and their decisions and solutions were often adopted by the Forest Service. For instance, by 1912, the Stanislaus National Forest had not reversed a single settlement brought to the supervisor by its local advisory board. The supervisor understood that stockmen needed to feel that the Forest Service took local opinion and traditions seriously (Rowley 1985: 80-81). Oftentimes, District 5 openly disagreed with and ignored instructions from the *Forest Manual* because of range conditions in California, concluding that "we have big men back in Washington that need only to be shown" (USDA Forest Service 1912: 164).

Grazing reconnaissance, the technical job in range work that sought to improve and develop grazing resource and its use, was initiated early in the DuBois era of scientific management. Reconnaissance work first began on the Modoc National Forest, largely because stock grazing was a principal industry of Northern California. At the 1912 supervisors' meeting, a set of questions and answers was prepared regarding grazing and related matters such as reconnaissance. From the questionnaire, District 5 officials learned that neither the district office nor supervisors were paying very much attention to regulations and instructions concerning grazing. Apparently neither District 5 nor Washington was very concerned about this lack of regard for reconnaissance work. In their view, intensive reconnaissance was impractical and of too small a value for administrative purposes to justify the cost (USDA Forest Service 1912: 162, 168, 180-182, 187-190).

To tighten up the situation and put grazing on a more scientific footing, in 1913, District 5 issued a standard outline for range reconnaissance that briefly described fieldwork methods and lightly touched upon the compilation of data. Grazing work plans were required of every forest, to put grazing administration on an efficient scientific course and to prevent changes in policy because of a change in personnel. Initially, these work plans were based on very superficial and inconsistent data collected by untrained district rangers and their assistants. This information was reported to the forest supervisor, who then routinely transferred it onto the grazing section of the annual report sent to Washington without any analysis (USDA Forest Service 1917: 1-2).

The implementation of the new grazing work plans became important in 1916, when Congress passed the Stock Raising Homestead Act, which authorized the sale of 640 acres of public domain land suitable only for grazing livestock to stock enterprises, provided they installed some range improvements. The effect of the act was the reduction of public grazing lands and a heavier demand for grazing privileges on California's nineteen national forests. The consequences of this legislation brought home to District 5 the necessity for better and more detailed data on California's forest forage resources (Rowley 1980: 91).

Once Congress authorized appropriations, District 5 initiated reconnaissance studies, also known as grazing surveys. District 5 began its grazing inventory work later than most Forest Service districts, and work did not really begin until 1916, when the first range examiner was assigned to the district. Thereafter, DuBois selected three national forests as beneficiaries of this work. The first scheduled range survey took place on the Warner Mountain Division of the Modoc National Forest. The range examiner and his assistant, along with a "special crew of properly qualified men of technical training and practical experience," were sent there. The Modoc was selected because of the intensity of use of the area, the dependence of adjacent property on national forest range and the importance of the livestock business to the community. When the Modoc work was finished, the California [Mendocino] and the Klamath National Forests were considered the next forests in line for grazing inventories. It was estimated that all available District 5 range survey funds would be needed on these forests for several years to come. While work progressed on these forests, the district office of Grazing planned to study conditions of the other forests and formulated a definite plan of work to complete studies for the entire district. There were specific problems to be resolved on the Lassen, Shasta, Stanislaus and Tahoe National Forests, but for the time being, the Southern California forests were ignored in this regard (USDA Forest Service 1917: 6-8).

In February 1917, DuBois sent out a district circular regarding range reconnaissance. In this circular, DuBois, similarly to his approach to fire planning, explained that grazing work plans were needed to systemize the haphazard records so far produced by the district. He requested all the information readily available regarding the utilization of areas, distribution of stock, methods of management and knowledge about improvement work – together with a thousand and one other details not recorded in previous inspection, supervisor and ranger reports, notes and maps. Once the information was gathered, it would be analyzed and then put into a definite working plan for the entire district (Ibid.: 9). The time was ripe for gathering this body of scientific work to support range resource use decisions as proposed by DuBois. But before he could implement his plan of attack, America's entrance into World War I intervened. If this information had been gathered, DuBois and Rachford would have learned that just as the nation went onto a war footing, District 5 had an overgrazing problem. The true condition of the range on some forests was completely obscured by periods of above-normal rainfall before 1917. These favorable growing conditions led to two fateful opinions. The first belief was that California's forest rangeland could sustain continued heavy use without harming the range. The second conviction was that close grazing helped eliminate possible fuels for fires. In essence, forest officials ignored common sense on both points. Rangeland was mishandled in the pre-World War I years, but forest officials could not be blamed. At the time, they were completely ignorant of the concept of rangeland as a biological system involving a complex interrelationship and interdependence of soil, water, vegetation and animal use. District 5's rangelands continued to be unwittingly mismanaged until Forest Service officials fully understood this point (Fox and Walker n.d.: 9).

Besides the overgrazing problem, forest officials also realized that grazing fees needed to be raised so they would be more in line with those charged by owners of private lands. Grazing officials concluded that the growing demand required higher fees, especially as the effects of the Stock Raising Homestead Act were felt and the pressure for forage heightened prior to the war. With a proliferation of permit demands, increased fees did not seem unreasonable, but would certainly disrupt the affable relationship that District

5 enjoyed with stockmen on its forests. World War I intervened before there were confrontations. With the war came a nationwide call for increased food production, followed by orders to fill all ranges to capacity. Controversial issues such as increased fees and overgrazing practices waited until after the war (Ayres 1941: 10-11; Rowley 1985: 92-94; Dana 1956: 229).

Prior to 1917, the prospect of going to war with half of Europe was a distant thought in the minds of most Californians. Instead, in the "age of the automobiles" many urban families with disposable income and more leisure time made the mountains and forests of California one of their first destinations. Increased recreation in California's national forests was due also in part to the growth of the California highways system, and use especially in the High Sierra grew in direct proportion to the development of the California highway system. This added rural mileage made recreational areas accessible even to the working classes. Greater recreational use was reflected in more human-caused fires, which concerned DuBois. In the summer of 1916, Chief Forester Graves directed the district foresters to report upon the number of recreation visitors and their activities on their districts – camping, fishing, hunting, motoring, hiking, etc. – on the nation's 136 million acres of forest land. The summary figures, which were considered low, indicated that just before World War I approximately two and a half million persons entered one

A "weenie" roast. Vacation camp for Pacific Electric Company employees, San Bernardino National Forest.

of the national forests for some kind of recreation (Waugh 1918: 23-24). Not surprisingly, Californians and their growing automobile culture led the nation at this time in seeking outdoor recreation. With more than 18 million acres on nineteen forests to serve them – many of which were very near to urban centers such as San Diego, Los Angeles and San Francisco – the transportation revolution caused a recreation revolution on National Forest land.

Pre-World War I California recreation took two forms – public and private. Public pleasure-seekers included tourists and campers. They engaged in a number of activities such as weekend or holiday picnicking, hiking on trails, picking berries and nuts, boating and sun bathing, and traveling to centers of resort activities. Due to ever-increasing road mileage, more and more people also took to the road to view California's scenic wonders, staying at hotels, resorts and "health" sanatoriums. To better serve the traveling public, in 1916, the Angeles, California, Cleveland, Stanislaus, Inyo and Mono National Forests published new maps with recreation information on them. The map for the Eldorado National Forest was stamped "compliments of the forest supervisor." It provided information on the history, administration, physiography, resources, climate, fishing, hunting, roads and rules for the prevention of fire. District 5 also began building automobile rest areas on Forest Service land. Many outdoor enthusiasts did not mind patronizing hotels part of the time, but automobile camps on national forests offered an inexpensive rest in a tent with a campfire. In addition to these automobile camps, there were also municipal playground camps and historic signage. For instance, by 1916, work had begun on the Lincoln Highway, the nation's first transcontinental highway. In response, Forest Supervisor Edward Kotok of the Eldorado National Forest posted a series of signs on the Lincoln Highway marking historic spots along the way to educate tourists on the history of the area. Additionally, a municipal playground camp named Camp Sacramento was located forty-five miles east of the state capital on the Eldorado National Forest (*Weekly Bulletin: Forest Service, District 5* 1916: 2, 6; Show 1963: 154; Supernowicz 1983: 168-170).

Municipal camps on many national forests established at this time included camps on the Sierra and Angeles National Forests. Little is known about the Sierra municipal camp, but the Los Angeles camp was the first and most fully developed of these facilities. The City of Los Angeles established it by permitting twenty-three acres on Seeley Creek Flats in the well-wooded mountain land within the Angeles National Forest. On this tract, the City erected sixty-one small summer bungalows, with a central clubhouse, cement swimming pool, tennis courts and other camp amenities. When in full operation, the camp provided an outdoor experience to about 300 persons (Waugh 1918: 19).

Cottages at the Los Angeles Municipal Camp in the Angeles National Forest. 1926

Source: Forest History Society

Public recreation in District 5 also included long and more serious horse pack trips into the backcountry of the Sierra and elsewhere. In order to help these recreationalists, in 1915, Washington reissued Hodge's "Handbook for Campers in the National Forests in California" (1908). The foreword to Hodge's booklet urged campers to use the ranger-constructed fireplaces in the attractive campsites. However, on the Angeles National Forest campers were required to obtain a permit before building any fires (Bachman 1967: 2). Despite these improvements, District 5 had no long-range plans to meet the needs of the camping public at this time, and neither did Washington for several years (Tweed 1980: 2).

On the national level, the subject of forest recreation was first elevated to formal policy discussion in the forester's annual report of 1912 and 1913, in which Chief Forester Graves noted that the construction of new roads and trails caused recreational use of the Forests to grow very rapidly (Ibid.: 2). The next year, the forester again mentioned public camps as part of general idea, but few improvements nationwide went beyond seeing that the Forest Service met the primitive camping necessities. In 1915, in response to the forester's message, the first prepared campgrounds in District 5 were built on the Angeles National Forest. "In those days," according to S. B. Show, "camping was in the canyons which were excessively dangerous and from which recreational fires started and spread. They were very simple campgrounds indeed, largely consisting of clearing the ground, developing springs so that the fire could be contained, perhaps putting in a garbage pit." Rangers and guards built them on what was called "contributed time" (Show 1963: 152-153).

On other forests, the main idea stimulating the creation of campgrounds was the need to separate grazing cattle from intruding campers. As early as 1912, District 5 acknowledged a tourist versus grazing problem on the Sierra, Sequoia and Kern National Forests, and sought solutions. On the Sierra, certain stockmen assessed themselves three cents per head to build fenced pastures so that the stock wouldn't wander into campgrounds. With tourists coming in at ever-increasing numbers, some supervisors cancelled a few grazing permits, much to the anguish of the permittees. In recognition of the importance of recreational use, all livestock were "kicked out" of the upper Kings and Kern river drainages, and such use was one among many reasons why sheep were no longer allowed on the High Sierra. Additionally, Chief Forester Graves approved this measure as a defensive move against the creation of additional national parks, a subject that is discussed further below. At this time, the Sierra Club actively campaigned against sheep in the Sierra, which interfered with their mass annual treks into the mountains, ate the feed desired by their pack animals and messed up their favorite camping spots. Other forests, such as the Eldorado and Stanislaus, took less drastic measures. They cleared specific "permanent" camping areas, locating them where campers tended to congregate. In addition to clearing the land, they also built a few rustic tables and loose stone fireplaces as amenities for these campgrounds (Ayres 1941: 12; USDA Forest Service 1912: 165; Show n.d. 77; Show 1963: 153-154; Fry 1963: 181-182).

On the other hand, there were those private pleasure-seekers who chose to build recreational residences on national forest lands under various agreements and terms. The area of greatest concentration of summer residences in District 5 naturally was the Angeles National Forest because of its proximity to a large and rapidly-growing urban center. Until around 1915, the location of these summer homes on the Angeles was very haphazard (Tweed 1980: 2). This situation changed that year, when Congress passed the Occupancy or Term Permits Act. This piece of legislation essentially allowed the granting of special permits for stores, hotels and other similar structures on national forests. These permits were for land not to exceed 80 acres, and leases could not run longer than thirty years. Thereafter, recreational residences were established on a plan-wise basis in District 5, which began to locate and survey specific residential tracts on various forests (Ayres 1941: 12). However, term permits were not in general use for summer homes until the 1960s.

The year 1916 was a major turning point for outdoor recreation activity on California's national forests. Once the Sierra Club and other preservation groups were defeated in the Hetch Hetchy Valley reservoir episode, they effectively campaigned to create a separate Bureau of National Parks in the Department of the Interior. These preservationists wished to set aside large tracts of land as national parks based on their scenic value and unique landscapes, precluding any timber, grazing, mining, hydroelectric or other uses. They enlisted the help of California Congressman John E. Raker, who wished to see the creation of Lassen Volcanic National Park in the area, which in 1914 and for several years thereafter experienced a series of major eruptions that devastated the surrounding area with ash and mudflows. At first Chief Forester Graves supported the idea of a Bureau of National Parks, but later he wished to incorporate these parks under the aegis of the Forest Service instead of a separate bureau. There already existed Yosemite, Sequoia and General Grant national parks in California, and Graves believed the Forest Service could easily manage them. But for several reasons, which included being hampered by the Forest Service's multiple-use mission, Graves could not parry the thrust of the preservationist argument that the Forest Service could not be trusted to keep these parks free of logging, grazing and even summer home development. The National Park Service (NPS) was created as a new agency in 1916. Lassen Volcanic National Park, which was created one month earlier, now fell under the administration of the new agency (Steen 1976: 118-119; Tweed 1980: 5-6; Strong 1973: 39-49).

Under the circumstances, it seems safe to say that the Forest Service's interest in outdoor recreation stemmed from the service's hope of preventing the creation of a new park bureau. Once the National Park Service was created, competition between the two agencies grew over the years. The Forest Service needed a strategy to stand its ground against Stephen T. Mather, the new head of the National Park Service, who "challenged the idea that the Forest Service should be engaged in recreation at all" (Lux, et al. 2003: 33). It turned out that Mather was every bit a zealous crusader for expanding the park system as Pinchot was for the national forests (Dana and Fairfax 1980: 109).

To stem Mather's criticism and the tide of transfers of Forest Service land to the nascent NPS, in 1918 Chief Forester Graves hired Frank A. Waugh to study recreation facilities on national forests and make recommendations on how to develop and improve them. Waugh's "Recreation Uses on the National Forests" concluded that recreation should be put on par with the other major uses of forest areas, such as timber, grazing and watershed protection. Waugh rejected the idea that the creation of the NPS made recreation on national forests moot. He also rejected the argument that all recreation areas be divested from the Forest Service and added to Mather's NPS. In his final analysis, Waugh made the case that the Forest Service should recognize the value of recreation use on its forests; that the agency should protect particularly scenic areas along with relics of historic and archaeological value as a management function of Forest Service administration; and finally that the agency employ personnel "suitably trained and experienced in recreation, landscape engineering, and related subjects" (Waugh 1918: 27-37). Essentially, Waugh's report turned the attitude corner on forest recreation within the Forest Service from a negative one to a more positive one.

Concomitant with the recognition of the importance of recreation management, the Forest Service identified wildlife as a significant resource needing management as well. The role of California's national forests in wildlife resource management began when a system of state game refuges were first established on national forest lands. Early on, the Forest Service had problems with the California State Game Commission, which often neglected these game refuges. In 1918, this problem was resolved when Chief Forester Graves came to District 5, met with state officials and then advocated a new game regulation policy. Under this policy, rangers were given the power to enforce state game laws on national forest lands. Following this arrangement, many rangers were very aggressive in enforcement and were quite successful in making cases (Show 1963: 151-152; Show n.d. 78; *Supervisors' News Letter, District 5* 1918, No. 2).

Despite differences, the California Fish and Game Commission worked cooperatively with District 5. Together they introduced, or reintroduced, game and fish on California's forests for the benefit of hunters, fishermen and visiting tourists. Cooperation began as a slapdash policy, and long-range planning or effort was still decades away. District 5 also introduced new game birds into game refuges where native quail were depleted, and officials consented to moving big game animals onto national forests to supplement shrunken herds such as elk on the Shasta National Forest. In addition, C. E. Rachford advocated a program to protect deer herds with the cooperation of the state. California's forests were natural deer refuges and were blessed with many open places called glades, on which occurs a good growth of browse. Typically each spring, sheep ate these grasses down, but when the sheep moved off, large herds of deer finished the grasses off. California's forests were also surrounded by a dense growth of live oak brush, which acted as the home for thousands of deer each summer. Finally, District 5 cooperated with the State Fish and Game Commission to work out programs to stock trout fry from California state hatcheries in barren or depleted California National Forest waters (Show n.d.: 78; Supervisors' News Letter, District 5 1918, No. 5).

World War I and Its Impact on District 5

In April 1917, President Woodrow Wilson went to Congress to declare that a state of war existed with Germany. With "Armageddon" on hand, Wilson led America on a crusade of moral idealism, a "War for Democracy." The president asked each American to sacrifice and embrace the war effort as that "final struggle where the righteous would do battle for the Lord."

To meet the increasing demands of warfare, America for the first time in its history was confronted with the problem of swiftly changing its economy to meet war conditions. Despite the rise of industrialism that swept America beginning in the 1890s, Americans still had not been made over by the machine age. Technology was not overwhelming. Yet overnight, an "innocent" America would be forced to make many rapid economic modifications. She did so with tremendous business and technological innovation, but often with little forethought of the eventual consequences of these policies. In 1914, the outbreak of the European war nudged a sluggish California economy, but after it became an American war in 1917, demands for food and war materiel accelerated the state's economy at a rapid pace.

Upon hearing Wilson's war message, District 5 mobilized its war effort quickly. Less than twenty days after Wilson's war declaration, newly-appointed Governor William D. Stephens authorized a state Council of Defense, which was composed of leaders from professional and industrial ranks, along with various state agencies. The Council's first action was to survey California's agricultural resources and ascertain the condition of food production in order to find ways to increase it. District 5 did not hesitate to cooperate with the Council in this endeavor, assigning forest officers to facilitate networking with county officials and local leaders to gather this information (*Weekly Bulletin: Forest Service, District 5* 1917).

Administratively, major staff changes took place as supervisors, rangers and others enlisted and went to war. During World War I, many Forest Service employees were assigned to the war effort through the 10th and 20th Engineers (Forestry). Among the first to volunteer was District Forester DuBois, who was furloughed to the war effort and became a lieutenant colonel. DuBois served as a consul in Paris. While DuBois was gone, Assistant District Forester Woodbury took up the slack as acting district forester. Woodbury ran the office, save for DuBois' occasional visit to his San Francisco home, when he sometimes met with District 5 staff and supervisors and gave advice on matters (*Supervisors' News Letter, District 5* 1918, No. 1; No. 7).

While the "boss" was overseas, District 5's publication *Weekly Bulletin* changed to the *Supervisors' News Letter*, a confidential weekly bulletin wherein supervisors could frankly discuss problems that confronted them. While the *Supervisors' News Letter* covered normal topics of interest to supervisors, there was an undercurrent of discussion regarding unionism, low wages and general dissatisfaction with the Forest Service. World War I tended to depress an already lowering morale among supervisors and rangers. The war furlough selection process upset some. The statutory roll and poor working conditions disturbed others (*Supervisors' News Letter, District 5* 1918, No. 6; Show n.d.: 74).

Meanwhile, District 5 staff under Woodbury concentrated on Forest Service issues related to resource management. Probably the most important wartime policy change on District 5 was related to grazing. Nationwide there was a "Food Will Win the War" drive that aimed at increasing food production to support the Allied Powers and our troops overseas. The possibility that increased grazing utilization could cause overgrazing and be injurious to forest reproduction disturbed Graves, who was also an Army volunteer and was given the rank of colonel. However, most of the District 5 staff and supervisors felt that "full grazing use and especially full use by sheep" was about the only means they had to offset a growing fire danger on California's forests due to a lack of fire guards. Chief of Grazing C. E. Rachford made this point to Colonel Graves (*Supervisors' News Letter, District 5* 1918, No. 3), and District 5 was given approval for full utilization in order to meet war conditions (*Supervisors' News Letter, District 5* 1918, No. 5).

From then on, District 5 sought every means possible to increase California's national forest forage crop, even encouraging Basque sheepmen to graze on rough, steep and inaccessible browse range. For instance, in 1918, District 5 successfully experimented with a system of sheep management on these areas, which heretofore had been considered worthless. The success of this World War I experiment depended on the grazer, but according to Rachford, it opened up "wonderful possibilities on the Plumas, Tahoe, and Eldorado Forests." Furthermore, Rachford stated that "it might be well for us to consider the diversion of thousands of head of sheep handled by the despised Basque which now cross the Stanislaus and Mono Forests to graze on public domain in the valley, to these unused areas." In addition, Rachford asked rhetorically, "Aren't we ready to admit that the grazing of sheep on most of Forest areas is a greater means of [fire] protection than the grazing of cattle?" (*Supervisors' News Letter, District 5* 1918, No. 5).

Statistics for the Sierra National Forest from 1909 to 1918 illustrate the trends for a steady increase of cattle, horses, sheep and goats on California's forests during the war years. In 1909, there were approximately 11,500 cattle and horses on the Sierra Forest. No sheep or goats were allowed. By the time World War I broke out in early 1914, stock grazing on the Sierra amounted to close to 16,000 cattle and horses, as well as 4,500 sheep and goats. By 1918, the very height of the American war years, there were close to 19,000 cattle and horses, and more than 55,000 sheep and goats on the Sierra (*Supervisors' News Letter, District 5* 1918, No. 9). Cattle and horse numbers on the Sierra National Forest had risen 40 percent, a significant amount, and sheep and goat numbers had risen an amazing 92 percent to meet the war demands for food production and fire protection needs.

Timber sales naturally increased during the war. One of the first changes in timber policy Woodbury made was to raise the limit of sales from a maximum of two million board feet per sale to three million. Additionally, because of the lack of personnel, District 5, which was always fairly liberal compared to other districts, granted supervisors almost *carte blanche* authority to handle this work (*Supervisors' News Letter, District 5* 1918, No. 6). Timber sales grew in professionalism and efficiency as well. Forest Service timber specialists, in an effort to improve efficiency, outlined detailed logging plans for timber operators, who oftentimes operated inefficiently (*Supervisors' News Letter, District 5* 1918, No. 8).

During World War I, airplane production became very important. As it turned out, only certain types of woods (Sitka spruce, Douglas-fir and Port Orford cedar) were suitable for struts, wing beams and other airplane components. California's national forests played only a small part in supplying this line of war materials because a sufficient amount of material was found in Oregon and Washington (*Supervisors' News Letter, District 5* 1918, No. 2).

During World War I, fire control was important as well. Many feared that with fewer fireguards, a major fire season might break out and the Forest Service would not have enough personnel to fight it. These fears proved groundless. Actually, human-caused fires on California's forests showed a reduction of more than 50 percent during the war. Two important fire law cases accounted for this reduction. In each case, the Forest Service won against heavy odds even though the prosecutions were based on circumstantial evidence. One case resulted in an eighteen-month jail sentence for the accused, and in the second case, a substantial fine was levied. Nonetheless, to meet the labor shortage regarding fire control, District 5 turned to stockmen in lieu of fireguards to handle fires – as well as women fire lookouts (*Supervisors' News Letter, District 5* 1918, Nos. 2, 4).

Lookout on the Tahoe National Forest. 1923

Mining on California forests also became important during World War I. At the turn of the nineteenth century, the American industrial revolution stimulated the need for many new metals and non-metals. At this time, modern society's needs demanded the production of new mineral resources such as copper, lead, zinc, manganese, chromium, tungsten, molybdenum and other metals. Non-metal products – such as asbestos, nitrates, mica and phosphates – also came into high demand. Eventually, these new metals and non-metals became important materials for modern society and manufacturing. Because of their strategic importance, demand for these metals in the twentieth century proportionately increased during periods of warfare,. The Spanish-American War (1898) first stimulated increased demands for these "strategic metals" by the munitions industry, but it was the outbreak of World War I that gave the greatest impetus to the development of the strategic metals industry. Demand first came during the European portion of World War I, which started in August 1914. World War I interrupted the supply of many strategic metals and their byproducts necessary for the war effort, and Europeans turned to America for relief. For instance, smelters in Belgium and Germany were no longer available to manufacture brass for cartridges from Australian zinc ores. Therefore, the British and the Germans turned to the United States for help. European demand for zinc, especially for high-quality brass production, resulted in monthly exportation of 15,000 tons of spelter (impure zinc often used as a cheap alternative for bronze in casting) from the United States (Godfrey 2003: Vol. 1: 129-130).

Demand for many strategic metals increased when America joined the conflict in April 1917. When the United States entered the "Great War," it was largely unprepared to fight the "new mechanized, metal-dependent warfare." Therefore, the federal government stepped into the mining industry to efficiently expand, control and coordinate the production and stockpiling of critical metals used in war. War production boards were created, which brought an unprecedented degree of regulation and control to the American economy and especially the mining industry. War agencies increased the levels of production in western mines, sufficient to mass-produce munitions, tanks and other vehicles, planes and military equipment for the American Expeditionary Force (AEF), and for British and French armies abroad (Ibid.).

Many of these strategic metals were located on California's national forests, and a mini-mineral rush took place as old lode and placer sites were reworked with few restrictions. These formerly unprofitable deposits suddenly became more valuable, especially as German U-boats sunk American merchant ships. One important mineral in demand was copper. Technological developments associated with the war, such as radio technology and electronics, produced a demand for copper, fostering increased copper production. Copper smelters were particularly damaging to forestland, such as on the Shasta National Forest. Another important mineral was tungsten, which was used as a steel alloy component in the war production of armor plating, rifle barrels, high-speed tools and other weaponry. When tungsten exports from British colonies were halted, the price of the mineral skyrocketed, leading to increased domestic production. Mines located on the Sierra, Sequoia and Inyo National Forests helped to meet the increased demand (Palmer 1992: 145-146).

Finally, there was chromium, which was also used to harden steel. As sources of this strategic mineral in Greece and Turkey were cut off, the chromite sites on California forests, such as the Los Padres, Eldorado, Shasta-Trinity and Plumas National Forests, became very important to the war effort (Palmer 1992: 145-146). Military demand for chromite led to much speculation. Unfortunately, some rangers dabbled in transactions that were risky and illegal, but also potentially profitable. Some rangers acted as agents for relatives who owned claims. In other cases, wives took out mining leases in their name to avoid having their husbands profiting directly during their official work. These rangers and wives yielded to the temptation to gamble in chrome for many reasons. Some said that they did not know that the rules of conduct prohibited a ranger from locating claims. Others justified the behavior because of the low wages they received. Still others were certainly victims of "chrome fever." If supervisors discovered this behavior on their forest, they were told to give a single warning to the ranger. If the ranger offended a second time, or attempted deceit, supervisors were required to ask for the man's resignation (*Supervisors' News Letter*, District 5 1918, No. 3). There is no record that any California forest supervisor had to take this last measure.

End of War and Reconstruction Questions

On November 11, 1918, the European war ended when Germany signed an armistice agreement in a railroad car in Compiegne Forest on the Oise River in northern France. The war for democracy ended in triumph, and now the way was clear for post-war planning in the Forest Service. In the *Supervisors' News Letter*, Woodbury warned that District 5 would face "stupendous changes" which would "come crashing about us even more rapidly now than heretofore" (*Supervisors' News Letter, District 5* 1918, No. 4).

In California, the Forest Service had a number of reconstruction problems to consider. First, there were important personnel questions. How should District 5 reintegrate forest officers now on military furlough, and what should be done with the personnel who filled in their positions while they were gone? What about employment for the partially- disabled returning soldiers? What should be done about better pay for forest officers who stayed from a sense of loyalty? Next, there were questions regarding resource management. For instance, what difference would peace make in timber sale policy or grazing practices? What difference would peace make to the successful anti-fire propaganda and the fire law campaign? What about administrative improvements such as road programs, spring construction work for new ranger stations and the possibility of moving District 5 into a new headquarters (Ibid.)?

Finally, there was the question of Bolshevism and its influence on District 5. In March 1917, Emperor Nicholas II, under duress, abdicated his Russian crown to a provisional government led by Alexander Kerensky. Within a short time, revolutionary leaders such as Vladimir Lenin and Leon Trotsky undermined the provisional government and set up a communist government based on Marxist ideology. The Bolsheviks (as the Russian communists were then called) wished to forcibly overthrow capitalism and bring about a new proletarian order based on Marxist theory and industrial working-class ideology.

Regarding rising talk about Bolshevism, the *Supervisors' News Letter* four days after the signing of the armistice editorialized (Ibid.):

> If we can succeed in resisting the people who want to do our thinking for us and who will try to put catch words in our mouths and catch ideas between us and the truth: if we can hear always the call of the men who fill graves in the mud of France and Belgium in order that the World may be a better place for all kinds of people to live in: if we can remember tolerance and patient work is necessary to find any solution for the problems which will arise when great industries and whole classes of people clamor insistently for the righting of their real or fancied wrongs: if we can do these things we can help public opinion to find and take a safe and just middle course between the tyranny of an autocracy of economic power and the tyranny of Bolshevikism.

Nonetheless, talk of Bolshevism plagued the lower ranks of rangers early in December 1918. This "virus" of "unrest, discontent, doubt, spirit of knocking, and destructive criticism" first burst forth on the Eldorado National Forest following the return of a winter improvement crew. The problem was openly discussed in the *Supervisors' News Letter*. District 5 staff chalked it up to a few agitators, who found that the men living together for extended periods under inclement weather in these winter camps were susceptible to this kind of talk. However, they also recognized that the lack of sufficient salaries was another feature of the unrest. The *Supervisors' News Letter* did not propose the elimination of free and open discussion of grievances, which everyone realized would further the spirit of discontent. Instead, they proposed two solutions: first, make sure everyone had plenty of work to keep them full of activity, including keeping everyone busy during protracted storm periods, and second, "get rid of the agitators" (*Supervisors' News Letter, District 5* 1918, No. 7).

The ideological discussion did not end there. The next week it turned toward the subjects of democracy and unionism versus autocracy. All the talk of fighting for "direct democracy" apparently inspired several unnamed individuals to propose selecting new supervisors and assistant district foresters by secret ballot. This idea was couched in the overtones of the principles of democracy. After all, wasn't America fighting a war to make the world safe for democracy? For many years previous, it was the practice of District 5 to consult with rangers more or less before appointing a new supervisor, but some now wanted a new way of selecting, judging, and training supervisors (*Supervisors' News Letter, District 5* 1918, No. 8).

Veteran Forest Service Supervisor Richard L. P. Bigelow defended the past record and District 5 selection process, as well as the honor of District Forester DuBois. Supervisor Bigelow stated that "our District Forester had proved himself, time and time again, as a leader of men, a man thoroughly qualified to pick his aides without the help of the Rangers and Supervisors.... Who ever heard of a successful business organization choosing their superintendents and foremen by the popular vote of the employees?.... Is it the best man qualified for the position or the best politician?....a vote of lack of confidence in my mind is a fine way to break up our prided esprit de corps and be a breeder of discontent amongst our force" (*Supervisors' News Letter, District 5* 1918, No. 10). Forest Service discussions of Bolshevism and democracy were part and parcel of the popular debate of the day. They continued in the post-World War I period, and came to a head during the 1919 Red Scare, which will be discussed in the next chapter.

War and the California Conservation Movement – A Prelude

Some scholars have labeled the 1910s as the root years of the Forest Service's "forest protection or custodial management era" (Williams 2000: 31). This label implies that the Forest Service simply held on to, looked after, or generally maintained national forests. During this period, this label may be a complete misnomer when it comes to the history of California's District 5. The Olmsted administration helped establish a permanent system of publicly-owned national forests in California; District Forester Coert DuBois actively put that system on a firm scientific basis. For most of the 1910s, whether it was timber or range resource management, or even fire control and protection, District Forester DuBois directed District 5 policy toward utilitarian conservation using scientific management principles. Most certainly, protection was a key element in the DuBois administration, which was exhibited by fire control developments such as the nation's first fire protection plans and manuals, as well as research to protect the state's forests from disease and insect damage. Yet the pre-World War I years are also highlighted by continued and increasing utilization of California's forest resources – in direct contrast to the preservation ethic of the newly-created National Park Service. Steady timber sales advanced by improved cruising and marking techniques and better timber reconnaissance, and expanded range use by local operators, fostered by the support and cooperation of District 5 with livestock associations, mark these years as well. During the 1910s, the responsibilities of District 5 even expanded to include utilization of California's forests for public and private recreation and as game refuges. Exploitation of timber, range, recreation and wildlife resources increased because of the development of Forest Service highways granted unprecedented access to them for the public.

Then came World War I. To meet the economic demands of warfare, District 5 suspended most conservation concerns it held in order to meet the emergency contingencies of the war. Timber sales rose dramatically, grazing expanded to the point of overgrazing, and mining of strategic metals for war production was allowed with little restriction. Fortunately, these demands on California's forest resources during World War I were minimal and created no long-lasting, injurious impacts to them. In California, World War I tested the two somewhat contradictory mandates inherent in Pinchot's philosophy of utilitarian conservation. Utilitarianism, the first operating principle of the phrase, centers on production of products and services (timber, forage, minerals, water, wildlife and recreation) for the public. On the other hand, conservation, the second operating principle of this term, calls for the wise or conservative use and protection of all forest resources for the future enjoyment and need of generations to come. America's call for the production of needed wood, forage and minerals during World War I at the expense of "wise use" served as a policy prelude for how California's forests would be needed and utilized in future national emergencies such as World War II and the housing crisis of the 1950s.

Chapter V: 1919-1932

Maturation of District 5 to Region 5 and the Great Depression

The Red Scare of 1919

In 1919, the Great Red Scare gripped the country. The Bolshevik Revolution of November 1917 awakened an exaggerated fear of radicalism in the United States, causing an extreme reaction as many Americans failed to distinguish between genuine revolutionaries and radicals advocating peaceful social changes. Some believed that the revolution in America would come in a matter of months, based on the growing militancy of organized labor, such as when 35,000 Seattle shipyard workers went on strike in early January 1919 for higher wages and shorter workdays. Many Americans also associated radicalism with terrorism, especially after a series of homemade bombs were mailed to prominent citizens such as capitalist John D. Rockefeller in April of that year. Then in early June, anarchists set off a series of bombs that exploded in eight different cities at the same hour. A new wave of strikes followed the bombings, which included a police strike in Boston and nationwide steelworker and a mineworker strikes. All of this labor unrest was a reaction to poor working conditions, low pay and rising inflation. Fortunately, the Red Scare ended as quickly as it began, but not before Attorney General A. Mitchell Palmer set up an antiradical division of the Justice Department under J. Edgar Hoover and then raided the homes of alleged anarchists, communists, socialists, "radical" organized labor leaders and even pacifists. The Palmer/Hoover raids yielded nothing in the way of dangerous revolutionaries, even though people were held incommunicado, denied counsel and subjected to "kangaroo" trials. At no time in America's history had there been such a wholesale of violation of civil liberties (Leuchtenburg 1958: 66-79).

In California, the Red Scare led to the passage of California's Criminal Syndicalist Act of 1919, a vague statute making it illegal to promote "any doctrine or precept advocating…unlawful acts of violence…as a means of accomplishing a change in industrial ownership or control, or effecting any political change." During the next five years, more than 500 individuals were prosecuted, tried and some convicted under this California statute. The absurdity of the law became clear when novelist Upton Sinclair, noted author of *The Jungle* (1906), a muckraking attack on Chicago's meatpacking industry, was arrested under it for reading the United States Constitution aloud in public. Nonetheless, the law remained on the books until 1968, when federal courts finally invalidated it (Rice, Bullough and Orsi 1996: 368; Rolle 1963: 508).

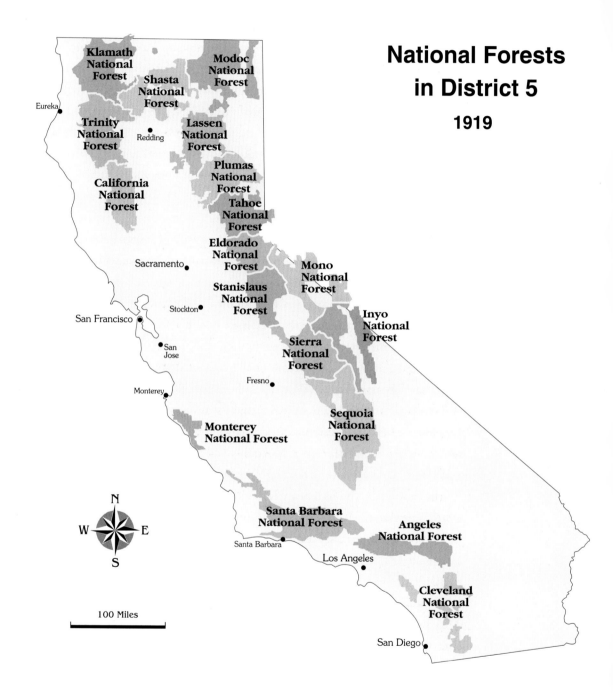

The Big Money: The Rise of Southern California

When domestic peace finally came to the nation following the Red Scare, California's economy was lifted to a new economic plateau. For most of the 1920s, the state witnessed phenomenal prosperity in many economic sectors. Much of the newfound wealth originated in Southern California, where a pattern of prosperity that began in World War I continued into the postwar era. California agriculture, which profited from fresh financial capital, lowering interest rates on farm mortgages, developing trucking transportation systems and the extension of railroads and advertising, became a highly organized and mechanized business. In addition, the completion of the Los Angeles-Owens Valley and the San Francisco-Hetch Hetchy aqueducts, and projected development of new sources of water for irrigation systems, such as the capture of Colorado River water into the eighty-mile long All-American Canal to the Imperial Valley, led to vastly increased growth in the citrus industry, especially the orange-producing counties. In the 1920s, new crops introduced into California's agricultural economy, such as grapefruit, dates and avocados, and perishable crops like tomatoes, melons and lettuce in the Imperial Valley, helped to contribute to this prosperity. The return to old crops such as rice in the lower Sacramento Valley and cotton in the San Joaquin and Imperial valleys also added to farmers' incomes. The profits of California's highly organized and mechanized agricultural system in the 1920s also depended on cheap transportation and low wages paid to pickers and field hands. Owners relied heavily on Mexican labor, although Filipinos, Japanese and African-Americans were also employed to deliver an uninterrupted flow of farm products to the country (Cleland 1962: 312-340 passim; Rolle 1963: 483-492 passim; Rice, Bullough and Orsi 1996: 409, 412-413).

Besides agriculture, the big money in California in the 1920s came from a real estate boom. Following World War I, the demand for real estate grew as an influx of people moved into Southern California for fun, sun and work. By the mid-1920s, Southern California found itself in the throes of an orgy of real estate speculation as farms, grazing land, orchards and chaparral-covered hilltops were subdivided and hawked by real estate promoters and salesmen as pieces of Eden, much as their predecessors had done forty years earlier. Populations of communities doubled, and then tripled, as one of the largest internal migrations in the history of the American people transformed landscapes from sleepy rural towns into energetic, busy, noisy, sprawling metropolitan centers with fifty-foot tract lots and square, flat-roofed, stucco houses (supposedly conforming to Spanish-Mexican tradition) piled upon each other on hillsides. Climate, scenery and boosterism, along with low gasoline prices and an automobile and highway revolution, had freed Californians from urban centralization. In the automobile age, Californians now worked anywhere and owned single-family homes with lawns, flowers and shrubbery in low-density suburban communities with modern supermarkets to serve them. A mode of economic decentralization developed as well, as businesses moved to the outskirts of cities and formed new manufacturing districts to accompany this population shift to emerging residential suburbs. Well-engineered concrete highways appeared to link these suburban regions to the cities like Los Angeles, which by 1925 was ranked first economically among California cities (Cleland 1962: 295-297).

As the real estate boom grew, ancillary demands such as energy for homes and manufacturing became almost insatiable. Because of the state's lack of coal, hydroelectricity became a necessity to the development of the state, and the struggle over public and private control of these power sources continued into the 1920s. In Northern California, the private Pacific Gas and Electric (PG&E) achieved almost a complete monopoly, in sharp contrast to the south, where municipalities such as the City of Los Angeles owned the distribution of electric power. By 1925, California's power plants produced forty times the amount of power they had produced in 1900. In order to continue its growth and meet demand, in 1930 Los Angeles obtained from the federal government the promise of 36 percent of the power from the immense Boulder Canyon Dam project upon completion. Los Angeles also had the right of first call for unused power in order to pump water through the Colorado Aqueduct (Ibid.: 301, 337-343 passim; Rolle 1963: 488-496; Caughey 1940; 554; Rice, Bullough and Orsi 1996: 396-398, 405-406).

The pervasive influence of the automobile on California culture and the growing needs of the automobile, set off by and coupled with the real estate boom, led to the search for oil in California. In the 1920s, discoveries of California "black gold" led to the rise of an oil industry, which played an important part in the state's later history. Petroleum was known to exist in California since the missionary days. However, it was not successfully exploited and utilized until the early 1890s. At that time, encouraged by market demand to substitute coal for oil, drilling operations began in Fresno and Kern Counties in the San Joaquin Valley. Prior to 1920, two-thirds of California's oil came from wells in this region, from which it was then shipped to major refineries along San Francisco Bay. Nonetheless, predominance in the oil industry passed to Southern California in the 1920s with the discovery of three major fields in the Los Angeles basin. In 1919-1920, Standard Oil discovered and tapped new fields at Huntington Beach. The next year, Shell Oil brought in the bonanza at the northern end of Long Beach near the prominent landmark known as Signal Hill. In the same year, Union Oil opened up the immense Santa Fe Springs field south of Whittier. To give an idea of the breadth of these discoveries and their capacity, by 1923, these three fields produced 70 percent of California's oil, which during the 1920s was valued at close to $2.5 billion. For a time during the 1920s and the early 1930s, California led even Texas in crude oil production (Cleland 1962: 298-301; Rolle 1963: 467-475; Bean 1968: 368-377; Caughey 1940; 551-553; Rice, Bullough and Orsi 1996: 404-406).

Besides residential districts and oil derricks, motion pictures became another new thriving Southern California industry in the 1920s, even though films had been made there as early as 1908. By 1916, most "features" were made in the Los Angeles area because of Southern California's climate and supply of cheap labor. Vertical integration of the movie industry – linking the big studios (e.g., Paramount, RKO and FOX), stars, directors, writers and artists to lavishly-built theater chains nationwide – began at this time too. Almost overnight, the sleepy little village of Hollywood attracted most of this business. In time, it became the Mecca for would-be stars and starlets. As time passed, a growing number of the middle class went to the movies in lush, studio-owned theaters to watch favorite silent-screen stars like Charlie Chaplin, Rudolph Valentino, Mary Pickford or Lillian Gish. By 1923, Hollywood studios accounted for 20 percent of all manufacturing products in California, and more than 20,000 actors and actresses worked before the cameras in 200 different studios – not to mention the thousands of agents, stunt men, writers, electricians, scene painters and many others. Even so, attendance increased after 1927, when Warner Brothers, one of the smaller studios, revolutionized the motion picture industry with the production of the first "talkie," *The Jazz Singer*, starring Al Jolson. As old squeaky voiced stars disappeared, new ones like Clark Gable, Spencer Tracy, Claudette Colbert and Joan Crawford replaced them. Not even the Great Depression seemed to slow down the flourishing factory of popular film, for movie matinees provided an escape to viewers with their silly boy-meets-girl plots, big-laugh comedies, song-and-dance extravaganzas and other daydreams (Cleland 1962: 295, 378-385; Rolle 1963: 540-546; Bean 1968: 381-391; Rice, Bullough and Orsi 1996: 407-409).

The Great Depression

On October 29, 1929 (Black Tuesday), the New York Stock Exchange had the most catastrophic day in the market's history. The collapse of the stock market that day led to the Great Depression – an ever-increasing downward economic spiral that adversely affected countless millions of people. As American cultural observer Alistair Cooke noted, if the roaring 1920s was the "promise fulfilled" after World War I, then the Great Depression was the "promise broken," for "within three months of the crash, men who worked in big factories, small men who had merrily played the market, were warming their hands before scrap wood fires….It was not just a blow to the extremes of the millionaire and the coal miner. It blighted everybody, even the very poor who had nothing to lose" (Cooke 1973: 327). Like most of the nation, California endured this wretched economic catastrophe until the election of President Franklin D. Roosevelt and the rise of the New Deal.

There is a myth that Southern California was a "balmy paradise of orange groves and Hollywood fantasy" during the Great Depression, and that somehow it was shielded from the economic disaster. In reality, Southern California felt the full impact of this financial downturn because the region's "chief products – specialty crops, tourism and movies – made this area of the state particularly vulnerable to the contraction of national income after 1929." For instance, as agricultural revenues sank from $750 million in 1929 to scarcely less than half of that value three years later, Southern California's mainstay rural communities were especially hard hit and sank into poverty. Unemployment also hit other areas of Southern California's economy as well. For example, when the oil industry could no longer sell its barrels of petroleum, major layoffs of workers throughout the business followed. Or, for instance, when unemployed people couldn't afford to buy new homes or make payments on existing mortgages, banks foreclosed on them. As a result, developers, real estate firms and construction firms collapsed by the hundreds. The nation's economic collapse also affected a disproportionate number of lower-middle-class workers in Southern California's service industries. Southern California also had the highest proportion of elderly in the nation, virtually all of whom were devastated by the faltering economy. In the end, statistics alone cannot convey the severity of the grief and anxiety over employment, loss of income and mortgage foreclosures felt by all the citizens and families of this part of the state, or express the human toll of the Depression upon minorities, which was incalculable (Rice, Bullough and Orsi 1996: 423-425).

Northern California suffered during the Depression as well, especially urban centers such as San Francisco, the home of many of California's largest corporations. By 1932, unemployment in San Francisco reached 25 percent, with no end, or hope, in sight. In Sacramento, a stunned state government faced the daunting task of supporting 1.25 million Californians on relief – an impossible mission considering that delinquent taxes had caused a drastic reduction of state revenues. California needed leadership at this hour of crisis, but instead elected former San Francisco Mayor James ("Sunny Jim") Rolph Jr. as governor. "In Rolph," as one source observed, "the state got one of its least competent governors in one of its most critical times of need." Governor Rolph suggested that the "best way to deal with it was for everyone to take a vacation to stimulate the economy through spending – advice he followed by going fishing." But as the Depression continued to envelope the state, the State Unemployment Commission took more direct action. To aid the transient and homeless men and women, the commission set up state labor camps. By 1931-1932, twenty-eight forestry camps and two highway camps were organized throughout the state in order to feed and house them and to give them jobs with some dignity. Workers in these camps constructed roads, cleared firebreaks and built campground facilities in state parks and later in national forests (Rice, Bullough and Orsi 1996: 422-426).

Public reaction to the crash and California's growing economic quandary was slow at first, but eventually the public became more and more radicalized. In response to these trying times, some Californians enthusiastically turned to social messiahs such as evangelist Sister Aimee Semple McPherson, who preached the "Four Square Gospel" at her Angelus Temple. Sister Aimee quickly established herself as the most successful of the lady ministers and even had her own radio station where she brought hope to the downtrodden and the lonely. Other Californians listened to and were consoled by Reverend Robert P. Shuler, another popular evangelist who operated a rival radio station. Many older Californians adopted new utopian economic philosophies to boost their spirits. First there was Dr. Francis E. Townsend, an unemployed Long Beach physician, who proposed a plan of federal taxes on business transactions that he thought could support and provide pensions for the elderly. Townsend supporters believed that their plan would bring general prosperity to all by restoring the "proper circulation of money." Thereafter, Townsend Clubs were formed to campaign for a decent living for old people. Then again, many Californians had a brief flirtation with Technocracy – a doctrine very reminiscent of turn-of-the-century Taylorism. Proposed by economic freethinker Howard Scott from Columbia University, this economic doctrine centered in Los Angeles. Technocrats, as they were called, advocated a new society based on scientific management. They believed that society should substitute the engineer for the businessman, "for the former would serve collective efficiency while the latter was motivated by private acquisitiveness." Both Technocracy and the Townsend Clubs emerged out of California to become powerful forces in national politics (Ibid.: 381-395 passim; Bean 1968: 409-414; Rolle 1963: 514-515; Kirkendall 1974: 26, 60-61).

Against the above economic, social and political backdrop of the promise of 1920s and the broken promise of the early 1930s, California's Forest Service history played out. Starting in 1919, a year of turmoil, District 5 would mature into California Region 5 and rise in the tide of the prosperity of the decade, only to founder like the rest of the nation after the 1929 crash.

Salaries and Working Conditions – 1919

In late February 1919, Colonel Coert DuBois unceremoniously returned to work after almost two years in the military with the 10th Engineers. Other district office staff, as well as a number of supervisors, rangers and others who served in the 10th Engineers overseas, returned as well. The discharged men were told to report to their respective forests for information regarding when to begin employment. In an effort to return to normal, DuBois immediately changed the confidential *Supervisor's News Letter* back to the *News Letter*. But in the years while DuBois was gone, much had changed (*Supervisor's News Letter (Confidential)* 1919, Nos. 11, 12; News Letter [District 5] 1919, No. 17).

At the close of World War I and during the year of the Red Scare, California District 5 faced turbulent employment problems as enlisted men returned to the Forest Service seeking their former jobs, and the Bolshevism threat spread among rangers who were dissatisfied with wages and working conditions. Discontent over salaries and workload continued to rankle many, and veteran District 5 personnel, such as Forest Assistant Edward N. Munns, characterized the causes for this disgruntlement as partly the fault of the varying standards in the Forest Service and uncongenial, monotonous work assignments, and partly the inability of qualified men to advance in the service because of the statutory roll, which included all positions in the federal government receiving less than $2,500 *(Supervisor's News Letter (Confidential)* 1919, No. 11). Nonetheless, during World War I, thirty-three rangers, assistant rangers and other staff, such as scalers, those who worked in timber sales or in improvement and other classes of work, resigned or were terminated – close to 25 percent of the 134 statutory ranger positions in the district *(Supervisor's News Letter (Confidential)* 1919, No. 12.5). Undoubtedly, some of these men were forced out because they were "agitators" during a trying time in American history.

While DuBois was gone, Assistant Forester Roy Headley tried to deal with the insurgency. His best effort was the creation of a new lengthy internal evaluation form, which he specifically devised to judge the work of supervisors and rangers. It was a direct response to the criticism leveled at the district office by rangers who desired a new "democratic" system of reviewing the achievements and failures of supervisors. The form included comment sections on the person's ambition, originality, personality, assumption of responsibility and physical stamina. It also contained sections on technical forestry skills and the person's grasp of the essentials of "national forestry" ideas and methods of public service in the management of timber, range, water, land, fire control and game management, as well as ability to work with local communities. In addition to the evaluation form, Headley also proposed a training school for new rangers, a common practice on other districts. Neither the questionnaire nor the proposed training school, however, ended ranger rank-and-file dissatisfaction with the San Francisco office. Every year, new and important work was heaped onto rangers by the district office through hundreds of circulars and written manuals. The annual workload built up so that it was beyond the reach and capacity of the average man to accomplish. In the end, rangers also simply wanted more respect for their views of the job, more effort from District 5 to learn those views, and they particularly desired better working and living conditions and increased salaries to combat inflation *(Supervisor's News Letter (Confidential)* 1919, Nos. 13, 15, 16).

In June 1919, the situation in District 5 reached crisis proportions. In a tell-all letter to Chief Forester Graves, District Forester DuBois described the devastating effect that low salaries and poor working conditions were having on his men and women. In a matter of only a few weeks, DuBois lost many key personnel, ranging from Miss Alma Krans, his chief clerk for the past six years, to experienced forest examiners and timber appraisers, all of whom left for more lucrative positions in the private sector. DuBois keenly felt their loss. "Evidence is accumulating every day," he urgently wrote to Graves, "which tends to show that unless the Forest Service can offer salaries to its employees which bear the proper relation to the cost of living, our organization is going to disintegrate rapidly…. It is discouraging to a Chief of Office to see an organization, which he has carefully built up through a long period of years, disintegrate within a few days for the lack of adequate compensation" (*News Letter [District 5]* 1919, No. 34). In the *News Letter*, he forthrightly and in frank terms addressed his supervisors on the subject. "Rangers' and Supervisors' statutory salaries," DuBois explained, "have been regarded as a tabooed subject, but I'm going to write about them anyhow. They are absurdly low. We've lost fifteen or more good men since January 1 on account of them. The good men that are hanging on are doing so against their judgment…hanging on just out of loyalty for the Old Service and what it stands for." DuBois encouraged his staff to hang on further and have confidence in Agricultural Secretary David Franklin Houston and Chief Forester Henry S. Graves, who were aggressively working with Congress regarding the statutory roll, and not "bawl…like a lot of I.W.W.'s" (*News Letter [District 5]* 1919, No. 34; Show n.d.: 84).

But apparently, both DuBois and Chief Forester Graves could not make the sacrifice that they asked fellow Forest Service staff to make. In November 1919, the district forester left the Forest Service after twenty years of tenure and moved on to a distinguished second career in the Consular Service under the State Department, where he served in various posts in Europe and around the world, until his retirement in the early 1950s (*News Letter [District 5]* 1919, No. 52; *Who Was Who* 1962: 239). Shortly after DuBois departed, Chief Forester Graves, also one of Gifford Pinchot's close and early associates, quit the Forest Service as well. In returning to Yale Forest School, Graves gave the low salary of his post as the reason why he relinquished his position (Show n.d.: 82), while in reality he left for health reasons. However, before Forester Graves left, he appointed Paul Goodwin Redington to fill DuBois' post (*News Letter [District 5]* 1919, No. 54).

Paul G. Redington, District 5's third district forester (1919-1926)

In mid-November 1919, Redington became California's third district forester. In 1904, Redington had joined the Forest Service at the age of twenty-six, after graduating from Dartmouth and Yale Forestry School.

District 5 Officers. Left to right: T. D. Woodbury, Forest Management; L. A. Barrett, Lands; Paul G. Redington, District Forester; Frank E. Bonner, District Engineer; S. B. Show, Research; J. W. Nelson, Grazing; Dr. E. P. Meinecke, Pathology; W. I. Hutchinson, Public Relations; A. W. Smith, Finance & Accounts; R. L. Deering, Operations. 1925

Redington rapidly rose through the ranks of the Forest Service. For five years, he served as supervisor of the Sierra National Forest and then left the Forest Service, taking a job as the city manager of Albuquerque, New Mexico. Unlike his two Pinchot-era predecessors, Olmsted and DuBois, Redington was no dictator. Rather, his management style depended greatly on finding good staff and then listening to them. Much to the consternation of some supervisors, over time he brought to District 5 several men who had served under him in the Southwest, and others from the Washington headquarters. Key posts included R. L. Deering as chief of operation, Paul P. Pitchlynn as general inspector, and Wallace I. Hutchinson as head of public relations. Redington also named S. B. Show to head a separate staff in charge of fire research. In time, Show became the most influential person in fire control on District 5, and would eventually be Redington's successor (*News Letter [District 5]* 1919, Nos. 54, 56; *Who Was Who* 1962: 440; Show n.d.: 85; Cermak n.d.: 217-219).

Assistant Regional Forester Wallace I. Hutchinson in charge of public relations

Naturally, one of Redington's first challenges was to sort out personnel matters, which needed his direct and immediate attention. District 5 had been through fifteen years of hiring, firing, promotion, transfer, resignation and assignment to specialist posts. By 1920, the makeup and composition of the field organization was a hodgepodge of forestry graduates and non-graduates who had come through to the service via the forest assistant and forest ranger examinations. Over the previous decade and a half, District 5 had drawn from this corps to fill its deputy supervisor and forest supervisor positions. DuBois, Redington's predecessor, attempted to professionalize his staff with technical men, but World War I halted this process, and staffing was in regression through non-replacement. In 1920, none of the deputy supervisors were considered technicals, and only nine of seventeen supervisors were trained as such. As Redington's successor S. B. Show remarked later, "all this meant the waning of hybrid vigor" (Show n.d.: 87-88).

Redington did not face this challenge alone. In 1920, William B. Greeley replaced Graves as chief forester. During his early years in office, Greeley and Redington struggled to resolve the low pay scale and statutory roll situation, but for many years they were unsuccessful. Redington frequently updated district personnel on the salary question through the *News Letter*. However, many good men and women continued to leave critical positions on District 5 because of the low pay, continuing inflation and the absence of a pension system. Some personnel transferred to the National Park Service (NPS), which offered a superior pay scale for comparable jobs, while others remained but continued to look around for better opportunities. The best Redington could offer his disgruntled clerical force in the district office to keep them on board was a separate rest room for the women of the office. Triumph over the salary question finally came in 1924, when the statutory roll was abolished. At this time, a modest but regular quota program of promotion for the rank and file was created. A new schedule of reclassified positions, grades and salaries was implemented as well. A retirement system soon followed (*News Letter [District 5]* 1919, Nos. 58, 60; 1921, Nos. 1, 2; 1923, No. 12; Show n.d.: 89-90).

By the end of 1924, San Francisco headquarters personnel included seventy-six employees, while District 5's seventeen forests employed 109 forest supervisors staff, and a field organization of 648 persons, including 394 protection guards and fifty-five cooperative guards under the supervisor's control (*California District News Letter* 1925: No. 7). In S. B. Show's candid estimation, when Redington inherited the organization, District 5 was missing that vital spark from the Pinchot days, and the morale of the rank and file was at its lowest ebb. Many field officers who had stuck with District 5 through the early 1920s were simply mediocre and/or incompetent. New recruits were also few because better jobs and pay attracted personnel to other positions. District 5's intake of men from the historic forest ranger examination continued, as forest guards tried to rise in rank and men continued to filter up from the re-established junior forester position into assistant ranger and then into district ranger posts, but for the most part, these men were average. Besides this problem, it was during Redington's administration that district ranger autonomy slipped away. Gradually, rangers were shoved "firmly and formally" into the job of fire control officer. Resource management duties pertaining to timber and range were to be done in whatever time the ranger had left over in the day. Increasingly, a corps of forest staff specialists took over these fields. In the winter of 1924-25, some personnel, including soon-to-be District Forester S. B. Show, rebelled openly against this state of affairs and invented the term "Operation Mechanic" to describe the lowly state that the district ranger's job had fallen into (Show n.d.: 90-95). Redington recognized the need for more organized training to create better field personnel, and in 1923, some serious training began on the Sierra National Forest, when a school was established there for rangers from different forests (*California District News Letter* 1923: No. 10). But Redington did not act further on the situation until 1925, when he authorized the creation of the historic Feather River School.

Feather River Training Camp Plumas National Forest, class of 1932

Headed by District 5 Inspector Paul Pitchlynn, the Feather River School, also known at the time as "Plumas College," was located at the Feather River Experiment Station on the Plumas National Forest. With its creation, District 5 began the first effort to continuously and systematically train men for higher administrative positions. Each fall some twenty-five to thirty forest officers, both district rangers and technicals, arrived at the school, which offered a tough, broadly-balanced scientific and practical curriculum covering all aspects of work on District 5 (*California District News Letter* 1926: No. 41). Older rangers were given this training as well as the younger men. S. B. Show, in his history of District 5, noted that assignment to the school was "not a pleasant vacation," and that half of the pioneer class was "promptly retired to private life" because of either an inability to learn, or character or personality defects (Show n.d.: 98). The ultimate objective of the school was the elimination of staff specialists on District 5. Shortly thereafter, District 5 took this message to the All-Service Conference on Organization held in November 1925. At this conference, Show and others firmly declared that the job of staff specialists was one of self-elimination through teaching the district ranger how to do the job. District 5 took the lead on this matter, setting this type of formal training in motion nationwide. After 1933, the Feather River School expanded its coursework, and became District 5's training ground for upper level positions on the district. Thereafter, passing this advanced course and graduation from Feather River School became virtually a requirement for any promotion into the ranks of deputy supervisor or supervisor positions (Ibid.: 96-98; Ayres 1941: 28).

Paul Pitchlynn, head of the Feather River School

Another key administration event in the Redington era was the reestablishment and restoration of attention to public relations, which had fallen by the wayside during World War I. In June 1920, Washington created the branch of public relations, and shortly thereafter District 5 responded with the creation of a district-level office of public relations. In July, Wallace Hutchinson was made head of the new office. District Forester Redington considered public relations a major activity for the district and strongly supported it, stating to all forest officers "all of us must put our shoulders to the wheel to see where and how we can be of fullest usefulness in this field…(*California District News Letter* 1923: No. 20).

One of Hutchinson's first actions as head of the office of public relations was to change the name of the *News Letter* to the *California District News Letter*. The revamped publication changed from an administration newsletter to a public relations vehicle for District 5. Newspapers throughout the state turned to it as a source of newsworthy items on a variety of subjects such as fire prevention week, timber sales and road reports, to name a few (*News Letter [District 5]* 1919, No. 84). The *California District News Letter* now contained feature articles on District 5 "happenings," such as an article describing District Forester Redington's weeklong outing with the Sierra Club in the backcountry of the High Sierra. Besides the thirty or so packers and cooks used to keep the hikers well-fed and comfortable on their journey through this rough and scenic region, more than 230 other people made the trip (*California District News Letter* 1920: 4 August). In addition to articles such as this one, many issues of the *California District News Letter* contained an occasional Gifford Pinchot speech or mention of one of his publications, or provided inspirational material to rangers in the form of quotations, or witty doggerels on Forest Service life and work (*California District News Letter* 1920: 22 September). Then there were the many human-interest items, such as a three-part series entitled "A Forest Ranger's Wife," which added flavor to the weekly publication (*California District News Letter* 1923: Nos. 33, 34, 35).

In addition to using the *California District News Letter* to inform the public on District 5 matters, Hutchinson wrote articles himself for national magazines, such as "California National Forests – Yours to Enjoy" (*California District News Letter* 1924: No. 15). He also often worked with Hollywood to get the Forest Service message across to the greater public. For instance, one issue of the *California District News Letter* contained quotations from popular film stars such as Charles Chaplin, Douglas Fairbanks, Mary Pickford and others, advocating reforestation efforts on Southern California's national forests (*California District News Letter* 1925: No. 18). Hutchinson also encouraged supervisors to get in touch with local editors and chambers of commerce and furnish them with readable articles concerning their particular forests, and he advanced ideas to them regarding how to write for outside publications (*News Letter [District 5]* 1921, No. 12; 1923, Nos. 2, 14). Revered veteran Sierra National Forest Supervisor Charles H. Shinn gave his all toward what he called "forest educational work," and tried to convince his colleagues to do the same (*California District News Letter* 1923: No. 42).

One problem that Redington addressed through the *California District News Letter* was the confusion and misunderstanding that existed in the public mind as to the essential differences between national forests and national parks. In an informative letter to California newspaper editors, Redington pointed out the basic dissimilarities between the two agencies, explaining that in California, there were seventeen national forests with a net area of just over 19 million acres, while there were four national parks in California with an area of approximately one million acres. According to Redington, "…the principle of use of resources is the vital distinction between National Forests and National Parks." Furthermore, he informed newspaper editors that:

> the purpose of the NATIONAL FORESTS is to protect and maintain, in a permanent productive and useful condition, lands unsuited to agriculture, but capable of yielding timber, or other general benefits, such as forage for livestock, water for irrigation, domestic use and power, and playgrounds for our citizen. All the resources of the National Forests are developed to the greatest possible extent consistent with permanent productivity under the principle of coordinated use." [On the other hand,] the purpose in creating NATIONAL PARKS is to preserve the scenery, the natural and historic objects, and the plants and wild life. The objects are the enjoyment of the people and the aiding of education and scientific study by keeping such areas unimpaired. National Parks are protected completely from any and all utilitarian and commercial enterprises save those necessary for and subservient to legitimate park uses
> (*California District News Letter* 1923: No. 46).

At the beginning of Redington's term of office, the Forest Service and the NPS worked more cooperatively. For instance, in the early 1920s, the Forest Service and NPS agreed to review national forests for sites best suited to transfer to national parks. The Forest Service also agreed to protect entrance areas to parks situated within national forests and to take special care with respect to logging and grazing near parks. This cooperative spirit continued until around 1925, when the NPS aggressively sought to transfer recreation areas in California's national forests to the Park Service. There were even renewed efforts to return the Forest Service to the Interior Department at this time, but these attempts were staved off.

Next to public relations, administrative improvements ranked as the next important concern of District Forester Redington. Accordingly, he focused his attention on moving District 5 out of the congested San Francisco headquarters. In July 1920, after six months in office, Redington engineered a move from the crowded offices in the Adams-Grant Building on Sansome Street to a more commodious and comfortable location in the Ferry Building at Market Street and The Embarcadero. The new location offered plenty of space for the district forester, operations, lands, public relations, products and

Ferry Building on the Embarcadero, San Francisco, CA. This was the fourth home of District 5.

Source: San Francisco Public Library

silviculture, grazing, law and research, and for stenographers, a receptionist, a conference room and even a small partitioned-off "ladies' rest and recreation room" (*News Letter [District 5]* 1920: No. 86). The new Ferry Building headquarters had a series of desks grouped in twos with no intervening partitions, which at the time seemed perfect, but by 1926, the big room had become a "Tower of Babel." With eighty employees, thirty typewriters and ten adding machines clacking away, a person had as much privacy as "Polar Bear in a New York zoo" (*California District News Letter* 1926: No. 20). At that time, District 5 office staff hoped to move into a new federal building reportedly to be constructed at the San Francisco Civic Center. However, when this plan became delayed as a consequence of the Great Depression, the California Forest Service headquarters was relocated to the Beehive Building at 85 Second Street in central San Francisco some time prior to March 1931 (*California Ranger Region Five* 1931: No. 17).

After finding a new home for district headquarters, Redington turned his attention to road building. When Redington took office after World War I, California's national forest road system was just beginning a major period of growth. In mid-January 1919, District 5 had held a conference to determine a five-year program for the expenditure of Section 8 road funds on big projects. Projects were segregated into four groups, each group containing certain counties. The southern forests received $600,000, more than half of the funding, for the Cuyama, San Gabriel and Bear Valley road projects, while the forests north of San Francisco were allotted only $81,000 (*Supervisor's News Letter, District 5* 1918, No. 13).

District 5 also had sufficient funds to construct minor roads as well, thanks to Congress. Toward this end, in 1919, Congress appropriated a half million dollars for minor roads and trails within national forests, of which $160,000 was expended in California (*News Letter [District 5]* 1919: No. 23). District 5's secondary low-gradient roads were designed and engineered for then current automobile speeds, which at that time was twenty to twenty-five miles per hour. They also had easy curvature and correct banking, and generally they were tied into the layout of the emerging California State road system. Many were built specifically for recreationists. California's burgeoning population and related recreation demand created a sudden need for a highway system in California's national forests. Seeing the dramatic direction that forest recreation was taking, one forest supervisor wrote: "If there is a real desire on the part of a large proportion of the people for recreation roads, then we must build them, as it is the people's money." Therefore, most new construction in the early 1920s took recreational development into account. When planning routes, District 5 designed many of its routes through venues with scenic values, and gave ample consideration for their relationship to campgrounds, resorts, and hotels (*News Letter [District 5]* 1919: No. 21).

By 1921, rapid development of automobile speed, mechanical efficiency and wide distribution strongly influenced Congress to pass the Federal Highway Act (42 Stat. 212), which expanded the nation's highway systems. The Federal Highway Act provided aid to the states in the construction of rural post roads and for other purposes. Besides authorizing funds to aid states in road construction, Section 23 of the Act also approved federal appropriations for the specific purpose of further development of a system of national forest roads and trails. Section 23 made separate appropriations for roads of general public importance and for roads of primary forest importance. Over the next decade, California's national forests received close to $16.7 million from the Federal Highway Act and its amendatory acts and obtained another one to two million dollars annually under Section 23 (Burnett 1933: 15-16; Smith 1930: 57, 177-180). Thanks to the passage of Federal Highway Act, many important Forest Service highways were constructed, such as the

Forest Service fire truck on California National Forest. 1923

proposed but uncompleted 800-mile Sierra Way (Burnett 1933: 16) and the Klamath River Highway, which was a "godsend to those sections of California having a small population, a restricted revenue, and large areas of public land" (*California District News Letter* 1923: No. 30). Another important Forest Service highway begun at this time was the Angeles Highway into the San Gabriel Mountains. The Automobile Club of Southern California first pushed for this highway in the pre-World War I period for its scenic and recreational values, but to no avail. Then in 1919, after several canyon forest fires destroyed the homes of important personages, such as Hollywood producer Cecil B. DeMille's residence in Little Tujunga Canyon, the Los Angeles Chamber of Commerce and California's congressional delegation heavily lobbied Agriculture Secretary Houston regarding the need for this highway, arguing that it was vital for firefighting purposes. Public opinion and pressure won over the Forest Service. Subsequently, work began on the highway. Using funds from the Federal Highway Act, as well as money from other Forest Service road building programs, the Angeles National Forest began work on what would eventually become the Angeles Crest highway system, which was not fully completed until 1956 (Hoffman 1968: 309-319, 337).

Besides changing the location of district headquarters and District 5 road improvements, District Forester Redington was confronted with one other major administrative policy issue. This involved the General Land Exchange Act of 1922 (42 Stat. 465). Reminiscent of the old forest lieu clause, which had caused considerable trouble in the past, the General Land Exchange Act made it possible for the Forest Service to exchange government land within national forests for private land of equal value within the boundary of a national forest. This legislation allowed the Forest Service to consolidate its holdings – a desirable action on California's national forests where a large percentage of private lands were intermingled with timber lands and lands valuable for recreation. This act also made for better administration and management in accordance with future planning initiatives because the process of a mutually desirable exchange was greatly simplified by this legislation to a matter of just signing papers, involving no additional legislation (Ayres 1941: 26; Smith 1930: 61-62).

Active Custodial Care of California Forests, 1920-1925

Though District Forester Paul Redington faced real administrative challenges – personnel matters, headquarters relocation, road building and simplified land exchanges – he also found that managing the timber, forage, water, mineral, recreation, and fish and wildlife resources on California's forests, as well as fire control and prevention issues, would prove very challenging during the first half of the decade.

Timber policy was certainly a prime concern for District 5 in the 1920s. California, unlike eastern states, had a significant amount of timber left – approximately 13.5 million acres of timberlands. In fact, at this time California's forests contained 15 percent of all the remaining virgin timber in the United States, and lumbering and wood-using industries ranked fourth among the California's industries, employing some 25,000 people annually (*California District News Letter* 1924: No. 15). The growth of this industry directly correlated with the phenomenal increase of wealth and population in California during the postwar period. This growth was reflected in every phase of California economic life, and the great postwar real estate boom stimulated the lumber industry to no end. The Forest Service did not stand in the way of growth; indeed, its utilization policies encouraged it. For example, before Chief Forester Graves left office in early 1920, he established a policy that "land chiefly valuable for the production of wood must produce wood when the adjacent communities need wood" (*News Letter, [District 5]* 1919, No. 31). To meet lumber demands, District 5 timber management policy sought and approved bigger and longer timber sales on its forests. For instance, in 1921, one of the largest California timber sales was made within the San Joaquin River watershed on the Sierra National Forest. Under this twenty-three-year contract, 600 million board feet of timber, mostly pine, was proposed for cutting, at a value of $1.8 million. Of this amount, 25 percent was to be paid to the counties within the Sierra National Forest for roads and schools, and an additional 10 percent paid to the state for road building purposes (*California District News Letter* 1921: No. 23). These early sales were largely designed as railroad logging operations. The long period was so that the lower-elevation trees would be cut first, then activities moved up the valley over two decades or more – a slow process. They were often touted as sustained-yield operations. As the operations moved up the valley, trees replanted would grow in the cut-over areas, ensuring a sustained yield over decades (Williams 2005).

Very few sales of the size of the Sierra sale had previously been made by the Forest Service in other states, and for good reason – America's forest heritage was dwindling away. In the early 1920s, many in the nation were concerned about America's declining forest resources. By this time, America had consumed, squandered or destroyed 60 percent of the nation's original timber wealth and was using timber at a rate four times as fast as it was being grown. National apprehension over this destruction was expressed on the floor of the Senate in 1920 with the passage of the Capper Resolution, named after Senator Arthur Capper of Kansas. Capper's resolution directed the agriculture secretary to report to the Senate on the extent of the forest devastation in the United States and its effect upon the price of wood products and the public welfare. Ultimately, the Capper Resolution favored federal control as the solution over state and private control of forestry and fire protection, and from then on, a debate raged in the nation. Dozens of hearings were held throughout the nation's timber regions on the subject of public versus private ownership. In 1923, President Warren G. Harding joined the controversy. Harding openly supported a "national conservation policy to protect and conserve the fast dwindling forest areas of the country," and he asked for greater cooperation among the federal government, the states and private owners to work out this policy (Clar 1959: 559; *California District News Letter* 1923: No. 7.

Regrettably, California's state authorities, including the governor and the legislature, did not have such an enlightened attitude on the subject. At the very time California needed to, and could, cooperate with the Forest Service in working out forestry problems, conservative state authorities announced severe budget reductions, which affected many state forestry activities. These budget reductions forced the state to discontinue fire protection and forest preservation efforts, ended maintenance for state nurseries and cancelled funds for Humboldt Redwood Park, which had been created as a county park in 1921 due to the lobbying of the "Save the Redwoods League," an organization incorporated in 1920 to "preserve the oldest trees in the world" (American Tree Association 1924: 68). To protest the budget reductions by the state legislature, ex-Governor George Pardee resigned from the California State Board of Forestry; and ex-Chief Forester Graves, who was then the president of the American Forestry Association, appealed to California Governor Friend W. Richardson to restore the budget cuts. District Forester Redington joined the fray. Redington warned Governor Richardson that California's forests were in great jeopardy if the budget cuts went through. Redington let it be known that federal forest officials would do all in their power to protect California's forests, but that the Forest Service had no authority to extend fire protection activities beyond national forest boundaries. Redington cautioned that California would "fail to share in the great national campaign so forcefully urged by President Harding for the more adequate protection of our dwindling forest resources." Apparently, this vigorous campaign worked, and the cuts in the budget were restored (*California District News Letter* 1923: Nos. 7, 9).

President Harding never had the chance to push further for forestry conservation. In August 1923, worried by growing scandals in his administration (e.g., Teapot Dome), the president decided to take a trip across the country and up to Alaska. While in San Francisco, uneasy and depressed, Harding, according to one source, fell ill of ptomaine poisoning, then of pneumonia, and died of an embolism there. "Silent" Calvin Coolidge, who "exalted inactivity to a fine art," succeeded Harding, restored the people's confidence in the Republicans that fall and won election (Morison 1965: 932-934).

With the Harding scandals behind them, national attention once again turned to forest conservation, and in the winter of 1923-1924, it reached a culmination when the Coolidge administration announced that a new "Forest Conservation Act" would be introduced at the next session of Congress (*California District News Letter* 1923: No. 49). In January 1924, a national forestry policy bill was introduced, but it did not become law until June 13, 1924, when President Coolidge signed the bill. Forestry leaders throughout the country acclaimed the legislation as the "greatest conservation measure passed by Congress in a generation…" (*California District News Letter* 1924: No. 24). Known as the Clarke-McNary Act of 1924 (43 Stat. 653) for its sponsors, New York Congressman John W. Clarke and Oregon Senator Charles L. McNary, this act provided for the protection of forestlands, reforestation of denuded areas, extension of national forests, and for other purposes, in order to promote the continuous production of timber. The Clarke-McNary Act also expanded the 1911 Weeks Act authority for federal-state cooperation in fire protection and forestry efforts, allowing for purchases of forest lands in watersheds, not just the headwaters of navigable streams (*California District News Letter* 1924: Nos. 1, 13; Clar 1959: 561-563; Smith 1930: 63). District Forester Redington certainly was gratified at the passage of the Clarke-McNary Act, for now there was no question that the State of California would pick up its burden of forest fire protection.

Meanwhile, in the wake of the Capper Resolution, District 5's cutting and marking policy underwent considerable study and re-evaluation under the newly named branch of forest management (1920-1935). For more than ten years, District 5 conducted a detailed study of cut-over lands on permanent plots on the Shasta, Lassen, Plumas, Tahoe, Stanislaus, Sierra and Sequoia National Forests in order to determine the effects of cutting on the growth of remaining trees and on reproduction. By 1922, the initial scientific results from these sites were ready, and timber specialist Duncan Dunning published the first results for the branch of forest management. Dunning's initial conclusion from the data was that in the absence of experience, the early days of timber sale practice on District 5 tended to be conservative. Cutting at this time favored sugar and yellow pines, and heavy stands of large mature trees were left in hope of increasing the proportions of these species in the stand and in hope of securing increased growth for a second cut. However, Dunning found that there was little correlation between this practice and probable growth and reproduction, which depended more on a tree's seed capacity, the relative value of the species, wind firmness and other factors (*California District News Letter* 1922: Nos. 19, 30; Woodbury 1930: 695). Dunning

Reading the life history of trees and forests from the annual rings of increment cores taken from living trees. Instrument invented by Duncan Dunning of the California District, U.S. Forest Service. 1925

published his final study in January 1924, which District Forester Redington sent to all supervisors, emphasizing a few of Dunning's conclusions. His final report confirmed his earlier conclusion that the past group selection system was a failure from a silvicultural standpoint. Instead of leaving groups of ponderosa pine, Dunning's conclusions advised supervisors to thin these groups in order to secure increased growth and for seed production. As a result, marking rules were redesigned to release well-established young growth from light and root competition brought about by the early practices (Ayres 1958: 29-30). Assistant District Forester T. D. Woodbury summed up the importance of the Dunning study when he stated, "Now for the first time we are able to handle cutting operations on the basis of facts obtained in our own region through careful research…. we administrative men who are responsible for growing the maximum amount of wood of the best species on our cut-over lands can now ascertain…the most productive trees to leave in order to accomplish this objective" (Woodbury 1930: 695).

Three significant changes in timber policy that were related to sustained-yield management grew out of Dunning's scientific study. First, District 5 began reserving between 15 and 20 percent of the volume of a merchantable stands on every cutting area for sustained-yield management purposes. This reserve was less than what was left when the Forest Service first began selling timber, and was prompted by both the 1920 Capper Committee report on the status of timber supply of this country and the Dunning study (Ayres 1958: 30; Woodbury 1930:694-695).

Second, District 5 revived the preparation of working plans on the district. Working plans were rejuvenated because it became apparent after the Capper Resolution that the Forest Service was making a mistake in selling timber without knowing the management objectives of the sale. These new working plans were not the big, bulky, unwieldy, European-styled, unused things which District Forester DuBois initiated in the 1910s. Instead, they were more modern ones that divided the forest into working circles, studied the problems in the circles and then drew up a policy statement for each circle. It is fitting that in 1923 the Plumas National Forest, which had the first untenable working plan in the 1910s, had the first viable one in California District 5. After 1923, management plans for California's national forests steadily went forward. They succeeded at this time because the pioneering work in District 5 was about finished, because the initial construction period on each national forest was completed, and because supervisors and staff were now prepared and willing to assume the technical problems connected with forest management (Ayres 1958: 48-49).

The third change to timber policy to come from the Dunning study was that it stimulated interest in reforestation on District 5 once again. In

Fire fighters on their way to the Forest Hill Divide Fire on the Tahoe National Forest. 1924

early 1924, Chief of Forest Management Woodbury recommended that District 5 begin a ten-year period of planting experiments under the branch of research, which was directed to improve the production and utilization of timber. Woodbury suggested that the Feather River Nursery provide trees for the project. Trees were to be planted in the northern forests on lands cut over in logging, in typical areas denuded by forest fires in the timber zones and in brushfields where there was no chance for natural reproduction. Woodbury's suggestion to confine the planting to the northern forests came just months before the fateful 1924 fire season, which included the worst fires in California history up to that year. They scorched more than a half a million acres in California's national forests, and after that season, Woodbury revised his plans to include Southern California forests as well (Ibid.: 72).

Preceding the disastrous fire year of 1924 were several years of innovation in fire control, beginning with the introduction of the use of Army airplanes for fire protection patrols in May 1919. The development of air patrols on California's national forests came about when the needs of Army Air Service and the Forest Service fortuitously crossed following World War I. The Army Air Service needed to keep the public aware of its pilots and planes, and the Forest Service was mindful of the potential of aircraft in fire control. As early as 1915, Eldorado Acting Supervisor E. L. Scott had asked Washington officials if they had ever investigated the use of airplanes or dirigibles in fire protection. The Forest Service had not. Nothing more happened until March 1919, when Agriculture Secretary Houston made a request to the secretary of war for air patrol service. At approximately the same time, recently discharged Colonel Coert DuBois had a chance discussion of the topic in an encounter with Major H. H. "Hap" Arnold of the Army Air Service in a San Francisco bar. These innocent occurrences resulted in the inauguration of the cooperative fire protection/air patrol service for California between the Army and the Forest Service (Cermak n.d.: 182-186; Ayres 1942: 15; Cermak 1991: 292-294).

Trial flights made in early May indicated that it was very difficult to locate most of the lookout stations, which pilots needed to help locate their position, because standard Forest Service colors did not show up well from the air. To remedy that problem, lookout roofs were painted red and white, and in some instances large crosses were whitewashed on open ground near some lookouts. With this problem solved, airplane patrols were inaugurated on portions of the Cleveland, Eldorado, Stanislaus and Tahoe National Forests on June 2, 1919. That day, Assistant District Forester Redington and Angeles Forest Supervisor R. H. Charlton took off from March Aviation Field outside Los Angeles on the inaugural patrol, which covered 600 miles. A few weeks later, as a courtesy for his influence in obtaining the project, District Forester

Colonel DuBois took off from Mather Field near Sacramento on the northern patrol, which went to Oroville and back. Thereafter, Army planes and personnel began to patrol set routes from bases located at Rockwell Field, near San Diego; at Red Bluff Field to fly over the California, Trinity, Shasta, Lassen, Modoc and Klamath National Forests; and at Fresno to cover the Stanislaus, Sierra and Sequoia National Forests. Unsurprisingly, California's press immediately picked up on the significance of these events, and newspapers contained vivid stories of the daring of these pilots and the perils confronting these heroes of the skies. One paper predicted that some day "large airplanes will be able to quickly get [fire] fighters to remote points.... Bombs containing gases which will put out fires can be dropped by the fliers, while the observers can keep the fire fighters informed as to the trend and direction of the flames and advise them as to the best disposition of their forces" (*News Letter [District 5]* 1919: Nos. 30, 32, 34, 47; Ayres 1941: 16). As a sideline, in the summer of 1920, District 5 also tried out a small one-man Goodyear blimp for fire patrol over the Angeles National Forest. Its slow cruising speed was good for observation, but blimps did not meet the planned need to transport equipment and firefighters, and the project was discontinued (Ayres 1942: 16).

By September, Major Arnold reported that the Army air patrol had discovered 118 fires, but that they had "first" reported only twenty-three. These statistics may seem paltry; however, these were days before radio

Stearman biplane used on Forest Patrol in Southern California. Paul Mantz, pilot (right) and Forest Inspector Paul Pitchlynn (left).

communication. Spotting a fire meant that the airplane had to land or drop a note from the plane to the ground. By 1920-1921, radios were tried and worked (Williams 2005). Nonetheless, these problems in accurately reporting fires from the air, because of the haze and smoke and in a timely fashion, led to criticism of the program. District Forester Redington vigorously defended the program against its critics in an article entitled "Airplanes and Forest fires," which was reprinted in the state forester's *Eighth Biennial Report* (1921) along with a "full page photo of a lonesome little biplane floating like a box kite against a broad expanse of timber covered mountains." Redington acknowledged problems with the service, but he argued that these mistakes could be rectified with better training and better maps. The State Board of Forestry endorsed the experiment and urged the War Department and Congress to grant a special appropriation to expand the cooperative project to other western states. In 1921, the California legislature, as well as lumbermen and foresters west of the Rocky Mountains, joined in the general plea to extend air patrols during the coming fire season. The project continued that summer, but in 1922 Congress failed to appropriate sufficient funds, and the cooperative program was discontinued. To meet the fiscal shortfall, the Southern Chapter of the Sierra Club took up the banner and recommended that Los Angeles County pay half the costs of the airplane patrol to protect the timber and brush cover that conserved water for the vital citrus and agricultural industries of Southern California. This plea fell on deaf ears, and regular patrols were discontinued. However, planes were still used in emergency scout duty and to transport supervisory personnel. This arrangement continued until 1925 (*California District News Letter* 1919: No. 59; 1920: 21 July; Cermak n.d.: 189-197; Ayres 1941: 16; Ayres 1942: 16; Cermak 1991: 296-302; Clar 1957: 458-461).

Besides the "Air Age" coming to California forestry, another major action in fire control and protection also came that year. Redington and District

Ranger patrolling the Tahoe National Forest. 1915

5 tackled the controversial question of "light burning" or "cooperative burns." This practice was a traditional way that many used to clear lands for agriculture, grazing and prospecting purposes, as well as to protect homes and property. As early as 1916, popular California author and naturalist Stewart Edward White led a vigorous and large-scale propaganda effort in favor of light burning. White believed that the practice was the answer to the growing destruction of pines near his cabin home on the Sierra National Forest due to beetle bark infestation. His arguments rested on the assumption that insects that attacked trees bred in dead and decaying wood and that light

Ranger C. E. Jordan and Ford Roadster fitted with flanged wheels for use on San Joaquin and Eastern R. R. Sierra National Forest.

burning would remove them. Finding the Forest Service unresponsive to his theory, White and other supporters of the cause launched a series articles attacking the Forest Service insect control methods. These articles appeared in magazines such as the *Lumberman*, *The Timberman* and *Sunset* (*California District News Letter* 1919, No. 58; 1920, No. 67; Ayres 1941: 81-82; Pendergrass 1985: 105-108). District Forester Redington tried to correct White's impression regarding insect control and responded to these attacks by citing principles of forestry science. Redington also opposed light burning because he believed that all fire needed to be kept out of the woods.

As will be remembered, Redington's predecessor, District Forester DuBois, instituted a "let burn" policy for California, a non-scientific policy based on personal opinions. This policy, espoused by Chief of Operations Roy Headley and tried out while DuBois was in the Army, believed there was an economic cost ratio of damage versus control and that certain fires should be allowed to burn if expenditures were disproportionate to the value of the resources under protection. S. B. Show, who had just been made head of District 5's research division, exposed this as errant thinking. Introducing analytical statistics to improve fire prevention, Show analyzed forest fire records for the preceding seven years and found DuBois' theory unworkable. Show's analysis demonstrated that personal opinions and judgments of Forest Service personnel experience with firefighting was not enough to fight fire in a state as diverse as California. His study also concluded that systematic analyzing of fire reports regarding fire behavior typology along with weather conditions and a matrix of average rates of spread could contribute as well. Show's results were presented at the supervisors' meeting at Davis in February 1919, where Show stated that there were two radical flaws in District 5 fire policy: "Detection was lacking at the beginning and end of the season; and the district as a whole was undermanned." Based on Show's report, Redington reoriented District 5 fire policy. The order went out to "get more and better men for fire guards and show more speed and efficiency in fire suppression." In February 1920, District 5's slogan became "put the fire out in the quickest possible time" (Ayres 1941: 16; *News Letter [District 5]* 1919: No. 58; 1920: No. 67; Pendergrass 1985: 102-103; Ayres 1942: 13).

District 5's new "put 'em out" fire policy conflicted harshly with the opinions of light burning supporters. The brewing conflict came to a head in late 1920, when the Society of American Foresters sponsored two public meetings on the subject. At these debates, Forest Examiner Show presented the Forest Service's viewpoint, while popular author White took up the light burners' views. Nothing conclusive came directly out of the Show-White public forums. Nevertheless, the controversy led to three important developments. First, the Southern Pacific offered a tract of 79,000 acres for an experimental burn to be observed by a distinguished California Forestry Committee composed of prominent men, including W. C. Hodge, a former staff member of District 5. From the test, the California Forestry Committee concluded that light burning cost more than the benefits it provided. Thereafter, light burning became official heresy within District 5. Second, District 5 began an active educational policy against light burning to counter its proponents. And third, District 5 authorized the San Joaquin Project on the Sierra National Forest, where White's cabin home was located. This project embraced the front of the sugar pine belt in the forest, and was the chief entomological activity of District 5 for the next few years. Before the San Joaquin Project folded in 1924, the project led to a number of important technical studies regarding how to maintain control of insect attacks, what classes of trees were subject to attacks and which trees were the most resistant. The San Joaquin Project was moved to a more serious infestation on the Modoc National Forest in the northern part of the State. By 1928, a western pine beetle infestation had reached epidemic proportions on the western portion of the forest (Ayres 1941: 82-83; Pendergrass 1985: 107-108; *California District News Letter* 1928, No. 37).

After and beyond the light burning controversy, District 5 witnessed several important events related to fire control. These events marked the growing desire among Californians and other western states to prevent forest fires. By the early 1920s, Chief Forester Greeley recognized the worsening fire situation throughout the West and called for a national fire conference to be held in Sacramento (Wilson and Davis 1988: 3). District 5 hosted this first national conference on fire control held in November 1921, which was also the first national Forest Service conference on any subject. The Mather Conference

Fire conference at Mather Field. 1921

(so-called because it was held at the Army Air Service base at Mather Field) achieved many positive results in fire control. It established forest fire control as a national priority. It set various fire control standards for reporting, atlases and terminology. It also endorsed fire research "as an essential tool in developing an adequate fire control organization." The Mather Conference was a milestone in national forest fire control history in this regard, and not surprisingly, District 5 emerged as a leader in fire control and prevention because most of the recommendations adopted by the conference were currently practiced on District 5. One negative result of the Mather Conference was that it essentially re-made district rangers into fire control officers (Cermak n.d.: 220-223; Wilson and Davis 1988: 5; Show n.d.: 93-94).

Following the Mather Conference, S. B. Show, together with Eldorado National Forest Supervisor Edward I. Kotok, were assigned to additional analytical research on fire protection in California. Show and Kotok's teamwork resulted in the 1923 landmark publication, "Forest fires in California, 1911-1920: An Analytical Study" (Wilson and Davis 1988: 3; Ayres 1942: 13). But before Show and Kotok could move farther along the path of scientific analysis of fire control, the devastating fire seasons of 1923 and 1924 delayed work their work.

Edward I. Kotok. 1937

In 1923, California, as a whole, suffered the worst fire season in the United States, with 2,349 fires throughout the state that year. California's national forests also had their worst season since 1917, when 1,492 fires burned close to 150,000 acres of government land and 25,000 acres of private holdings. Fires on the Angeles, Santa Barbara and Cleveland National Forests accounted for 125,000 burned acres, and all but 25,000 acres of that amount resulted from two large fires on the Santa Barbara National Forest. Northern California was not exempt from destruction. The California National Forest (Mendocino National Forest) led the list of northern forests with the most destruction, when flames swept across 20,000 acres. There were also major fires on the Plumas, Shasta, Lassen and Klamath National Forests. In the final analysis, lightning caused nearly 50 percent of the fires on District 5, and the remainder was due to carelessness. Of the human-caused fires, tobacco smokers caused fully 50 percent of them. Fires caused by campers, railroads, brush-burners, lumbering and incendiarism made up the rest (*California District News Letter* 1923: No 52; 1924: No. 15).

The 1923 California fire season was heated, but it paled in comparison to the following year. Two years of continuous drought had left the forested and brush-covered areas of the state in an extremely inflammable condition. The disastrous fire season of 1924 resulted in the devastation of about one-half of one percent of national forest land in California (*California District News Letter* 1924: No 37).

District 5 supervisors, rangers and other fire personnel, in cooperation with the general public, put up a heroic fight that season, but they were no match for the flames caused by a severe drought, coupled with record heat and winds. In early 1924, Redington, Show and Kotok realized the seriousness of the situation and began preparations immediately. Hutchinson and District 5's public relations department quickly got the message out to the public to be vigilant against carelessness. They used "cartoons, editorials, films, lantern slides, even talks broadcast over the newest fad, radio…to sell the message." Other measures taken by District 5 included signing up user groups and associations interested in forest fire protection. Groups such as lumber companies, livestock associations, conservation groups and local water and irrigation companies that had a stake in fire prevention were contacted and enlisted in the cause. Chief Forester Greeley advised Redington to prohibit smoking, camping, hiking and hunting in California's national forests except in designated grounds and suggested that the hunting season be cancelled and all matches and firearms be prohibited as well. There were even calls for federal troops for fire patrol duty in Southern California forests (Cermak n.d.: 236-240). District 5's informal historian R. W. Ayres wrote that it was "the damnedest season" he had ever seen, and when the season closed, "it left behind a vast area of blackened forest, and what was worse, broken men…" Every phase of suppression failed – from manpower, transportation, supply, equipment and communication to overall organization (Ayres 1942: 16; Benedict 1930: 709).

While these fire prevention efforts were noble, they did not prevent the holocaust that ensued. Starting in early June and for the next five months, unstoppable large fires burned relentlessly until rain finally came in late September and early October for most of California's national forests. but the rains did not come to the Cleveland National Forest until early November. In September, Chief Forester Greeley visited fourteen of California's national forests to get a grasp of the situation and even lent a hand in fighting a couple of the big fires. The most intense conflagration and the one which received the most publicity was the San Gabriel Fire on the Angeles National Forest because of its intensity and because of its proximity to nearby urban areas. After Greeley left and the season passed into history, the physical damage was assessed. The season resulted in 1,932 fires, which burned 762,000 acres and caused $1,275,000 of damage. That record stood for many years thereafter. The relentless character of the season also took its human toll as well. At first, most Californians accepted the daily reports of the forest fires with only a passing glance. Most did not feel any sense of danger. But as flames started near or reached the outskirts of their community, then concern, anxiety and even panic set in. With day after day, week after week and month after month of fighting, Forest Service employees were stressed and exhausted as well. For the first time in history, District 5 asked for and received fire protection personnel from other districts, and twenty-five men came from District 3 to help the dog-tired men. During the season, five men lost their lives, one from overexertion, and four fighters were killed on the line by accidents (*California District News Letter* 1924: Nos. 37, 46; Cermak n.d.: 240-244; Ayres 1942: 16, 18). Though District 5 did a remarkable job of containment, the 1924 fire season may have prompted the president of the Mission Indian Federation of San Jacinto, California, to write President Coolidge suggesting that the control of California's national forests be put back into the hands of the Indians. President Coolidge replied, stating that the "administration and protection of the National Forests presented a great many problems, which called for the services of men of wide experience and training," and encouraged all Indians to join the Forest Service and help out (*California District News Letter* 1924: No. 46).

The excessive losses due to the 1924 fire season pointed to unrevealed weaknesses in District 5's fire protection program. To investigate the program, a Board of Fire Review was convened. The board, composed of supervisors and District 5 staff as well as state officials and other prominent and knowledge individuals, studied the worst of the fires to discover what exactly had gone wrong. The board did a thorough investigation, soundly reviewing and critically analyzing everything connected with forest fire fighting from transportation, fire line and camp organization to communication and personnel training. Essentially, they determined that District 5 was simply unprepared for and overwhelmed by the adverse weather conditions – excessive drought, high winds and low humidity – and that "mediocre management and poor leadership throughout the District 5 organization," such as "shortcomings in recruitment, training, supervision, and work planning," contributed to the problem as well. The board equally noted problems on both Northern and Southern California forests. Even so, it proposed that fire protection of the Angeles and other southern forests be given the highest consideration – and even recommended asking Congress for a $1 million matching fund under the terms of the Clarke-McNary Act for fire protection in Southern California. This was the first time that fire control in Southern California's forests was recognized as a "problem with national implications" (Ayres 1941: 17; Cermak n.d.: 260-265; Benedict 1930: 709; Ayres 1942: 18).

Meanwhile, in 1925, the State of California revised and tightened up its fire laws to make them more conducive to future fire protection. Changes included protection requirements for logging procedures for private operators, prohibition of smoking in the woods and banning the use of wood in steam engines instead of oil when practical. Use of oil instead of wood was promoted to reduce the risk of fire – wood combustion creates more sparks. A corresponding action was taken by District 5, when Chief Forester Greeley made protection clauses part of all Forest Service timber sale contracts. Finally, in May of that same year, District Forester Redington, in one of his last administrative actions, re-established the San Bernardino National Forest from parts of the Angeles and Cleveland National Forests, an action that became effective on September 30, 1925, and was in line with the Board of Fire Review's recommendation. The board believed that such a division would improve administration and fire protection for this area, which at this time was the most intensively used mountain recreation region in the United States (*California District News Letter* 1925: No. 19; Ayres 1941: 17; Ayres 1942: 19).

Not unlike fire management, range management was ablaze with change in the early 1920s, especially after the federal government proposed raising grazing fees on national forests. First proposed before World War I, that conflict naturally postponed this controversy, but after the war ended in November 1918, a range war broke out in the American West over the issue. In 1919, Congress pressed for increased rates, and some members muttered that forest ranges were no more than a subsidy for livestock interests. By the end of World War I, Congress had also provided that a percentage of collected fees be returned to the states for schools and road construction, and argued that below-cost fees deprived communities of potential funds for these objectives. The question also soon arose as to how range values should be determined by the Forest Service. Should fees be tied to range value or to market values? In 1925, this grazing question and others reached the heights of a Senate investigation. At that time, Christopher E. Rachford, Washington inspector of grazing, and former District 5 chief of grazing, reported, and Chief Forester Greeley emphasized, the "need to raise rates to a commercial level in line with pricing policies for other forest resources." Rachford's report considered it necessary that national forest grazing fees be raised 60 to 70 percent to equal comparable commercial rates (Rowley 1985: 112-128; Dana and Fairfax 1980: 136-137; Ayres 1941: 11).

In the early 1920s, there was a general deterioration of the range and the decline of grazing on forest ranges. A number of factors contributed to this decline. There was a major drought in 1919, followed by a short economic depression in 1920. Then there was the disastrous drought of 1923-1924, which was also marked by great forest and range fires. Raising grazing fees on national forests just elevated the ire of California livestock owners, who were in difficult and dire economic times. Even so, Rachford, who was quite familiar with conditions in California, recommended raising fees on California's national forests. His report recommended that rates be based on the average rental value of private lands over the past ten years. California forest supervisors were instructed to work closely with livestock associations on this matter because these rates were expected to go into effect very soon (Fox and Walker n.d.: 9; Ayres 1941: 11; *California District News Letter* 1923: Nos. 51, 52).

Notwithstanding the growing grazing crisis in California in the early 1920s, business seemed to continue at its usual steady, if not somnambulate pace, on each of the national forests. Annual District 5 range reports provided statistics on achievements in water development, revegetation of depleted ranges, improvements in range and stock handling and reconnaissance work, as well as reduction in the loss of livestock from predators, disease and poisonous plants. Furthermore, each year's reports often just stated, "You will notice the conditions surrounding grazing affairs this year are almost the same as last year and several years prior to that." By the end of 1923, there were 195,000 cattle and 467,000 sheep grazing on California national forest lands, representing 13 percent of all the beef cattle in California and 19 percent of all the sheep in the state. The Modoc National Forest led in cattle numbers grazed on national forests, followed by the Sequoia and then the Lassen National Forests, which had half as many each. The Modoc National Forest also led in sheep numbers, followed by the Plumas and the Tahoe National Forests, respectively. Grazing management of these numbers of livestock was fundamentally nonexistent, for by the end of 1923, there were no completed studies of carrying capacity on any California's national forests – a key element in range management. Finally, in 1924, grazing reconnaissance studies initiated on several of California's national forests prior to World War I were completed. They revealed badly damaged ranges from overstocking and fire. Based on this information, District 5 staff recommended that livestock numbers be significantly reduced. The proposed range management plan recommended cutting cattle numbers by 20 percent and reducing sheep numbers by 50 percent. Interestingly, when implemented, it reduced cattle numbers by only 13 percent and sheep by a mere 12 percent (*California District News Letter* 1924: Nos. 4, 15; 1925: No. 7; USDA Forest Service 1919-1925, various).

The lack of implementation of real livestock reduction on California's forests may have been due in part to recovery of the range following the 1924 fire season or to the debates raging in Congress that focused on the issue, for before a new schedule of grazing fees on national forests was to take effect on January 1, 1925, Agriculture Secretary Henry C. Wallace deferred the rate increase, citing a recent depression in the livestock industry. Subsequently, bills were introduced into Congress to weaken the authority of the Forest Service in grazing matters. Moreover, some organizations, such as the Society of American Foresters (SAF), proposed that grazing matters be completely removed from the Forest Service and placed under a separate agency. Ultimately, in a decade noted for extensive congressional investigations, action on the fee system and the administration of rangelands by the Forest Service was delayed even longer when Oregon Senator Robert N. Stanfield opened hearings around the West on the subject. The upshot at the end of the hearings was that Congress took no action (Ayres 1941: 11; Dana and Fairfax 1980: 137-138; Rowley 1985: 124).

In the early 1920s, management of general water conservation issues and mining resources on California's national forests was less controversial than either grazing or fire control. Nevertheless, it was a time of active custodial care and better understanding of the California's forest water resources. For instance, in 1919, Forest Examiner Edward Munns produced a path-breaking document on the problems of floods in California titled "The Control of Flood Water in Southern California." At this time, floods occurred often in Southern California, when deposited detritus in streambeds heightened them to flood stage. These floods frequently redirected stream channels through valuable fertile lands, damaging them, as well as transportation systems, highways and municipalities downstream. To control flash flooding in Southern California, the Munns study advocated the construction of check dams, which Munns considered the most effective means of prevention (Munns 1919: 423-424, 429). In 1925, District 5 forest water policies regarding flood control garnered the support of the California Development Association (CDA), an organization that appreciated the "dependency of California's agriculture, industry and recreation upon the establishment of a substantial and intelligent forest policy." In their words, California's forests were "highly important in their control of the flow of water from mountain slopes and constitute one of the most effective means of checking floods and holding back reserve supplies of water for slow distribution through the long dry season." In addition, the CDA realized the importance of California's forests for preventing erosion, particularly in Southern California, and California's national forests' role in conserving water to be used for power purposes and not just for irrigation (*California District News Letter* 1925: No. 51).

Regarding mining, the only important change happened with the passage of the Mineral Leasing Act of 1920, which removed phosphate, sodium, potassium, native asphalt, sulfur and fuel minerals from locations under the Mining Law of 1872. Such deposits were thereafter subject to exploration and disposal only through a prospecting permit and leasing system. The act also specified royalty rates, lease size and lease term for each kind of leasable mineral. After World War I, and following the passage of the Leasing Act, District 5 officers continued to aid and cooperate with the mining industry in every practical way. There undoubtedly were some problems on California national forests when mining operations conflicted with roads and recreation, or produced pollution. But any difficulties were usually ironed out on the forest level and not referred to the district office. That there were few complaints against any District 5 actions in the 1920s can also most likely be attributed to an industry slump in the American West. There was little demand for strategic metals following World War I, and most mining in the West during the 1920s, including California, was done by small-scale operations, involving very little capital investment (Friedhoff 1944: passim; Dempsey 1992: 101-102; Godfrey 2003: passim).

Recreation management in District 5 also seemed less controversial than the management of other resources. Prior to World War I, the term "forest recreation" was almost unknown. Then, with the steady proliferation of automobiles, the revolution in California's highway system, the production of cheap gasoline due to the Southern California oil boom in the 1920s, and with a decade of steady and rapid population growth, by 1925, rangers were truly confronted by a new human use of California's national forests – an explosion of travel by public and private recreationalists to District 5's scenic forests. Californians had discovered that the state's national forests had uses other than the production of timber. Ever-greater numbers of visitors took to the highways to vacation in California's rugged and scenic forests. For instance, by 1921, travel to Lake Tahoe via Placerville (one of seven possible routes to this destination) had risen to 150,000 people annually, far out-doing the 90,000 visitors to Yosemite National Park that year. By 1924, thanks to Forest Service road construction of approximately 7,500 miles of improved roads, California's forests became the playground of all of America. In 1924, more than 4.3 million visitor days were logged, and 88 percent of all visitors were motorists. Of this number, there were 2,420,00 transient motorists; 768,000 picnickers; 618,000 campers; 430,000 hotel and resort guests; and 100,000 or more summer-home owners (*California District News Letter* 1921: No. 40; 1923: Nos. 14, 15; 1924: No. 15). To aid travelers, District 5 distributed guides to California's forests prepared by various automobile clubs and groups. These maps and guides included descriptions of all improved Forest Service campgrounds, along with slogans on fire prevention and camping rules and regulations (*California District News Letter* 1922: No. 19; 1923: No. 14).

There were negative impacts from this recreational explosion. One immediate adverse effect of passable access to remote parts of California's forests was the ever-increasing numbers of deer killed by automobiles, particularly on the Modoc National Forest (*California District News Letter* 1922: No. 50). Some forests, especially in Southern California, began also to show signs of overcrowding. The fiscally conservative administrations of Harding and Coolidge failed to provide adequate appropriations to keep up with the

Hunter's Camp and five-point Mule Deer, Modoc National Forest. 1926

growing recreational demands on the nation's forests and the increasing need for improved facilities. As overcrowding became an issue, it also raised the specter of the loss of important wilderness in California.

In 1890, John Muir said, "The clearest way into the Universe is through a forest wilderness,"(Browning 1988: 225) but it was not until the late 1910s that the preservation concept of wilderness on national forests came about, when Arthur H. Carhart, a Forest Service landscape architect, advocated the principle. Forest Service employee Aldo Leopold, later famous for authorship of the *Sand County Almanac* (1949), picked up the mantle thereafter and taught, wrote and fought for wilderness in America. As early as 1920, in an article for *Journal of Forestry*, Leopold suggested establishing wilderness areas of at least 500,000 acres in each of the eleven western states. Four years later, on June 3, 1924, the first Forest Service wilderness area was created on the Gila National Forest in New Mexico (Williams 2000: 58-59; Shepherd 1975: 230-231), due in large part to Leopold's efforts. In October 1925, Leopold wrote an article in *American Forests and Forest Life* titled "The Last Stand of the Wilderness," which gave pause to some of California's forest officers, like Tahoe National Forester Supervisor R.L.P. Bigelow. The *California District News Letter* noted Leopold's piece, and Bigelow stated, "Mr. Leopold's convincing presentation of this most important subject should start us all figuring as to how some of our forest areas can be left in a virgin state for the benefit of future generations." Bigelow thought at that the very least, wilderness should be considered in the preparation of recreation plans for each forest (*California District News Letter* 1925: No. 45).

But before California's national forests would get their first wilderness areas, District Forester Redington announced on January 15, 1926, that Chief Forester Greeley had selected him to serve in Washington as assistant forester in charge of the public relations work of the Forest Service (*California District News Letter* 1926: No. 2). Redington accepted the offer, which placed him in line as the apparent heir to Greeley's position. A month later, Redington named Stuart Bevier Show, or "S.B." as many called him, as the fourth district forester for California (*California District News Letter* 1926: No. 6).

Stuart B. Show, District/Region 5's fourth forester (1926-1946)

Active Administration and Custodial Care of California Forests, 1926-1929

S. B. Show's term in office (1926-1946) spanned two vital decades of forestry history, from the pre-Great Depression years, through the Great Depression and the New Deal, and then continuing into World War II and its immediate aftermath – the longest tenure of any California district forester. Born and raised in Palo Alto, California, Show received his bachelor's degree from Stanford, where his father was a professor of history, and his master's degree in forestry from Yale University. In 1910, Show joined the Forest Service and spent his "rookie days" as a technical assistant on the Shasta National Forest. By the time Show joined the Forest Service, Chief Forester Gifford Pinchot had already been dismissed and replaced by Henry S. Graves. All the same, throughout his formative career as one of the West's leading figures in forestry, Show's writings and career indicated that ideologically Show was a utilitarian conservationist, much like his hero, Pinchot (Show 1963: i).

Show made his reputation in the Forest Service very early on through his contribution to fire protection research. In 1915, while working on the Shasta National Forest, Show branded light burning as an impractical method of forest management. He and Shasta Forest Supervisor R. F. Hammett wrote several papers on the subject, in which they contended that light burning sacrificed "long-range forest values, reproduction, and soil protection for the immediate reward which the preservation of mature forests offered." Clearly, Show was the ideological cousin of former District Forester DuBois, who promoted Show in 1917 to serve as the assistant head of the Feather River Experiment Station on the Plumas National Forest. In 1919, with the assistance of his brother-in-law Edward I. Kotok, Show issued a series of bulletins, which expounded on DuBois' *Systematic Fire Protection in the California Forests*. They strengthened District 5's position in the light burning controversy and also led to the end of Headley's economic fire fighting theory. Show felt that any policy that emphasized low costs as the fire objective was very dangerous, and he advocated larger expenditures for initial attacks to offset escalation of suppression costs (Pendergrass 1985: 103-109; Show 1963: i).

In 1926, Show made several important changes which built a "fire control tradition" for District 5 during the last half of the 1920s. But Show also realized he had greater responsibilities as well, which included improving the organization of District 5 and its personnel and enhanced management of

District 5's many forest resource programs. While forest fire research continued to draw his attention, he focused his initial efforts on these objectives.

At the time Redington left for Washington, D.C., the District 5 force was comprised of mostly non-technically trained personnel, who in Show's opinion were mediocre at best. There were eighteen forest supervisors (ten non-technical), eleven deputy supervisors (ten non-technical), ninety-one district rangers (eighty-seven non-technical), fourteen assistant rangers (mostly technical) and sixteen project sales officers, which were both technical and non-technical. Given the unevenness of his staff's ability, Show began a rebuilding program. In his *The Development of Forest Service Organization, Personnel and Administration in California,* Show stated that at this time in his career, he made a critical decision. Rather than continuing his predecessor's policy of promoting average middle-aged men in their forties to the upper-level positions, who in his opinion were too "settled in a firm mould from which little now could be reasonably expected," Show placed his hopes on training and advancing his youthful assistant rangers because they still had "an unknown ceiling." Despite his technical training, Show also realized that District 5 needed generalists, "men who could and would give balanced attention to all aspects of the job." Show had little faith in government by specialists, nor in the accepted opinion that "expertise" in an area such as grazing or timber automatically qualified a person as supervisor (Show n.d.: 102, 106-114).

The answer to Show's dilemma was "trained brains," which he hoped the Feather River School curriculum would produce for District 5 in the future. In his words:

> What we were trying to buy was a mind of at least medium intelligence, with some training in the processes of learning, analysis and application of facts, reasonably articulate, coupled with the basic qualities of character and with personality characteristics qualifying him to deal effectively with people outside and inside the Service – in total a reasonably normal, balanced and sound person, accepting and able to get along in a real, sometimes troubled and always competitive world (Ibid.: 114-115).

Of course the old line, non-technical supervisors opposed this "new look." They believed that "practical experience which had once qualified them was still valid as a basis for entrance and that further practical experiences on the job was superior to the – as they saw it – impractical training at the [Feather River] school, which, moreover took men from useful work on districts and forests." Eventually, these supervisors either accepted the program or retired. Show looked to Paul Pitchlynn, the demanding headmaster of "Plumas College," for tough-minded appraisals of school enrollees in order to find promising individuals to fill the upper ranks. During Show's administration, the failure rate of the Feather River School soon rose to an average of 30 percent (Ibid.: 115-116).

Besides Show's search for general managers versus specialists and technicians, national administrative changes came as well. In 1928, Chief Forester Greeley resigned and in May of that year named Robert Y. Stuart, a World War I friend, to the position of chief forester. A native of Pennsylvania, Stuart joined the Forest Service in the summer of 1905, before graduating from Yale University. Stuart soon rose through the ranks, serving in the early part of his career in Montana, northern Idaho and then Washington, D.C., until World War I, when he volunteered for staff duty and went overseas. After the war, Stuart left government altogether and went to work for Gifford Pinchot, who was then commissioner of forestry of Pennsylvania. In 1927, Stuart returned to the Forest Service as an assistant forester in the branch of public relations. He replaced former District 5 Forester Paul G. Redington, whom Greeley appointed as chief of the Bureau of Biology, thus moving Redington out of the line of succession. The following year, Greeley recommended Stuart as the new chief forester. The serious-minded Robert Stuart saw the Forest Service through the initial years of the Great Depression, until his untimely tragic death following a fall from a window in his office on the seventh story of the Atlantic Building, Washington, D.C., in October 1933 (*California District News Letter* 1928: Nos. 19, 25; Williams 1993: 5; *California Ranger Region Five* 1933: No. 48; Clar 1969: 237).

With new national leadership, an important period of change came to District 5. Fire protection and forestry research were emphasized during the remaining pre-Depression years over timber management, while grazing management was put on hold. New emphasis was also placed on watershed management and hydropower development, and a "war" broke out between the Forest Service and the National Park Service over recreation issues, which led to the founding of California's first primitive areas.

With Show as district forester, it was natural that fire protection and forestry research became key elements in District 5's program. During the years 1926 to 1929, District 5 accomplished a great deal in enhancing fire prevention and control on the district. The year 1926 was another bad fire season due to drought. Northern California national forests suffered the most in this season, losing approximately 132,000 acres to forest fires on the Klamath, Shasta, Lassen and Plumas National Forests (Cermak n.d.: 287). But under Show's direction, there were few failures in the field as supervisors and rangers attacked the fires at the very onset. At the end of the season, Chief Forester Greeley concluded that "a season like 1926, occurring fifteen years ago, would have brought an appalling disaster," and complimented District 5's "grit," "staying power" and handling of the situation (*California District News Letter* 1926: Nos. 26, 37). This praise did not satisfy Show, who continued to drive home fire prevention on the district. Increased forest use caused increased numbers of fires, and starting in 1927, all campers were required to carry shovel and axe when visiting California's national forests. Coming on the heels of the 1926 fire season, the district forester stated that the purpose of this regulation was to "further safeguard the 19,000,000 acres of national forests from damage and destruction resulting from unextinguished camp fires.... [which cost] serious loss of timber, valuable watershed cover, and other needed forest resources, and cost the Forest Service many thousands of dollars to extinguish." Newspapers such as the *San Francisco Chronicle* actively supported this requirement, but apparently the Forest Service was not very vigilant in checking campers for these items, and soon the policy was dropped (*California District News Letter* 1927: Nos. 23, 36).

Clearly, from the 1924 fire season onward, it appeared more and more likely that despite all of the Forest Service's methods of fire protection, each year District 5 would need to expect a loss of 100,000 acres or even more from forest fires. Some national forests escaped with very little acreage burned each year. For instance, the Inyo, Mono and the Eldorado National Forests were termed "asbestos" forests because fire was seldom a problem on these eastern Sierra Nevada forests. On the other hand, each year it seemed that Southern California forests raged with blazes (Cermak n.d.: 313). Nonetheless, Show pressed onward with his fire prevention and control crusade. He supported and participated in forest fire research, which came into being during the early years of the Show administration.

In 1926, Show and District 5 were at the forefront of fire research in the nation – a status that was only enhanced by the establishment of the California Forest Experiment Station at University of California, Berkeley, that year, under the direction of Edward Kotok. Despite the inherent regional and personal nepotism in his appointment – Kotok was Show's brother-in-law – he was highly qualified. Kotok served in District 5 forestry in a number of positions: forest examiner on the Shasta (1911-1916), forest supervisor on the Eldorado (1916-1919) and fire chief in the District office (1919 to 1926). When Kotok left District 5 to assume the directorship of the California Station, Jay H. Price succeeded him as chief of fire control. Among the staff Kotok assembled for the station was Duncan Dunning, District 5's expert on reforestation (*California District News Letter* 1926: No. 26; and Cermak n.d.: 295).

The proximity of Berkeley to the San Francisco headquarters provided an excellent opportunity for close cooperation between District 5 and the California Forest Experiment Station. To work out a coordinated program, a series of forest research conferences were held in the Ferry Building, headquarters of District 5, which resulted in an annual investigative meeting. Chaired by District Forester Show, these conferences were attended by all of Region 5's department heads (operations, forest management, grazing, products and public relations) as well as by representatives from other federal agencies engaged in forestry research or investigative work in the California Region. Papers were presented on a variety subjects related to forestry research at these conferences, and current research projects were discussed openly, all of which help set policy and project goals for the California Forest Experiment Station (*California District News Letter* 1927: No. 4; USDA Forest Service 1927: passim; 1930a: passim). It is interesting to note that the establishment of the California Forest Experiment Station predated the passage of the 1928 McSweeney-McNary Act, which legitimized experiment stations such as the California Forest Experiment Station. Nevertheless, the McSweeney-McNary Act opened the way for a larger and more adequate development of forestry research in California. The act established a ten-year forestry research program and survey of forestry resources in the Forest Service, and also provided appropriations for broad-scale research by the experiment stations. The California Forest Experiment Station's initial budget amounted to $200,000 per year (Williams 2000: 40; *California District News Letter* 1927: No. 43; 1928: No. 21).

One immediate product of this funding and the collaboration between District 5 and the California Forest Experiment Station was the publication of a new study by Show and Kotok, one that had a profound effect on fire protection in California. Titled *Cover Type and Fire Control* (1929), their study thoroughly discussed the relationship between fire behavior and fuels types. Although *Cover Type and Fire Control* proved useful, Show and Kotok were cognizant of the need to test their theories. What they needed was an experimental fire laboratory where they could try out corrective measures in actual fire fighting conditions. Recognition of this need led to the creation of the Shasta Experimental fire Forest, whose objectives were to "determine the best methods of fire control for a selected area by trying out both accepted and new methods…" (*California District News Letter* 1926: No. 23; Ayres 1942: 13; Wilson and Davis: 1988: 5; Cermak n.d.: 296-300).

By his own admission, S. B. Show clearly focused his attention on fire protection and forestry research during the years 1926 to 1929. He realized that he needed to improve his knowledge of California's national forests and learn to "see with eyes and interpret with the minds" of his staff (Show n.d.: 120). In fact, he left timber, water, grazing and recreation management to veteran staff, men like Assistant Forester T. D. Woodbury. During Show's first administrative years, Woodbury, after twenty years of directing District 5's timber management program, was promoted to assistant regional forester. Under him, District 5 developed a silvicultural cutting system based upon intelligent research that yielded reasonably satisfactory net increments in cutover areas. At the same time, Woodbury's program left sufficiently high-grade timber of the better species to furnish ample seed for restocking, thereby guaranteeing a profitable second cut. Up until about 1950, national forests were dependent upon natural regeneration, since reforestation efforts prior to that time were unsuccessful. Therefore, it was important to save existing young growth on timber sales. Distinct 5 also developed and enforced a protective code based upon stringent timber sale contract clauses to minimize fire, logging and insect and disease damage losses on cutting areas. Through practical national forest timber sales, District 5 expanded its annual cut from 52 million board feet in 1908 to 333 million board feet in 1928. District 5 also made fair progress in preparation of sustained-yield management plans under Woodbury's direction. By 1929, nine sustained-yield management plans had been prepared and approved on District 5, which involved a total of 816,000 acres of government and private lands containing a combined allowable cut of 191 million board feet (Woodbury 1930: 697-699).

Sometimes the timber industry growled at District 5's policies. Such was the case in 1926. Wallace Hutchinson had just released a District 5 news item critical of loggers titled "Less Destructive Logging Methods Urged by U.S. Forest Service." This release drew the ire of one logging industry leader, who responded immediately. In a piece entitled "You've Gotta Stop Kickin' My Dog Around," he roundly condemned District 5 for Hutchinson's statements. In the end, the industry leader feared that the "constant propaganda" regarding fire protection as the key to forest perpetuation was leading many in California to believe that the forests were fast disappearing and that the "future of forestry lies in raising of cellulose." District Forester Show defended Hutchinson and the points raised in Hutchinson's news release. Show declared, "if I were to write it myself I should phrase it differently…. [But] I would suggest the real answer to the growing feeling that the forests are disappearing is not to suppress actual facts, for I believe it to be a fact that under current logging methods in California some 40 percent of the private cutover land is left unproductive…. the real solution is to actually leave the cutover lands in good condition" (*California District News Letter* 1926: No. 41). The industry leader's fears may have derived from declining timber production in the state. By the end of the decade, California had dropped from the fifth to sixth leading producer in the country (*California District News Letter* 1929: No. 6).

Progressive grazing management did not match progress in sustained-yield timber management. The controversy over the reduction of grazing livestock and increased fees came to a head in 1926-1927. In February 1926, the Forest Service instituted a new grazing regulation, whose purpose was twofold: to help stabilize the use of national forest lands by the Western livestock industry, and to give livestock owners a greater say in the settlement of grazing matters. This new regulation had three major components. The regulation's first part authorized a ten-year grazing term permit, which could not be revoked except for violation of the contract by a livestock owner. Usually specified in these term permits were livestock reductions to protect range, timber or watersheds, or to provide forage for other users. The next part of the regulation encouraged individual grazing allotments wherever practicable under local conditions of range use. Third, the new Forest Service regulation provided for local grazing boards covering either a single or a group of national forests. The purpose of the grazing boards was to study and settle grazing questions as far as possible (Ibid.).

Regarding grazing fees, in February 1927, Agriculture Secretary James T. Jardine announced that there would no increases that year and that the schedule of fees recommended by the Forest Service would be implemented on a graduated scale beginning in 1928 and extending to 1931 (*California District News Letter* 1927: No. 5). These increases never came about because by this time, a movement was underway to transfer all public grazing lands to the states. At a time when even the future of the Forest Service's future looked

Herbert Hoover just after landing a 6-lb. Steelhead from the Klamath River, Klamath National Forest

dim under the Herbert Hoover administration, creating hostility among Western livestock owners would have been politically unwise. By 1929, when the subject of grazing fee increases had been fully exercised by all, the stock market crashed, ushering in the Great Depression and even further delays (Rowley 1985: 144-145; Dana and Fairfax 1980: 138-139).

Neither the 1926 grazing regulation nor the postponement of increased grazing fees seemed to affect grazing policy on District 5, probably because Show only took a perfunctory interest in the subject. Nonetheless, statistics from the annual Region 5 grazing reports provide telling information about the grazing situation on California's national forests. First, annual report figures indicate that though the useable range on California's national forests remained the same (approximately 11 million acres), cattle numbers dropped, while sheep numbers stayed about the same. For instance, in 1926, 173,591 cattle and horses and 422,200 sheep and goats actually grazed on District 5's forests. By 1929, those numbers had dropped to 155,603 cattle and horses and 428,230 sheep and goats. Both numbers were well within the established carrying capacity and range allowance set for District 5 forests. However, as time passed, there was only an 8 percent increase in term permits for cattle from 1925 to 1929, and only a 2 percent increase in term permits for sheep. California's livestock owners clearly were avoiding the new ten-year permits because of the possibilities of stock reduction (USDA Forest Service 1926-1929, various).

If grazing management was not a high priority for S. B. Show, then neither was watershed management – at least until 1928, when the St. Francis dam collapsed on the Angeles National Forest, causing so much damage. The collapse called into question the safety and supervision of Forest Service dams on California's forests. As it turned out, the St. Francis dam was constructed at a very early date under the authority of an Interior Department easement, and the Forest Service had no responsibility for checking the suitability of the design, foundations or performance. Most dams on California's forests came under the authority of the 1920 Federal Water Power Act (41 Stat. 1063). To correct the problem associated with the St. Francis dam collapse, Region 5's engineering staff assumed supervision of all dams on national forests not built under the Federal Power Commission (FPC) (*California District News Letter* 1928: No. 28; Smith 1930: 52, 175).

St. Francis Dam after collapse in spring of 1928. Angeles National Forest

In fact, after the passage of the 1920 Water Power Act, California's national forests experienced unparalleled expansion of hydroelectric facilities. From 1923 to 1928, twenty-two dams of major size were built on District 5 forests. All were closely supervised at all stages by the engineering staff of the San Francisco office. There were, for instance, the Florence Lake and Shaver Lake dams on the Sierra National Forest that controlled water for Southern California Edison Company's Big Creek power project. Other important dams in the High Sierra included the Bullards Bar dam on the Tahoe National Forest and the Almanor dam on the Plumas National Forest. In 1928, there were five different dams recently licensed and being built under the FPC. The two largest were on the Mokelumne River, which were being built by the

Pacific Gas and Electric Company. Moreover, in "Water Powers of California" (1929), Region 5 Engineer F. E. Bonner estimated that slightly more than 90 percent of the undeveloped waterpower resources in the state of California were situated wholly or partly within the national forests. Bonner also affirmed that the Sierra National Forest, which embraced all the San Joaquin and the major part of the Kings River power resources, possessed greater hydropower potential than any other forest, although the Plumas National Forest was not far behind (*California District News Letter* 1928: No. 28; 1929: No. 3).

Kern River hydroelectric power plant #3, in Southern California. Sequoia National Forest. 1930

Besides the huge power resources of California revealed in Bonner's report to the Federal Power Commission, watershed management, which was designed to preserve or to restore watershed conditions, to stabilize soils and vegetation, or to maintain high quality of water yield, was also in the back of Regional Forester Show's mind. These concerns were brought out in three District 5 research publications, two produced by the California Forest Experiment Station. "First Progress Report on Work in Southern California," written by Charles J. Kraebel in 1927, was prompted by the major problem affecting all of Southern California from San Luis Obispo County to San Diego – water conservation – or how to find a type of forest cover that permitted maximum run-off, maintained soil in place and in and of itself was the least susceptible to destruction by fire. A tall order indeed, and one that Kraebel proposed could be resolved during the next ten years through cooperative research studies by the Forest Service, the State Forestry

Department and the two major counties currently affected by the problem, San Bernardino and Los Angeles. According to Kraebel, this research effort needed to focus research on cover types, erosion and runoff, fires, forestation, water use, climate and planting for the answer (Kraebel 1927: 1-12).

Kraebel's paper was followed by Edward Kotok's paper, "Forests and Water" (1928), which emphasized that the protection of brushfields offered the greatest hope for soil building, the restoration of tree-like forests and the replacement of original forest barriers which Kotok theorized originally stood throughout Southern California. According to Kotok, a high coniferous forest of pine, spruce, fir and cedar once covered the mountain ridges and the plateaus and extended through the winding canyons and slopes to the coastal floor. The lower slopes, according to Kotok, supported hardwood, woodland forests of oaks, sycamore and walnuts. But through centuries of fires, these forests were lost and supplanted by xerophytic plants, which were better able to withstand extremes of drought. Kotok deplored the loss of the original forests, but "nature has substituted the brushfields, which serve a most useful purpose (Kotok 1928: 1). Kotok's faith in brushfields and Kraebel's desire for research were emphasized and supported by W. C. Lowdermilk in his paper, "Role of Chaparral and Brush Forests in Water and Soil Conservation in California" (1928). Lowdermilk, a District 5 silviculturist, lamented the condition of Southern California's watershed situation. Southern California's stream flows were now fully appropriated, ground water supplies were steadily lowering because of pumping for irrigation, and importing water under the 1922 Colorado River Compact was years away. Lowdermilk lectured Southern Californians to start considering the ability of chaparral and brush forests in nearby mountain areas to store native water from winter precipitation. To Lowdermilk, these vegetation types could absorb enough capacity for future use (Lowdermilk 1928: 1-4).

Finally, the last subject of resource management that Show had to learn about early in his administration, but one that he later enjoyed discussing and writing about, was recreation. Recreation permits and summer cabin building continued at an unrestricted pace. Each year, national forests became more and more overcrowded, and Congress failed to provide appropriations to meet this demand. But in the early years of Show's administration, it was outdoor recreation and wilderness that took front stage.

Aristotle once drew the distinction between three sorts of human activities: work, recreation and leisure time. Recreation in the Greek philosopher's eyes was an activity that one pursued in order gain refreshment after work or to make new work easier to do (Kneipp 1930: 618). By the 1920s, the American concept of recreation considered the activity as one to be done for its own sake. The growth of California's car culture, coupled with the wealth and prosperity of the 1920s, allowed Californians to adopt that philosophy with a vengeance. In response to this need in California and elsewhere in the nation, in 1926, Chief Forester Greeley ordered an inventory of all undeveloped or unroaded national forest lands larger than 230,400 acres (10 townships). Greeley wanted to withhold these areas against unnecessary road building and forms of special use of a commercial character that would impair their wilderness character. Thereafter, in 1926, the Forest Service created seven new "recreation areas" in California for public use. These areas were set aside because of their outstanding scenic and recreation values. Three of them – Mount Shasta, San Gorgonio Range, and Laguna Mountains – came in the early part of the year. By the end of 1926, four new areas were added: the Salmon River Alps in the Klamath, Shasta and Trinity National Forests, which included the headwaters of the Trinity and Shasta rivers; Echo

Public campground on shores of Gull Lake, Mono National Forest. 1934

Source: Forest History Society

Lake and Desolation Valley, a popular and well-known summer outing region on the Eldorado National Forest; the Lakes Basin on the Plumas National Forest, noted for the fishing waters of Long, Gold and numerous other lakes; and the High Sierra vacationland of Reversed Creek on the Mono National

Forest. In total, more than 275,000 acres of national forest land in California was set aside at this time, primarily for recreational use (*California District News Letter* 1926: No. 42).

The designation of these recreation areas was prompted by worries of losing valued recreation areas to development. In the 1920s, the use of powerful road graders and bladers for road construction had allowed permittees to push many roads and developments deep into the mountains and deserts of each national forest. Unless this process was controlled, Show and others feared that they would "end up with the California mountains lacking areas that people valued because they didn't have roads and structures in them." The Forest Service was alarmed that unless this road construction was halted, "they would work into every corner of the mountains and destroy a kind of country, a kind of forest recreation, which had been a cherished part of the American way." District 5 officials made several attempts to stop this road-building process. First, District 5 tried to block owner-built roads, or what some called a "road," to their land, but lost in the courts. Next, District 5 resorted to regulating these owner-built roads by requiring permittees to obtain road-building permits loaded with stipulations, but it was ineffectual as well. Finally, District 5 tried to stop heredity ownership of summer home permits with the introduction of the non-transferable life estate permit. This permitting system helped the Forest Service cancel some use in areas it wanted to restrict for specific recreational use (Fry 1963: 157-159).

Beyond this altruistic desire to prohibit road building in natural areas, the designation of these initial recreation areas by District 5 in 1926 was most likely also prompted by serious threats by the National Park Service to transfer and acquire land valued for recreation purposes that was currently under Forest Service administration. According to David C. Swain in his book *Federal Conservation Policy, 1921-1933* (1963), "The Forest Service was fearful that national parks might win guardianship of these areas that were valuable primarily for their natural beauty, and it was to reduce Forest Service vulnerability on this point that the Forest Service acted." In an interview on the matter, S. B. Show agreed, stating that it "certainly was an expected result of the wilderness area program" (Fry 1963: 159-160; Swain 1963).

By 1927, California national parks were the most popular in the country, which gave NPS head Stephen T. Mather and the Park Service much national power and persuasive force. Yosemite led the list of parks in the United States with close to one-half million visitors annually, followed by Sequoia with just over 100,000 people, General Grant with just under 50,000 visitors and then Lassen Volcanic with just over 20,000 people. Muir Woods, which was a national monument at this time, attracted more than 100,000 visitors as well (*California District News Letter* 1927: No. 45). Park Service Director Mather was an "extremely aggressive, strong man," in Show's opinion, "who clearly believed in his service and assignment, and he was very skilful and effective in obtaining public support" (Fry 1963: 181). Mather's covetous eyes turned to California in the mid-1920s.

During the early 1920s, the Forest Service at first exhibited an attitude of cooperation with the NPS, "wishing to see areas of outstanding interest set aside under a separate administration [e.g., NPS] so they might be given the kind of attention they merited." This accommodating attitude continued until at least 1925, when, through the initiative of the Forest Service, several scenic areas were transferred to the NPS. They included the transfer of Lava Beds National Monument on the Modoc National Forest to the Park Service, the extension of Sequoia National Park to include the Kern River Canyon area of the Sequoia National Forest, and the transfer of California's highest peak, Mount Whitney, from the Inyo National Forest. Chief Forester Greeley approved these actions because he felt that they were proper additions to the national park system (Fry 1963: 180-184). In fact, the NPS and the Forest Service even cooperated in building a trail to the summit of Mount Whitney (14,494 feet), the highest trail in America. This trail was dedicated in September 1930 and eventually became part of the High Sierra trail system (*California District News Letter* 1928: No. 31; 1930: No. 37). But at the same time, Chief Forester Greeley worried about NPS raiding national forest lands. Therefore, according to Show, Greeley struck an informal deal with NPS Director Mather "that if the Forest Service got out of the way" on these transfers, then "the Park Service would drop additional pressures for additional major transfers of areas to park status" (Fry 1963: 180-184). This deal held as long as it was to the advantage of the National Park Service, which turned out not to be very long.

As soon as Show became District Forester in 1926, Mather increased pressures for the transfer of additional areas of California forest recreation land to the Park Service. First, NPS Director Mather requested an expansion of Lassen Volcanic National Park. Show opposed much of the expansion, but the "high brass" in Washington failed to sustain his opinion, and the land transfer occurred. Next, in 1927, Director Mather and the Sierra Club wanted not only Kings Canyon to become a national park but also the transfer of 30,000 acres from the Sierra and Mono National Forests to Yosemite Park. Behind the scenes, Chief Forester Greeley, District Forester Show, and Forest Supervisor M. A. Benedict actively conspired to resist Mather's efforts. They stacked hearings on the subject with opponents of the projects, mainly irrigators and ranchers. Irrigators that opposed the proposal included the City of Los Angeles, which of course at this time was reaching out far and wide for large quantities of water (*California District News Letter* 1927: No. 33; Fry 1963: 184-190). The war between the Forest Service and NPS was on, and relations changed from bad to worse as District 5 fought back in several other ways. These battles, however, were fought on the District 5 Office level or higher. For the most part, lower-level Park Service and Forest Service employees continued to cooperate and did not join the fray.

The NPS/Forest Service conflict was not helped by the State of California. In 1928, the state passed a $6 million park bond, which established a statewide park system and development program. Will Colby, "one of the old Bull Moosers of the Sierra Club," charmed the first California State Park Board. But to the chagrin of the Forest Service, in 1929, Colby filed a report demanding the immediate transfer of parts of several national forests to the State of California, such as the Mount San Jacinto area in Southern California. To stop Colby, District Forester Show got the word out that if the State of California tried to conduct land raids, as the NPS had done, he would fight them vigorously and in no uncertain terms. Fortunately, cooler and more rational heads prevailed, and a plan of joint administration and cooperation between the Forest Service and the California State Park Board was worked out for the San Jacinto area, setting an admirable example of how public agencies should act toward each other (*California District News Letter* 1928: No. 47; Fry 1963: 164-166).

Meanwhile, in the winter of 1927-1928, Chief Forester Greeley sent out a letter to all the districts to prepare proposals for "a system of wilderness areas," as they were originally called, "through which roads, buildings, and formal recreational developments would be barred." According to the letter, "large areas of presently or imminently commercially-exploitable timber" were to be excluded from qualifying, and grazing, mining and waterpower development on recommended lands was allowable (Fry 1963: 16).

Show and his chief of lands, Louis A. Barrett, accepted the wilderness idea without reservation. They studied the Greeley letter carefully, and ingeniously and much to their credit, they picked apart the letter in order to obtain a formula. In their approach, Show and Barrett were as broadminded and flexible as they could be in putting the greatest acreage under protection within these wilderness areas. For instance, they noted that Greeley's letter said nothing about inclusion of private lands and that Greeley had not set a minimum or maximum size limit. Furthermore, they consulted closely with Sierra Club leaders such as Will Colby, Duncan McDuffie, Walter Huber, Francis Farquhar and Walter Starr, as well as individual forest supervisors. All were enthusiastic about setting the course of this new Forest Service policy, which concerned many, many people. Finally, District Forester Show personally set out on a number of fact-gathering horse pack trips to potential wilderness areas, including two lengthy trips to the Marble Mountains and the Trinity Alps in Northern California. Barrett was unable to join Show on these trips, but instead handled meeting arrangements and prepared maps and reports. Cautious not to include essential areas where they might need to build fire control roads in the future, they defined four acceptable criteria for wilderness. First, they selected areas that would have no more than a relatively small amount of private land. Second, they chose areas that only had scattered and patchy stands of inferior tree species. Third, they opted for areas that would be of the best quality and largest size possible. And fourth, they sought to have at least one wilderness area on each of California's national forests (Ibid.: 159-163).

Using the above criteria, Show and Barrett in mid-January 1929 proposed the first fourteen wilderness areas for District 5, encompassing one and a half million acres. Some of this was land previously designated as recreational area land in 1926 but which Show and Barrett recommended be redesignated as wilderness areas. As a general rule, they selected areas located in the higher scenic mountain regions, where fire hazard was limited and where there would be no necessity to build roads for forest administration purposes. In conformity to the Greeley letter, these wilderness areas were not to be developed by road building or opened to any form of permanent recreational occupancy under permit. However, under Show and Barrett's proposal, much of this land was still subject to grazing, and in years to come, Show and Barrett believed that some timber cutting and water development could be allowed on selected wildernesses (*California District News Letter* 1929: No. 3).

Apparently, Show and Barrett felt that Washington's approval of these early California wilderness areas was not needed, but in due course

Washington approved most of them with modifications under the 1929 Regulation L-20, which Agriculture Secretary Jardine modified and better defined. Under the 1930 amendment, wilderness areas were re-designated as "primitive areas" under this regulation.

L-20 was the first regulation regarding wilderness management. Essentially, it provided for a series of areas to be known as primitive areas, wherein the Forest Service was to "maintain primitive conditions of environment, transportation, habitation, and subsistence with a view to conserving the value of such areas for purposes of public education and recreation." The purpose of a primitive area was to "prevent the unnecessary elimination or impairment of unique natural values, and to conserve, so far as controlling economic considerations will permit, the opportunity to the public to observe the conditions which existed in the pioneer phases of the Nation's development, and to engage in the forms of outdoor recreation characteristic of that period, and promoting a truer understanding of historical phases of national progress."

L-20 went on to state that if "doubt existed as to the highest form of service, ordinarily it will be resolved in favor of maintaining primitive conditions." Utilization policy was clarified in March 1932 with another amendment. At this time, primitive areas were raised to the level of other forest uses. The 1932 amendment also affirmed that primitive areas were not just natural areas under another name and that in primitive areas, as elsewhere in national forests, the principle of highest use would prevail. This principle was sufficient justification for partial or complete restriction or postponement of the utilization of timber, forage or reservoir sites "where such utilization would nullify the value and service of the primitive area to a degree exceeding the benefits or advantages accruing from such utilization." Fire prevention administration and road and trail construction were to be confined to the bare minimum. Special permits were also to be confined, but as a general rule no hotels, resorts, permanent commercial camps, summer-home communities, individual summer homes, or commercial enterprises would be authorized within designated primitive areas" (USDA Forest Service 1976: various). With promulgation of L-20 and later amendments, Show and Barrett were forced to conform to stated national policy, and the original fourteen wilderness areas they proposed were either sanctioned, modified or eliminated based on L-20. The description below summarizes these changes.

On the southern national forests, Show and Barrett carved out five wilderness areas. They were the 27,000 acre Agua Tibe, or Tibia, Primitive

Area on the west end of the Palomar Mountain in the Cleveland National Forest (increased to 34,553 net acres in 1931); the 19,000 acre San Gorgonio Primitive Area covering the San Bernardino and San Gorgonio Ranges (reduced to 13,083 net acres in 1931); the 7,500-acre Telegraph Primitive

Campers at Boiling Spring Camp Ground, Cleveland National Forest. 1923

Area around that peak (never officially approved); the 22,000-acre San Jacinto Primitive Area (reduced to 16,645 net acres in 1931) covering the country east of Idyllwild – all on the San Bernardino National Forest; and the 52,894-acre Ventana Primitive Area, which included wild mountain lands at the north end of the Santa Barbara (later Los Padres) National Forest (*California District News Letter* 1929: No. 3; USDA Forest Service 1976: various).

The majority of wilderness or primitive areas set aside by Show and Barrett were located in the High Sierra. In total, there were six areas designated as primitive areas in these scenic mountains. Going from north to south, there was the 12,000-acre Murphy Hill Primitive Area, which surrounded Campbell, Morris and Lotts lakes west of Belden on the Plumas National Forest (never officially approved); the 41,700-acre Desolation Valley Primitive Area north of Echo Lake and west of Lake Tahoe on the Eldorado National Forest; the 23,000 acre Hoover Primitive Area west of Mono Lake on the Inyo and Mono National Forests (increased to 25,656 net acres in 1931); the 87,000-acre Mount Dana-Minarets Primitive Area between Tioga Pass and the Devil Post Pile country on the Mono and Sierra National Forests (reduced to 82,181 net acres in 1931); and the 700,000-acre High Sierra Primitive Area, the largest wilderness area created at this time, which took in the High Sierra crest in the Inyo, Sequoia and Sierra National Forests (increased to

825,899 net acres in 1931). This rugged mountain wilderness ran seventy-five miles from the Mammoth Lakes region on the north to Mount Whitney on the south. Finally, there was the 97,020-acre Emigrant Basin Primitive Area established on the Stanislaus National Forest along the north boundary of Yosemite National Park (*California District News Letter* 1929: No. 3; USDA Forest Service 1976: various).

In Northern California, Show and Barrett established three wilderness areas. Going from west to east, they were the immense 200,000-acre Middle Eel-Yolla Bella Primitive Area at the head of the Middle Eel River on the California and Trinity forests (reduced to 107,195 net acres in 1931); the 130,000-acre Salmon-Trinity Alps Primitive Area, which encompassed the headwaters of the Trinity and Salmon rivers in the Klamath, Shasta and Trinity National Forests (increased to 221,370 net acres in 1932); and the 75,000-acre South Warners Primitive Area around Eagle Peak in the South Warner Mountains on the Modoc National Forest (reduced to 68,242 net acres in 1931) (*California District News Letter* 1929: No. 3; USDA Forest Service 1976: various).

All in all, Show and Barrett had quietly tried to turn close to one percent of California's national forests into designated wilderness areas – largely without any public comment or interference from Washington – with a stroke of the administrative pen. Luckily, L-20 supported many of their early designations. By 1931, Washington approved of most of their recommendations with few exceptions, and in April 1931, four other primitive areas were added to the Region 5 under regulation L-20. They included the 16,443-acre Caribou Peak Primitive Area and the 15,495-acre Thousand Lakes Primitive Area on the Lassen National Forest, the 5,000-acre Cucamonga Primitive Area on the San Bernardino National Forest, and the 234,957-acre Marble Mountain Primitive Area on the Klamath National Forest. In January 1932, the 74,160-acre San Rafael Primitive Area on the Santa Barbara National Forest and the 36,200-acre Devils Canyon-Bear Canyon Primitive Area on the Angeles National Forest were created as well (USDA Forest Service 1976: various). By the end of 1932, eighteen primitive areas were established on California's national forests, encompassing an amazing 1.9 million acres.

Marble Mountain Primitive Area on the Klamath National Forest. 1929

District 5 Becomes Region 5

On May 1, 1930, District 5 changed into Region 5. At that time, Washington renamed all USDA Forest "districts" as "regions" in order to avoid confusion with ranger districts. That important administrative event came under Chief Forester Stuart. Under Chief Forester Greeley, district autonomy held at its traditional level, and the Washington office staff had been, as Show put it, "firmly and knowledgeably controlled in expanding its directive powers." Chief Forester Stuart did not cramp Region 5's autonomy in choice of people and projects either, although the creation of the Bureau of the Budget in the Hoover administration set up a "new hurdle" in the battle to obtain appropriations. To mark the change from District 5 to Region 5, or just Region 5, the *California District News Letter* was renamed in December 1930 and became the *California Ranger Region Five* (*California Ranger Region Five* 1930: No. 1; Show n.d.: 111-112).

In 1930, the Forest Service celebrated a quarter-century of progress and achievement in the development of forestry and the conservation of America's natural resources. This 25th anniversary date was celebrated on February 1, 1930, with a national coast-to-coast radio broadcast on the National Broadcasting Network (NBC). Chief Forester Stuart led the radio program, which included several significant speakers, such as former Chief Foresters Gifford Pinchot and Henry S. Graves. At Region 5 headquarters, all listened intently and with enthusiasm. Certainly it was a time for reflection by District 5 "old- timers," many of whom had served in California forestry even prior to 1905. They included District Officers W. I. Hutchinson (1901), R. W. Ayres (1902), Joseph Clinton Elliott (1902), L. A. Barrett (1903) and T. D. Woodbury (1904). Woodbury was one of the last survivors of the first crew of the California Inspection District 5 – full of youth, hope and ambition. He and "Uncle Joe" Elliott, a senior lumberman, were the only remaining members of the original forest management team that started out under "Fritz" Olmsted more than two decades earlier (*California District News Letter* 1930: No. 6; *California Ranger Region Five* 1930: No. 1; 1931 No. 47).

Then there were old-time District 5 forest supervisors such as Tahoe National Forester Supervisor R.L.P. Bigelow (1902), whose work and long-term service was acknowledged. Much has already been said regarding Bigelow's career and personality. On the other hand, little has been said about lower-level forest officers such as District Ranger Jacinto Damien Reyes. Known to his fellow officers simply as "J. D.," Reyes spent thirty-one years as a district ranger on the Cuyama Ranger District on the Santa Barbara National Forest. Born in 1871 to Angel Reyes, one of the first settlers and ranchers in the Cuyama Valley fifty miles north of San Buenaventura, Reyes rode the range as a *vaquero* with his father, developing at that time the valuable experience and native resourcefulness that he later used when the General Land Office in 1900 placed him in charge of the 400,000-acre Cuyama District, the same

District ranger Jacinto Damien Reyes, Santa Barbara National Forest.

Source: Ventura Historical Museum

ranger district that he managed with dedication for the next thirty-one years. During the early days of the Forest Service, Ranger Reyes actively rode the territory each day, constructing trails, telephone lines, lookout houses or whatever was needed. Each summer, Reyes fought major fires on his and other ranger

Picnic party at Happy Camp lookout, Modoc National Forest. 1927

districts. He was proud that he had never lost a man, or had one seriously burned on a fire under his charge. Reyes himself was off duty only once in his three-decade-long career. Returning home from a fire, Reyes' mule bucked him off, and he was laid up for several weeks. At the barbeque held for his retirement in 1931, more than 500 persons attended. The guests provided testimony to Reyes' conscientious dedication to his work and his contributions to the community. Stories were exchanged about Reyes' adventures with various forest supervisors and district foresters; the times he escorted Presidents McKinley and Roosevelt when they toured the Cuyama District; and the first time Reyes flew in an airplane over his ranger district, viewing all the trails he had ridden and all the fire battlegrounds he had worked on (*California Ranger Region Five* 1930: No. 48; Reyes and Hogg: 1930).

Nostalgia soon gave way to tallying up the accomplishments of District 5 and educating the public regarding progress made by the district over the years. First, the *California District News Letter* carried a feature article detailing District 5's accomplishments in operations (fire control, plans and training), lands, forest management, engineering and public relations for several weeks in 1929 (*California District News Letter* 1930: Nos. 14-19). Then in 1930, to educate, celebrate and publicize Region 5's past accomplishments, purpose and current progress, Region 5 and its leadership produced several public relations programs. In that year, for the first time, Region 5 began to issue a periodic "Accomplishment Report" for the California national forest region. Over the years, these photo-illustrated accomplishment reports took many forms, sizes and shapes, but each issue usually covered a variety of subjects such as timber, forage, mineral, recreation, water management, fire control, road-building activities and research. They almost always presented a positive, progressive image of the Forest Service (USDA Forest Service 1930b: passim).

In addition to beginning the annual accomplishment report series in 1930, Region 5, in cooperation with the California State Chamber of Commerce, issued the *Forestry Handbook of California*. The *Forestry Handbook* was a joint educational effort among Region 5, the California Forest Experiment Station and the Sierra Club, whose purpose was to provide the public with knowledge concerning the conservation of California's natural resources. Revised and reprinted several times thereafter, the *Forestry Handbook of California* discussed at length conservation and forestry, forest types and trees in California, lumbering and reforestation, forest and water conservation, grazing and forestry, wildlife of the forest, the natural and man-made enemies of the forest – fires, insects, and disease – recreation on California's forest playgrounds and good woodsmanship and behavior in the forests (USDA Forest Service 1930c: passim). A year later, the Forestry Handbook was complemented by the highly popular pamphlet *The Forest Rangers' Catechism: Questions and Answers on the National Forests of the California Region*. Based on an idea originating with Sierra Forest Ranger Frank M. Sweeley and prepared by R. W. Ayres and Wallace Hutchinson, *The Forest Rangers' Catechism* described in a question-and-answer format what national forests were, how they were administered, how Region 5 derived revenues and allocated expenditures for improvements, and provided to the public direct and important information on how Region 5 fought forest fires and managed its resources (water, timber, forage, minerals, recreation, fish and game). Other subjects covered in *The Forest Rangers' Catechism* included the question of light burning, reforestation, forest research, public relations and the differences between a national forest, a national park and a national monument, along with practical information such as how to contact the Forest Service or how to obtain a Christmas tree from California's national forests (Ayres and Hutchinson 1931: passim). In the same year, Region 5 also published *Federal Activities in the National Forests of California Region*, the first comprehensive look at the state of affairs on California national forests and examination of the value of national forest resources to the state. Most likely, this publication was produced at the request of Chief Forester Stuart, who was seeking to create a more far-reaching national program of forestry and probably presented reports such as this to key members of Congress. According to *Federal Activities in the National Forests of California Region*, the accumulated value of these resources was far in excess of the devastated resources that the General Land Office assumed control of starting in 1897 (*California District News Letter* 1929: No. 46).

The turning point in the conservation of California's resources, according to *Federal Activities in the National Forest of California Region*, was the pragmatic desire of Californians to preserve watersheds and water supplies for farming and urban centers. In 1931, water on California's national forests served three major uses. First, with a population of more than three million people, as compared to 1.2 million in 1890, California's forests provided domestic and municipal water supply for cities, towns and settlements. These communities over time had invested more than $100 million in water systems to bring water from the national forests to their communities. The Angeles, Cleveland, Eldorado, Inyo, Klamath, San Bernardino and Santa Barbara National Forests continued to be most valuable for the watershed protection they afforded. Second, of the five million acres of irrigated land in California, fully three million acres were dependent on California's national forests for their water supply. With the average cost of $10 per acre, this amounted to a total $30 million annually for the water supplied by the national forests, which helped produce irrigated crops (fruits, grains and vegetables) valued at $250 million each year. The California, Mono, San Bernardino, Shasta National Forests and, to a lesser extent, the Lassen National Forest continued to be most valuable for the conservation of water for irrigation they afforded. Third, in 1931, hydroelectric plants on California's national forests produced 72 percent of the total electrical energy generated in the state and 18 percent of the hydropower in the entire United States. The Eldorado, Inyo, Lassen, Plumas, San Bernardino, Sequoia, Sierra, Stanislaus, Tahoe and Trinity National Forests served as important hydroelectric power locations and in 1930 had three or more Federal Power Commission licenses underway under the 1920 Water Power Act. Furthermore, from 1922 to 1929, California's national forests accumulated almost $300 million in receipts from hydroelectric power of which more than $81 million was turned over to the State of California under Section 17 of the 1920 Act (USDA Forest Service 1931: 1-2, 7-12, 34).

Despite the devastation of timber resources left by nineteenth-century miners, railroads and loggers, by 1931, the forests of California contained one-fourth of the timber found on the Pacific Coast and an estimated 14 percent of the remaining timber in the United States. Commercial forest area within California's national forests amounted to more than 8.6 million acres, or 49 percent of the total commercial timber in the state. More than 5,000 people were employed by operators utilizing national forest timber, which had a total stand of 102 billion feet. Annually, they produced lumber and timber products valued in excess of $10 million. While government timber sales on California's forests were minimal at the turn of the century, by 1929 they amounted to more than 450,000 (M.feet B.M.) and thereafter dropped as the nation's financial system devolved into economic depression. California national forests noted for timber reserves and production included the Eldorado, Klamath, Lassen, Modoc, Plumas, Sequoia, Shasta, Sierra, Stanislaus, Tahoe and Trinity National Forests (USDA Forest Service 1931: 2-3, 7-12).

Stock raising in California was the oldest industry in the state, and after the Forest Service assumed control of the range in California's national forests in 1905, many important changes took place. After that date, the Forest Service stopped range wars between conflicting industries (cattle versus sheep) by establishing equitable range boundaries and limiting the number of stock to the forage capacity of the range. Thanks to the Forest Service, depleted ranges were under restoration by the curtailment of stock numbers, proper seasonal range use and by reseeding overgrazed areas. Additionally, the Forest Service secured better use of the range through the construction of driveways, drift and division fences and corrals, the development of springs and watering places, and the destruction of predatory animals and the elimination of range-destroying rodents, along with the eradication of poisonous plants. In 1931, within California's national forests, there were 10.5 million useable government grazing lands, along with another 1.4 million acres of private lands on which stock grazed under Forest Service permit. Foraging on these lands were 1.8 million cattle and 4.1 million sheep. Important forage range was found on the California, Eldorado, Inyo, Lassen, Modoc, Plumas, Santa Barbara, Sequoia, Shasta, Sierra, Stanislaus, Tahoe and Trinity National Forests (USDA Forest Service 1931: 3, 7-12).

Though Californians had always recreated on national forest lands, the building of roads and trails, the establishment of free public camps and the issuing of permits for summer home sites increased recreation on them at an amazing pace. For instance, in 1916, the first year that statistics were kept, there were just over 700,000 visitors to national forests in California, of which close to 430,000 traveled by automobile. By 1923, there were close to two-and-one-half million transient visitors, or tourists, to California's national forests, of which close to two million stayed for a period of time as visitors. By 1930, the number of forest visitors almost reached three million, and there were just over 14 million transient visitors. To accommodate these campers, by 1931, the Forest Service had built 500 improved campgrounds and several hundred unimproved campgrounds. To accommodate those who wished to return annually for health, rest and recreation, the Forest Service allowed more than 6,000 summer homes to be built on lots varying from one-quarter to one-half acre in size, depending on location and demand, and charged a lot fee from $15 to $25 per annum. The greatest number of these summer homes at this time were located in the national forests of Southern California. Finally, by 1931, there were recreation areas totaling 72,000 acres and another two million acres established as primitive areas in sixteen inaccessible parts of the mountains. Although all of California's national forests offered recreation, several forests attracted more visitors than others. Owing to its proximity to large centers of population, the Angeles National Forest was used for recreational purposes more than any other national forest not in just California, but in the United States. Next in line was the Cleveland National Forest, where the 11,500-acre Laguna Recreation Area offered valuable summer vacation ground, followed by the San Bernardino and the Santa Barbara National Forests, in that order. Important High Sierra mountain vacation retreats included the Eldorado, Inyo, Plumas, Sequoia, Sierra, Stanislaus and Tahoe National Forests. In Northern and Central California, recreational use on the Shasta, Lassen and Trinity National Forests was increasing year after year (Ibid.: 4-5, 7-12).

Fish and game were also now considered important resources on California's forests. Since the Forest Service began in 1905, the protection and conservation of California's game animals, birds and fish were important management duties of the agency. A large part of the large game habitat in California existed within thirty-two state game refuges containing more two million acres within the California national forest system. In the early 1920s, there were only an estimated 250,000 deer and 500 antelope in the state. By 1931, deer and antelope populations increased slightly to 260,000 deer and

1,200 antelope. In that year, more than 100,000 hunters entered the forests and bagged an estimated 24,000 deer and an unknown quantity of antelope. Conservation of deer became important after 1924, and the California National Forest contained a larger deer population (estimated at 25,000 in 1930), more than any other forest in California. Large game animals such as bear, elk and mountain sheep, however, were dwindling at an alarming rate. In the early 1920s, there were an estimated 12,000 bears, 150 elk and 10,000 mountain sheep left in the state. By the end of the decade, there were only an estimated 9,000 bear left, 87 elk and barely 600 mountain sheep. Clearly, these animals were endangered species that needed attention. Regarding fishing, California's national forests continued to provide many species of trout in their streams and lakes, including rainbow, eastern brook, golden, cutthroat and steelhead. Each year, forest officers cooperated with the State of California Division of Fish and Game, as well as local sportsmen's organizations, assisting in planting fry in national forest waters to maintain these populations. In 1930, they planted an estimated 4.6 million fry (Ibid.: 5-7).

All in all, fiscal contributions to the State of California by 1930 from various Forest Service funds were very significant indeed. For instance, between 1907 and 1930, California's national forests had receipts of more than $16 million and expenditures of close to $31 million, excluding expenditures on forest roads built by the Bureau of Public Roads, which totaled approximately another $7 million. From 1906 to 1931, California's counties received more than $4 million of receipts from the 25 percent fund, with the leading benefactor counties being Plumas County (close to $600,000) followed closely by Tuolumne (close to $450,000) and Fresno (over $400,000) counties. Of the counties hosting national forest system lands, Orange and San Benito counties received the least amount, just under $2,000 each. Additionally, from fiscal year 1912 to 1931, the Forest Service expended more than $1.5 million from national forest receipts for roads and trails in the State of California (Ibid.: 36-37).

Great Depression and Region 5, 1929-1932

Unlike the Great Red Scare, which absorbed the country in 1919, the stock market crash a decade later did not at first fully grab the attention of Californians, but as the dire economic effects of the crash snowballed and California's economy turned sour, the populace began to worry. Anxiety turned to despondency as California felt the full impact of the financial downturn.

Though forest-dependent industries were curtailing or were in the process of closing their operations soon after Black Tuesday (because, for example, businesses could no longer get bank loans, and workers were having their paychecks cut), Region 5 and the Forest Service first really felt the effects of this national economic crisis in their budgets as declining revenues from resources resulted in decreasing budgets and programs. The principal revenue items declining were timber sales and grazing fees. After 1930, all of the Forest Service regions were in the red, with Region 5 in the worst condition. During the first quarter of fiscal year 1931, California's national forest receipts fell off more than $250,000. This figure was only the beginning of the slide in Region 5's revenue, which by the end of the year declined to $500,000 (*California Ranger Region Five* 1931: Nos. 7, 36; 1932 No. 37). By 1931, financial matters turned from bad to worse. The situation became so bad that President Hoover recommended a work reduction and a furlough system to keep everyone working. He ordered that government employees spread their work over a "five day week," so that those employed by the federal government not be deprived of all income and thrown into the bread lines (*California Ranger Region Five* 1932: No. 3) A six- or five-and-one-half day work week had been standard for many years. By July 1932, the general Forest Service budget situation became so critical that Chief Forester Stuart asked the various regions to reduce the volume of mimeographed work and use of paper. In the spirit of cooperation, Region 5 reduced the size, editions and pages of the *California Ranger Region Five*, deleted all personal items, left out all inspirational and similar material, and included thereafter only official material necessary to the prosecution of Service work (*California Ranger Region Five* 1932: No. 34). Still the Depression deepened. By August 1932, receipts for Region 5 were 46 percent lower than the previous year. The biggest drop, $1.5 million, was in timber sales (*California Ranger Region Five* 1932: No. 37). On the eve of the 1932 election, President Hoover extended the "furlough system" as the Great Depression deepened (*California Ranger Region Five* 1932: No. 3)

Meanwhile, as unemployment figures rose, Region 5 began to experience a steady rise in crime rates on its forests. As the itinerant unemployed roamed California searching for work, many intentionally set fires either out of anger or frustration, or to get work on fire fighting crews. These arsonists, when caught, were fully prosecuted. For instance, one desperate couple got a year in jail for deliberately setting a fire on the Trinity National Forest. Other crimes occurred as well. For instance, in 1931, a former employee of the Forest

Service in Montana was caught in Southern California for "check kiting," or, in other words, for passing bad checks. Using his "natty Forest Service uniform and forged credentials," the young man passed more than a thousand dollars' worth of bad checks and even bought a new coupe in Los Angeles with his

Loading fire fighters on the Plumas National Forest. 1934

"earnings." Apparently, he issued the checks to cover a meal or some small article of purchase, but always collected a considerable amount in change. Interestingly, he always obtained a receipt for the amount of the bill, for his "expense record." There were more serious crimes as well. Many built stills in national forests to earn income and a livelihood from mash and moonshine. Bootlegging turned to murder on the San Bernardino National Forest when a forest guard was shot and killed on duty when he unexpectedly came upon a man in the woods while working on his still (*California District News Letter* 1930: No. 35; *California Ranger Region Five* 1931 No. 44; 1932 No. 3).

As unemployment in California took a turn for the worse in 1931, Region 5 did not stand by but instead set out to alleviate some of the unemployment. In December of that year, Region 5 established and operated unemployment work camps on two of Northern California's national forests in cooperation with the State of California. These work camps were the forerunners of the Civilian Conservation Corps (CCC) camps of the New Deal and were established to give productive work during the harsh winter for single, unemployed men in return for food and housing. Region 5 furnished equipment, bedding and supervision costs, while the State of California paid for subsistence costs and local public agencies provided medical attention and transportation

Loading fire fighters onto bus on the San Bernardino National Forest. 1928

for the men to the camps. One camp of fifty men from San Francisco was located on the Stanislaus National Forest. The other camp was located on the Sequoia National Forest and was filled by men recruited from the Fresno area. Regional Forester Show pointed out that more camps were contemplated, particularly in Southern California, and would be established at an early date (*California Ranger Region Five* 1931: No. 2).

The men (there were no women camps) at both camps were put to work building firebreaks, clearing roadsides of inflammable material, falling and burning snags and removing other burnable material from valuable stands of timber. The men at these camps were very different from the average "pick-up fire fighter" that supervisor and other field officers were used to selecting for work. According to Regional Forester Show, they were different:

> ...not only in the level of intelligence and education, which is higher, but the attitude toward the work of any kind is vastly different. College students, trained mechanics and even professional men are found among the unemployed of the city. At the first mention of a possible job the selecting officers are swamped with appeals for work of any kind. It is not a case of calling for volunteers but of handling the numbers who are ready and anxious to trade their place in the bread line for a man's job in the woods, and who beg not for a dole but for a chance to work to help pay for the food and lodging…(Ibid.).

In the spring of 1932, with the closing of the camps, California Governor "Sunny Jim" Rolph Jr. thanked Regional Forester Show for his assistance. Rolph thought that this social experiment proved successful and that its accomplishments were "due in large measure to the splendid cooperation of the Forest Service." The Governor was also pleased with the project's contribution to forest conservation in California and the betterment of fire protection on the national forests, and hoped to adopt a similar program the next winter (*California Ranger Region Five* 1932: No. 22).

Custodial Care of California's National Forests and Conservation

On the eve of the New Deal, California's national forests were largely in good condition. After more than twenty-five years of labor, the Forest Service had kept its bargain to protect and conserve California's various forest resources for the future. By the early 1930s, California Region 5 understood how best to utilize them for the public good. Through viable "wise use" policies, Region 5 continued to foster conservative utilization of timber, forage, water and mineral resources. As time passed, Region 5 better understood how to protect these resources for the future enjoyment and use of generations to come, through such programs as sustained-yield management of timber and controlled grazing measures. The Forest Service was not always successful in these endeavors. Even so, its work clearly was one of "active" custodial care, particularly in the area of fire prevention and control. Region 5's determination and innovation worked to defend California's national forests from its ever-growing seasonal nemesis – fire. They had the faith; they just needed to develop the appropriate religion to fight what they saw as a fiendish devil.

By the early 1930s, Region 5's policy regarding the conservation of the recreational values of California's national forests had also matured. California's car culture, prosperity and population growth placed an almost insatiable demand on the national forests, which Region 5 could barely meet. Tourists and campers no longer were considered just a nuisance, but important Forest Service clientele. This circumstance convinced Region 5 at a very early date to place managing recreation values on par with other forest resources such as timber and water. Furthermore, whereas Region 5 officials had once disagreed with John Muir's appreciation of wilderness, resulting in a fractionalized conservation movement in the Pinchot years, now S. B. Show and Region 5 attentively embraced the precept, if not the concept, of wilderness with the establishment of California's first primitive areas. Admittedly, Region 5 was dragged into this state of affairs through battles with the National Park Service over the recreational hearts and minds of Californians. Even so, by the early 1930s, thanks to the leadership of men like S. B. Show and Louis Barrett, Region 5 had set aside more primitive areas than any other Forest Service Region in the country.

Finally, the Forest Service had other bargains to fulfill, such as the protection of wildlife. Not much has been said on this subject so far because the consciousness of most Californians had not been raised to a level of concern. As the New Deal years unfolded, awareness of endangered species such as the California condor began to prick not just the conscience of Californians, but that of the nation as well.

Chapter VI: 1933-1941

A New Deal for Region 5

A New Deal for California

By the time of the 1932 election, which pitted Republican President Herbert Hoover against the Democratic candidate, New York Governor Franklin Delano Roosevelt (FDR), Californians were in dire straits. As one writer noted, the economic disaster had "cut through the underpinnings of California's economy like a scythe," hurting the state's "real industries" such as agriculture, oil, and real estate. Some residents were turned into virtual paupers, while thousands were forced to stand in bread lines, reduced to selling apples and pencils, and dependent on state and municipal relief work (Watkins 1973: 363, 365). Californians blamed Hoover for the Depression, and faulted his economic recovery program as well. Not surprisingly, President Hoover lost not only his home state of California but even his home county of Santa Clara (Bean 1968: 417). Ironically, one of Hoover's proposals, the Reconstruction Finance Corporation (RFC) – created by Congress in early 1932 as a means to promote economic recovery – funded the construction of many of California's most notable large-scale public works projects during the next decade, including the construction of the Oakland/San Francisco Bay Bridge and the Colorado River Project (Kirkendall 1974: 17-18; Rice, Bullough and Orsi 1996: 428-429).

The Roosevelt landslide, however, did not translate into Democratic control of California's governorship. In 1933, noted muckraking author Upton Sinclair decided to run for governor as a Democrat, based on his program to End Poverty in California (EPIC). In his utopian novel, *I, Governor of California, and How I Ended Poverty: A True Story of the Future,* Sinclair espoused putting all of the unemployed to work in state-aided cooperative farms, factories and other enterprises. The national Democratic administration disassociated itself from Sinclair's socialistic ideas, and conservative New Deal Democrats threw their support behind the Republican candidate, Governor Frank F. Merriam, who won handily in California's 1934 gubernatorial election (Bean 1968: 415-416, 419-420; Rolle 1963: 514-517).

With Roosevelt and the Democrats controlling the presidency and other national offices, the federal government played a greater role in California economic life. Roosevelt's New Deal programs passed during the "first hundred days," and thereafter, clearly involved a strong commitment to capitalism. The New Deal both maintained and changed America's capitalistic system, providing a "new deal" for workers, farmers, the elderly and the impoverished by rejecting the theory that poverty was the consequence of the defects of individuals.

Roosevelt created programs to put the country on a better economic footing, such as the National Industrial Recovery Act (NIRA), which set up the adoption of codes of "fair competition" for all businesses. Also, for the first time, labor unions were given the right to organize and bargain collectively. NIRA also set up the Public Works Administration (PWA), a multi-billion-dollar public works agency that funded large projects involving large capital expenditures. The PWA was designed to pump money into the economy to stimulate its recovery. Among the large-scale public works projects funded by the PWA in California was Orange County's development of Newport Harbor (Kirkendall 1974: 38-46 passim; Rice, Bullough and Orsi 1996: 427).

California farmers were not left out by the New Deal. The Agricultural Adjustment Act created the Agricultural Adjustment Administration (AAA), which involved itself in production controls, encouraged farmers to play a role in its administration, advocated disposal of surpluses abroad at low prices and sought to protect farmers from foreclosure on their properties (Kirkendall 1974: 38-46 passim). New Deal financing also set into motion the Central Valley Project (CVP), which transported the abundant waters of the Sacramento Valley to the arid San Joaquin Valley. The federally subsidized irrigation project was put under the charge of the Bureau of Reclamation, with construction beginning in 1937 and completed in the 1950s. The CVP construction project created thousands of jobs, relieving the impact of the Depression on the state, and became one of the most enduring legacies of the New Deal in California (Rice, Bullough and Orsi 1996: 430).

Despite these accomplishments, California farm workers suffered for most of the decade because of the migration of destitute outsiders – the dispossessed, exhausted refugees of the Dust Bowl. Among the newcomers were the 350,000 farmers from the Middle West, the so-called "Okies" and "Arkies" from Oklahoma and Arkansas, poignantly portrayed in John Steinbeck's well-known novel, *The Grapes of Wrath*. In migrating to California, they hoped to find a better life, only to be taken advantage of by farm owners who worked them for long hours at starvation wages and housed them in unsanitary hovels and crude tar-papered shacks. By 1936, migration to California, especially the farming region of the central valley area, overwhelmed the system. Desperate state officials set up guard posts at all rail and highway entrances to the state in Oregon, Nevada and Arizona, and closed off entry to any hitchhikers, boxcar riders and "all other persons who have no definite purpose for coming into the state." Refugees from Mexico continued to easily slip over the international border, congregated in the vicinity of Los Angeles and worked long hours for little money, alongside other foreign groups such as the Japanese and Filipinos (Watkins 1973: 363, 365; Rolle 1963: 512-514).

Then there were relief programs for the unemployed, which were designed to bring immediate assistance to the millions of native-born unemployed and to restore their morale and health. First was the Federal Emergency Relief Administration (FERA), which made grants to the states for relief. Under Governor Merriam, a California version of that agency was quickly organized to administer these relief funds, called the State Emergency Relief Administration (SERA), which lasted until 1935 (Clar 1969: 219-234). Second came the creation of the Civil Works Administration (CWA). Opposed to simply giving out handouts, FDR and fellow New Dealer Harry Hopkins created the CWA to employ millions of people to help them survive the winter of 1933-1934. CWA workers were put directly on the federal payroll. The CWA took half of its workers from the relief rolls and the other half from people who simply needed jobs. In California, the CWA employed more than 150,000 Californians in a wide variety of activities, such as building airports, bridges, roads, schools and other public structures. Both the CWA and FERA programs were temporary measures.

After their expiration, Congress established the Works Progress Administration (WPA) in 1935 to supersede them, which in 1939 changed its name to the Works Projects Administration. For more than a decade thereafter, nary a school, post office or other public structure was not built in California without some WPA funding. Furthermore, under the WPA, the Federal Writers Project engaged unemployed writers and historians to produce histories of California's counties, the Federal Theater Project hired unemployed actors and musicians to present plays in California's theaters, and the Federal Arts Project put artists to work painting murals in practically every public building in the state (Leuchtenburg 1963: 121-122; Kirkendall 1974: 38-53 passim; Rice, Bullough and Orsi 1996: 427-428). The California Division of Forestry was also the beneficiary of many substantial WPA projects, which included the design and construction of ranger residences, lookout stations, barracks, warehouses, bridges and other structures and facilities (Clar 1969: 218).

A closely related program to FERA, CWA, and PWA was the popular Civilian Conservation Corps (CCC), which combined relief and conservation and put millions of unemployed young men to work in forests and parks (Kirkendall 1974: 38-53 passim). This relief program was closest to Roosevelt's heart. FDR felt strongly that the creation of a civilian forest army would benefit the character of city men, while at the same time put the "wild boys of the road" and their energy to good purpose in the national forests (Leuchtenburg 1963: 52). The CCC will be discussed in greater detail later in the text.

Newly-elected California Republican Governor Merriam was not a "do-nothing" conservative who held back the New Deal from Californians. He realized that he would not have been elected without the support of conservative New Deal Democrats. Therefore, the Governor brought his program into conformity with some of the New Deal's early policies and made use of the New Deal "alphabet" relief programs such as FERA, CWA, WPA and the CCC, as well as recovery and finance programs such as the NIRA and the PWA. Though he generally resisted liberal social legislation, Merriam's "pragmatic conservatism" and political skills "brought the New Deal to California, hesitatingly perhaps, on tiptoe, and a little shamefacedly, but he brought it nevertheless." Even so, Governor Merriam was no friend of labor, for on July 5, 1934, or "Bloody Thursday," he sent in the National Guard to break the San Francisco longshoreman's strike, killing two strikers and injuring sixty-four, thirty-one of whom were shot by National Guard troops (Watkins 1973: 365-369; Rice, Bullough and Orsi 1996: 433).

Despite his support of some of the New Deal programs, Merriam was destined to serve only one term as California's governor. In the 1938 election, he lost out to the active Democrat leader of the California's legislature, Culbert L. Olson. Governor Olson's was the first Democratic administration in California in the twentieth century. Everyone expected that Olson would inaugurate a "New Deal in miniature" for the state. In spite of having the support of the Roosevelt administration, Olson accomplished next to nothing during his term of office as a New Dealer, largely because of his administrative shortcomings but more importantly because conservative Republicans controlled the state senate and blocked even his mostly modest reforms. Governor Olson achieved significant humanitarian reforms for the mentally ill and prisoners, vigorously protected civil liberties and minorities, and raised the standard of living conditions for migrant farm workers. But Olson met the determined opposition of pressure groups such as the Associated Farmers, who resented his "meddling" with their seasonal labor system. Then, before long, Californians, as well as the nation, turned their attention away from economic, political and social reform to the problems of foreign affairs and national defense as war clouds gathered in Europe and Asia. The approach of World War II eventually lifted California out of the Depression and placed the state on a path of unprecedented growth and prosperity in the post-World War II period. But, ironically, Governor Olson's "New Deal for California" would prove meaningless, as he participated in one of the most tragic events in California history, the removal and incarceration of Japanese Americans in 1942 (Bean 1968: 420-423; Watkins 1973: 376; Rice, Bullough and Orsi 1996: 433-436; Rolle 1963: 517-519).

A New Deal for Region 5

From his window in the Ferry Building, Regional Forester Stuart Bevier (S. B.) Show witnessed the growing daily effects of the Depression. By the end of 1932, Region 5 began to experience a wave of incendiary fires – obviously set for jobs, according to Show, because the "same gangs of homeless, jobless men would show up at those fires." As winter approached, these seasonal workingmen went back to the big cities such as San Francisco. In desperation, they slept in doorways, pilfered food from fruit and vegetable stands, walked the streets or lined up each day in a soup line, across from the Ferry Building, in the rain or otherwise, for a bowl of stew. In April 1932, the twenty-eight cooperative work camps for the transient unemployed that operated under the state emergency program of Governor "Sunny Jim" Rolph Jr. closed because of a lack of money. The Forest Service had spent $14,000 and contributed the time of its staff and use of its equipment to this project. Despite the closure, Regional Forester Show required his supervisors to prepare advance plans of work for when and if an expanded program might be needed (Show 1963: 1-6).

Meanwhile, Democratic Senator Royal S. Copeland from New York called for a congressional investigation of forestry under Senate Resolution 175. The central purpose of Copeland's investigation was to outline a coordinated plan that would "insure all of the economic and social benefits which can and should be derived from productive forests by fully utilizing the forest land" (U.S. Senate 1933: 1). Regional Forester Show and his staff responded to the Copeland resolution by preparing a lengthy, detailed and very comprehensive report on California's forestry situation. Region 5's report, dated September 1932 and simply titled *Copeland Resolution Report: Senate Resolution 175*, described the present and potential forest land and timber resources of California's national forests, covered Region 5's forestry practices and progress in all resource areas from silviculture to recreation, provided a description of conditions and forestry practices on California state and private forests, and, as part of the larger national report, finished with a telling summary of current forest devastation and deterioration of forest lands in the west and in California. In this lengthy report, Show called for action to stop further devastation by improved silvicultural practices, reforestation of barren and devastated lands, and enlargement of intensive management areas through land purchases, especially regarding recreation and fish and wildlife game refuges and fish preserves. Finally, the Region 5 report called for aid to California in the form of direct federal expenditures (USDA Forest Service 1932).

Meanwhile, FDR won the election that November, and as the nation waited for his inauguration on March 4, 1933, many wondered if the country would make it until the swearing in. By March 4, thirty-eight states had closed their banks, and in the other states, banks operated on a restricted basis (Leuchtenburg 1963: 39). The Forest Service was going broke as well. Receipts from all the national forests for the fiscal year ending June 1932 were down 50 percent from the previous year, due to declining timber sales and falling grazing fees (*California Ranger Region Five* 1932: No. 37).

Once in office, President Roosevelt moved quickly on forestry and other issues, assuring the people that he would use the power of the government to help them. First, on March 9, FDR called Congress into special session. Listening to the pleadings of mayors, county and state officials for federal assistance, on March 14 the President asked four of his cabinet members to consider the idea of a conservation corps as a relief measure (Leuchtenburg 1963: 52). At this time, Agriculture Secretary Henry Wallace may have submitted for the president's consideration what became known as the *Copeland Report*, which embodied many of Region 5's national forestry ideas. The monumental, two-volume *Copeland Report*, which Chief Forester Stuart had placed before Congress earlier, had four main findings: first, that practically all of the major problems of American forestry centered in, or had grown out of, private ownership; second, that one of the major problems of public ownership was that of unmanaged public lands; third, that there was a serious lack of balance in constructive efforts to solve the forest problem between public and private land ownership; and fourth, that the forest problem ranked as one of the nation's major national problems. The report recommended greater public ownership of forest lands and more intensive management of public lands in order to stabilize permanent forest industries, provide employment for two million men, increase taxable property and maintain a balanced rural economic and social structure by utilizing land productively for the purpose for which it was best suited (*California Ranger Region 5*: 1933: Nos. 19, 49). The document also called for comprehensive management plans for all national forests, including plans for administrative facilities and lookouts, roads and trails, and recreation facilities.

Subsequently, on March 21, less than a week later, FDR sent an unemployment relief message to Congress promoting a civilian conservation corps of 250,000 men (no women), which in his words was "to be used in simple work, not interfering with normal employment, and confining itself to forestry, the prevention of soil erosion, flood control and similar projects.... It will conserve our precious natural resources. It will pay dividends to the present and future generations. It will make improvements in national and State domains which have been largely forgotten in the past few years of industrial development" (*California Ranger Region Five* 1933: No. 17). Not seven days after Roosevelt's message, Agriculture Secretary Wallace on March 27 transmitted the *Copeland Report* to the Senate (later printed as *A National Plan for American Forestry*). On March 31, Congress passed the bill that created the Emergency Conservation Work (ECW) program later called the Civilian Conservation Corps (CCC). FDR signed it that same day and immediately named Robert E. Fechner, a labor leader from Boston, to head the newly established CCC (*California Ranger Region Five* 1933: No. 19).

Region 5 and Roosevelt's Peacetime Army – The CCC

To perfect plans for the CCC, Chief Forester Robert Y. Stuart immediately called all the regional foresters to Washington, D.C., for a series of meetings that started on April 3. One of the most "dynamic and demanding of the group" was S. B. Show of California. According to one observer, Show "thought and [had] written more about technical forestry problems and their solutions than any of his associates.... And it was no secret that Show was a champion of extended federal control of forest land as well as being a critic of destructive private timber harvesting methods" (Clar 1969: 239-240).

In his unconstrained recollections of the first Washington meeting on the CCC, S. B. Show related that when the regional foresters convened, Show and the other regional foresters found Robert Fechner, whom they called "Uncle Bob," a "lovable, cooperative and amiable elderly gentlemen, quite unhampered by any knowledge of conservation, but disposed to accept the regional foresters as the fountain heads of wisdom, which the Regional Foresters didn't discourage." In no time, a meeting was held with FDR – who "imagined himself as a great forester" – along with his devoted aide, Louis Howe, to discuss the CCC. The president had specific notions regarding the CCC. FDR wanted most of the enrollees to come from the big cities, where he judged the problems were most acute. Enrollees were to be unskilled. Work was to be done by hand. FDR also firmly insisted that the Army run the camps – after all, the Army was without jobs too – and FDR specified that he wanted 500,000 young men with 281 men per camp – an Army company. Even so, Regional Foresters Show and Evan Kelly of the Northern Rockies tried to ameliorate some FDR's ideas. Show and Kelly fought for and achieved the inclusion of the concept of local experienced men (LEMs) to supervise the men in the field, along with other "facilitating" personnel. LEMs were to be drawn from the men whom the Forest Service normally hired as fireguards and workers on construction projects. Show and Kelly argued that they would be needed in order to teach the enrollees "useful handicrafts and keep them from injuring themselves." Show and Kelly also effectively made a case for the purchase of trail-building equipment as opposed to hand labor. Not only would this type of equipment increase efficiency, but also training enrollees on this machinery would prepare them for civilian employment. This concept was not part of Roosevelt's plan, but Show and Kelly convinced Fechner on this point, who then put it across to the president (Show 1963: 12-19).

During the whole time, according to Show, "nobody in the Washington office, including the Chief [Robert Stuart] had anything whatever to do with all this development." Show scrambled to put together a map, a list of camps and their locations, and a detailed work program for each camp for the next six months. The president wanted this information on his desk by Monday morning. Starting on late Friday afternoon, Show worked through the weekend conferring with R. L. Deering (Chief of Operation) and Jay H. Price (Chief of fire Control) back in California by telegraph and telephone. Fire protection naturally and inevitably got the great number of the early projects, since more study had been done on protection needs on Region 5 than for timber, range, recreation or wildlife. Stenographers beat their typewriters far into Sunday night, but by Monday morning Show delivered to FDR's desk a massive load of impressive documents on schedule. Word later came back that FDR was "pleased with the breadth and completeness of the plans." By taking such initiative, Show managed to secure 165 CCC camps for Region

5, far more than other regions. Later, Show joked that it was "true that Forest Supervisors later had difficulty in recognizing projects on their own forests" from the documents he submitted that day. Show just figured that adjustments in men and camp projects could be made after the program got started (Show 1963: 20-36).

By early April, young men between the ages of nineteen and twenty-eight, some with dependents, were being selected from Eastern and Western cities and sent to Army conditioning camps throughout the nation (*California Ranger Region Five* 1933: No. 20). California's initial quota included 11,500 men. These men were sent to Forts McArthur and Rosecrans and March Field in the south, and to Forts Scott and McDowell and The Presidio in the north. At these military installations, the men were clothed, fed and "hardened" with exercise for a period prior to their transfer to CCC field camps. All national forest, state, county and private CCC projects in the state of California, as well as all requisitions for men for such work, were to be submitted to Regional Forester Show for approval (*California Ranger Region Five* 1933: No. 21).

The Army was placed in charge of all CCC camps themselves, and controlled the enrollees when they were in camp. Alternatively, the Forest Service handled all the men while they were on the job, including their transportation between camp and work points. In Region 5, working relations between the Army and the Forest Service were generally good. Regional Forester Show and the other regional foresters saw to this point early on by working closely to reach certain understandings with the Army at the beginning of the program. Before the regional foresters departed Washington, D.C., they drew up a long list of questions about Army-Forest Service relations, such as management of spike (temporary work or side) camps, camp overhead, discipline and the proportion of camp strength available for Forest Service projects. The latter issue presented a persistent relations problem between the Army and Region 5 because Army officers in California tended to assign way too many men to maintaining the camps for surprise inspections. These men spent the entire day picking up cigarette butts, raking the grounds and keeping everything nice and neat, while the Forest Service needed them in the woods. Show often threatened to call and complain to General Malin Craig, commanding general of the Ninth Corps [at The Presidio in San Francisco], which got results in some cases. On several occasions, Show snidely put local commanders, such as Major H. H. "Hap" Arnold in charge of March Field, on the defensive by saying, "You mean you can't run this little old camp without using 85 men?" Before long, Regional Forester Show persuaded the Army to reduce camp teams from twenty-four to twenty-one men (Show 1963: 37-48).

To their credit, the regional office recognized the many opportunities that the CCC provided to California forestry very early on, and took great advantage of them. Working with his supervisors, Show immediately set to work to get the CCC underway. Upon his return to California from Washington, D.C., Show gathered his supervisors together and asked each of them to revise

Teaching Civilian Conservation Corps workers basic forestry techniques, Lassen National Forest. 1933

Source: Forest History Society

the camp work programs he had "imaginatively" created while in the nation's capital. Show set up a review board composed of himself, Deering, Price and Chester Jordan. Working three shifts a day, they listened as each supervisor presented his proposals for work projects. This camp work program revision process brought in a fair number of new range, recreation and water development projects. While holding off the political pressure of California congressmen for projects and camps in "their" districts, and for hiring unemployed constituents, Show moved to set up supply centers that he strategically located to serve the camps; centralized purchases of heavy machinery at rail delivery centers at Redding for Northern California, and at Sacramento, Fresno and Los Angeles; trained trail-builder operators; and then hired worthy and unemployed persons as "facilitating" personnel (Show 1963: 48-55). In the end, according to Show, "neither Washington nor the politicians moved fast enough to cramp the Region's autonomy in choice of people and projects." In the first hectic days of the CCC, Regional Forester Show felt like the freedom to act that the Forest Service had enjoyed during the time of Gifford Pinchot had returned. But then rules began to be made by the Washington office (WO), which cramped and slowed action, diverting time and attention from productive work (Show n.d.: 133-134).

On May 1, 1933, one of the biggest moments in California forestry history took place at the Ferry Building when Regional Forester Show, Major General Craig and representatives of National Park Service (NPS), State Division of Forestry and county and private forestry agencies met to carry out planning for Roosevelt's "peace time army." A total of 166 camps were eventually authorized for California national forests, along with another thirty-three camps on State-private lands, twelve camps on national parks, and five on state parks, under the authority of Governor Rolph. Additional camps under the administration of the Indian Service, Department of the Interior, were located on the Hoopa Indian Reservation in the Trinity/Klamath rivers region, and in Southern California at the Mission Indian Agency. With each camp consisting of 212 men, including officers, there were close to 45,000 men in the California CCC during the initial period. The first CCC camp in California was on the Angeles National Forest. That camp was occupied on May 13, 1933, when 187 men left Fort MacArthur and arrived at the Piru Canyon CCC Camp (*California Ranger Region Five* 1933: Nos. 23, 24, 25, 30).

Meanwhile, in the pre-New Deal period, Region 5 had built perhaps three ranger stations and two or three guard stations a year, and had only occasionally built a residence, a warehouse or an office, because of a lack of funding. Local design of each individual structure was acceptable, and various officers had come to fancy themselves as architects. Now, however, the scope of the CCC program outlined by Show and the forest supervisors, in terms of financing, available labor and plans for the construction of hundreds of needed buildings demanded that the scale, rate of progress and type of plans be radically changed.

To meet the demand, Show went outside the Forest Service for experienced labor and sought to hire from the unemployed ranks registered civil engineers to run construction projects, seasoned logging superintendents and even recent forestry graduates from the University of California, Berkeley. Additionally, Region 5 hired architects for the first time to design the desired administrative improvements, and office staff as well to handle the mounting paper work. The architects and staff worked day and night on the designs and specifications but accomplished the job of designing, drawing specifications, obtaining bids, awarding contracts and setting delivery schedules for construction projects that were programmed for the next six months under the CCC. From the outset, Chief of Lands Louis A. Barrett, who was assigned the vast job, and Show decided that functional types of buildings – ranger stations, guard stations, offices, and large and small warehouses – should be standardized and shipped ready-cut (Show 1963: 56-60; *California Ranger Region Five* 1933: No. 33; Show n.d.: 132).

By July 1933, complete sets of plans and specifications for nine different types of buildings were mailed off to each forest. A month earlier, however, CCC camps in California's national forests and elsewhere were reduced, mainly because the total number of camps applied for by all agencies was greatly in excess of the men available. In the revision schedule for the first CCC period, which ended in October 1933, authorized CCC camps in California's national forests were decreased from 165 to 131 – still the largest number in the country, even though only fifty-three camps were fully occupied at this time. They included the following: Angeles (6); Cleveland (8); Inyo (3); Klamath (2); Lassen (1); Mendocino (1); Modoc (3); Mono (1); Plumas (2); San Bernardino (15); Santa Barbara (3); Sequoia (1); Shasta (1); Sierra (1); Stanislaus (3); Tahoe (1); and Trinity (1) (*California Ranger Region Five* 1933: Nos. 28, 29). For the second CCC work period (October 1933 through March 1934), the number of California's national forest camps dropped to ninety-four (Show 1963: 81).

The men who occupied California's CCC camps came from all over the country. This circumstance was due largely to Show's lengthy work project list, which far exceeded available California manpower. Initial enrollees in California's camps included boys (young men eighteen to twenty-five years old) from as far away as the New York Bowery. Enrollees from Ohio, Indiana and Kentucky, inducted under the Army Fifth Corps, represented the largest group, and a total of ninety-one companies were stationed in Central and Northern California. A large number of men also came from New York, followed by numbers from Nebraska, the Dakotas, Minnesota and Missouri under the Army's Seventh Corps. The men from the Seventh Corps were mostly stationed in Southern California (*California Ranger Region Five* 1933: No. 25; Cole 1999: 17).

Many of the enrollee ranks were dispossessed, underfed and undernourished young men. Some had never ever seen a toothbrush. Some were anxious to learn how to drive a car. Others were hysterically scared of coyotes. Whatever their malady or desires were, LEMs took each bunch of raw kid enrollees under their wing, and soon living in a CCC camp "began to get the wrinkles out of their bellies" (Show 1963: 56-68). In California, the majority of enrollees were Caucasian, but there were African-Americans, Japanese, Chinese, Filipinos and Hispanics in the camps. The law establishing the CCC contained an anti-discrimination clause based on race, but segregation and discrimination were central features of the agency's policy toward African-Americans, including the CCC in California. For instance, African-Americans

Santa Ana Cone ECW Camp F-153. Enrollees working on water ditch abutments, San Bernardino National Forest. 1933

were integrated into all-Caucasian camps only when there were not enough men to form separate companies. Once assigned, they worked side by side with other enrollees and were well treated, though some ended up in menial jobs such as part of the kitchen crew or similar duties. With the exception of Native American, who were under the direct control of the Indian Service, scholars have not fully researched the treatment of other minorities in California's CCC camps (Cole 1999: 12, 17-20).

Understandably, running the CCC and other New Deal agency projects took up much of the time and created the headlines of the Show administration in the New Deal years. Wallace I. Hutchinson, as head of public relations, saw to it that each issue of the *California Ranger Region 5* publicized Region 5's CCC developments and accomplishments in a section called "Chips from the E.C.W. Camps." In addition to Hutchinson's CCC-related public relations highlights, the Forest Service also promoted CCC activities with a thousand-foot film reel on the subject, and the Fox Film Corporation of Hollywood planned a picture on CCC activities as well (*California Ranger Region Five* 1933: Nos. 26, 31-37, 41). More will be said about the role the CCC played in Region 5 when discussing the various resource programs associated with timber, grazing, water, recreation and wildlife.

Region 5 and Other New Deal Agencies

In addition to the CCC, Show and Region 5 took advantage of other New Deal emergency relief and recovery programs. Immediate work relief was provided by the Civil Works Administration, and the Forest Service helped many get through the winter of 1933-1934 using this program. Unlike the CCC, no camps were established under the CWA program, and all work was arranged so that men could leave at the end of the day for their homes. Region 5 put in several extensive proposals, and they got much of what they asked for from the CWA. By the end of 1933, Region 5 was assigned 8,500 men. Research employed 500 of them on erosion control on the Devils Canyon and San Dimas projects, and the other 8,000 were used on miscellaneous forest projects. Though the CWA proved immensely popular, it also was very costly and the program was widely damned as a "boondoggling and leaf raking" one. Roosevelt discontinued this "temporary" agency in the spring of 1934. FERA thereafter took up the relief burden and continued the CWA's unfinished work projects (*California Ranger Region Five* 1933: No. 1; Leuchtenburg 1963: 122-123; Show 1963: 78), but apparently, Region 5 made little use of FERA funding.

In addition to working with the CWA, Region 5 worked with NIRA. After passage of the National Industrial Recovery Act in June 1933, large NIRA allotments were made available to the regions for useful projects. Region 5 immediately sought NIRA funding for various major road-building projects. This was direct money to the Forest Service without the Army holding any of the purse strings (Show n.d.: 135). In August 1933, Region 5 Engineer Bruce B. Burnett's *Forest Highways in California with Connecting State and County Highways* laid out the history of Forest Service highway funding in California, as well as the state of affairs of forest highways and the state highway system and the estimated cost of completion of Class 1-3 highways. Burnett anticipated that it would cost approximately $28.5 million to complete the 2,291 miles of the California forest highway system. To speed construction, Region 5 naturally sought NIRA funding from Agriculture

Secretary Wallace. In answer to Region 5's request, in September 1933, Secretary Wallace approved an additional $2.3 million of NIRA funding for thirteen California forest highway projects to improve the California National Forest Highway system (*California Ranger Region Five* 1933: No. 41).

For the remainder of the New Deal, Region 5 continued to seek NIRA funding. Show, R. L. Deering, and Jay Price wrote proposals for timber, range and recreation projects, and obtained most of what they sought. One project where NIRA funding played a key role and received a great deal of publicity was the Ponderosa Way firebreak, which ran almost 700 miles along the Sierra foothills. There were even plans to extend it an additional 225 miles around the Sacramento Valley and south along the broken timberline on the eastern slopes of the coastal ranges. Show first conceived the Ponderosa Way firebreak in 1929, and may have been inspired by the firebreak that was built in the winter of 1914-1915 by the Sierra and Sequoia National Forests (Cermak n.d.: 124, 366). But at the time Show envisioned

CCC crew building firebreak on the San Bernardino National Forest. 1933

Source: Forest History Society

it, it appeared too grandiose of an idea and was not taken seriously. The idea was renewed in S. B. Show and Edward I. Kotok's study, *The Determination of Hour Control for Adequate Fire Protection* (1930). In this study, Show and Kotok recognized the need for the establishment of an integrated statewide fire detection system, and the Ponderosa Way was considered an essential part of that system. Then, in July 1933, Show revitalized the idea once again, and successfully sold it as a CCC-labor and a NIRA- funded project (Clar 1969: 244-245, 252-258).

In a letter to the nine forest supervisors in charge of national forests fronting upon the Sacramento and San Joaquin Valleys, the regional forester described the project and how the CCC could be used to construct it, and surveying on the "world's longest firebreak" began immediately. Local views in regard to location were obtained, but Region 5 was not governed by them, and most specifications for firebreaks were based on fire behavior. The western extension never materialized, but Forest Service CCC labor constructed close to 70 percent of the firebreak, which was used not only as a firebreak but also included a continuous truck trail on or near the line for access to remote areas (Show n.d.: 136; Pendergrass 1985: 114-115; Cermak n.d.: 366-367). Because the Ponderosa Way crossed many deep and rough canyons, a major job for Region 5 engineers involved the design, location and construction of many bridges. Though Forest Service and state CCC enrollees who participated in the Ponderosa Way project had become fairly skilled by this date at road and trail building, they did not have the skills necessary to construct the required steel bridgework. Consequently, Forest Service personnel were assigned to these bridge projects and liberal amounts of NIRA funding made possible thirty-six NIRA camps for them (Show 1963: 78-83; Show n.d.: 136).

Bridge across Sacramento River built by CCC under regional office supervision, Shasta National Forest. 1934

Rise of the California Forestry New Deal, 1933-1938

By August 1933, thanks to the relief and recovery measures of the CCC, CWA and the NIRA, things were looking up for the Forest Service. Another indication that times were growing better under the New Deal was that national forest receipts for the fiscal year ending June 30, 1933, had increased. Timber sales, waterpower and special uses were still in the red, but they were offset by an increase in grazing fees (*California Ranger Region Five* 1933: No. 37). Then came the shocking news on the morning of October 23, 1933, that Chief Forester Stuart had died. Stuart's death came just before an important meeting with the agriculture secretary on timber conservation provisions of the NIRA code for the lumber industry and in the midst of a transition period in the nation's forestry policy (*California Ranger Region Five* 1933: No. 48).

Within a week of Stuart's untimely death, President Roosevelt approved the appointment of Ferdinand A. Silcox to succeed the deceased Stuart. Chief Silcox was a contemporary of Frederick "Fritz" Erskine Olmstead, Coert DuBois and others who came west when the reserves were transferred from the General Land Office (GLO) to the Forest Service. But after serving as District 1 Forester under Chief Forester William B. Greeley and in the 20th Engineer group during World War I, Silcox transferred to the Labor Department after the war to act as an arbitrator in shipyard labor problems. There, he met Rexford Guy Tugwell, a Columbia University economics professor, who became a leading spokesman and prophet of the New Deal (*California Ranger Region Five* 1933: No. 49; Show n.d.: 138).

Inevitably questions arose regarding the appointment of Chief Silcox, especially because he had been out of the service for so many years. Some regional foresters, such as Show, criticized the decision, particularly the lack of consultation with Forest Service leaders prior to it. They judged that Silcox was unqualified because he was far out of date on the state of affairs in the field (Show n.d.: 137-138). But Silcox's appointment was secured thanks in part to political lobbying efforts by Assistant Agriculture Secretary Rexford Guy Tugwell (*California Ranger Region Five* 1933: No. 1). Tugwell felt that Forest Service leadership had "drifted from the liberal and articulate faith of the founding fathers toward the kinds of conservative outlooks and practices which the New Deal was out to destroy or curb" (Show n.d.: 138). Fortunately, in the early period of Silcox's administration, former California District 5 Forester Paul Redington transferred to the Forest Service from his position as chief of the Biological Survey and helped Silcox in the initial years of his administration (*California Ranger Region Five* 1934: No. 15; Show n.d.: 140). To educate himself on the current state of affairs, Silcox immediately made quick trips to various regions and attended the Regional Forester meetings. In August 1934, he toured the California region with

Forest Supervisor J.E. Elliott and family at Rae Lake, Sequoia National Forest. 1937

Regional Forester Show. At that time, Silcox met most of the supervisors, who pronounced him a "regular fellow and a real leader" (*California Ranger Region Five* 1934: No. 38; Show n.d.: 140). The appointment of Ferdinand Silcox as chief forester came at a point in time in Forest Service history – a time of new life and interest for a chief forester, and a time of cruel irony in the wake of Stuart's death. For many years, the service under Chief Forester Stuart had been starved of funding and its activities restricted. But now the Forest Service had come into a period of sympathetic public interest and expansion, with the CCC working in the woods to build the long-overdue administrative improvements from public works funds (*California Ranger Region Five* 1933: No. 1).

By 1935, Chief Silcox, in what could be considered a power struggle with his regional foresters, reorganized the head office staff. At that time, without consulting with his regional foresters, Silcox created a staff of assistant chiefs for operation, state and private forests, lands, research and an at-large assistant chief. It was Show's opinion that this general staff reorganization had the potential to block off access by regional foresters to the chief. He also feared that they would accumulate power as a group and become the *de facto* seat of decision and policy making in the Forest Service, with the chief relegated to ratifying decisions. Under Silcox, the Forest Service also moved toward greater departmentalization. Regional Forester Show believed that "these steps weakened Forest Service autonomy; imposed time-consuming, basically unproductive chores of dealing with new and not always well-informed or well-disposed new agencies and people; [and] built up [the] volume and complexity of formalized and largely meaningless paper work. Government by clerks was becoming an unpleasant reality." Naturally, this "creeping centralization and formalization" did not sit well with regional foresters like Show. Men like Show had little understanding or sympathy for being regarded as "executors of ideas developed at a higher level" (Show n.d.: 155-156, 158-160).

In the meantime, Region 5 underwent several significant administrative events of its own starting in 1933. First, the regional headquarters' location changed frequently during the New Deal period. In December 1933, Region 5 moved from the Ferry Building, "where the salt air and smell of tar was swell and everybody had about as much privacy as a carload of spring lambs" to the Wells Fargo Building at the corner of Second and Mission Streets, a move most likely prompted by an increase in staff due to the administration of the various New Deal programs. By March 1935, thanks to an internal spasm of growth, the regional office had more than 170 people, who by that date were scattered from the second to the seventh floors of the Wells Fargo Building. In October 1934, the *California Ranger* expressed the hope that the region would eventually be moved into the new Federal Office Building, an addition planned for the Civic Center of San Francisco. That desire was not fulfilled. Nonetheless, by April 1935, the "brother-can-you-spare-a-dime" aura outside the Wells Fargo Building had made many regional office workers very self-conscious going in and out of the building. Subsequently, the staff moved and temporarily occupied part of the Wentworth-Smith Building at 45 Second Street. Then, in June 1935, the regional office found a more permanent home in the Phelan Building located at 760 Market

Phelan Building in San Francisco–sixth home of Region 5. 1927

Source: Forest History Society

Street. The California Ranger's editorialized that "now that we're here it's not so bad, but two moves in one New Deal is like a game of checkers." The regional office remained in the Phelan Building for a time, but in the spring of 1941, the regional office was planning on moving the next year into the soon-to-be-constructed eighteen-story Appraisers Stores and Immigration Station between Washington and Jackson Streets (*California Ranger Region Five* 1934: No. 46; 1935: Nos. 17, 31, 32; 1941: No. 17).

Of course, administrative improvements occurred on the field level as well, thanks to the CCC and other New Deal programs such as the WPA. New supervisors' headquarters and ranger stations, and other physical improvements such as barracks, warehouses, supply depots and other structures were built in a massive Region 5 building program. Lands were purchased for needed administrative sites, and the liberalization of cost limits resulted in a new set of buildings being constructed at some ranger stations. Furthermore, between 1933 and 1939, five of the California national forests received completely new supervisors' headquarters. In some towns, the land was donated to the government for these headquarters, and WPA labor was used in their construction (Ayres 1941: 31).

Moreover, there was a revolution in Forest Service architecture. In June 1933, the California region appointed landscape and park engineers George Gibbs and L. Glenn Hall to develop administrative site plans, such as ranger stations, and to assist in the development of campgrounds. Gibbs, who was from Palo Verde, California, had worked closely with the noted landscape architect Frederick L. Olmsted. Among his assignments, Gibbs was hired to prepare general plans for large recreation areas such as the one at Kings Canyon. Glenn Hall, who was from Los Angeles, had twenty years of experience in landscape architecture, city planning and civil engineering throughout the country. He became a specialist in campground planning for Region 5 and produced standardized landscape plans in these early phases of landscape architecture, which the Forest Service put out as a handbook (*California Ranger Region Five* 1933: No. 29; Tweed 1980: 17; Lux 2005).

At the same time that Gibbs and Hall were working on plans for campgrounds, the region hired David Muir as a landscape engineer in its engineering division to work on roadside treatments in order to beautify Region 5's highways. The California national forest highway system was expanding rapidly, thanks to CCC labor. To illustrate how Region 5's transportation had evolved over time, Louis Barrett once remarked that in the first half of his years of service (1902 to 1918), he traveled 24,270 miles by horse, 7,900 miles on foot and only 6,530 miles by automobile. In the last half of his service (1919 to 1936), Barrett estimated that he traveled only seventy miles by horse, 1,900 miles on foot, and covered 51,355 miles by automobile (*California Ranger Region Five* 1937: No. 3).

Ford and burro transportation on the Inyo National Forest. 1919

In 1937, Region 5's transportation system continued to grow in importance. That year, there were 2.5 million passenger cars registered in California, an increase of almost 332 percent since the passage of the Federal Highway Act of 1921. Many of these motorists came to California's national forests for recreation (*California Ranger Region Five* 1939: No. 19). Nevertheless, the initial forest highway system planning for Class I and II highways developed in the early 1920s was now totally out of date. Furthermore, the Class III highway system in Region 5 was simply an aggregation of projects thought to be of community interest and were distributed such that each national forest and most national forest counties had at least one. Finally, a very large part of California national forest mileage consisted of simple, single-lane dirt roads designed to serve fire control, and only incidentally designed to serve other resources such as timber and recreational use. When such unimproved dirt roads provided access to attractive areas or viable stands of timber, the resulting heavy use meant that the roads became hopelessly overloaded. Prior to New Deal planning, the only cohesive forest highway concepts in Region 5 were those designed specifically to keep the traveler in attractive mountain country for recreational enjoyment, such as the Sierra Way or the Angeles Highway into the San Gabriels. This situation changed in 1937, when Bruce B. Burnett was assigned to design and initiate an all-purpose transportation plan for California's national forests. A transportation conference was held in 1938, where traffic maps with overlays of resource maps were presented, including data on timber, mining, recreation, grazing, fish and game, water, agriculture and population. This data brought out the point that traffic plans and patterns within each forest should take into account the development of all of these resources. Thereafter, the basis of all-purpose transportation planning in Region 5 included resource classifications to determine as far as possible the highest and proper use of roadways without abusing national forest lands (Show n.d.: 194-199; *California Ranger Region Five* 1938: No. 37).

From this point onward, Forest Service highway planning in California became more cohesive and took into account several resource variables. Major projects, predominately sections of federal and state highways, continued to be built in the interests of the whole, including approach and access roads to California's national parks. However, other projects, predominantly Class III highways, were programmed and spread out throughout Region 5, resulting in greater cooperation between the Forest Service and counties, especially the neglected "cow" counties in Northern California. The Forest Highway program of the 1930s became a potent force in maintaining cordial and friendly relations with county citizens and governments. Each mountain county earnestly desired to capture a share of the floating recreational dollar. They, along with loggers who by this date had largely adopted truck logging from woods to mill and then on to market or main shipping lines, recognized their mutual interest with the Forest Service "all purpose" road plan. Both groups were amiable and cooperative with the Forest Service in seeking funding for a rational system of forest roads that included large land exchanges and acquisitions necessary for highway construction (Show n.d.: 194-201, 204-205).

While this highway construction took place and while the regional office moved around downtown San Francisco, various departments tried to operate efficiently. In public relations, Wallace Hutchinson faced an interesting dilemma. An unanticipated consequence of emergency programs such as the CCC, CWA and NIRA was that many loyal friends and supporters of the California Region believed that all forestry questions were now solved and their active support was no longer required. In other words, an apparent result of the CCC wealth was the slackening off in the strength of Region 5 public support.

This diminishment was especially hard felt in Southern California, where groups like the Angeles Protective Association, whose primary purpose was "to provide tough, skilled fire-fighting leaders on the line," had drifted into a "sort of social club." Many former Region 5 supporters were simply aging as well and did not have the energy and drive any longer (Show 1963: 102-105).

Hutchinson and the region worked hard to abate these feelings. To increase public contacts, Hutchinson encouraged supervisors to engage in "show-me" trips with key congressional, state, county and community leaders. Hutchinson also arranged for a barnstorming tour whereby Forest Service officials toured thirty-five to forty towns from Eureka to Calexico, telling the Forest Service story. This project was supplemented by a Hutchinson letter to 2,000 "key men" telling them about current Forest Service work and inviting them to comment (Show n.d.: 145-146; Show 1963: 106-108). Hutchinson also gave emphasis to the theme of conservation at every turn. The region worked this theme into press and radio programs, education programs in schools, and exhibits at state and county fairs. Region 5 especially supported and participated in "Conservation Week" campaigns in California, which were designed to promote and arouse public interest in fire prevention, tree planting, water and soil conservation and roadside beautification. During these weekly activities, Region 5 distributed more than 100,000 pieces of literature to its forest officers and another 40,000 to schools to get the message out (*California Ranger Region Five* 1935: No. 44; 1936: No. 15; 1937: 21). Region 5 even experimented with motion picture production. Starting in 1934 and for the next six years, Region 5 produced "popular" reels under the direction of Frederick E. Dunham. They had catchy and not-so catchy titles, such as *Winter Sports in the National Forests, Top of the U.S.A.*, and *Lest We Forget Chaparral*. Dunham and Region 5 also completed a series of training pictures for the CCC that included *Mustard Sowing for Erosion Control, Hold That Silt, Bridge Building by the CCC* and *Eyes of the Forest*. In 1940, when production was shut down due to a lack funds, Dunham was busy completing a firefighting picture for the California CCC (*California Ranger Region Five* 1940: No. 33; USDA 1937: 15-16).

Besides combating the decline of public support, Regional Forester Show also had several personnel issues that needed attention. Starting in 1936, all employees in the Civil Service of the United States who had rendered thirty years of service were required to retire. In Region 5, this requirement forced the retirement of many key Region 5 employees. including Chief of

Recreation and Lands Louis A. Barrett and Tahoe National Forest Supervisor R.L.P. Bigelow. Both men had surpassed the thirty-year limit several years before. Upon leaving the service, Barrett remarked that he had done everything from "sweeping out the office around 8 AM to turning out the light about midnight when he was 'all in' and ready to call it a day." Bigelow said that he was leaving the "Forest Service financially with about the same worldly goods" as he entered it. Consequently, several Region 5 Forest Service veterans formed an association (*California Ranger Region Five* 1936: Nos. 20, 24), and Region 5 became ever more aware of its own history. A year later, Region 5 initiated a "History of National Forests" project under the supervision of R. W. Ayres. Three forest histories were to be completed each year, with the entire series to be finished by 1940 (USDA Forest Service 1937: 15). There were other personnel issues, including how best to handle the increased workloads and salary reclassification issues. Collaterally with these issues came a newly-created division of personnel management whose responsibilities grew exponentially through the years (Show n.d.: 150-153).

The professionalization of Region 5's ranger corps continued through the operation of the Feather River School. Although there was a lapse in instruction for two years, in 1935 the school reopened with Assistant Regional Forester Paul Pitchlynn as its headmaster once again. In this ninth year of operation, the Feather River Training School had thirty-two trainees from sixteen national forests, the State Division of Forestry, the National Park Service, the Indian Service and the Los Angeles County Forestry Department. In addition to the regular session of the Feather River Training School,

Training course for Assistant Supervisors, University of California, Berkeley. Back row, (L to R): Ben Hughes, R. O.; P. P. Pitchlynn, Dean of class; DeWitt Nelson, Shasta; E. E. Bachman, Tahoe; J. P. Farley, Trinity; J. S. Everitt, San Bernardino. Front row, (L to R): Guerdon Ellis, Angeles; W. T. Murphy, Santa Barbara. 1934

starting in 1932, an advanced training course for assistant supervisors in the broad phases of national forest administration was also offered in order to develop higher leadership for Region 5's future openings. The first part of this advanced course was held at the University of California at Berkeley, while the second part took place in the field and was held on any one of a number of selected national forests (*California Ranger Region Five* 1935: No. 46; 1936: No. 17; Show n.d.: 149-150).

Multiple-Purpose and Multiple-Use Management of Region 5, 1933-1938

In 1932, the *Copeland Report* presented the term "multiple purpose use." The phrase was used in the section of the report written by California Regional Forester Show and Senior Forest Service Economist W. N. Sparhawk in an article they wrote entitled "Is Forestry Justified?" The article correctly stated, "Multiple-purpose management for production, conservation, and utilization of timber, forage, water, wildlife, and recreational values was first developed and is now generally found on the national forests." The term clearly expressed Show's viewpoint toward the various resources that he managed in Region 5. Thereafter, the expression "multiple purpose use" appeared in Chief Robert Stuart's 1933 annual report and then appeared frequently in other Forest Service reports and communications (Wolf 1990: 24-26).

In 1936, Regional Forester Show formally took his concept of multiple-purpose use and applied it to the management of resources. By this date, a chapter in Region 5 history was closing and another opening. Under Show all the necessary development work for fine administration in Region 5 was accomplished thanks to the achievements of the CCC and other New Deal agency programs. With that behind them, Region 5 was able to better plan for the handling of resources and seek the broader objectives of sustained-yield management, improved grazing administration, recreational development, wildlife management, superior fire protection and erosion and flood control (*California Ranger Region Five* 1933: No. 1; 1936: No. 10). The first formulated example of Show's multiple-purpose management principle came when he proposed a detailed plan for the Kings Canyon area of Sequoia National Forest. In his plan, the regional forester proposed a High Sierra Primitive Area, with the rest of the area open to logging, dams, reservoirs and every other use from summer homes to mining. Show devised this multiple-purpose management plan in response to a NPS proposal to establish a Kings River

Canyon National Park, about which more will be said later (*California Ranger Region Five* 1933: No. 28; Wolf 1990: 27).

By 1937, Region 5 entered into a planning phase and began to develop "multi-land use" plans as well. There were individual timber, grazing, recreation, wildlife and transportation plans, but none evinced an integrated multiple-use approach, possibly with the exception of Burnett's all-purpose transportation plan. Nevertheless, by 1937 the term "multiple use" entered Region 5's vocabulary. Interestingly, C. B. Morse, regional office division of lands and recreation, quoted the following paragraph in the *California Ranger*:

> It is becoming increasingly apparent that we must value our forests not only as a source of our supplies of timber, but also for their many other uses – as food and shelter for our game and fur-bearing animals, as regulators of the water-flow of the streams in which we fish, and as attractions for the tourist and other recreationists who delight in the great outdoors. Our forest areas must be developed and protected from fire in the interests of these "multiple uses." Forest management plans must recognize both the industrial and recreational uses of forest areas. By far-sighted planning we must make these various interests harmonize as much as possible, secure greater returns from our forest areas, and make them the greatest single drawing-card for our rapidly increasing tourist trade (California Ranger Region Five 1937: No. 43).

While this quotation may seem to be descriptive of Region 5, it actually came from the 1936 Report of the Forest Branch, British Columbia, and Morse suggested that the paragraph regarding the "multiple use of forest areas" could likewise be applied to Region 5. From this point onward, Region 5 absorbed the phrase into Region 5 planning language, although no one provided actual specifics describing how "multiple-use" management worked, or how it was applied. For example, in 1940 at a national conference on planning, Regional Forester Show championed the theme of "protection, conservation and wise of use of natural resources." In his presentation, he used various maps to illustrate the "significance of multiple use" and stressed the "use of planning for the use of all forest lands to insure the best benefits to each class of users" (*California Ranger Region Five* 1940: No. 33). Fundamentally, however, even though the terms "multiple-purpose use," "multi-land use," and "multiple-use" were bandied about since 1933, they were internal Region 5 phrases that would not become public nomenclature until well after World War II. Even so, the following paragraphs demonstrate the evolution of multi-land-use management planning during the New Deal for the many resources of Region 5.

In 1933, according to Chief of Forest Management T. D. Woodbury, timber business in California hit an all-time low. Timber sales receipts on Region 5 amounted to just over $250,000 and were less than any year since 1919, when receipts were just over $190,000 (*California Ranger Region Five* 1933: No. 44). Using the records of the Plumas National Forest, the most consistent timber sale forest of Region 5, Woodbury found that less timber was cut on Region 5 in 1933 than any other year since 1914, when

Assistant Regional Forester T.D. Woodbury talking with the woods crew at Big Springs ECW Camp F-48, Lassen National Forest. 1933

the outbreak of World War I caused a temporary depression (*California Ranger Region Five* 1934: No. 17). Some of the heavy producers in the 1920s "Golden Era" of prosperity, notably the Stanislaus and Sierra National Forests, had by 1933 dropped toward the bottom of the list and were producing at the same level as the Angeles and Inyo National Forests. From the standpoint of timber cutting, in 1933 the Lassen, Plumas, Klamath and Shasta were the best producers in Region 5 (*California Ranger Region Five* 1933: No. 10).

At the same time that timber sales reached this nadir, President Roosevelt and the Washington office made it be known to leaders of the lumber industry that they expected them to adopt measures for the prevention of destructive exploitation of forestland if they expected any relief under the National Recovery Administration (NRA). The conservation of natural resources was a declared policy of NRA, and the Roosevelt administration favored a sustained-yield policy to ensure the perpetuation of the basic forest resource for the general public. Furthermore, there was a strong call by the Forest Service for the acquisition of all private forests by the federal government. The recently submitted *Copeland Report* had revealed that 90 percent of the total area of devastated and poorly stocked forest land, and 95 percent of the current devastation, was on privately-owned lands (*California Ranger Region Five* 1933: No. 34).

By February 1934, the lumber industry acquiesced to forestry-minded President Roosevelt's sustained-yield policy, and adopted NRA codes of forestry practice. Under these codes, private timberland owners and operators committed themselves to "conserve the forests and regulate their cutting practices in such a manner as to bring about a system of 'sustained yield' timber production." The haphazard lumber industry, according to the *San Francisco Chronicle*, was based on the speculative exploitation of timber provided by a bountiful nature. Henceforth, the newspaper editorialized, they would "be replaced by a permanent industry based on timber growth according to definite plans" (*California Ranger Region Five* 1934: No. 12). In order to cut timber on a national forest, a lumber company was now required to obtain a sustained-yield certificate from the Lumber Code Authority of the NRA (Article X), pledging themselves to the use of sustained-yield practices before getting any federal timber sale contract (*California Ranger Region Five* 1935: No. 45). The NRA lumber codes were short-lived. On May 27, 1935, the Supreme Court, in a 9-0 decision that became known as the "sick chicken" case, struck down the National Industrial Recovery Act (Leuchtenburg 1963: 145). All the same, starting in 1934, Region 5 diligently practiced an "active" sustained-yield policy and exercised that policy in selective stand improvement and cutting practices, in conservative timber marking, in reforestation efforts and finally through timber land acquisitions.

For more than a quarter of a century, the California Region had mainly confined itself to protecting the national forests from fire and other enemies such as insects and disease, for which the CCC insect and blister rust control programs (1933-1938) provided added information. The region had also prevented devastation indirectly through timber sales based on sound silviculture practices of the day and had improved stands by conducting some thinning and sanitation, which encouraged reproduction. Even so, the treatment of stands was too often left to sale operators and the activities of individual lumbermen. However, with additional manpower available through

New Deal agencies, Region 5 began stand improvement cuttings, which were designed to better the growing conditions for the most valuable species on the best and most accessible sites on cut-over areas in the national forests. The principle object of this work was to increase the production of the most desirable species for market. To accomplish this end, NIRA-funded camps were established on six forests – the Shasta, Lassen, Plumas, Eldorado, Stanislaus and Sierra National Forests (*California Ranger Region Five* 1934: No. 26).

In 1935, the Branch of Forest Management changed its name to Division of Timber Management (1935-1974), and the division continued to advocate a sustained-yield policy to private lumber operators even following the Supreme Court NIRA decision. First, the timber division conducted several logging and milling studies to demonstrate the principle of sustained yield to private owners of timberlands (USDA Forest Service 1936: 5). In line with this policy, T. D. Woodbury and forest supervisors began to "sell" the idea of selective cutting to timber operators on the national forests, starting on the Lassen National Forest with the Fruit Growers Supply Company of Susanville, California (*California Ranger Region Five* 1935: No. 45). They also pointed out to lumber operators the financial and silvicultural advantages of judicious tree selection. In this respect, they carried out economic studies designed to show the practicability of better silviculture procedures such as sustained-yield practices on private land, with the idea of interesting lumber companies to continue with the work on their own (USDA Forest Service 1937: 7).

The timber division also swung toward more conservative timber marking practices at this time, leaving a larger percentage of the stand behind in reserve than when the first timber on Region 5 was sold thirty years before (Ayers 1958: 32). Region 5 was successful in this regard because technology was rapidly changing the logging industry, making it possible to conduct the selective logging practices the Forest Service desired. Dolbeer donkey engines (and other makes) used to yard logs in the woods went out, and "Cats," or Caterpillar Tractor Company tractors, came in. With their Caterpillar treads, timber operators were far less damaging to the forests than their donkey engine logging predecessors. "Cat bosses" supervised the logging, and felled trees were taken out with minimum damage to reproduction and reserve stands. In addition, Cat operations also laid out roads for the tractors, taking advantage of all openings for use as roads or turning places (*California Ranger Region Five* 1936: No. 35; Ayres 1958: 37).

Reforestation also became a prime factor in the New Deal's sustained-yield forestry effort. With CCC labor available, the California Region immediately began large planting programs on each of the forests, and pine nurseries were established to supply this effort. In 1936, Agriculture Secretary Wallace evoked the relationship of forestry to agriculture in a publication titled *Forests and Farms: A Pictorial Presentation of the Social and Economic Services of the National Forest to Agriculture*. In it, Wallace related the "irreduc-

Region 5 publications such as *Forests and Farms* emphasized the link between the Forest Service and the Agriculture Department. 1936

ible inter-relationship between forestry, wild life conservation, recreation and agriculture." These "man-made forests" brought new life to cut-over lands in Northern California, and pine nurseries such as at Susanville held the future for California reforestation efforts (USDA California Region 5 1936).

Timberland acquisitions were the last component of Region 5's sustained-yield program. Region 5's timberland acquisition agenda included plans to acquire various redwood forest tracts from private landholders and to buy up tax delinquent cut-over pine regions as well. Region 5 acquisition of these lands came under the Weeks Act (1911). Show prepared plans to establish several redwood national forests in California as early as 1933. The "Redwood Empire" along the coast of Northern California was foreign land to Region 5's silviculturists, timber salesmen, cruisers and scalers. Few of them knew more about this part of California than the average tourist and traveler who visited the area along Highway 1 (*California Ranger Region Five* 1934: Nos. 47, 52).

There were at this time only a scattering of redwoods within the boundaries of California national forests, and they were chiefly on the Monterey division of the Santa Barbara National Forest (Ayres 1941: 34), which in 1936 was renamed the Los Padres National Forest because the Santa Barbara Forest was often confused with the county of the same name (*California Ranger Region Five* 1936: No. 3). The National Forest Reservation Commission, authorized by the Weeks Act, approved Show's redwood project in April 1934, asking Region 5 to prepare acquisition zone reports prior to the purchase of individual tracts (Ayres 1941: 34). In response to this action, in September 1934, the California legislature supported California Governor Merriam and passed the National Forest Land Acquisition Act. This legislation expressly gave state consent to the acquisition of private lands by the federal government for national forest purposes and empowered the Forest Service to proceed with its plans to establish a new national forest in California's redwood region. By this date, many large redwood property owners also supported the idea. They had given up hope of any economic recovery for the area and just wanted to get out. They saw Forest Service interest as an avenue of financial escape. Even so, many California county officials – in this case in Humboldt, Del Norte and Mendocino counties – opposed the creation of a redwood national forest within their boundaries. Much like the opponents of California's early northern forest reserves and national forests, county officials feared the possible loss of revenue by the removal of these lands from their tax rolls (California Ranger Region Five 1934: No. 43; Show n.d.: 144; Ayres 1941: 34; USDA 1936: 8).

Despite county opposition, in July 1934, Region 5 made field examinations of the redwood country, and several redwood purchase unit proposals got underway. Ultimately, the Forest Service hoped to acquire two redwood national forests of approximately 100,000 acres each under the Weeks Act and the National Forest Land Acquisition Act, which set no limit on the area that might be acquired by the Forest Service. In 1935, J. K. Brandeberry was assigned to study and prepare a management plan for the units, and his report called for two acquisition zones: one in Sonoma and Mendocino counties, of 650,000 acres; and one in Humboldt and Del Norte counties, of 250,000 acres.

However, to their frustration, Region 5's proposal to the National Forest Reservation Commission was delayed by challenges from the Sierra Club and the Save-the-Redwoods League. These groups were suspicious of the Forest Service's motives and expressed doubts that the Forest Service was the best agency to administer these lands. The Save-the-Redwoods League accepted Brandeberry's report, but Sierra Club resistance, along with strong local opposition, continued to block the way for the purchases. Other private land purchases were made elsewhere under the Weeks law, such as on the Tahoe National Forest, where more than 51,000 acres of cut-over pine was purchased from the Hobart Estate, and on the Sequoia National Forest, where 21,500 acres of big trees were purchased from the Sanger Lumber Company. Lamentably, however, by 1941, only 20,000 acres of redwoods were purchased for the redwood acquisition in the Del Norte area north of the Klamath River. After that date, World War II interfered with the land purchase program. The opportunity soon faded into history, but not before the Redwood Experimental Forest was established in 1940 (USDA Forest Service 1936: 8; Show n.d.: 145, 156-157, 186; Ayres 1941: 34; *California Ranger Region Five* 1939: No. 5).

One positive case in point happened in the Lake Tahoe area. Prior to 1915, only about 12 percent of the Lake Tahoe watershed was public land, which is one reason it never became a national park. The entire shoreline was in private ownership, and the forested area back from the shoreline belonged to timber and flume companies, and almost all of the meadows were in private ownership. During the Depression years, the Forest Service purchased the Tahoe Flume & Lumber Company holdings after the forest had been cut over. This action started a process whereby today the watershed is about 80 percent national forest and 7 percent state lands, thereby avoiding much development and giving the public shoreline access where none existed before (Leisz 2005).

Fire control and protection were less controversial than timber management in the New Deal years, and unlike the 1920s, fires did not dominate the administrative or physical landscape. As noted earlier, by 1933 the Ponderosa Way firebreak was underway. The firebreak was one of several that were a part of a California master fire plan developed by former Forest Supervisor G. M. Gowen. Before Gowen transferred to the California Forest and Range Experiment Station, he developed a sound method of selecting lookout stations to form an integrated system that covered the known areas where fires had started in the past (Show n.d.: 136-137). However, the idea was not new. Regional Forester Coert DuBois, who developed a fire plan in 1911 using the Stanislaus National Forest as a model, had designated key mountaintops as permanent lookout stations (Lux 2005). The CCC thereafter provided the available labor force to build the required lookouts, telephone communication system, fire roads and structures called for in the fire detection master plan. There were few problems with lookout construction and fire trail building. On the other hand, the telephone system required a complete overall, which included relocating lines, limiting loads, installing metallic circuit pole lines and replacing ground circuit lines where interference problems existed (Show 1963: 84-93).

Meanwhile, the CCC was built into a first-strike firefighting force. Each camp had an organized fifteen-man fire crew of tough youngsters fully equipped to make the initial attack on any fire and thereafter take care of any immediate follow-ups. During the first period of the CCC, the fire crews provided a potent

A CCC fire suppression crew on the California National Forest. 1933

Source: Forest History Society

suppression net. But as the numbers of CCC camps on California's national forests dropped from 167 in 1933, to 58 in 1935, and then to 37 in 1938, the CCC became less and less a major factor in the first attack and reinforcement for ongoing fires. To some degree, such losses in personnel were offset by increased truck trail mileage, which resulted in better and easier placement of first attack forces to fire areas. By May 1936, Region 5 became so confident in its firefighting abilities that the office of fire control established in 1930 went out of existence as a separate division and was absorbed by the Division of Operation as the Forest fire Control Section (Ibid.).

As the New Deal began, Region 5 had just faced the severe 1932 fire season. During the season, the Matilija Fire, an isolated event, burned

219,000 acres on the Santa Barbara National Forest, and the Eel River Fire consumed 24,000 acres on the Mendocino National Forest. These two fires accounted for nearly 90 percent of the acreage burned that year on California's national forests. For the next six fire seasons, from 1933 to 1940, only 277,000 acres burned on Region 5 timberlands (Cermak n.d.: 343-344. 348-357),

Metal lookout towers replaced many early wood platform lookouts, such as this one on the Mendocino National Forest.

although there were fires notable for their size and destructiveness. They included the Nelson Point Fire on the Plumas National Forest (1934), Red Cap Fire on the Klamath National Forest (1938), Bear Wallow Fire on the Trinity National Forest (1938), Deer Creek Fire on the Shasta National Forest (1939), San Joaquin Fire on the Sierra National Forest (1939), Log Springs Fire on the Mendocino National Forest (1940) and the Keenbrook Fire on the

Malibu Fire, near Angeles National Forest. 1935

San Bernardino National Forest (1940) (Show 1963: 93-99; *California Ranger Region Five* 1936: No. 23). Nevertheless, by May 1941, with the growing threat of World War II, the build-up of the defense industry in California and the waning of the CCC and the New Deal, the firefighting forces of Region 5 changed back to the available force that existed prior to the inauguration of the nationwide CCC program (*California Ranger Region Five* 1941: No. 23).

CCC crew fighting a fire near Mammoth on the Inyo National Forest. 1931

During the same period of time, technology marched on, and recent innovations were applied to the Region 5 fire protection and control program. In the period 1933 to 1940, a new air war was conducted on fire in Region 5. During this time, airplanes continued to be used as an aid in forest fire suppression. They were used for scouting large fires or general patrolling, and Region 5 contracts usually required one plane for Southern California and two for the northern part of the state (Cermak n.d.: 410). Besides using aircraft to scout for fires, Region 5 had experimented with the use of air-dropped fire retardants as early as the 1920s. In the fall of 1925, L. W. Hess of the district office first experimented with the use of fire "bombs" on a landing field with varying results – and ultimately concluded that the available substances tested did not have the fire suppression qualities they needed. A decade later, Region 5 revived and reinvestigated the idea. In the winter of 1935 and through the fall of 1936, aerial experiments took place – this time at the Oakland airport, where a pilot dropped and splattered water, flour and cement in bombs or containers onto the runway (Ayres 1942: 21). From the test results, Region 5 decided to conduct a more exhaustive study, and in the summer of 1937, the investigation continued of fighting forest fires with airplanes. That summer, even more types of bombs and bombing materials were dropped in aerial tests conducted in the San Francisco Bay region (Show 1939: 6). Encouraged by the results, in 1938, the Forest Service acquired its first official airplane – painted Forest Service green – and a pilot, Harold C. King. Both were detailed to the Aerial Fire Control Project on the Shasta National Forest. The chemical section of the project tested fire retardants in the laboratory and the field, while the aerial section developed equipment and devices necessary to release retardants from airplanes. The latter group also worked out sighting devices to secure accurate placement of retardants, flight and approach techniques, design of containers suitable to hold fire

Fire patrol plane at San Diego. Longacre, Wisler, DeLapp & Pilot Robinson. Cleveland National Forest. 1934

retardant chemicals, and the effectiveness of aerial treatment ongoing fires (*California Ranger Region Five* 1938: No. 43). In 1939, this project, which for the most part was largely a failure, was moved to the state of Washington (Cermak n.d.: 415-417). However, in a 1940 patriotic-toned publication for the *California Conservationist*, Regional Forester Show extolled the role the airplane played in Uncle Sam's forest fire fighting forces in California (Show 1940a: 1-4). Region 5, by the 1940 fire season, had leased five airplanes and stationed them around the region. They were used to transport key firefighting personnel when necessary and to scout fires, and at least one cargo plane was leased specifically to drop thousands of pounds water and food supplies to fire fighters by parachute over fires in inaccessible areas. Compared to ground transportation, the airplane could clearly get supplies and personnel to fires more quickly (*California Ranger Region Five* 1939: Nos. 24, 48).

Fire research also made progress during the New Deal years. Basic and long-term watershed and fire research was handled by the San Dimas

Experimental Forest, which was established 1932 in Big Dalton and San Dimas canyons in Los Angeles County. A year later, a CCC camp was established there, and its workers built most of the experimental forest facilities (Cermak n.d.: 380-381). In the north, the CCC helped newly-hired research technicians at the California Forest and Range Experiment Station to move ahead with important studies – integrated detection systems, communications, and firebreak systems, fire road systems and initial attack strategies – that had been at a standstill because of a lack of funds. At the station, the CCC contributed to and accelerated research in studies regarding fire line production rates, fire behavior and fire planning. In these studies, Region 5 worked closely with the station in fire management studies, a collaboration that eventually led to several landmark publications and helped cement the working relationship between the station and Region 5. The station's first publication, *Region 5: Fire Control Handbook* (1937) provided a working tool for fire control people based on fire research results and actual practical experience. The second publication, *A Planning Basis for Adequate Fire Control on the Southern California National Forests* (1938), emphasized, "person-caused fires in flash-fuels and brush zones presented the dominant Southern California fire problem." Carelessness by smokers was blamed for many of these fires, and Show and Kotok proved it in this publication (Wilson and Davis 1988: 5; Cermak n.d.: 381--382). Thereafter, Region 5 publicized the results of this study in many public relations pieces, including one article entitled "Put It Out." This article clearly admonished smokers by blaming them for 43 percent of all human-caused forest and brush fires in the state (USDA California Region 5 1940: 1) and may have led to the Smokey Bear "anti-smoking" campaigns of the future.

Besides fire control and protection, Region 5 range management also provides an interesting story. Since the founding of the Forest Service in 1905, forage resources and grazing on California's national forests had declined significantly. Several factors caused this decline and a loss of forage capacity on California's eighteen national forests. First, there was a steady decrease in actual forest acreage due to eliminations and the patenting of claims within the boundaries of many national forests. Second, recreational use of mountain areas and closing of tracts for watershed protection, such as for the Central Valley Water Project, impinged on grazing acreage totals. And third, fires, such as the recent Matilija Fire on the Santa Barbara National Forest, also contributed to the loss of grazing areas. By 1932, these factors, combined with the subnormal rainfall and poor economic conditions, resulted in California's national forests supporting only 152,000 head of cattle as compared to 178,000 in 1909, a decline of 15 percent in twenty-three years. To compound matters, the issue of computing grazing fees for livestock had not been resolved, and there were also serious overgrazing problems on some national forests (*California Ranger Region Five* 1932: No. 15; USDA Forest Service 1932-1941: 1933).

Relief and recovery came to the livestock industry with the New Deal. First, in 1932-1933, a new method of computing grazing fees was devised that suited both the Forest Service and livestock owners. The combination of drought and the Depression brought a recommendation that the national forest grazing fees schedule be tied to the fluctuations in livestock prices. As one historian put it, this was the "most pronounced social-welfare gesture of the New Deal toward range users in the forests." The new system ended charging for ranges on a competitive basis, especially in rough economic times, and was a major departure from the previous view that producers who could not pay should not remain in the livestock industry (Rowley 1985: 150-151).

With the fee schedule settled, the important Region 5 issue during the remainder of the New Deal years was overgrazing. In 1933, each of Northern California's national forests had substantial areas of overgrazed lands, including the Klamath (15,000 acres of government land and 14,000 of private lands), Shasta (34,000 acres), Modoc (125,000 acres), Lassen (9,000 acres) and Trinity (53,000 acres) National Forests. Serious overgrazing was also a problem in the High Sierra on the Plumas (14,000 acres), Tahoe (12,500 acres government land and 73,000 acre of private lands) and Eldorado (7,000 acres) National Forests. In contrast, no overgrazing problem on Southern California national forest ranges existed except for the Santa Barbara National Forest (10,500 acres) and a few watering places on the Angeles, San Bernardino and Cleveland National Forests (USDA Forest Service 1932-1941: 1933).

Overgrazing was in part a problem of mismanagement by Region 5 that could be corrected, in part related to not having control over large tracts of private lands intermingled with national forest lands such as on the Klamath and Tahoe National Forests and in part due to demand for range that was greater than availability, a condition especially felt on the Lassen and Modoc National Forests. In 1933, Region 5 had reduced forage capacity due to fire protection on the Modoc National Forest and estimated that further reductions of between 10 and 25 percent would be needed over the next few years to bring an end to overgrazing. Up until this date, there had been comparatively few limits placed on ranges in Region 5, and forced reductions had been only occasionally applied in the past. Naturally, these forced reductions on the Modoc National Forest stirred up a controversy between stockmen and the Forest Service, which ended up in the so-called "Modoc Grazing Wars" a decade or more later (USDA Forest Service 1932-1941: 1933; Rowley 1985: 198; Swift 1936: 8).

Livestock reductions were the only solution to the overgrazing problem, and in 1934, Region 5 reported that it was "evident that the degree of stocking to prevail in this region for the next ten year period must be conservative if the damage caused by overgrazing is to be remedied" (USDA Forest Service 1932-1941: 1934). Of course, some conservation groups, like the Izaak Walton League, sought the complete abolishment of sheep and cattle from California's national forests. On this point, the Los Angeles chapter of the League petitioned Region 5 to immediately and absolutely abolish all sheep and cattle grazing permits in California's national forests. The League's petition stated that sheep grazing was a costly mistake and a man-made hazard to forests and watersheds, that cattle grazing had been too heavy, the privilege abused, and that both types of livestock hindered wildlife and were in direct conflict with recreational use. In their words, "forest travelers by trail and saddle find meadows grazed absolutely bare, cow-slips in the most desirable camping spots, no feed for the saddle stock which suffer the most, and public pastures too few and far between; all bring great mental anguish and quandary as to whom do our National Forests belong" (*California Ranger Region Five* 1934: No. 42).

Of course, the Forest Service was not about to completely eliminate grazing – one of the many resources it was mandated to protect – on the national forests. But with the introduction of CCC labor and NIRA funding, various range improvements were made that did help grazing management immensely and eased some of the problems and conditions found objectionable by the Izaak Walton League and other conservationists. Unlike the areas of timber and fire management which had multiple projects during the very first CCC period, Region 5 range improvement projects did not really begin until the second CCC period (USDA Forest Service 1932-1941: 1933). Once begun, these projects included many kinds of improvements, such as construction of drift fences, stock driveways, erosion control dams, stock bridges and trails, holding corrals, the development of springs, windmills, wells, reservoirs, and water tanks, the reseeding of range, the eradication of poisonous plants, and rodent control. But because of the large program of improvement work carried out under NIRA and CCC funding in the early New Deal years, range administrative personnel had little time for direct grazing supervision, or to develop range management plans. Those tasks were entrusted to local ranchers and operators (USDA Forest Service 1932-1941: 1934, 1935).

That situation changed in September 1935, when former Modoc National Forest Supervisor F. P. Cronemiller was appointed to the position of assistant regional forester in charge of the Division of Wildlife and Range Management (*California Ranger Region Five* 1935: No. 44). Under Cronemiller, grazing management took a turn for the better. One solution to the overgrazing problem that Cronemiller explored was the acquisition of private land located within forest boundaries through the Agricultural Adjustment Administration. In 1934, Region 5 estimated that there were approximately one million aces of private land within the forests of Region 5 that were alienated because of their grazing value, and proposed that the AAA be induced to purchase the land. In 1935, the California Region reiterated its proposal and submitted to the AAA an acquisition project of approximately 90,000 acres of strictly grazing land for the Lassen National Forest (USDA Forest Service 1932-1941: 1934, 1935). However, special AAA activities on California's national forests seemed to be limited. A few grazing men rode out to look at submarginal lands with a few muddy water holes to determine their potential grazing values, and in 1937, a cooperative AAA range conservation program was funded to study the carrying capacity on the state's ranges, which was part of a larger nationwide western range inventory survey. Region 5 contributed to the project by assembling and making available data already on hand and conducting additional range surveys through the AAA range program. But, in the end, the AAA purchase program failed to materialize (*California Ranger Region Five* 1936: No. 51; 1937: No. 2; USDA California Region 1937: 10-11). Another solution was greater cooperation and understanding of the cattlemen's viewpoint. For example, when Modoc cattlemen continued to

Assistant Regional Forester Fred P. Cronemiller. 1937

protest protection reductions and attributed unsatisfactory range conditions in some areas to winter over-browsing by deer, Cronemiller responded to their concerns and pushed for greater coordination of grazing with recreation and wildlife (USDA Forest Service 1932-1941: 1935, 1936, 1937; California Ranger Region Five 1936: No. 34). Though these solutions ultimately failed, the overgrazing conditions on the California's national forests eventually were reversed. This happened as a result of abundant rainfall over the next few years, which favored plant growth on forest ranges, and of the accruing effects from the yearly range improvements made by the CCC on each forest. Over the next few years, Region 5 reduced overgrazing conditions to what it called a few "sore spots," and by 1938, Show's annual grazing report made no mention of overgrazing conditions whatsoever (USDA Forest Service 1932-1941: 1936, 1937, 1938).

Save for the above work, Regional Forester Show and his forest supervisors demonstrated very little interest in grazing matters, particularly in grazing research. Acting on this instinct, in 1938, Show turned all range research over to the various experiment stations. Prior to this date, the regional office had cooperated with the research branch in utilization studies, which aimed to develop simple, clear-cut, workable standards for range utilization, but by 1938, Show decided to halt further regional grazing research projects. Show's action was based on Lloyd W. Swift and A. Fausett's *A Report on the Grazing Administrative Studies in the California Region*. The Swift and Fausett report examined the history of Region 5 grazing administrative studies. Finding them wanting in every category, they concluded that continued range research on the regional level was futile. According to Swift and Fausett, grazing administrative studies failed because Region 5 forest supervisors clearly lacked any interest in grazing studies, they were not very research-minded and they simply did not have the time to properly plan and administer these studies. Their report stated that the stations could provide the required attention, had adequate time and finances, and competent personnel to assure proper work, and Show concurred (USDA California Region 1937: 10-11; Swift and Fausett 1938: various).

To a lesser degree than range management, watershed management caused concerns during the New Deal period. By the 1920s, the Forest Service and Region 5 keenly understood the multiple-use aspects of nature's water as a resource for irrigation, hydroelectric power and recreation, and grasped how to store water and save soil on watersheds, too. California's great water projects, such as the Owens River aqueduct, the Colorado River diversion and the Central Valley Project were well underway. In the rest of the nation, severe flooding had occurred on the Ohio, Mississippi and Missouri rivers, and these disasters ultimately led to the passage of the Omnibus, or Upstream, Flood Control Act of 1936, which recognized that flood control is a national rather than a local problem. In California, national forests for the most part helped to prevent flooding in the state, and CCC erosion and flood control efforts contributed to an overall good record for Region 5. Even so, floods did occur in California that made national headlines during the decade of the 1930s. For example, the 1933 New Year's Eve La Crescenta-Montrose Flood outside Los Angeles took the lives of thirty-four people and wreaked property damages that included the destruction of 200 homes. The catastrophe was blamed on an intense rainstorm aggravated by the destruction of forested watersheds by the disastrous fire season of 1932 (USDA Forest Service 1940d: 18). In December 1937, there was the Downieville Flood, whose high waters following some of the heaviest downpours on record in Northern California caused more than $1 million of damage (*California Ranger Region Five* 1937: No. 4). Then, in the spring of 1938, flood waters partially or wholly inundated nearly all of Southern California's valleys and towns. The Southern California floods were set off by five days of steady downpour, and the resulting floodwaters took the lives of 210 men, women and children, injured hundreds, damaged an area of 30,000 square miles and brought an estimated property loss of $50-60 million. The flood caused damage and losses to Southern California's Angeles, Los Padres, San Bernardino and Cleveland National Forests totaling $1.6 million. An estimated 750 summer homes were destroyed at that time, mostly on the Angeles and San Bernardino forests; 90 percent of the public camps on the Angeles were ruined beyond repair; and hundreds of miles of roads throughout all four southern forests were wiped out (*California Ranger Region Five* 1938: No. 17). Then there was the Sacramento Flood of 1940, which covered approximately 200,000 acres of farmland and caused a loss of $15 million (Fox 1948: 35-40).

These periodic floods indicated that the State of California, its counties and Region 5 still had a great deal of work to do in watershed management. To combat the destructive potential power of future flooding in all parts of California, including along the coastal area from Santa Barbara to San Diego and the area between Los Angeles and San Bernardino, which suffered from frequent flood damage, Region 5 conducted a number of watershed surveys, watershed restoration of burned areas, erosion control projects and other improvements under the Omnibus Flood Control Act. One such project was the Los Angeles River Flood Control Project. Under this joint project with the Army Corps of Engineers, Soil Conservation Service (SCS) and Los Angeles County, Region 5 used CCC and WPA labor to intensify its forest fire control program, stabilize slide areas, revegetate denuded lands and construct channel barriers in the Arroyo Seco drainage on the Angeles National Forest for better watershed management and upstream flood control (*California Ranger Region Five* 1940: No. 46).

Not surprisingly, mineral management within Region 5 also became an important resource issue during the New Deal, when many of California's forests were inundated with new mining claims. By the early 1930s, almost everyone in the United States was in the economic grip of the Great Depression, and mine owners and miners alike enjoyed very little relief from the Hoover administration. When large mining enterprises collapsed and machinery ground to halt, armies of unemployed miners sought work elsewhere. Consequently, several thousand of these penniless men migrated to California, to try to eke out a living from reworking, again, California's stream and bench gravels (Friedhoff 1944: 33).

Though the Hoover administration did not help them, the New Deal did. In time, Roosevelt put Hoover's Reconstruction Finance Corporation (RFC) to work and the RFC helped get the gold and silver industry back on its feet once again by advancing money to California's banks and financial institutions, which in turn loaned it to mining companies in order to finance any number of projects. Two additional actions taken by Roosevelt and Congress also helped reinvigorate California's mining industry. First, in 1933, Roosevelt decided to raise the official price of gold to $32 per ounce, thereby making gold mining of marginal claims a worthwhile pursuit. And, second, in 1934, Congress passed the Silver Purchase Act, which nearly tripled the price of silver at home and stimulated the metals mining industry nationwide. These actions caused a new mining boom that enticed many unemployed miners to stake out new claims and to rework old ones. Small-scale operations sprang up as miners began to rejuvenate and update former mining and milling operations and, in some cases, even begin new ones. Lode mines produced once again, along with other mining operations such as dredges, hydraulic mining, drift mining, small scale placer mining and "sniping," a term used to describe small-scale migratory mining operations (Godfrey 2003: 61-62).

Because the principal mineral belts of California were located outside the boundaries of the national forests, as of 1932 there were few viable working lode mines or placer mines in Region 5, and even fewer individual prospectors. However, by 1937, mining figures jumped dramatically for the region because production had become important on the Plumas, Inyo and Klamath National Forests, thanks in part to the passage of the Gold Reserve Act (1934). These three national forests accounted for 45 percent of the national forest gold lode production. If the Tahoe National Forest were included among them, the four forests accounted for 90 percent of the total placer gold production. On the other hand, the four southern national forests – the Angeles, San Bernardino, Los Padres and Cleveland – contributed less that 1 percent of national forest gold production, yet contained more than 15 percent of the located and patented mineral acreage on California's national forests. For example, the Angeles National Forest, from which the total gold production was about $300 a year, became one of the leading mining claim forests of California (Friedhoff 1944: 22-23). On the Angeles, the ranks of the unemployed employed methods similar to those used by earlier Californians to search for color, such as panning and sluicing in Texas, Bouquet, San Francisquito and San Gabriel canyons. So many people worked claims in the latter canyon that shantytown communities, so-called "Hoovervilles," grew up along the canyon walls – until the Great Flood of 1938 wiped them out. Though panners found less than $2 per day, far below average daily wage, they thought it was better than having no income at all (Angeles National Forest Grapevine 1990: 4).

Even so, with the economy failing, a new kind of gold rush was taking place on Southern California forests such as the Angeles, one not necessarily after mineral wealth. Locators, recreational miners and "mining" homesteaders on the Angeles and other Southern California national forests accounted for the discrepancy between the lack of actual gold production and the increase in mining claims. This discrepancy pointed out major problems with the functioning of the mining laws in the national forests of California (Friedhoff 1944: 19).

First, for every old-time prospector, there were probably a hundred individuals who were "locators," or people who filed a claim on any area of land that appealed to them as having real estate value. At least 90 percent of the locations each year were worthless, with not even a potential value for minerals. Nonetheless, it cost almost nothing to locate a mining claim, and after two years, they could simply abandon the claim, sell their cabin, and move on to another location. Additionally, with improvements, many placer claims along streams were used as yearlong residence sites. Some were even developed into costly mountain estates with as much as $150,000 of expenditures for buildings alone. The "locator" and "sniper" influx during Great Depression on California's national forests was reminiscent of a similar problem in the 1920s, when people inspired by the outdoors movement and recreationists seeking summer homes flooded into California's national forests by automobile. At that time, opportunists caused a mining claim boom on certain national forests when they attempted to locate claims to cover road surveys. Claims were made along streams where highways made ingress and egress easy for campers and tourists. Then locators built shacks as close as possible to the highway and waited for the federal government to pay them when they located claims upon lands ripe for highway rights-of-way acquisition (Ibid.: 33-34). Compounding this problem were the many mining locations made by large lumber companies and/or power companies as a self-defense mechanism against someone blocking their expansion programs (Ibid.: 51).

Panning for gold, Plumas National Forest. 1964

Then there were the "recreation miners," people who enjoyed getting out of the cities and playing at placer mining. Handbills were circulated throughout Southern California hawking this new form of recreation. They exclaimed in big bold letters: "Gold Strike – Come out and be a 49er – Dig gold and pan it yourself – Organize your family and friends into a gold digging outfit and spend your weekend in healthful and profitable pleasure." Once at one of these gold panning camps, usually illegally located on Forest Service lands, real estate speculators encouraged recreational miners to purchase nearby property so they might be nearer to their new hobby, and perhaps might even erect a hunting or fishing cottage in a nearby resort community (Ibid.: 37-38).

Finally, there were those who used the national forest mining laws as a substitute for the Homestead Act of 1862, which ended when in 1935 President Roosevelt withdrew all public lands from further public entry. According to one report, many people became imbued with the erroneous idea that:

> every citizen of the United States was entitled, under the mining laws, to a twenty-acre claim in the National Forests. As a result, not only people in distress but also others with moderate incomes moved into the forests and filed on or purchased claims, constructed cabins and settled down for life. The latter class consists generally of elderly people of small incomes who find the mountain life most satisfactory and who are beating the high cost of city living. They have no interest in mining (Ibid.: 35).

These "mining" homesteaders picked a building site (generally adjacent to a highway and stream), then staked out a placer claim, built a cabin and garage – never considering the idea of mineral discovery. When the Forest Service questioned them, they universally produced their location notice, which they considered a "sacred document" and declared that their residence sites should be considered mines. Some claimants even drove cuts and short tunnels into granite hillsides as development work on their claims in order to "verify" the ruse (Ibid.: 40-41).

Summer home "improvements" on patented Oak Bar placer claim, Klamath National Forest. 1935

The influx of these Great Depression "homesteaders" onto national forests, along with "locaters" and recreational miners, presented Region 5 with several problems. The primary problem was how to protect the rights of legitimate prospectors and miners to conduct bona fide mining operations, while defending national forest fishing streams and camping places against claim locators and homesteaders. A secondary problem was how to accomplish forest development without the necessity and expense of purchasing rights-of-way for highways and power transmission lines across fallacious mining claims. Finally, by the late 1930s, the mining situation called into the question issues related to multiple use. Was mining an area the best use of the land among competing uses, such as recreation? (Ibid.: 48-50). This multiple use question, along with the mining situation, continued well after the 1930s and would not fade away for decades to come.

After the 1920s, conflicts between recreationists and miners, as well as with timber companies and grazers, were inevitable because recreation on California's national forests had grown at a phenomenal rate. By 1932, California led all states in recreational use, and in the tough economic times of the Great Depression, camping was seen as one way to enjoy a cheap vacation (*California Ranger Region Five* 1932: No. 23). Recreational management within Region 5 involved a wide spectrum of recreation resource uses and areas, ranging from highly developed and used areas near large urban centers, which included recreation residences, to natural wilderness utilized by a just few determined hikers.

Ever since the San Francisco office opened in 1908, the regional administration had preoccupied itself with timber, grazing and watershed management as well as fire control. Then, in the 1920s, several fundamental shifts of recreational policy and agency priorities took place. First, under the Redington administration, Region 5 began to emphasize and develop private facilities for recreation, such as recreation residences. Following World War I, there was a push to privatize the national forests, and the Forest Service reacted by directing forest supervisors to get important citizens involved in the national forests, suggesting that summer homes was a good way to do so, and tracts were established on many of the forests (Leisz 2005). Region 5 permitted private facilities such as summer homes and resorts in order to serve not just the public's needs, but its own needs as well. By 1923, individual forests such as the Inyo and the Sierra National Forests had rangers specifically assigned to recreational survey work. Fundamentally, however, Region 5 viewed recreation primarily from the standpoint of sanitation and fire prevention (Bachman 1967: 2-4; Tweed 1980: 13-14). The early S. B. Show years largely continued this tradition, but soon Region 5 realized that recreational demand by Californians was greater than anticipated and needed more attention. Therefore, in the fall of 1926, Show reassigned mineral examiner W. H. Friedhoff's responsibilities to include recreation, the only man on any district headquarters staff in the nation with explicit recreational responsibilities.

When evidence of campground and picnic area deterioration due to excessive and unregulated use surfaced in 1928, Show employed forest pathologist Dr. E. P. Meinecke of the USDA Bureau of Plant Industry as a consultant to study the problem in California's national forests. Meinecke reported his findings that year. He recommended that roads and trails be built with log rails and barriers to control auto and foot traffic and that visitors be "trained" to stay on them. He also recommended that parking areas, or spurs, be provided and their use be required. Furthermore, Meinecke suggested that stationary fireplaces and tables be installed to protect vegetation and improve site appearance. Until implementation of New Deal forest improvement programs, the Depression, as can be expected, slowed or even prevented the campground construction program on California's forests. Nevertheless, in 1932, Meinecke published his report as "A Campground Policy." The regional office approved the report, although some in the lands division believed that the public would never stand for such regimentation (Bachman 1967: 4-5; Tweed 1980: 13-15).

However, from the moment FDR took office until the outbreak of World War II, recreation largely dominated many of Region 5's activities as the administration increasingly focused its attention on recreational management. Under the New Deal, a frenzy of recreational development occurred, checked only by America's entrance into World War II. The modest level of recreational development that persisted in the 1920s and early 1930s, which included several major campgrounds in Southern California national forests, blossomed into a New Deal boom of recreational facilities – but with one major philosophical adjustment. The Forest Service accentuated community over individual rights, and public over private interests in accordance with the 1915 Term Permit Act. The CCC-built public works program evinced these values and philosophy as thousands of campgrounds, picnic areas, trails, lodges and other public recreational-related facilities were built nationwide. As the number of these public facilities increased during the New Deal era, surveys of new recreation residence tracts on California's national forests declined and virtually stopped by World War II (Tweed 1980: 22; Lux, et al.. 2003: 35).

Recreational planning began in March 1934, when the California Region hired six junior foresters and trained them to become recreation planners. At Pinecrest campground in the Stanislaus National Forest, they were given two weeks' training in laying out camp and picnic sites, then each one of them was sent to a different national forest. The early CCC-built recreational facilities for Region 5, designed by these junior foresters, started out very simple and did not compare well with NPS recreation projects funded and developed by the CCC, largely because Forest Service projects were tied to designs from an "earlier regime of scanty funds." Much of the work lacked good design, but they were an improvement over that of previous years. As time passed, Region 5 made a greater commitment to well-designed recreational planning. First, they hired experienced landscape architects such as Gibbs and Hall, much like the NPS did, but initially Gibbs and Hall's assignments were specific to various projects. Then, from 1937 to 1939, Region 5 employed six additional landscape architects, who were each assigned to an individual forest. By 1938, Region 5 used the architects exclusively for campground and other recreation site planning. Working together, these landscape architects eventually standardized facilities, a process that would not come to the rest of the service for another twenty-one years. Chief Silcox supported the regional level program, but by the end of the CCC period and the approach of World War II, Region 5 dropped all landscape architects from the program with the exception of one individual, who became the regional architect (Bachman 1967: 5-6; Tweed 1980: 17-18, 20).

Besides supporting good recreational design, Chief Silcox also instructed his regional foresters to lend greater attention to the "social" functions of the forest in their recreation programs. For instance, the San Gabriel Mountains, which protected the coastal plain from hot and desiccating winds and served as a force to store precious water, were now viewed sociologically as "raw nature" behind a great city providing recreation to thousands who "may momentarily evade the tribulations of a work – grief-burdened world and come into intimate communion with the good earth – nature in the raw" (*California Ranger Region Five* 1937: No. 32).

Meanwhile, in 1935, a broad reorganization study of the entire Forest Service took place that directly affected recreational resources on Region 5. This reorganization effort was part of a battle between the Agriculture and Interior departments. In an effort to create a Department of Conservation, Interior Secretary Harold L. Ickes tried to transfer the Forest Service away from Agriculture Secretary Henry Wallace. This interdepartmental war was fought on many battlefronts, but the central issue focused on which department could best administer, regulate and/or manage various resources on public lands. For instance, the Forest Service attempted to take over grazing management on all public lands, until the Interior Department created the U.S. Grazing Service to counter the Forest Service (Williams 2000: 75). On the other hand, the National Park Service, as it did in the 1920s, attempted to usurp and annex Forest Service recreational lands for its park system.

The pages of the *California Ranger* closely followed this quarrel, gladly enumerating the points made by Wallace against Ickes' arguments. Wallace took the stand that forests were growing crops and that the Agriculture Department had all the technical services needed to manage forests. The agriculture secretary evoked Gifford Pinchot's missionary spirit for utilitarian conservation but for the first time affirmed the principle of multiple-use management as a national policy. Unlike Pinchot, Wallace considered recreation and wildlife on an equal footing with other forest resources. Wallace stated, "The whole idea has been to devote the land and all its resources to its highest public use; to fit national forest lands for such uses as their character, that of their resource, and the needs of the public, will permit. To do this, multiple-use is necessary." The *California Ranger* bolstered the agriculture secretary's opinion with a *San Francisco Chronicle* editorial that found the Agriculture Department's record friendlier toward conservation than the Interior Department history, citing the Pinchot-Ballinger affair and the Teapot Dome scandal under Interior Secretary Albert B. Fall in order to denigrate the Interior Department (*California Ranger Region Five* 1935: Nos. 36, 40). One battle of this war broke out in California over the formation of Kings Canyon National Park. Regional Forester Show used the concept of "multiple use" to philosophically attack the NPS on the formation of the park by labeling it a "single-use" agency.

Proposals to add the Kings Canyon region to Sequoia National Park started in 1911 but were dismissed until 1933, when powerful California Senator Hiram Johnson renewed interest with a proposal to create a Kings River Canyon National Park. Some in the state, such as the California State Chamber of Commerce, quickly opposed the formation of this new park (*California Ranger Region Five* 1933: No. 28; Wolf 1990: 26-27). But in response to Senator Johnson's proposal, Show sent noted landscape and park engineer George Gibbs to Kings River Canyon to make a reconnaissance of the area preliminary to the preparation of a detailed plan for the development of the recreational resources of the area. Show hoped to have the reconnaissance report completed in the next two or three years (*California Ranger Region Five* 1933: No. 29).

The Kings Canyon recreational plan called for by Show was completed in 1936, a year after the Forest Service-Interior Department transfer fight had broken out. By this date, the interdepartmental conflict had turned into a slugging match, and against this backdrop, Show fought to keep the Kings Canyon region as a Forest Service recreational area under his multiple-purpose plan.

First, Show strategically, but privately, went to both the California Chamber of Commerce and the Sierra Club to garner their support for his Kings Canyon recreational plan over Senator Johnson's bill to create a new national park of the area. President Roosevelt had given specific orders instructing all Forest Service members not to appear in public meetings on the subject, orders likely given because Senator Johnson played such an important role in the Roosevelt New Deal program. Therefore, Show's action was directly insubordinate to FDR's orders. The regional forester realized this, but felt strongly that it was his duty to inform the public and that the president's directive asked him to violate his public duty. Show's defiance had consequences and resulted in the Federal Bureau of Investigation (FBI), or "snoopers" as Show called them, tapping the official Forest Service telephones at the regional office and at Sequoia National Forest. Fearing that the FBI might raid his office files, Show instructed Public Relations Officer Wallace Hutchinson to take them to his home, where they eventually were destroyed. Furthermore, Show learned of two very sharp notes scribbled out by Roosevelt to Secretary Wallace asking why he and Chief Silcox had not "called off" Show from "pushing" against the Kings Canyon Park proposal. One note may have even suggested Show's transfer elsewhere (Show 1963: 195-200; Wolf 1990: 27).

Meanwhile, Show fought on. At issue to him was the current and future use of Kings Canyon water, which Show felt would be needed for upstream storage to even out the flow of water for irrigation and power development below – both important uses in that part of country. Of course, the California State Chamber of Commerce and power companies such as Southern California Edison supported Show and lobbied vigorously against the park proposal (Show 1963: 203-206). Conversely, the Sierra Club declined to support

Show's "multiple-purpose" management plan. The Sierra Club felt the Forest Service could not guarantee the future protection of the scenic values of the canyon (Wolf 1990: 27). In the meantime, in January 1939, the California state legislature passed a resolution by a substantial margin in both houses petitioning President Roosevelt and Congress to oppose the proposed Kings Canyon National Park (*California Ranger Region Five* 1939: No. 10), and Show even went so far as to enlist commodity groups such as the timber industry to support his multiple-purpose management proposal (Wolf 1990: 27).

Yet the efforts of Show to undercut the Roosevelt administration's decision to create a Kings Canyon National Park failed. In February 1939, California Congressman Bertrand "Bud" Gearhart introduced a bill to establish the John Muir-Kings Canyon National Park and to transfer thereto the lands included in the General Grant National Park (*California Ranger Region Five* 1939: No. 13). In May of that year, new legislation establishing the Kings Canyon National Park was reported favorably, but the bill also permitted construction of projects for water conservation, irrigation or hydroelectric development throughout the area upon the recommendation of the Army Corps of Engineers (*California Ranger Region Five* 1939: No. 28).

However, in July 1939, the park bill passed the House, after an uproarious and bitter debate, rejected the provision permitting irrigation and power dams. Interestingly, following the debate, Chief Silcox toured Region 5 with Regional Forester Show (*California Ranger Region Five* 1939: No. 35), perhaps to keep Show in line before the Senate took up the measure. Afterward, the Senate passed the measure, although it was still opposed by the California

San Joaquin Power Company hydroelectric plant at Big Creek, Sierra National Forest. 1940

legislature, the California Farm Bureau and seventy-two other organizations. In 1940, Kings Canyon National Park was created (*California Ranger Region Five* 1940: No. 13).

The fight over the establishment of Kings Canyon National Park was very much reminiscent of the Hetch Hetchy Valley conflict of an earlier era, only this time the preservationists won out over utilitarian conservation. Initially naming the park after the celebrated California naturalist John Muir symbolically acknowledged Muir's contributions to the preservation ideology and perhaps signified the power of preservationists in the Roosevelt administration. Show did not accept defeat graciously. He attributed losing the battle to a last-minute shifting of support by Agricultural Secretary Wallace and to a highly injudicious personal attack on Congressman Gearhart by California Congressman Al Elliott, which influenced the final debate. Ultimately, in his personal recollections, he condemned FDR for his strong-arm tactics, which in Show's opinion had gone beyond the limits of decency and dignity. The regional forester also vilified Ickes, who Show felt had used his ambition and determination to ignore and override the public interest. In Show's opinion, it was never "a question of jurisdiction; it was case of trying to search out wherein the major public interest lay." However, the larger question raised by the conflict according to Show was, "What proportion of the land with resources of public value can and should be withdrawn from multiple use for a single use?" (Show 1963: 208-210).

Meanwhile, at the same time as the Kings Canyon conflict, the issue of wilderness designation arose. In November 1935, Chief Silcox created a Division of Recreation and Lands in the Washington office, and in May 1937 appointed forester and wilderness advocate Robert Marshall, who had written the recreation section of the *Copeland Report*, to head the division. As division head, Marshall worked tirelessly to establish a secure position for recreation in the multiple-purpose program of the Forest Service (Tweed 1980: 24; Williams 2000: 77-78). Marshall's career was short-lived, but before his untimely death from a heart attack thirty months after his appointment, he promulgated, and the agriculture secretary approved, the "U" regulations of 1939. These regulations strengthened and refined areas to be designated for protection for their wild and scenic beauty. According to the U-1 regulations, the agriculture secretary could designate, modify or eliminate unbroken tracts of 100,000 acres or more as "wilderness areas," and under U-2 regulations, the Forest Service could create, modify or eliminate other areas of 5,000 to

100,000 acres as "wild areas." These designations were in addition to primitive areas, created under Regulation L-20 (1930), which remained in full force and effect, but were now managed as wilderness areas under U-2a. The latter U regulation protected primitive areas established by Region 5 in the late 1920s and early 1930s, with the exception of 432,000 acres of the High Sierra Primitive Area on the Inyo and Sierra National Forests, which in 1941 was transferred to the NPS (USDA Forest Service 1976: 1-2, 6).

Before any wilderness areas or wild areas could be added to Region 5 under the new U regulations, World War II put a damper on all recreation in the country. However, the stage was set in California for a vastly increased role for forest recreation in the postwar period. The mountains continued to be the summer playgrounds for vacationing, hiking and camping Californians, but much of the postwar growth would center on winter sports on the national forests of California.

Winter sports gained the attention of Forest Service recreation managers as early as 1938. In that year, the first ski-training course for Region 5 forest officers was held on the Tahoe National Forest. The experiment turned out so successfully that it was decided to hold other training schools, and the High Sierra ski school for staff officers, supervisors, district rangers and forest guards was born. For the next few years, the Mono National Forest hosted the school, and in one year, Alf Engen, considered by many experts to be the most outstanding all-around skier in America, offered extremely valuable training to Region 5 personnel in skiing techniques necessary for the proper administration of national forest winter sports areas and for carrying out other wintertime duties. In addition to ski technique, training was given in making snow surveys and first aid in relation to winter sports accidents (USDA Forest Service 1940a: 1-3; *California Ranger Region Five* 1941: No. 7). Even so, no major winter sports developments took place until after World War II (Coutant 1990: 2).

As recreation's star rose in importance in Region 5's multiple-purpose management program, so did fish and wildlife management, but at a much slower pace. Prior to the New Deal, active wildlife conservation policies were limited and mostly devoted to the protection of habitat. Little protection of wildlife occurred, since national forests were not closed to sport hunting and fishing. But with the closing of the public domain following the Taylor Grazing Act of 1934, active management of game and fish became more prevalent. Once it did, Region 5 learned that several species were in need of protection. Two such species were California's bighorn sheep and the California condor.

Bighorn sheep, one of the most majestic and reclusive of animals on California's national forests, thrived on diverse habitats that ranged from high mountains to deserts. The sheep preferred precipitous mountainsides above 3,000 feet that support chaparral, oak woodland and conifer vegetation. Here, they could rapidly outdistance pursuers. In the 1930s, thriving populations were still found in the San Gabriel Mountains and elsewhere, but Region 5 realized that with growing pressures on their habitat they would need protection and effective management, and therefore Region 5 began cooperative efforts and programs with the California Department of Fish and Game and with various wildlife organizations (Angeles National Forest Grapevine 1989: 1).

In the 1930s, another important cooperative wildlife effort between the federal government and groups such as the Audubon Society sought permanent protection for the California condor. In 1935, noted naturalist Aldo Leopold brought the plight of the California condor to the attention of C. E. Rachford, chief of the grazing division in Washington. After reading an article written by Ernest I. Dyer that described a breeding remnant of condors located on one the national forests of California, Leopold declared that if the Forest Service was "really serious about taking a hand in wildlife, then I think the local administrative plans ought to take account of such a national resource." Rachford immediately wrote Regional Forester Show regarding Leopold's letter, stating that "fortunately, most of the areas known as condor country are within the boundaries of national forests, and it would seem possible to have certain areas set aside within such Forests and close them to hunters, keeping them under Federal control as sanctuaries." At that time, Forest Service officials knew very little about the condor's habits and territory. Apparently, there was a constant market for the rare condor eggs, which brought upwards of $1,000 each from collectors, a worthy sum that drew many to hunt and raid nests for profit during the Depression. At the time, Region 5 officials believed that the bird ranged from the Angeles to the Sequoia national forests and westward to the Santa Barbara National Forest (renamed the Los Padres National Forest in 1936). They also erroneously believed that the bird was holding its own and gradually increasing in numbers (USDA Forest Service 1935: various)

Thus began Region 5 efforts to protect the endangered birds. Forest Service officials soon learned more about the black plumaged California condor, with its eight- to nine-foot wingspread. In 1935, it was obvious Region 5 needed more and better information, so lookout men on all of its Southern California national forests were instructed to fill out reports if they spotted the bird. From these observations, Region 5 learned of three roosting areas on the Santa Barbara National Forest. The largest (thirty birds) was near Sisquoc Falls in Santa Barbara County, another nesting area was located in the Whiteacre Peak-Hopper Mountain area in Ventura County (thirteen birds) and a third roost (two birds) was located near Huffs Hole in San Luis Obispo County. In 1936, Region 5 suggested closing off these nesting and gathering grounds from public access for the perpetuation and protection of the condor. It was also suggested that Region 5 halt further road building into at least the Sisquoc River country until a survey could better determine the nesting habits of the condor. Interestingly, Region 5 awareness was also drawn to similar problems with potentially endangered species such as antelope on the Lassen National Forest, golden trout on the Inyo National Forest and the California fisher on the Sequoia National Forest (Ibid.; Robinson 1940: 1-3, 15).

In 1937, as the first measure of California condor protection and perpetuation, the Forest Service set aside 1,200 acres on the Los Padres National Forest as the Sisquoc Falls Sanctuary. This area was thereafter posted and all use prohibited. Two years later, all federal, state and county lookout stations again were instructed to record all condor observations. From them, Region 5 learned that the flight zone of the bird extended north to Monterey Bay, south to Santiago Peak in Orange County in the Cleveland National Forest and sometimes as far south as San Diego County. Additionally, in the same year, the Audubon Society premiered a full-length movie on the California condor, for which the Forest Service provided valuable production assistance (*California Ranger Region Five* 1939: No. 32).

With the above data gathered, in 1940 Associate Regional Forester Cyril S. Robinson produced the first comprehensive Forest Service study of the California condor. Robinson's research paper described the bird itself, which once ranged as far north as Salinas Valley and Pfeiffer's Point below Monterey on the coast. His report covered the condor's nesting and mating habits, its feeding habits and food preferences, its natural enemies, the man-caused losses of this rare bird and the range of the condor on national forest lands. Ultimately, Robinson concluded that the condor was in grave danger of extinction because of the peculiar habits and characteristics of the birds themselves and because of other dangers, ranging from the "curiosity and cupidity of those who care little for preservation of the condor" to harm from fires and other natural hazards (Robinson 1940: 1-15, 18). For the next few decades, the California condor would be watched, studied and protected by Region 5 in what appeared at times to be a futile attempt to perpetuate the species.

Waning and Demise of the California Forestry New Deal, 1939-1941

As the prospects for survival of the condors waned, so did the New Deal and so did the Forest Service's decentralization and the independent power of regional foresters, epitomized by S. B. Show. The first blow to the Forest Service came on December 20, 1939, with the death of Chief Silcox (*Who Was Who* 1962: 1125). Under Silcox's supervision, many CCC and WPA projects not only helped millions of unemployed in the nation but also provided the administrative improvement infrastructure that the Forest Service badly needed headquarters, ranger stations, roads, trails, recreation and fire control facilities, to name a few. President Roosevelt named associate chief of the Forest Service Earle H. Clapp (1939-1943) as acting chief, but Clapp was never officially made chief, apparently because the president blocked his appointment (Williams 2000: 79) because of Clapp's public opposition to the transfer of the Forest Service to the Interior Department (Steen 2004: 244-245).

During his tenure, Acting Chief Clapp struggled with various issues such as the continuation of the CCC; working with the Joint Congressional Committee on Forestry (JCCF), established in 1939 to hear testimony on the condition of forest lands in each of the western states; growing centralization of power to the Washington office; and mobilizing the nation's resources behind the impending war. In Region 5, all of these issues were very important.

By 1939, the CCC was on the decline. Although Roosevelt proposed a permanent CCC in 1937, it was rejected by Congress (Cermak n.d.: 427). A renewed effort was made in 1939 (*California Ranger Region Five* 1939: No. 11), but by the latter date President Roosevelt had clearly lost interest in his "pet" New Deal project. With war clouds looming on the horizon, Roosevelt turned his attention away from domestic issues to larger global ones. While Congress did not officially abolish the CCC until June 1942, for all intents and purpose, for the last three years of its existence, the California CCC was moribund compared to its heyday years of 1933-1935 (Cermak: 427-428; Pendergrass 119-120). During its existence, California's federal, state and county CCC program employed more than 135,000 persons and salvaged the lives of many young men. By the end of the program, the accomplishments of the California CCC included thousands of miles of firebreaks, truck trails and minor roads, and many more miles of telephone lines. CCC enrollees planted more than 30 million trees, not to mention permanent improvements to the forests such as the construction of hundreds of lookouts, ranger stations, barns, warehouses, storage sheds, pastures, dams and stream improvements, which added much to Region 5's infrastructure. One of the key areas of benefit to Region 5 was the great increase in technical work and research conducted at the various forest experiment stations. Beyond these accomplishments, the California CCC also spent 1.6 million man-days fighting and preventing fire. As a token of Region 5's appreciation, Regional Forester Show made sure that each enrollee left the outfit with an impressive certificate stating how that individual had worked to protect and improve the country's resources (: 108-119; Pendergrass 1985: 116-117).

Regarding the JCCF, hearings began in California in December 1939, mainly because the state held extensive forested areas and because the population was dependent upon these forests for watershed cover for irrigation and flood control, as well as for timber, forage and recreation (*California Ranger Region Five* 1939: Nos. 2, 3). Following the hearings, Clapp met with the regional foresters and thereafter addressed a confidential letter to them regarding two groups of problems, the first of which centered on Forest Service morale. Many in the Forest Service in 1940 expressed confusion over the service's objectives; some felt uneasy regarding their future, with the constant threat of government reorganization; and others were uncertain about their individual jobs and service responsibilities because New Deal emergency work had given them little or no time for needed progressive forestry work. The second group of problems was broadly associated with the national forestry problem – the differences between public and private management of forest resources – and the numerous and frequently opposing forces that were then testifying before the JCCF. These competing groups advocated different programs, which tended to nullify each other and prevent any real progress on major issues such as forestry regulation and the extension of public ownership over private forestry. Clapp's answer to both problems was a "nation-wide educational campaign designed to obtain real public understanding of the Forest problem in the United States, and to stimulate aggressive public action to safeguard its own interest." Clapp hoped that if people became interested in the forestry problem they would naturally want to do something about it, and that this interest would reach Congress and the JCCF. In many ways, Clapp was carrying the torch left behind by deceased Chief Silcox. Silcox had campaigned for public ownership of all forests since the day he took office and during his tenure had fought for the conservation of private forestlands. Acting Chief Clapp also hoped that this education crusade might unite his organization as well (USDA Forest Service 1940b: various; 1940c: various).

After meeting with Clapp, Regional Forester S. B. Show and Edward I. Kotok, Director of the California Forest and Range Experiment Station, were bought into the education program. In support of Clapp, they began a major public relations effort to educate the public and the JCCF on California problems caused by private forestry and the lack of regulation. For instance, in October 1939, the California station produced a lengthy report entitled *The Forest Problems of California*. This report detailed difficulties associated with private ownership, which controlled half of the commercial forestland in California (5.5 million acres of pine timberland and 1.4 million acres of redwood). In comparison, public ownership (federal, state, county and municipal forests) controlled the other half of California commercial timber. However, though publicly-owned forests in California accounted for 40 percent of the total annual growth of commercial timber, it furnished only 12 percent of the annual cut, mainly because cutting was regulated and public timber was generally located in more inaccessible districts (USDA Forest Service 1940d: 3-16).

The Forest Problems of California also closely described the national forests in terms of their extent, and the problems affecting them, particularly regarding protection, use and proper development of their four major resources: water, timber, forage and recreation. The crux of the water problem on the national forests rested in the condition of the vegetative cover, which in turn depended on effective fire protection and prompt restoration of cover where natural processes were slow to recover. But in 1940, 15 million acres of water-yielding protection forest were in private holdings, which, according to *Forest Problems of California*, prevented a well-managed forested watershed. The most critical problem in timber on the forests pertained to the difficulty of properly utilizing the 90 billion feet of government timber and adjacent private holdings through sustained-yield management plans. Many public units were intermingled with private holdings, which, according to *Forest Problems of California*, hindered this type of utilization and presented problems in fire, insect and disease control, as well as reforestation (USDA Forest Service 1940d: 16-24).

The problems associated with grazing on national forests in 1940 lay in the deterioration of the range due to overgrazing and to the lack of technical manpower to permit proper periodic inventory of public range resources and determination of proper use. Once again, *Forest Problems of California* concluded that the lack of control of private holdings was problematic in range-improvement practices, including developing water holes, springs and meadows, and controlling conflicts between use of range by domestic livestock and wildlife. Finally, regarding recreation, *Forest Problems of California* found that just as in the case of watershed, timber and forage management, private ownership was a detriment to better management. In the case of recreation, lack of control of natural units due to private ownership of key tracts contributed to improper use of public lands. Many of the most desirable and accessible recreational areas, and main points of entry to them, were in private title (Ibid.).

On the other hand, Region 5 produced its own synopsis of the forest problems of California entitled *The Forest Program and Its Application to California Problems and Conditions* (USDA Forest Service 1940e: various). This publication reiterated many of the arguments laid out by *Forest Problems of California*. Regional Forester Show followed up this Region 5 publication with other Region 5 "educational" efforts. In a confidential letter to all forest supervisors, and in coordination with Clapp's forestry education program, Show enlisted the entire region – from regional officers to district rangers – in a grand strategic campaign to support Clapp's forest program and win over public opinion. First, Show instructed all Region 5 personnel to begin indirectly or directly contacting California congressmen, to hold group meetings with the public and give talks, and to use the newspaper and radio media to get the Forest Service message across. To this confidential letter, Region 5 compiled and attached a list of important individuals and organizations within zones of responsibility for future contacts. The list read like a "who's who" of important national and state business, outdoor, labor and women's organizations. They ran the gamut from the Agricultural Extension Service Farm Advisors at the top of the list to the Western Pine Association at the bottom. Furthermore, Show "militantly" divided the state into two educational zones – Northern California (depicted as less populous, with many timber producers with private holdings, who were emotionally opposed to the Forest Service) and Southern California (described as more populous, respectful of the Forest Service's objectives and personnel, and as consumers of water and power). Show firmly instructed his staff to avoid the appearance of exerting pressure, and Show instructed them not to disclose any confidential material recently sent to them. All active campaigning was to begin January 1, 1941 (USDA Forest Service 1940d: 3; 1940f: various).

Despite such thorough preparations, Clapp's education program was unsuccessful. By 1941, the political climate in America had greatly changed since the beginning of the New Deal. Though Acting Chief Clapp drove the service into a "New Crusade" for conservation, President Roosevelt had dropped conservation and the Forest Service from its once favored spot on his agenda. Without FDR's leadership, the program languished in Congress. Furthermore, the lumber industry, which had come hat in hand to the Forest Service for help in the Depression, no longer sat "cowed and silent in the storm cellar," according to Show, "ready to accept, albeit reluctantly, almost any measure of Federal action." Now the lumber industry increasingly arrayed themselves against the Forest Service and its quest for public ownership of private forests. From the viewpoint of business, particularly of the lumber industry, the Forest Service was repositioned as being too radical. At this time, California's lumber industry made a progressive shift of support to state forestry, which it deemed more compliant and manageable (Show n.d.: 211-215).

Earle Clapp had badly misjudged the timing of his conservation crusade, and Region 5 blundered by following suit when it aligned its reputation with him. Part of the problem was the growing centralization of power to the Washington office, which started under Chief Silcox. Washington's staff of assistant chiefs had grown under Silcox and then Clapp, which facilitated a shift in power from regional control to Washington when regional foresters failed to "hang together" on key issues. Moreover, Acting Chief Clapp, who was absorbed by regulation, took little part in the day-to-day direction of the service, and Clapp increasingly delegated service direction to his staff. One major area of staff encroachment was in curbing regional forester selections of officers from supervisors on up. Clapp and Washington staff believed that provincialism existed in western regions such as California, and he strongly intended to break it up. He implemented policies that required that all future supervisors have a forest school degree, that functional experience and repute were required for supervisors of forests with specific emphasis (e.g., grazing in Modoc and timber in Stanislaus), and that broad geographical experience was now considered more important than actual knowledge of regional conditions and problems. Of course, these dictates flew in the face of Show's Feather River advanced training course at Berkeley for higher leadership.

Another major area of centralization occurred in 1936 when the Washington staff introduced the general integrating inspection (GII). Under this system, a two-man team visited and thoroughly inspected all aspects of a region's program every few years. In California, Assistant Chiefs Earl Tinker and Carter in 1937 conducted the first GII for Region 5. Initially, Show welcomed the GII because it presented the first comprehensive depiction of the Region 5 program as a whole and provided general and specific comments and recommendations regarding administration and operations, personnel, improvements, forest, grazing, water, wildlife and recreation management, as well as fire control, land acquisition, state and private forestry, research and public relations. The 1937 Tinker-Carter GII was followed up in 1941 by the Granger-Forsling GII. Regional Forester Show accompanied both inspections and supported them because he felt that they allowed higher officers in Washington to attain a fairly broad and up-to-date level of knowledge. He hoped that this data would offset the results from functional inspections made by what he called "third and fourth stringers." Nevertheless, the GII system also gave the Washington office greater power in the overall scheme of things, and the inspections and other events nibbled away at the Pinchot-era concept of Forest Service decentralization (Show n.d.: 216-233). In the meantime, Show faced the breakup of his old-time regional office staff through retirements and promotions. When key men like Assistant Regional Forester T. D. Woodbury retired in September 1941 (*California Ranger Region Five* 1941: No. 40) and when others moved on, such as Edward Kotok, who left California for Washington to serve as assistant chief forester in charge of state and private forestry (December 1940) (*California Ranger Region Five* 1940: No. 2), Show found himself as the last survivor of the bygone Pinchot era. In his recollections, Show cynically declared that the Forest Service had become no more than a "bureaucracy" that was "ruled by clerks" (Show n.d: 246).

On the Eve of World War

On April 9, 1940, German blitzkrieg troops crossed the Danish border, overran Denmark in a matter of hours, conquered Belgium in eighteen days and then swept into France. In three weeks' time, they drove the British off the continent and were at the doorstep of Paris. France surrendered, and later that summer the Germans rained bombs on Great Britain and, together with fascist Italy, invaded the Balkans, Greece and North Africa. Meanwhile, half a world away, the Japanese viewed the Allied reversals in Europe and elsewhere as a source of opportunity. They tightened their relationship with the fascist Rome-Berlin axis and invaded northern Indochina. When Germany surprised the world by invading Russia in June 1941, Japan felt it no longer needed to fear interference from Siberia and promptly moved into southern Indochina. Facing little opposition, they completely occupied it by July.

The impending entrance of the United States into the world war meant major shifts of emphasis for Region 5, which began as early as June 1940. At that time, Region 5 began to cooperate with the military and to work out new emergency plans for firefighting labor with them, with industry and with individual employers. There was also a slowdown in various Region 5 programs such as public relations, training and testing, largely because of a lack of money (Show n.d.: 251-252). At that time and up to the declaration of war in December 1941, the *California Ranger* called for America to take the lead in assuring world peace through equitable distribution and planned conservation of natural resources. Thus began a discussion regarding how natural resources could be used for national defense. After all, according to one article, in the "perilous situation" confronting the country, the "strongest nations, the mightiest people, are those with the greatest natural wealth…" (*California Ranger Region Five* 1940: Nos. 2, 33; 1941: 31). Looking back at World War I in conjunction with the growing Japanese threat that might cut tree rubber supplies from the Far East, in March 1941 Region 5 turned to research on the domestic guayule shrub, a substitute resin supply which was being grown in the Salinas Valley (*California Ranger Region Five* 1941: No. 17). Other articles, such as "If War Comes to the Forest," assured the public that national defense fire protection plans were in place in case of a war emergency (*California Ranger Region Five* 1941: No. 42). But as prepared as most Californians were for war in their minds, no one was quite prepared for the Japanese attack on Pearl Harbor on December 7, 1941.

Conservation Anniversary

In 1891, the Forest Reserve Act set aside the first timberland reserve, marking America's awakening to the importance of conserving forests and other natural resources. For many years, Americans had exploited these resources, but by 1941, Americans were celebrating their conservation. Aware of the mistakes of the past and the misuse of California's unique environment and abundance, the Forest Service had diligently conserved eighteen pieces of paradise – California's national forests. Thus, fifty years after the Forest Reserve Act, the most acute forest conservation problems in California had to do with privately-owned forests and not with the national forests.

The Forest Service's view of multiple-purpose management had also significantly grown since Agricultural Secretary James Wilson's letter of February 1, 1905, authorized the protection and conservative use of timber, water, range and minerals on the nation's forest reserves. By the eve of World War II, the Forest Service's conservation policy now centered on the management of lands for the greatest number of uses. Still, partly in accommodation to pressures from preservationists and partly by its own volition, Region 5 firmly added scenic wilderness and wildlife protection to its conservation management mandate. Attention to these resources balanced the Forest Service's spectrum of concerns sufficiently to avoid much controversy with preservationists during the New Deal era. It was true that during the New Deal, disputes surfaced between utilitarian conservationists and preservationists, such as over the establishment of Kings Canyon National Park; but the brunt of that dispute was an interdepartmental struggle between two government agencies. In the end, while S. B. Show realized that part of the general public still looked upon all-purpose use of the national forests as something detrimental to conservation, he considered them the "last survivors of an outmoded era when lands were set aside and preserved for this or that use; not put into circulation for this use and that use" (Show 1940b: 7). Conservation implies some use, yet Show could not account for the defense demands of World War II and California's post-war growth pattern that would upset his pre-war view of balanced multiple use of California's national forests in favor of timber management over that of other resources. As will be seen, not many others predicted these circumstances either, so it's hard to fault Show for this.

Chapter VII: 1941-1945

Region 5 at War

California At War

On the eve of the Japanese attack on Pearl Harbor, America was divided between those who wished to avoid intervening in what they saw as a European war and those who saw its inevitability and wished the United States to become involved. While the United States was largely unprepared for a large-scale war with Imperial Japan, the bold attack on Pearl Harbor quickly provoked and united a divided isolationist America and turned it into a tremendous war production engine. With a formal declaration of war against Japan, and then against the other Axis powers, Germany and Italy, America was thrust into World War II. California played a key role in that war production machine. Eventually, World War II lifted the state of California out of the Great Depression and set it on a course of exceptional growth, development and prosperity that lasted well into the postwar years.

Naturally, California's location, geography and climate aided the state's contribution to the war effort. After all, California was America's window on the Pacific and an expected target for any invading enemy. Once secure from Japanese invasion, California's important naval ports from San Diego to San Francisco became important staging areas for troops, ships, planes, tanks, supplies, food and other materiel to the Pacific battlegrounds. Naval facilities on Treasure Island in San Francisco Bay, and the Terminal Island installation in Los Angeles, expanded and grew out of the war effort. In addition to these and other facilities, California's many sandy beaches, now empty of tourists, soon became the training grounds for thousands of marines and soldiers as they prepared for assaults on the Pacific islands, while the state's sagebrush and creosote- bushed deserts were used to prepare armored divisions for action in North Africa. Furthermore, California's ideal flying weather led American air power to expand and build new bases throughout the state. Major air war activity was concentrated at March Field, McClellan, Mather, Travis, and George Army air bases (Rice, Bullough and Orsi 1996: 472), as well as air bases at El Toro (Marines) and North Island (Navy). Meanwhile, modern wartime economic demands led to the expansion of many California manufacturing and industry interests. Of the $29 billion in war contracts let by the federal government, California secured $18 billion for new war plants, which comprised 45 percent of the national total (Nash 1973: 200).

But it was the overriding demand for airplanes and ships that stimulated California's economy the most. Aircraft construction, which up until 1939 had been very modest and employed approximately 20,000 workers, leapt forward and became a giant California industry centered in Los Angeles, Orange and San Diego counties. In the spring of 1940, President Franklin D. Roosevelt issued a call for the construction of 50,000 warplanes a year, and after Pearl Harbor, the president requested even more planes. Prior to that date, many of California's aircraft manufacturers had already contracted for construction of planes for various foreign governments, most notably for Great Britain. But with the declaration of war and $10 billion of federal funds, California's aircraft industry rose to employ a workforce of 280,000, which built more than 100,000 planes in 1943 alone. In fact, more than 60 percent of all federal monies spent in California were related in some way to aircraft. Manufacturing plants for pioneering firms such as Douglas and Lockheed, and new firms such as North American, Northrop and Hughes developed into enormous complexes that spread out over thousands of acres in Los Angeles, Long Beach, Santa Monica and El Segundo. The aircraft industry thereafter became a critical part of Southern California's postwar economy and the base for the aerospace industry that developed after World War II (Rice, Bullough and Orsi 1996: 451-452, 473, 475; Watkins 1973: 435-436; Nash 1973: 201-203).

California's shipbuilding industry underwent a similar wartime expansion. Seagoing tugs, tankers, amphibious landing craft, PT boats and freighters were needed immediately. The federal government's largess in this case went to the San Francisco Bay area rather than Southern California. Soon shipyard employment jumped from 4,000 to 260,000 people as the yards in Sausalito, Vallejo (and Mare Island), Alameda, Berkeley, Oakland and San Francisco built hundreds of transports and warships. To meet the demand for cargo carriers, President Roosevelt turned to Henry J. Kaiser and his corporation. Though Kaiser's corporation, located in Oakland, had no experience in shipbuilding, in the 1930s it had constructed major projects such as Hoover and Parker dams on the Colorado River, the Bonneville and Grand Coulee dams on the Columbia River and San Francisco's Bay Bridge. In Richmond, just across the Bay from San Francisco, Kaiser built a whole new shipyard facility that relied on innovative techniques of management, design and prefabricated construction to produce the ugly, but reliable, "rust bucket" known as the Liberty ship. This vessel played an important part in winning the war by carrying supplies to all theaters of war. All in all, the round-the-clock Richmond plant operation launched a new freighter every eight hours and produced one-fourth of the Liberty ships in the United

States (Rice, Bullough and Orsi 1996: 451-452, 473; Nash 1973: 201; Watkins 1973: 436-437).

Besides the expansion of manufacturing, Californians witnessed the creation of a vast new scientific-technological complex, making the state one of the leading research centers in the nation. California's nucleus of distinguished scientific schools and universities, ranging from the radiation laboratory at University of California, Berkeley, in the north, to the California Institute of Technology in Pasadena in the south, played a significant wartime research role. Using an atom smasher at the Berkeley laboratory, able scientists such as E. O. Lawrence, J. Robert Oppenheimer, and Enrico Fermi diligently concentrated their work on nuclear research. Their investigations into the world of elements and isotopes led to the fabrication of the atomic bomb, which ushered in a whole new age for the world. Further to the south, scientists at the California Institute of Technology experimented with rocket motors and propellants; this became the nucleus of the Jet Propulsion Laboratory, famous in the postwar decades for missile research that took Americans into orbit (Nash 1973: 207-208; Rice, Bullough and Orsi 1996: 477).

The contribution to the war effort of California's farmers, cattlemen, lumbermen and miners was more down to earth, but no less noteworthy. Increased prosperity, population growth and war demands expanded agricultural output in the state from $623 million in 1939 to $1.75 billion by 1945 (Rice, Bullough and Orsi 1996: 451-452, 472-476). The many new laborers migrating to California to work in its war industries, as well as troops training in California for oversea assignments, needed to be fed. This demand boosted California's production of dairy products, fruits, nuts, vegetables and cotton. California's cattlemen prospered as well; as the Depression ended, meat consumption increased because an increasing number of Americans moved into urban centers. Before long, demand far exceeded meat supplies, and the federal government was forced to impose price ceilings. Nevertheless, California's livestock owners met the demand and did not complain, largely because they were making good profits from their livestock (Nash 1973: 199). California's petroleum industry also played a crucial role in the war effort. To meet the demand of high-test aviation fuel and fuel oils necessary for the war against Japan, California oil production increased by 50 percent between 1941 and 1945 (Rice, Bullough and Orsi 1996: 451-452, 475). Miners also strained to increase their output. Wartime needs for new alloys such as ferro-manganese, ferro-chrome and ferro-silicon were met by the

California mining industry, which continually explored for new sources for strategic metals such as chrome and tungsten.

Important social consequences followed the extraordinary growth, development and prosperity in California directly attributed to World War II. First, the war proved to be an important turning point for women and minorities in California. As fathers, husbands and brothers left for the war, women replaced them in the workforce and sought jobs in shipyards, aircraft assembly lines and other manufacturing industries. Fully 40 percent of the employees in California's aircraft industry were women. Many more took jobs as chemists, engineers, railroad workers and lawyers, and in service industries such as banking, retail sales and education. African-Americans also benefited. Of the two million new residents who arrived in California, a large percentage were African-Americans, who settled in Los Angeles, Oakland and Richmond to work in the shipyards and military installations. Due to a booming wartime economy, many Mexican-Americans moved from unskilled to semi-skilled and skilled labor jobs into the mainstream of California society, resulting in a growing Mexican-American middle class. Mexicans arrived in greater numbers as well. With a shortage of harvest workers owing to Californians entering military service and other reasons, the federal government made arrangements with Mexico to import harvest labor. The United States provided transportation, health care, decent wages and housing to Mexicans who signed up to work in California's fields. By 1944, more than 26,000 Mexicans signed on to this program, which later led to the *bracero* program. The importation of Mexican harvest labor, the migration of African-Americans to California and the entry of women into the workforce caused acute problems in some communities. For example, serious housing problems arose as communities, such as Richmond, grew overnight from 20,000 to more than 100,000 persons. Crowding of two or three families into single-family houses and apartments was not uncommon. California's swelling urban population also led to overcrowded schools, and rising demands on municipal services such as water, sewer, gas, electricity, telephones and transportation, which stretched the budgets of many urban centers (Rice, Bullough and Orsi 1996: 477-479, 482-484; Watkins 1973: 438-440).

Blatant racial discrimination, segregation and conflict also raised their ugly heads on a number of occasions during World War II, such as the Los Angeles "Zoot-suit" riots in the summer of 1943. Sailors and other military personnel, along with newly arrived workers from the South, objected to the fancy clothing, called "Zoot-suits," that was favored by many Mexican-American youths, and ran them down and beat them. The Zoot-suiters retaliated when they found unwary sailors wandering the streets. The Port Chicago or Mare Island Mutiny occurred in 1944, shortly after a massive munitions explosion at the Port Chicago naval base on nearby Suisun Bay atomized two vessels, demolished the base, damaged the adjacent town and killed 320 men, more than 200 of whom were African-Americans. At Mare Island, fifty African-American service workers refused to load munitions ships at the Mare Island Naval Depot because of the dangerous working conditions (Rice, Bullough and Orsi 1996: 474, 482). In 2000, the Mare Island Naval shipyard would become the home of the Region 5 regional office, marking the end of its ninety-year relationship and identity with San Francisco.

More significant was the removal and detention of people of Japanese ancestry from the entire West Coast shortly after Pearl Harbor, which was considered a war zone at the time. On February 19, 1942, President Roosevelt signed Executive Order 9066, which authorized their removal and relocation. The order created the War Relocation Authority (WRA), and 110,000 people of Japanese descent – even though two-thirds of the evacuees were American citizens – were ordered to assembly centers with little more than the clothes on their backs. After weeks of waiting, they were transferred to ten different "non-strategic" locations in the country, including Tule Lake in Northern California and Manzanar in the Owens Valley, both of which had the capacity to hold 10,000 evacuees (Watkins 1973: 429-432). Most of the detainees remained in relocation centers until October 1944, but when the war ended in August 1945, nearly 44,000 remained in the camps for awhile thereafter, fearful of returning to their homes and the lingering hostility of many Californians. The removal of Japanese-Americans and their detention has been called "our worst wartime mistake, a blunder resulting from fear, confusion, and racial prejudice" (Rice, Bullough and Orsi 1996: 479-480).

Region 5 Publicizes the War Effort

With the declaration of war on December 9, 1941, the city of San Francisco and regional office (RO) personnel feared an imminent attack by Japanese aircraft. Spooked by a "sighting" of a formation of enemy aircraft 100 miles off the coast, "reportedly" approaching the Golden Gate Bridge and the Monterey Peninsula, they sounded the civil defense sirens. The city went dark, searchlight beams prodded the clear sky overhead, and sailors, soldiers and marines rushed to their battle stations – but nothing happened that night (*California Ranger Region Five* 1941: No. 2). Thereafter, San Franciscans read their morning newspapers and went about their daily work.

So did the RO, but with one important change: Regional Forester S. B. Show put the RO on an immediate war footing, starting with double-shift staffing of some seventy-five fire lookout stations for enemy aircraft detection (USDA Forest Service 1942: various). Shortly thereafter, Show suspended the publication of the *California Ranger* and replaced it in May 1942, with a new "confidential" added to the masthead of the *California Region-Administrative Digest*, an action very reminiscent of World War II. Issued by the Division of Information and Education under Assistant Regional Forester W. I. Hutchinson, the *California Region-Administrative Digest* was limited to information of administrative character, which would be useful and necessary to office and field personnel in the efficient transaction of business. This step was taken in response to Forest Service policy regarding release of public information concerning defense activities. Published by Region 5 throughout the remainder of the war, the *California Region-Administrative Digest* provided current and important official information contributing to resource protection, limited information on Forest Service war activities and contributions

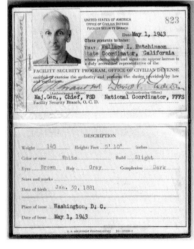

Civil defense identification card of Assistant Regional Forester Wallace I. Hutchinson. 1943

to the war effort, publicity and promotion of government-sponsored activities, and routine notices regarding personnel changes and promotion of safety. Hutchinson acted as the censor for the region, and all informational material with possible military significance had to be cleared through him (*California Region – Administrative Digest* 1942: No. 1; USDA Forest Service 1942: various).

To keep the public informed on Forest Service activities, Show turned to newsletters that were sent to a select list of 1,200-1,500 "key" people in California, many of whom were on the 1941 "contact" list for the Joint Congressional Committee on Forestry (JCCF). The first of these periodic letters, issued on January 30, 1942, discussed the difficulties ahead in addressing forest, water conservation and range problems, as well as fire protection from arson and sabotage, with limited budgets and personnel. Optimistically, Show was prepared to meet these challenges and others as well (USDA Forest Service 1942: various). His letters outlined Forest Service participation in war and defense activities and were designed to make the homefront strong and were also designed to keep the public's eye on the future problems of the Forest Service in California. For instance, in the January letter, Show wrote, "As foresters and public servants, we are not forgetting that in the near future a comprehensive forest program will be needed to check wasteful exploitation of our forests, both public and private, and to insure the proper management of forest, watershed and range lands." In his May 1942 letter, Show continued this theme, stating, "As the tempo of conflict increases, we shall give greater time and energy to the Nation's drive for victory. But in so doing, we must at all costs keep on top of our long-time job – to conserve and protect our valuable natural resources for the crucial days that will follow the writing of the peace" (USDA Forest Service 1942: various).

Show and Forest Service officers also kept the public informed on Region 5's war effort activities through articles in various publications. In 1942 alone, the indefatigable Show published numerous articles on various aspects of Forest Service management during time of war. For example, regarding the general war effort, he composed "The National Forests in Wartime" and "Forest Service War Activities." He also penned "Forest Protection in the War Zone" and "We Must All Prevent Forest fires," about fire protection, and finally, he wrote "Your Forest Playgrounds in Wartime" regarding recreation. Each article was positive and uplifting, and kept the public generally informed of Region 5 wartime activities (Show 1942a; Show 1942b; Show 1942c; Show 1942d; Show 1942e).

Defense and War Actitivies of California Region 5

Supporting this public facade was Region 5's direct work with the armed forces and other military organizations during the war. Region 5's war program was a multi-faceted one, and over the next few years, Region 5 played a key part in the Forest Service's war effort plans in six different areas, the first of which was assistance to farm families for the war effort.

After the Pearl Harbor attack, President Roosevelt appointed the War Production Board (WPB) to coordinate procurement programs for the armed forces and to allocate materials between civilian and military needs. The WPB-Forest Service program provided assistance to farm families, fostered cooperation with livestock owners in production campaigns and helped resolve many farm labor problems. Russell W. Beeson from the RO was assigned to this national defense activity and the California WPB. Under his direction, Region 5 participated in WPB meetings throughout the state, explaining farming priorities and seeking solutions. The Forest Service was also placed in charge of handling all initial applications for farm dwellings costing more than $500, other farm structures of $1,000 and industrial off-farm structures of $5,000. Applications were necessary to begin construction, and if the WPB denied the application, the Forest Service handled the appeal from the prospective builders (*California Region – Administrative Digest* 1942: No. 1).

Region 5 personnel also worked closely with the WPB's "Food for Freedom" Program. This program was designed to stimulate greater food production of fruits and vegetables through better land use, and more meat production through improved range management. Special instructions were sent to all California Region 5 forest supervisors regarding cooperation with

Farmers' market, Fresno, California showing produce grown with water from Sierra National Forest. 1943

livestock owners in the production campaign. In the field, each national forest took action, either by attending meetings, by personal contacts with livestock owners or by completing range studies, registering ranchers, making small increases in grazing allotments, stocking ranges to capacity and opening new ranges. For instance, in 1942, the Los Padres National Forest worked out a cooperative range management plan with stockmen and Army officials in order to open up the quarter-million-acre Hunter Liggett military reservation for livestock grazing. In the same year, the Sequoia National Forest increased its grazing range by using recreation areas for grazing while the war lasted. In 1943, the Modoc National Forest opened new community cattle ranges. Because of recruitment problems in most forest areas due to war industry competition, there was a shortage of rural fire protection forces, so Region 5 personnel helped recruit, organize and train some 25,000 farmers and farm laborers for this important task. Forest officers also served on deferment boards and handled cases for farm workers who were considered essential to the war effort. All of these activities slowed down around June 1944 and then ended immediately after September 2, 1945, when Japan formally surrendered (USDA Forest Service 1956: various).

In the battle for production of strategic metals, another aspect of Region 5's multi-faceted war program, providing critical raw materials proved very important. Region 5's war effort in this regard began with the survey of strategic materials on California's national forests. In March 1942, Region 5 initiated a statewide survey of both developed and undeveloped strategic metals and recommended a plan of access roads for working mines and other roads to develop new strategic mineral deposits. Region 5's mining war effort helped the country in other ways, too. It instructed all forest officers in the identification of strategic and critical minerals, and asked them to watch for mineral development possibilities and to offer assistance to bona fide operators by providing requested technical information concerning land status, topography and road and trail locations. Region 5 also worked closely with the state Bureau of Mines and the state Mining Board in recording known deposits. Thanks to this combination, new manganese, tungsten, chromium and zinc deposits were located on California's national forests, including a 130-mile belt of rich tungsten deposits on the western slope of the Sierras from Mariposa to Kern counties (USDA Forest Service 1956: various).

Once strategic minerals were located, the Forest Service worked to develop the deposits, a goal which the Region 5 achieved in several ways. Region 5 forest officers encouraged and assisted prospectors and mining operators to develop these deposits by simplifying procedures for permits granting access to or across public lands, and by clearing the way for any immediate work and installations in order to meet emergency conditions. Region 5 also built access roads to key strategic metal deposits located on California's national forests, such as an access road to chrome deposits on the Klamath National Forest and a road and camp development permit to a large body of tungsten on the Inyo

National Forest. In the fall of 1942, the divisions of engineering, recreation and lands reviewed fifteen strategic mineral road applications and approved ten, located on the Trinity, Klamath and Plumas National Forests. Region 5 also cooperated closely with mining companies such as the U.S. Vanadium Corporation in laying out town sites for their new operations (USDA Forest Service 1942: various). Finally, Region 5 participated in the conservation of critical minerals. Scrap metal drives were conducted on each forest. One forest reported collecting 1,200 tons from old mills and mines; another forest collected 200 tons from abandoned mines; and yet another, fifteen tons from auto graveyards. Collected metals were shipped to the nearest industrial center for recycling (USDA Forest Service 1956: various; *California Region – Administrative Digest* 1942: No. 27).

California's national forests also collected thousands of pounds of old scrap rubber for reprocessing. Region 5's participation in the Guayule Rubber Project – finding substitutes for imported raw materials – was the foundation of their third contribution to the war effort. With the Japanese invasion of Indochina, the United States was cut off from natural rubber supplies, and a substitute source was immediately needed. Attention was immediately drawn to the guayule plant (*Parthenium argentatum*) as a substitute for the imported raw natural material. For close to decade, the War Department had known that the United States was dependent on overseas sources for its rubber supply and saw guayule, a plant resembling sagebrush and native to northern Mexico and western Texas, as an alternate rubber source. In 1930, Dwight D. Eisenhower, then a young major in the Army, had studied guayule production possibilities for the Army and had even recommended that a 400,000-acre guayule farm be established with govern-

The Guayule Rubber Project was but one Region 5 contribution to the World War II effort.

ment aid somewhere in the American Southwest. For many years thereafter, the Intercontinental Rubber Company had a mill at Salinas, California, which processed guayule from shrubs grown in California and Arizona, but produced most of its rubber from guayule at Torreson, Mexico. With the outbreak of war, the Department of Agriculture acquired the Salinas property and, with emergency funding, established a guayule production facility to extract rubber through a process of grinding, washing and flotation. On February 6, 1942, Region 5 was assigned to oversee the operation (USDA Forest Service 1946a: passim; USDA Forest Service 1946b: passim; USDA Forest Service 1946c: 11; USDA Forest Service 1956: passim; USDA Forest Service 1990: 395-396).

The first urgent task was to increase nursery production of guayule seedlings. Within two months, a force numbering at times up to 3,000 persons worked on the project. Some installed the ninety-four miles of overhead irrigation pipe. Still others constructed seed-treating plant. Many women and high school girls recruited from the surrounding towns and farms were engaged in pulling weeds. Soon thereafter, some 21,000 pounds of seed (all that was available) were planted on 520 acres of nursery beds at Salinas. Other nurseries were established at San Clemente, Oceanside, Carlsbad and Indio, California, to meet the wartime emergency, and the Japanese-Americans relocated to Manzanar planted 200,000 seedlings there and 50,000 at the Parker Relocation Center in southwestern Arizona (Ibid.).

Congress initially authorized 75,000 acres for plantations but in October 1942 increased that to 500,000 acres. The seedlings grown from the Salinas and other nurseries were used in a field planting program on leased land in areas such as Bakersfield, and the Tracy-Newman area. Mills, as well as farm labor camps to house Mexican laborers and their families, were built at these plantations. Housing was built to a higher standard than most area migrant camps. This action spawned not just harassment by the Associated Farmers of California, who accused the Forest Service of wasting taxpayers' money on sinks for the migrant workers, but also a House Agriculture Subcommittee investigation that gave the critics of the housing program a scorching rebuke. In any event, by the spring of 1943, the War Department discovered that synthetic rubber production could be rapidly expanded to assure a continuing supply of rubber to meet military and civilian requirements. Though there was some sacrifice in quality for tire manufacturing, synthetic rubber eventually replaced the slow-growing guayule. All totaled, 2,947,273 pounds of guayule rubber were produced during the war years. The project was liquidated in

January 1946, and most of the fields were destroyed unharvested at that time (Ibid.; Granger 1965: 55-58; California Region – *Administrative Digest* 1942: Nos. 4, 7, 18, 21, 23, 24; 1943: No. 12; 1944: Nos. 9, 14; 1945: Nos. 20, 42).

The fourth major contribution of Region 5 to the war effort was the assistance it provided in defense operations through manning the Aircraft Warning Service (AWS). The AWS was initiated in the late 1930s on a limited basis, but within hours of the Japanese attack on Pearl Harbor, the partially-completed system went into operation, and soon thereafter, Region 5, working with the Army's 4th Fighter Command, assumed direct administrative control of all AWS aircraft warning posts in California, including those located on state lands. Fears of Japanese invasion or attack prompted this program to go rapidly forward, and additional components of the system were constructed to fill gaps in coverage. By September 1943, there were a total of 265 posts on thirteen of California's national forests, many of which were manned on a twenty-four-hour basis – many times by husband and wife teams who were above or below draft age. However, a year later, the AWS became less important because of the introduction of radar posts in 1943, and fear of Japanese attack subsided once the American Navy checked the Japanese advance in the Pacific war theater. By December 1943, the only remaining active posts in California were those located within a sixty-mile radius of San Diego, Los Angeles, Santa Barbara and San Luis Obispo. On June 1, 1944, the AWS was terminated, closing a most colorful mission assigned to Region 5, to operate the twenty-four-hour, yearlong AWS in wild mountain country and desert. These observers, the "skywatchers of the hinterlands," worked through trying experiences of snowbound posts, gales, rattlesnakes and even a few scares and harrowing experiences dodging friendly fire and bombs dropped by overeager pilots in training – yet no AWS observer in California ever saw an enemy aircraft. At the time of the program's termination, Region 5 still operated eighty-three aircraft warning posts (USDA Forest Service 1956: passim; *California Region – Administrative Digest* 1944: No. 23; Cermak n.d.: 479-485; Brown 1943: passim; Williams 1998).

Coupled with Forest Service assistance in defense operations was Region 5's continuous mission to prevent fires caused either by enemy sabotage, by friendly forces or by natural events. Unbeknownst to most of the public, the Forest Service and U.S. Army were secretly engaged in direct combat with the Japanese military along the Pacific coastline. After General Jimmy Doolittle's raid over Tokyo, the Japanese developed and implemented a plan

Ranger Art Gifford and Ann Hanstead plotting location of fire. Mohawk Ranger Station, Plumas National Forest. 1945

that they hoped would set America's western forests afire in retaliation, or at the very least intimidate America. Between mid-1944 and March 1945, the Japanese sent aloft from Honshu and other locations close to nine thousand firebomb-carrying balloons. Using the strong stratospheric wind currents, these silk-paper, hydrogen-filled balloons flew at heights from 25,000 to 35,000 feet over the Pacific toward America. The thirty-foot-diameter balloons took some eighty to 120 hours to reach the shores of North America, where they were detonated by an automatic altitude control switch that was activated when the balloon descended to a certain elevation. Most balloons carried four ten- to twelve-pound incendiary bombs, one thirty-pound anti-personnel bomb and a flash bomb to destroy the balloon itself to prevent recovery by Americans (*California Region – Administrative Digest* 1943: No. 23; 1945: Nos. 22, 24; 1946: Nos. 4, 11; Cermak n.d.: 485-487). By the end of the war, the Klamath National Forest supervisor reported that his organization had spotted many Japanese balloons passing over his forest. Furthermore, numerous suspected balloon cases were reported to the nearest Army control center by Forest Service personnel all over Region 5. During the war, remnants of twenty-three balloon bombs were found in Region 5, and another three after the war. In California, none caused any fires, serious damage or loss of life. All in all, the Forest Service was credited at the end of the war with finding 334 bombs nationwide, but during World War II, it censored all information related to them in hopes that the Japanese would decide that the bombs were not reaching the United States and would abandon the project (Cermak n.d.: 489; USDA Forest Service 1956: passim). Bombs landed from Alaska to Mexico and as far east as Ohio. The bombs failed because the best air currents carrying them were during the rainy season, and damp conditions prevented fires from starting. The only casualties to occur happened in Oregon, when a family found one bomb and triggered it accidentally (Williams 2004).

Ironically, though no fire damage occurred from the Japanese balloon attacks, America's military caused considerable harm to California's forests during training exercises or through carelessness by individuals unfamiliar with the dangers of fire from a simple discarded cigarette. Considering the hundreds of thousands of soldiers who trained in California, it was inevitable that the military would cause fires, which resulted from tracer bullets, hand grenades, artillery shells, tank shells, bombs and rockets. Southern California, where many training facilities existed, suffered the most. Altogether, American military forces caused at least sixty fires in that part of California. The most serious fire disaster in Region 5's history up until that date occurred on the Cleveland National Forest in October 1943. Known as the Hauser Creek Canyon Fire, it resulted in the death of nine marines and one Army cavalryman, with fifteen more men, including two Forest Service officers, seriously burned, resulting in one death when without warning the fire swept over a ridge, trapping the men in Hauser Canyon. Many others suffered burns of a lesser degree (*California Region – Administrative Digest* 1943; Cermak n.d.: 496).

In the meantime, the fear of general "natural" forest fires was uppermost in the minds of most Region 5 officers. With many regular officers off to fight in the war, with experienced fireguards, lookouts and firemen gone into defense industries and with the loss of CCC camps and their forest fire fighting capability, Region 5 was forced to reshape its firefighting organization in order to meet wartime needs. Regional Forester Show first decentralized fire suppression into four zones: Southern California, South Sierra, North Sierra and Northern, with each zone headed by a forest supervisor who acted as a zone coordinator. Next, in order to relieve some of the workload on the typical ranger whose time was consumed with timber sales and other administrative matters, Show created the fire control assistant (FCA) position in order to handle routine fire control duties on some of California's forests (Cermak n.d.: 469). Then, to meet wartime manpower shortages, Region 5 developed cooperative fire control measures with various groups in order to find sufficient personnel to fill out Region 5's normal regular forces. Eighteen sources were considered for recruitment into Region 5's infantry of seasonal firefighters (Cermak n.d.: 471). The final statewide coordinated forest fire control plan included lumber operators and their employees, American Legion posts, volunteer rural firefighter groups, conscientious objector and delinquent youth camps, inmates from San Quentin and Folsom prisons, and university, junior college and high school students, to name a few participants. The inmates were highly motivated by patriotism, or perhaps by boredom or a chance to escape, but nevertheless became effective fire line workers, and use of inmates became an important source of firefighters thereafter and up to the present day (Show 1942c; Show 1942d; Show 1942e; 1942f; USDA Forest Service 1956: passim; *California Region – Administrative Digest* 1942: No. 13; Cermak n.d.: 472-473).

There were many problems with this statewide program, the biggest of which was coordination and direction of the efforts of all these organizations willing to help in time of emergency. Another problem was mobilizing and training these inexperienced firefighters each fire season, and meeting equipment needs. This problem was solved when the federal Office of Civil Defense (OCD) assumed the role. One result of this cooperative effort was

Firefighting techniques: (clockwise) flame thrower used on the Cleveland National Forest

Hand crew working fire line.

Hose crew suppressing a fire.

Firefighting techniques: (clockwise) Hauck torch used for creating a backfire to check the advance of a wildfire.

Felling snags with hand saw after a fire.

Water pack pump. Eldorado National Forest

Fire engine and crew mopping up fire on Modoc National Forest.

the formation of the Forest fire Fighters Service (FFFS) (Cermak n.d.: 471-472; USDA Forest Service 1943a: 8-10). Meanwhile, Show increased the use of firefighting equipment to fill the gap left by departed Forest Service and experienced seasonal firefighters. Tank truck attacks on fires had begun in the late 1930s, and between 1940 and 1945, their use nearly doubled. The use of bulldozers on large fires and new fireline tactics helped as well. By the end of the war, bulldozers also became a very important part of large-fire suppression. Besides motorized equipment, hand tools remained important. Scraping tools (shovels, rakes, the Pulaski and the McLeod tool – a combination rake and hoe), cutting tools (double-bit ax, brush hook and pruning shears), timber falling and bucking tools (falling saw, wedge, sledge and saw oil), hand water-application equipment (backpack pumps) and backfiring equipment (fuses, Hauck torch and flame thrower) were some of the more common tools and small equipment used in firefighting at this time (Cermak n.d.: 470-471; USDA Forest Service 1943a: 24-25). Eventually, concern for fire prevention caused Show to restore fire control back to its former divisional status in April 1945 (*California Region – Administrative Digest* 1945: No. 16).

Besides attacking fires, Region 5 also initiated several forest fire prevention programs. For instance, since careless smokers caused more fires than campers, brush burners, railroads and lumber companies combined, the Forest Service invented the "fag bag," a red cloth sack bearing a fire warning, big enough to hold a pack of cigarettes and a box of matches. The "fag bag" became an institution in Southern California where, as early as the summer of 1941, it was extensively used on the Angeles National Forest. Region 5's fire prevention campaign was complemented with posters, press, radio and meetings on the subject (Show 1942c; Show 1942d; USDA Forest Service 1956: passim; *California Region – Administrative Digest* 1942: Nos. 15, 20). One such Region 5 publication was entitled *The National Forests of California: Their Social and Economic Resources and the Need for Protecting Them From Fire*, which in illustrated form brought home the message that the protection of California's forest was one of the most important duties a citizen could perform in time of war (USDA Forest Service 1943b: passim). One meeting between Forest Service officials and a newly-created War Advertising Council led to a campaign of captioned posters picturing the enemy against a backdrop of burning forests labeled "Carelessness – Their Secret Weapon – Prevent Forest fires." Other fire prevention posters followed. For instance, in 1944, Walt Disney Studios designed a set of fire prevention posters featuring the lovable deer, "Bambi," which sparked the idea to use a symbolic animal for fire prevention – the bear. After some evolution of the concept, the first Smokey Bear character was drawn, which was named after "Smokey Joe" Martin, assistant chief of the New York City fire Department from 1919 to 1930. The Smokey Bear fire prevention campaign continued after the war, eventually becoming one of the best-known advertising symbols in America (Cermak n.d.: 483-484; Williams n.d.).

Finally, Region 5 decided to close off strategic areas on several national forests due to fire hazards such as numerous small areas surrounding powerhouses, dams, bridges, railroads and other strategically sensitive areas on the Lassen, Plumas, Sequoia and other national forests. Closures of larger strategic areas occurred on the four Southern California national forests in the spring of 1942, due to the lack of protection facilities and personnel needed for potential outbreaks of human-caused fires. However, by June, Congress provided additional fire protection funds, and Region 5 was able to accommodate

Angeles National Forest was closed to visitors during WWII.

the usual number of visitors to campgrounds, resorts and picnic areas and to offer them a reasonable amount of freedom from war restrictions (Show 1942c; Show 1942d; Show 1942e; 1942f; USDA Forest Service 1956: passim; *California Region – Administrative Digest* 1942: Nos. 13, 15, 20). Fortunately, Mother Nature was kind to Region 5 during the four fire seasons of World War II. Spring fire weather was near normal for these years, and Southern California enjoyed wetter than usual conditions during the summers. Nevertheless, an average of 69,000 acres of national forest land burned each of these wartime fire season years (Cermak n.d.: 499).

The fifth area of importance to Region 5's war effort program centered on the region's special work to increase production of timber for defense purposes. For twenty years prior to World War II, California had been a prodigious user of lumber and was the leading state in the nation in the quantity of lumber consumed. In fact, during 1941, California used nearly twice as much as Washington, the second-leading state. Furthermore, other states depended on California's timber supply (Show 1944: 4-5). Before World War II began, private lands supplied the bulk of California's timber milled annually by lumber companies. However, as the war progressed and public lands could not meet war demand, the timber harvest on California's national forests grew steadily to fill the gap (Cermak n.d.: 468). Wartime demands for lumber ranged from material for military and civilian housing, where it was used because of its low cost and ready availability, to lumber used for shipping, in shipyards and airplane factories and on flatcars and transports. Docks of wood supported by timber pilings were stacked high with boxing, crating, dunnage and blocking to assure the safe arrival of weapons of war to the front (Show 1944: 5). During the war, the Army estimated 800 uses for lumber and the Navy listed 400 uses, and shipping lumber rose from 5.5 billion board feet in 1941 to 16.5 billion board feet by 1943 alone. Furthermore, as early as November 1942, a Senate War Investigating Committee predicted an acute lumber shortage if timber capacity did not increase. The Senate report estimated that 31.1 billion board feet of lumber would be required for military and civilian use in 1943 against an anticipated production of only 29 billion feet. According to the report, unless additional production of boxing and crating lumber was obtained, transoceanic shipment of essential war equipment vitally needed at the fighting fronts would be seriously delayed (*California Region – Administrative Digest* 1942: 34; 1943: No. 33: 1944: Nos. 5, 27)

Critical shortages of war timber resources made it imperative that the greatest possible production from the forests, consistent with proper forest conservation practice, be obtained. To meet this objective, Region 5's division of timber management steadily increased timber sales on the forests in many ways. The region as a whole made stumpage available for cutting as soon as possible, it conducted surveys and studies in timber use to expand timber sales, it completed appraisals in record time for possible new mill establishments and it surveyed and built timber access roads to speed logging operations. Twelve of California's national forests, many of which had sold only a small amount of timber in the past, showed increased timber sales activity during the war. In 1943, the Plumas and the Lassen National Forests expected record cuts, and one southern forest, the San Bernardino, where timber had been withheld from sale for many years, considered an application for ten million board feet for war boxes and crate use. At this time, Region 5 also set up a special, intensified timber sale training course at the Plumas Feather River School. The course brought timber appraisal work up to date and instructed additional men so as to relieve the Division of Timber Management from the added pressure of increased timber business.

Men from the Eldorado, Lassen, Modoc, Plumas, Sequoia, Shasta, Stanislaus and Trinity National Forests attended the course (USDA Forest Service 1956: passim; *California Region – Administrative Digest* 1943: Nos. 13, 16, 33). With increases in experienced staff, Region 5's timber receipts and the regional cut rose significantly. During 1943, the volume of timber cut under commercial sales and land exchange swelled to 376 million board feet, with a stumpage value of close to $1 million (*California Region – Administrative Digest* 1944: No. 5). In 1944, timber management activities on the forests were even more numerous than the year before – Shasta National Forest cruised, log graded, appraised and marked ten million board feet on one sale; timber sales on the Tahoe National Forest totaled fifty million board feet that season; and the Plumas National Forest completed four appraisals for thirteen million feet and sold twenty-three million feet during the second quarter of 1944 (USDA Forest Service 1956: passim). Furthermore, beginning in 1944, the immense Douglas-fir timber stands of the Klamath and Trinity National Forests became accessible for logging, and soon thereafter came under production (Cermak n.d.: 468). By mid-1944, Region 5 broke all previous records for timber production when the region's total timber cut reached 511 million board feet valued at $1.6 million. The Lassen National Forest led all the forests in both total cut and value, followed by the Plumas and the Eldorado National Forests. Region 5's previous high year was 1930. just before the effects of the Great Depression were seriously felt by many, when 460 million board feet valued at $1.3 million were cut (*California Region – Administrative Digest* 1944: No. 29).

Even so, by January 1945, national military and essential civilian requirements for container lumber were still greater than the supply available. The result was an unprecedented demand for ponderosa and sugar pine, and white fir would soon come into its own to meet the nation's lumber requirements. Good conservation principles were set aside when Region 5 made several war emergency sales on the Stanislaus and the Trinity National Forests to keep some operators and mills going once their stumpage was cut out (USDA Forest Service 1956: passim).

This unprecedented demand marginally declined following victory over Germany in Europe in May 1945 (V-E Day) and again following victory over Japan in August of that year (V-J Day). In celebrating, the *California Region – Administrative Digest* editorialized, "Let us do our part to keep the United States of America from ever becoming a 'have-not' nation." But Bennett O. Hughes, the new assistant regional forester in charge of timber management and private forestry, would need to confront that problem again as a housing crisis developed in California following World War II (*California Region – Administrative Digest* 1944: No. 45; 1945: Nos. 19, 33).

Besides the production of timber for defense purposes, the assistance in defense operations, the finding of substitutes for imported raw materials, the providing of strategic and critical raw materials, and the assistance given to farm families for the war effort, Region 5 participated in many different, but no less important, defense activities. Of course, service in the Armed Forces by Region 5 personnel was one major contribution, as hundreds of forest officers joined the war effort, including S. B. Show's son, who was commissioned a second lieutenant after graduating from artillery school (*California Region – Administrative Digest* 1943: No. 5). By the end of the war, 460 Region 5 employees had been given formal military furlough to join the armed forces. Other personnel were informally furloughed to the armed services, and others left to work in civilian war industries. Region 5 also took part in sabotage prevention programs, in which the Forest Service assisted the Federal Bureau of Investigation (FBI) in checking aliens. Personnel from Region 5 participated in war bond campaigns that exceeded quotas for the region and took a prominent place on many county draft boards as well (USDA Forest Service 1956: passim). One difficult duty for Forest Service personnel was air-ground rescue work, searching for airplanes that crashed on training missions. Too often the task turned from rescue to recovery of the bodies of crew members killed in crashes in the mountains (*California Region – Administrative Digest* 1943: Nos. 26, 29).

Due to a workforce shortage, more women were hired and were gradually able to play a more important part in Region 5's war effort. Each national forest had its own version of "Rosie the Riveter," and women who worked on forests were often nicknamed. For instance, women working on the Shasta National Forest were known as "Shasta Susies" (Pendergrass 1985: 156). During World War II, women were most often employed on fire or AWS lookout work and graduated to other jobs from there. For instance, in September 1942, a class of twenty women was trained on the Lassen National Forest to replace men for lookout relief and fire camp work. By early 1943, the Trinity National Forest supervisor felt women could work into his "he-man's outfit and do a bang-up job of it," and subsequently he employed women not just for lookout duty but also in a variety of jobs. They acted as district dispatchers, firefighters, fire camp cooks, truck drivers and

Women's volunteer motor corps on the Trinity National Forest. 1943

radio dispatchers. They even repaired and maintained tools, and wrangled horses and packed mules. By late 1943, the Eldorado National Forest had five women lookouts, four camp cooks and eighteen AWS observers, a task assumed by many Forest Service wives, while the Sequoia National Forest relied to a great extent on women to help out in their fire organization. These included one woman patrol, six suppression crew members, three fire lookouts and nine combination fire lookouts and AWS observers, a task many Forest Service wives assumed (*California Region – Administrative Digest* 1942: No. 20; 1943: Nos. 4, 32, 33, 44; Cermak n.d.: 473-474). With manpower at a premium, women sometimes temporarily took over the duties of rangers and supervisors who entered military service (Pendergrass 1985: 156), and by 1944, an all-woman tanker crew, including its foreperson, was formed on the Angeles National Forest. But even though all of these women performed well and proved they could do the job, it would be two decades or more before women began to be accepted into the fraternity of fire control staff (Cermak n.d.: 473-474).

Conscientious objectors also worked for Region 5 and were housed in Civilian Public Service (CPS) or conscientious objector (CO) camps on various national forests. The Selective Service Act of 1940 provided that any person, by reason of religious training and belief and who was conscientiously opposed to participation in war in any form could be assigned to non-combatant service. By the end of 1945, there were fifteen Mennonite, seven Friends and five Brethren CPS camps organized on California's national forests.

They occupied rehabilitated CCC camps and worked on a variety of projects such as manning fire suppression camps, hazard reduction, campground development, small timber surveys, telephone line maintenance and equipment repair. They also collected tree seeds, stocked fish, eradicated poisonous plants and conducted insect control work. The CPS program, however, was not at all well regarded by local communities or by forest supervisors. Local communities resented the CPS camps because they had young men and women in active war service. Especially hated were three government camps for those individuals who objected to serving in a camp controlled by one of the historic peace churches. These individuals listed their denomination only as War Resisters League, and most of the assignees were political rather than religious objectors. The most notorious government CPS camp in California was located on the Trinity National Forest. Enrollees were formerly located in Upper Michigan under the control of the U.S. Fish and Wildlife Service, but when the group outlived its welcome there, it was transferred to the Minersville CCC camp. Many considered them the "dregs" of all the CPS camps in the nation. Shortly after arriving, enrollees burned down their mess hall and damaged other buildings and equipment. Subsequently, they regularly put dust in their camp truck transmission and sugar in the gasoline tanks to disable the vehicles. In fact, they did everything possible to irritate and antagonize everyone with whom they came into contact, particularly Forest Service personnel. In the end they did as little productive work as they possibly could do (Price 1948: 1-9; Cermak n.d.: 474). Recalcitrant objectors at the Glendora CPS camp, which operated on the San Dimas Experiment Station, even struck in protest when their discharges were delayed (*California Region-Administrative Digest* 1946: No. 35).

Finally, there was the issue of Region 5's involvement in the War Relocation Administration (WRA) and Japanese internment camps. As will be remembered, in 1942, due to war hysteria, racism and economic greed, 110,000 people of Japanese ancestry in California were forced to leave their homes to be relocated to camps bounded by barbed wire and guard towers. Region 5 assisted the WRA by allowing the use of unoccupied CCC camps for Japanese assembly points. Once at the WRA internment camps such as the Tule Lake Relocation Center in Northern California and the Manzanar Relocation Center in Southern California, Forest Service personnel worked directly with WRA officials and provided aid to the camps on various agricultural projects such as the guayule emergency rubber project. Each Christmas holiday, the Forest Service furnished trees to the camps to cheer these American internees (USDA Forest Service 1956: passim).

Other Actitivies of California Region 5, 1941-1945

The war effort took up most of the time of Region 5 personnel, and they barely maintained many of their non-war duties and activities. The first war year, 1942, was so dominated by the urgencies of the war effort that established constructive projects such as the ranger and advance schools were dropped, and Region 5 undertook no new projects. By 1943, the unusual and unexpected had become routine, and Region 5 was able, according to Regional Forester Show, to "reinstate attention to creative work" (Show n.d.: 253). At this time also, FDR appointed Lyle F. Watts as chief forester. Watts served as chief of the Forest Service during the remainder of World War II. By the time of his appointment in late 1943, the war effort was under control, and he and his staff turned to planning what the national forests and the Forest Service would look like after the war. Chief Watts supported forest productivity to augment rural incomes and maintain payrolls in small communities and to sustain a tax base to support local government functions. Under his administration, the regions were also encouraged "to hire university forestry graduates to help develop forest road systems, and intensively managed, sustained-yield forests" (Williams 2000: 82).

The most essential study at the time in Region 5 was an overdue reanalysis of range conditions and problems. Show, with a sense of guilt for his lack of interest in the subject, promulgated this study, which made a real attempt to look at grazing administration on a planned and uniform basis. A regional grazing review board, which included Regional Forester Show and F. P. Cronemiller, assistant regional forester in charge of the Division of Wildlife and Range Management, held long public hearings on each national forest. During hearings, they dug into range problems with a searching and brutal frankness that long characterized Region 5 fire program reviews. Their final report covered the major trends in livestock use from 1909 to 1944, an analysis of the allotment situation including a description of major allotment problems and an administrative action program (Show n.d.: 254; USDA Forest Service 1945: passim).

The regional grazing review board found widespread biological illiteracy among forest officials, failures to recognize type and soil deterioration caused by overgrazing, paralysis of will in establishing and enforcing essential discipline in use of the range, and putting off of decisions and actions in the "rosy hope that the future would take care of itself." According to their final report, half of the grazing units were found to be in an unsatisfactory condition (Show n.d.: 254, 260-262). One major problem noted was competition for forage by deer, especially on the Modoc, Mendocino and Inyo National Forests. While domestic stock numbers declined after a peak following World War I, deer numbers had grown dramatically since 1924, which was seen as a major cause of overgrazing. The inescapable conclusion was that the Forest Service had no real control of ranges until it controlled deer populations, which needed thinning. The game versus livestock subject was a sensitive one among state game officials, sportsmen groups and outdoor columnists in the West. A second factor causing overgrazing during the previous decade was the growing practice of ranchers sending cow and calf herds to national forest range, while holding beef animals on valley pastures or feedlots. This practice was good animal husbandry, which sought an efficient method of meat production, but the practice also took advantage of the free-range use by young animals allowed by the Forest Service rulebook. Region 5 had not accounted for this practice in the determination of range capacity, which essentially fostered a pattern of sanctioned heavy overuse. The report estimated that the range on some national forests, where this method of animal husbandry was practiced, actually was being used as heavily as when the Forest Service first took over the General Land Office forest reserves in 1905. The plain bald fact was that the Forest Service lacked control of range use in Region 5. The grazing board concluded that most of the range problems were essentially due to "yielding to users [sic] wishes and convenience" and that grazing management had a long and hard road ahead (USDA Forest Service 1945: 4-8; *California Region – Administrative Digest* 1943: No. 49).

The aggressive action and a tighter grazing policy needed would have to wait for a younger generation of forest officers, because during mid-to-late World War II, many critical old-line staff in the regional office retired, including Paul P. Pitchlynn and Robert W. Ayres. Pitchlynn had headed the Division of Personnel Management for decades and also served as the "dean" for the Feather River School. Ayres, of the Division of Information and Education, typified old-time forest officers and had served under every district and regional forester of the California Region. He had the broadest understanding of the organization's history of anyone in Region 5 besides Regional Forester Show. Then there was Jesse W. Nelson, who had started out as a fire guard on the Yellowstone Timberland Reserve under the General Land Office (GLO) and worked his way to head Region 5's Division of Grazing and Wildlife in 1935. Nelson not only helped establish the early national forests but also helped develop lasting conservation policies and held the confidence of the livestock industry and the public he served (*California Region – Administrative Digest* 1943: Nos. 31, 38; 1944: No. 16; California Log 1954: No. 5).

At the end of the war, many forest supervisors voluntarily retired under the service's policy of retirement at the age of sixty-two. They included, but were not limited to, Forest Supervisors Maurice A. Benedict of the Sierra National Forest, Dave Rogers of the Plumas National Forest and Roy Boothe of the Inyo National Forest. Forest Supervisor Maurice Benedict was the first to go and closed his career of thirty-six years of service after working on various national forests, including the Plumas, California (Mendocino) and Sierra National Forests. Forest Supervisor Rogers joined the Forest Service in 1905 under Gifford Pinchot and in 1910 was placed in charge of the Plumas National Forest. For the next thirty-five years – a record that was said to be unsurpassed in the national forests of the United States – Rogers managed the Plumas, which became one of the leading timber-producing forests in the state of California. Forest Supervisor Roy Boothe joined the Forest Service in 1907, trained under the venerated Sierra Forest Supervisor Charles H. Shinn, and after fifteen years there was appointed supervisor of the Inyo National Forest. Under his leadership, the Inyo National Forest increasingly became an important source of water supply for the Los Angeles area, and because of its alpine scenery and good fishing, the Inyo attracted an ever-increasing number of recreationists (*California Region – Administrative Digest* 1944; No. 16; 1945: No. 21).

As these men left the service, Region 5 also underwent a massive field study on workloads, and Show and his regional officers tackled the vexing issue of true time requirements for multiple-use management. They concluded that the job on many districts remained far higher than the officer time available. As Regional Forester Show put it, "Overall the district ranger job had, in fact, become one of general management plus a strong but dwindling component of doing – the 'little supervisor' theory of three decades earlier came to pass.... The day of the one man district was over." Division of overloaded districts was clearly needed. This conclusion resulted in the redrawing of ranger district, forest and inter-forest boundaries and in relocating district and forest headquarters. Instinctive resistance to

S.B. Show at Mohawk Ranger Station, Plumas National Forest. 1945

change was expected and encountered. In the southern zone (Los Padres through Cleveland National Forests), the problems were relatively simple and resistance from forest supervisors was relatively low and non-combative. On the other hand, the southern Sierra zone (Stanislaus through Sequoia National Forests) brought to the fore the vexing question of forest headquarters relocation, which resulted in a few shifts of established district headquarters from "back in the brush" into major city or towns in use areas. Fresno replaced Northfork (headquarters for the Sierra National Forest since the horse-and-buggy days), and Stockton was proposed to replace Sonora as the headquarters for the Stanislaus National Forest. The northern Sierra zone (Lassen through Eldorado National Forests) brought about the first real question of inter-forest boundaries in Region 5, which involved the Plumas National Forest with its neighbors to both the north (Lassen National Forest) and the south (Tahoe National Forest). However, the situation was eventually worked out amicably. The northern zone, that far-flung half moon of five national forests (Modoc through Mendocino), was by far the most complex and difficult zone to change. Existing forest headquarters towns – Alturas, Yreka, Weaverville and Willows – were county seats of purely local significance. The final solution came later, when forest headquarters were reestablished in larger key cities such as Eureka (the dominant business and political center of the redwood region) and Redding (a strategically-located hub that fed all or the main parts of four forests) (Show n.d.: 265-274).

Another change of address for the RO occurred as well. In the spring of 1942, construction of a new federal building in San Francisco began on Sansome Street between Washington and Jackson streets. The new $4 million. fifteen-story federal structure, known as the U.S. Appraisers Building, was completed in August 1944, and Region 5 established quarters therein with other tenants that summer. This out-of-the-way location resulted in a decrease in the tabulation of visitors but would be Region 5's home for decades to come (*California Region – Administrative Digest* 1944; No. 31; 1945: No. 13).

Meanwhile, Region 5 sensed its early formative history coming to an end and a new era of forestry beginning. In 1942, Region 5 and the Angeles National Forest celebrated the golden anniversary of federal forestry in California and the creation of the San Gabriel Reserve (1892) along with the birth of early California conservationist Abbot Kinney. Three years later, Region 5 marked the fortieth anniversary of the Forest Service (1945) as a special occasion. Region 5 reprinted Gifford Pinchot's remarks before the American Forest Congress in January 1905 (which put forth the objectives and organization of a proposed new Forest Service), along with Agriculture Secretary James Wilson's famous letter on conservation use of forest resources (*California Region – Administrative Digest* 1942: 28; 1945: Nos. 5, 8). Perhaps Hollywood sensed some nostalgia of these men for the "by-gone days" of the Forest Service when, in 1942, Paramount Pictures filmed *The Forest Rangers*, a Technicolor fire-thriller romance starring Paulette Goddard, Fred MacMurray and Susan Hayward. *The Forest Rangers* was partially filmed on the Angeles National Forest, and Miss Goddard was made the first woman Honorary Forest Ranger in "recognition of her service to forest conservation and the ideals and objectives of the Forest Service as portrayed by her splendid acting in *The Forest Rangers*" (*California Region – Administrative Digest* 1942; Nos. 11, 14, 16, 17). Nonetheless, recently-appointed Chief Forester Lyle F. Watts, in a message to the service and nation titled "Life Begins at Forty," honored the accomplishments of the first forty years of the Forest Service but reminded everyone that the Forest Service must "carry forward a program of forest improvement and development that will make the Nation's forests play their full part in helping to meet the postwar emergency and future needs" (*California Region – Administrative Digest* 1945:

U.S. Appraisers Building on Sansome Street in San Francisco, California. This building was home to the Regional Office until 1998.

Source: San Francisco Historical Society

Movie stars like Paulette Goddard were used during World War II to advertise Forest Service campaigns. Acting chief of the Forest Service Earle H. Clapp stands at her right and Perry A. Thompson, who later became California's regional forester, is on her left. 1942

No. 8). Following V-J Day in August 1945, Region 5, though elated with prospects of peace at last in the world, was uneasily placed on the cusp of change for the future.

The Demise of Pinchot Conservationism

The demise of Pinchot conservationism came with World War II. Thereafter, a major transition period for Region 5 occurred. There were many factors behind this shift. First, there was the evolving loss of autonomy of Region 5, mirrored by a power shift to the Washington office (WO), first manifested in 1936 when Washington staff introduced the two-man general integrating inspection (GII). Regional Forester Show's struggle and defeat by the Roosevelt administration in the Kings Canyon National Park controversy (1939-1940) marked a new plateau in this power shift. This centralization of power within the Forest Service's WO continued and accelerated during World War II as Region 5 increasingly took orders from the WO and other higher authorities. World War II also created a gulf in program continuity as staff energy and Region 5 resources were guided to meet emergency war demands. At this time, Region 5 started its shift toward overlooking basic conservation principles in many areas such as timber, in order to meet national emergencies and needs. After forty years of essentially conserving timber for future use, Region 5 officials optimistically allowed steadily increasing timber sales for the next three or more decades.

The retirements of key Pinchot-era RO staff and forest supervisors by the end of World War II and the failure of many Region 5 personnel to return to their former jobs after the war created a severe loss of institutional memory and experience, as well as a break in Region 5's conservation spirit. The most acute blow came when Regional Forester S. B. Show decided to retire in August 1946, a year after peace came to the nation, to join the staff of the United Nations to work on international forestry issues and advise nations concerning forest policy (*California Region – Administrative Digest* 1946: Nos. 14, 35, 40). The retirement of Regional Forester Show after twenty years of service, as well as the loss of other Region 5 officers from the Pinchot era, created a break in values and traditions from the early Forest Service years. Their replacements were not schooled by decades of conservation work but were men largely trained to harvest timber, one of Region 5's major wartime activities. Instead of defining and defending the general public conservation interests in a balanced way, Region 5 increasingly found

Fisherman on North Fork of Feather River, Plumas National Forest. 1945

itself supporting the needs and views of individual special interests, such as the timber industry, before those of the general public. Finally, Pinchot's utilitarian conservation of timber, forage, water and minerals resources had given way to multiple-use management that now included recreation and wildlife management. However, after World War II, Region 5, like much of the Forest Service, had to learn to "harmonize a mix of uses while preserving the biological integrity and esthetics of the forest…[and to] reconcile competing demands…[and] set priorities when production and preservation came into conflict" (Hirt 1994: xix).

Chapter VIII: 1946-1954

Golden State of Managing Growth and Multiple Use

Postwar California

The migration to the Golden State after World War II, which literally transformed California, can only be compared to the early Gold Rush days of a century earlier. From 1940 to 1945, California's population increased from 6.9 million to 9.2 million (*California Region-Administrative Digest* 1946: No. 29), but only a decade later, the state had grown to more than 13 million people. This astonishing population influx was stimulated by migration from other states of people seeking to enjoy California's mild, healthy climate, casual lifestyle and economic opportunities. During World War II more than 700,000 soldiers, sailors, marines and airmen passed through California, and of these, more than 300,000 servicemen and women elected to be discharged there. Many of them settled in the state, seeking an elusive Eden-like lifestyle. Once settled, these new residents told their relatives and friends of the delights of living in California, and in turn many of *them* packed their bags and headed west. Though San Francisco listed some population increases, Southern California became the destination of the vast majority of this national population shift. This explosion in population impacted the state in virtually every way (Rice, Bullough, and Orsi 1996; 488; Cermak n.d.: 520; Watkins 1973: 450; Rolle 1963: 567).

Expanding economic opportunity, which matched the growth in population, was the major magnet for newcomers to California. The war had given great impetus to existing manufacturing and also helped to diversify industrial development in the state. A postwar boom in manufacturing followed the war years, and more than 7,000 manufacturing plants were established in California between 1947 and 1954. Much of this growth came from federal expenditures that were closely related to the Cold War and America's involvement in the Far East, such as the Korean War. California took full advantage of this federal largess. California's salubrious climate, available space, established factories and science-oriented universities all combined to broaden the state's economy even further. Federal funds poured into California's defense industries, and the so-called military-industrial complex dominated planning, research and manufacturing in Southern California, transforming that part of the state from a leading aircraft manufacturer to a research and development complex for missiles and space vehicles. Furthermore, after the war, secondary cities grew as well, such as Oakland, Stockton, Fresno and San Diego (Rice, Bullough, and Orsi 1996; 488, 497-498; Cermak n.d.: 522; Watkins 1973: 462; Nash 1973: 214, 230-231; Rolle 1963: 568-569).

The postwar population boom had other economic consequences as well. Despite the glamour of the aerospace industry, agricultural commodities (cotton, vegetables and fruits) were not to be eclipsed in importance to the state's economy. California's farm economy kept pace with manufacturing, and California continued to be the West's leading farm state. California farmers in the Imperial and San Joaquin valleys achieved record agricultural production levels. They emphasized mechanization of their operations, they developed new strains of vegetable and fruits, they used new fertilizers and other chemicals and they took advantage of the cheap stoop-labor of migratory field workers during the harvest season. The *bracero* program, which originated during World War II, was continued after the war and formalized into law in 1951 (Watkins 1973: 462-463; Nash 1973: 214, 230-231).

California's burgeoning economic and population growth had many social consequences that centered on three interrelated areas – suburbanization, transportation and recreation. With the rise of the new economic West came a proliferation of suburban communities that ringed important metropolitan areas such as Los Angeles and San Diego. These sprawling suburbs were very unlike the small towns and cities that were the norm in the pre-World War II era. Most lacked downtown areas, but were instead no more than clusters of residential areas linked by satellite shopping centers with drive-in fast-food restaurants, movies, and shops of all kinds to satisfy the needs of suburban dwellers. Reliance on the automobile and gigantic transportation construction projects – virtually a freeway fever – spawned these suburban communities. Congested freeways covered Southern California like the "latticework crust on a pie," and in the freeway age, California living became unthinkable without the automobile. The postwar population boom and the suburbanization of California created an incredible demand for single-family dwellings. In response to the situation, homebuilders developed mass-production techniques resulting in identical "tract" houses, which became the home of two-thirds of the state's new residents. Real estate developers in counties such as Los Angeles, Orange, Ventura, Riverside and San Bernardino in the south, and the counties south of San Francisco in Northern California rapidly subdivided agricultural land to meet demand. The new "California living" style emphasized ranch-style houses with outdoor patios, barbeques, clipped shrubbery and kidney-shaped swimming pools as integral parts of the property. As people migrated to the new "bedroom" communities and pushed into vital agricultural land, many inner-city areas declined; old buildings

Traffic jam on Laguna Cuyamaca Highway in Laguna Mountain Recreation Area. Cleveland National Forest. 1951

and homes were razed and covered with asphalt to serve as parking lots for commuting workers (Rice, Bullough, and Orsi 1996; 489-493; Watkins 1973: 452-460; Watkins 1973: 452-455; Nash 1973: 219-222). California's unplanned population growth had other consequences, such as impacts on the environment caused by ever-greater demand on resources for recreation. Californians took to the highways in caravans of Airstream trailers and other recreational vehicles, or RV's, placing greater and greater pressure on recreational outlets and California's magnificent scenic beauty. Simply put, more people meant more outdoor recreation by campers, backpackers and sightseers in their off-road vehicles, and more visitors to California's beaches and state parks, not to mention Yosemite, Sequoia, Kings Canyon and Lassen national parks, as well as California's national forests (Rice, Bullough, and Orsi 1996; 489-493; Watkins 1973: 496-497; Nash 1973: 225).

The Late Forties and Region 5 Postwar Inspections

In October 1946, Perry A. "Pat" Thompson, a veteran member of the Forest Service and chief of the division of fire control in the Washington office (WO), was appointed as regional forester of the California Region. This son of a "back country" newspaper editor grew up in the big-timber country of northwest Washington. He worked his way through college as a forest guard, survived overseas service in World War I and eventually returned to the Forest Service to become an assistant forester at Missoula, Montana. In 1939, Thompson was promoted to chief of personnel management for the Forest Service, and then became chief of fire control for the service in World

War II, when Japanese incendiary bombs menaced western forests (*Administrative Digest-California Region* 1950: No. 46).

Thompson replaced Regional Forester S. B. Show, who had served twenty tumultuous years. S. B. Show was "feared" by some, but "admired" and "respected" by many, as evidenced by the toasts at his well-attended farewell. Show had "transformed forest fire control in California from a job into an art form," and according to one source, Show "administered Region 5 as a man's world; expected men to be men, to live by their word, to work hard and to uphold his beloved Forest Service." Remarkably, during the first forty years of its existence, the California Region had had only four district and regional foresters (Olmsted, DuBois, Redington and Show). Over the next decade, Region 5 would have two regional foresters, Pat Thompson (1946-1950) and then Clare W. Hendee (1951-1955). Both Pat Thompson and Clare Hendee were not California sons, and both regional foresters could be viewed as transition leaders for Region 5. Neither Thompson nor Hendee was in place long enough in Region 5 to place their mark upon the region as Show had readily done (*California Region-Administrative Digest* 1946: Nos. 44, 45; Cermak n.d.: 524-526).

Perry A. Thompson, Region 5's fifth regional forester (1946-1950).

S.B. Show just prior to his retirement. 1946

Though Regional Forester Thompson was appointed in October, he did not take office until early December 1946. By that time, three important Region 5 administrative changes had taken place, or were about to take place. First, on June 1, 1946, a new Division of Wildlife Management was created, and F. P. Cronemiller, who had previously handled wildlife incidental to range management, was placed in charge. The division's initial purpose was to concentrate on the problems associated with overuse of range by deer, such as on the Modoc National Forest, and the failure to obtain cooperation with the State of California in management of the herds (Show n.d.: 262-263; Loveridge and Dutton 1946: WL 1-3). Next, on October 7, 1946, a new Division of State and Private Forestry was established. Region 5 was encouraged to take this action because in 1945 the State of California passed a forest practices law aimed at conservation efforts through sustained-yield management in redwood forests as opposed to logging on a liquidation basis. The California legislation also provided incentives for hazard reduction on private lands. Under Governor Earl Warren, the State of California also expanded its forestry service to include conservation and reforestation activities (Loveridge and Dutton 1946: SF 1-5, PF 1-4; *California Region-Administrative Digest* 1946: No. 14). Third, a new national forest was born by combining parts of the Siskiyou National Forest of Region 6 with parts of the Klamath and Trinity National Forests of Region 5, thus creating the Six Rivers National Forest. Though the proclamation would not come until June 3, 1947, Show and Region 5 had pushed for the creation of this forest in the north coast of California as early as 1935. The Six Rivers National Forest became the eighteenth national forest in Region 5. With its headquarters in Eureka, it was logically named after the six well-known rivers within it (Conners 1998: 1-12; *California Region-Administrative Digest* 1946: Nos. 46, 50, 51-52).

Like all new administrators, Regional Forester Thompson was confronted with personnel problems. Retirements continued and key supervisors had to be replaced periodically. With their retirements came a distinct "loss of devotion" to public service in Region 5. Newer personnel appeared not to be living up to the public servant image of the Forest Service, as had been exemplified by men like long-time supervisor of the Tahoe National Forest, Richard L. P. Bigelow, who had recently passed away (*Administrative Digest-California Region* 1948: No. 6). One friendly critic of the new forest officers commented, "The newer men do not admit to themselves that they are servants of the public and guardians of the public's interest and property. The newer men all possess college degrees in forestry, seem to have a superiority complex, generally, that makes them unable to establish proper relationships with their employers, the public" (*Administrative Digest-California Region* 1947: No. 38). Apparently, this comment, making Region 5 officers appear "smug, high-hat, and bureaucratic," was only the tip of the iceberg of public perception of the Forest Service in California at this time, for the popularity of the Forest Service of years ago had diminished among certain sections of the public. Region 5's advocacy of regulation to prevent destructive cutting on private lands, its vigorous efforts to acquire additional lands to place them under forest management, its range reductions to protect watersheds over the protests of permittees, its support for protecting the public interest in wildlife, and similar moves had been outstanding steps in meeting its overall public responsibility, but these actions made Region 5 officials unpopular to many Californians. Furthermore, a trend toward bureaucracy seriously affected Region 5, leading it away from intimate contact with forest users, which was the cornerstone of the public support that had allowed the Forest Service to weather many a political storm. In short, forest officials no longer seemed to live up to the image presented by Saturday matinee idol Roy Rogers as a U.S. forest ranger (Loveridge and Dutton 1946: PR 1-3; *California Region-Administrative Digest* 1946: No. 6). Though Region 5's Division of Information and Education (I & E) had excellent and numerous friendly personal relationships with newspaper editors, radio program directors, movie publicists and various chambers of commerce members, most people's perception of Region 5 was that it only fought fires. Many people wondered aloud, "What else does the Forest Service do, what does it advocate, and why?" (Loveridge and Dutton 1946: PR 3-5).

Regional Forester Thompson took several steps to meet this growing public relations concern. First, in 1948, he announced the appointment of nine leading Californians to serve as an advisory council for Region 5. The region's first advisory council was made up prominent personages from a broad spectrum of the public, including an elected official, a cattleman, a lumberman, an agriculturalist, a labor union representative, a water and power manager, a newspaper editor, a Sierra Club member and a bank representative. At the advisory council's first meeting, Regional Forester Thompson provided them background data on California's forest resources, each of the national forests, and Forest Service organization. The advisory council members thereafter gave a sympathetic hearing to administrative problems such as questionable mining claims on national forests or cooperative plans for increasing timber production through building new access roads, and reported and offered suggestions on possible solutions. This communication link with the broader segments of society, which later included "show-me" trips, was revolutionary for its time and continued for decades thereafter (*Administrative Digest-California Region* 1948: No. 42; 1950: No. 38; 1951: No. 19).

Well maintained home with flower gardens on mining claim near Hayfork on the Trinity National Forest. 1953

Next, I & E prepared several new publications, which sought to provide the public with a better understanding of Region 5's job, problems and accomplishments, alongside a conservation education theme. Charles E. Fox's "Know Your National Forests: A Story of Conservation Through Wise Use" (1948), met this need. Fox, an educational advisor from the Washington office for the California Region, aimed his booklet at high-school- level audiences. The booklet provided them with the basic facts about California's national forests and the Forest Service's multiple-use policy regarding timber, cattle, recreation, wildlife, watershed and fire control (Fox 1948a: passim). Region 5 followed up on this publication with several other educational publications, including "Where Rivers Are Born: The Story of California's Watersheds," which showed how the treatment of watershed affected California's water supply (Fox 1948b: passim), and "Green Gold: Resources of the National Forests of California," which described the forest situation and watershed values of California's national forests to Southern California residents (*Administrative Digest-California Region* 1948: No. 24). Finally, Regional Forester Perry Thompson joined in the public relations campaign. On the centennial celebration of the discovery of gold in California, Regional Forester Thompson pointed out how those early days were the antithesis of conservation and that California's second-growth ponderosa pine were a testament to the fine achievements of the Forest Service in managing the state's eighteen national forests. Thompson concluded that "while California celebrates one hundred years under the spell of gold, she is also looking ahead to the future well-being of her extensive but not inexhaustible supply of 'green gold' growing tall and trim on the timbered slopes of the Golden State" (USDA Forest Service 1947c).

Meanwhile, Region 5 underwent its first thorough and systematic WO inspection and was given direction regarding timber, range, water, recreation and fire control management on the various California national forests. The basic objective of all Forest Service inspections at this time was to determine the extent that Forest Service responsibilities to the public were being met – and if not, what should be done about it. At this point in time, there were five categories of Forest Service inspections: general integrating, functional, boards of review, investigations and fiscal or audits. The general integrating inspection (GII) constituted the primary instrument for program control of the WO over each region and was a long-range planning tool.

Ideally, the GII was carefully planned to supply the WO with the type of information necessary for an overall appraisal of progress in a region and for the translation of regional experience into policy. Upon arrival at the regional headquarters, inspectors contacted the regional forester and worked out a schedule to cover the area desired by the inspectors. At the conclusion of the inspection, a full report covering all aspects surveyed, plus recommendations for necessary changes in program or action emphasis, was submitted to the WO. However, prior to completion of the GII report, the regional forester was given a chance to review the findings and submit a rebuttal on controversial points. If possible, a conference was held between the inspectors and the regional forester, at which time final recommendations were reviewed and corrective action agreed upon. Thereafter, the chief forester signed off on the report, which was bound and placed in an inspection report library for both current and historical reference (Blanchard 1949: 42-43).

Under the strong-willed S. B. Show, two incomplete GIIs of Region 5 took place – the Tinker-Carter GII (1937) and the Granger-Fosling GII (1941). In the summer of 1944, E. W. Loveridge and W. L. Dutton began a third GII, which they completed in December 1946, several months after Regional Forester Show left to work for the United Nations and Regional Forester Thompson was placed in charge (Loveridge and Dutton: 1946). The Loveridge-Dutton GII raised several organizational questions for Region 5 that occurred under the Show administration but were not addressed until after the Thompson administration.

At about the same time, the postwar GII program for individual national forests began. The Cleveland, Inyo and Stanislaus National Forests were the first to be substantially reviewed, in 1945. Thereafter, each year one or more forests were scheduled for a GII. The first round of GIIs for Region 5 was not completed until 1951, and included the Klamath and San Bernardino National Forests (1946), the Angeles, Eldorado, Los Padres and Sequoia National Forests (1947), the Lassen, Mendocino, Shasta, Sierra and Tahoe National Forests (1948), the Plumas, Trinity and Six Rivers National Forests (1949), and the Modoc National Forest (1951). However, Region 5's record hit about 70 percent of the normal inspection frequency, when compared to the composite average of 80 percent for the other regions for the period prior to 1951. Moreover, many of the inspections completed in this first round were considered deficient because they barely rose above a sketch of conditions on each forest (Ibid.: 31-32).

In addition to these inspections of the "mainland" national forests, in 1950 Region 5 also conducted a significant inspection of Hawaii and the Pacific Islands. Forestlands in Hawaii were set aside as the property of the government as early as 1846 under the leadership of King Kamehemeha III. Between then and 1903, various laws aimed at prevention of forest destruction and its consequent diminution of water supply were passed, much like those in California. Then in 1903, the Territorial Forest Service was created to operate under the supervision of a seven-man board of agriculture and forestry. The board was a policymaking and advisory group and lasted until at least 1950, when the Territorial Forest Service became the Territorial Division of Forestry. Prior to 1950, the Hawaiian Board of Agriculture and Forestry received federal allotments under the Clarke-McNary Act (1924) for various activities. The first Hawaii inspection by Region 5 personnel occurred in 1931 (Price Inspection). Under the Clarke-McNary Act, such inspections were to be made annually, and a second "official" Hawaii inspection of Clarke-McNary activities was scheduled for 1940, but the impending war prevented it. A second Hawaii inspection did not happen again until 1950, well after World War II. The obvious reasons for these infrequent inspections by Region 5 were the cost in time and money, and the interference of World War II, as well as the relatively small area under Forest Service control and the lack of issues that competed well for management attention. However, Region

The forestry program in Hawaii became part of Region 5 in 1959.

5 was not made directly responsible for Hawaii and the Pacific Territories until 1959, when a Pacific Division was created. Region 5 thereafter changed its name to the USDA Pacific Southwest Region, and the California Forest Experiment Station became the Pacific Southwest Region Station (Price 1931: passim; Branch and Murray; 1950: passim; Branch and Wilsey 1956: passim; *Administrative Digest-California Region* 1950: No. 20).

As time passed, keeping up this pace of inspections on the mainland, along with overseas obligations, became increasingly difficult. From the late 1940s to approximately 1970, when GIIs were discontinued, the time and effort put on inspection preparedness and execution by inspectors and the regions alike were enormous. During this period of time, the regional office also increasingly lost more and more discretion to the WO as the Forest Service in general became increasingly more centralized in its policy planning.

Another important tool for long-range planning, which was developed in conjunction with the inspection system, was the annual program of work, or the non-recurrent work program. Starting in the postwar period, the WO submitted to the field a document indicating particular fields of work upon which the chief desired special emphasis to be placed for the year, as well as the program for individual functional divisions. Upon receipt of the chief's annual program of work, the regional forester prepared his own letter to each forest, indicating the contents of the chief's program, WO plans and policies, and how to localize and implement the announced objectives. Thereafter, each forest supervisor drew up a highly specific plan for implementing the WO and regional policies on his forest. District rangers were informed in this way of the specific jobs that they must supervise in order to put the chief's "special emphasis" program and individual functional programs into action. This system of long-range planning, along with the GII system, was an outgrowth of Earl C. Loveridge's *Job-Load Analysis and Planning of Executive*

Work in National Forest Administration (1932). Loveridge had clearly taken his organizational ladder ideas and systematic planning principles from Frederick W. Taylor, who had developed them at the turn of the twentieth century for industrial operations and had taken them far beyond Taylor's time and motion studies (Blanchard 1949: vii, 23, 25, 29).

The Early Fifties and Multiple-Resource Area Planning

In November 1950, Perry A. Thompson announced his retirement from his position as regional forester, and Clare W. Hendee, assistant regional forester in the Rocky Mountain Region, succeeded him. Regional Forester Hendee had fifteen years' experience in forest resource management in the Midwest and then a few years in the Pacific Northwest and the Rocky Mountain West. With a degree in forestry, Hendee started in the Forest Service in 1931, at the height of the Great Depression. Thereafter, he served successively as forest ranger, forest supervisor and forest supervisor on the Ottawa National Forest in Michigan (1931-1939), supervisor on the Superior National Forest in Minnesota (1939-1944) and then supervisor of the Mt. Hood National Forest (1944-1946). From 1946 until his move to Region 5, Hendee had been in charge of recreation and lands management for the Rocky Mountain states (*Administrative Digest-California Region* 1950: No. 46).

Clare W. Hendee, Region 5's sixth regional forester (1951-1955).

As an outsider to the California Region, Hendee spent much of his short term as regional forester just meeting with the forest supervisors from California's eighteen national forests and learning about conditions in each forest (*Administrative Digest-California Region* 1951: Nos. 5, 15). For instance, in the summer of 1951, he escorted Chief Lyle F. Watts on a "off the highway," three-week tour of Region 5. This trip was probably as educational for Hendee as it was for Chief Watts, who, during a 1949 visit to California with Regional Forester Thompson, declared that each forest should be managed on a "multiple use" basis (*Administrative Digest-California Region* 1949: No. 3; 1951: Nos. 29, 30).

During Hendee's five-year tenure, organizational issues continued to plague Region 5 and were raised by the 1951 E.W. Loveridge and M.M.

Nelson GII – Region 5's fourth GII. Gordon D. Fox, who conducted the first functional operation inspection of Region 5, reiterated these questions in 1952. Altogether, Fox felt, and the previous GIIs suggested that it was a "fair statement" to say that Region 5 had far to go to obtain management efficiency when compared with other Forest Service regions. In Fox's opinion, the large regional office (sixty staff officers with a GS-9+ rating) talked more about management than it practiced. Fox believed that Region 5 had become too centralized and that its personnel numbers were above the recommended staffing level – 40 percent for regional office staff and 60 percent for supervisor's office staff. Furthermore, Fox believed that supervisor staff levels were over-financed and ranger levels under-financed. Under Hendee, these organizational problems were addressed and fixed (Fox 1952: 1-2, 8, 10).

Public relations issues also concerned the region during Hendee's time as regional forester. As noted earlier, the 1946 Loveridge-Dutton GII had pointed out the growing unpopularity of the Forest Service in California among important and articulate groups and individuals because of several unpopular actions. Five years later, the 1951 Loveridge-Nelson GII pointed out that Region 5's public relations situation in California was still unsatisfactory in many regards, even though I & E had been placed under new leadership when the venerable W. I. Hutchinson retired the previous year. According to the Loveridge-Nelson GII, certain groups such as organized labor, women's organizations, water users, agricultural interests, religious organizations, and the logging and lumber industry were still wholly ignorant of the Forest Service's mission. Region 5 had also, according to the Loveridge-Nelson GII, failed to gain the attention of California's twenty-five congressmen through "show-me trips" and to hold the attention of both regional and forest advisory councils. But in all fairness, the Loveridge-Nelson GII pointed out that public relations was not a program in itself, but part of every job and needed to be addressed as such (Loveridge and Nelson 1951: PR-1-PR-3).

In the years following the Loveridge-Nelson GII, Regional Forester Hendee turned the situation around. Five years later, when Dan Parkinson made a functional inspection of I & E, he found that much had been done to correct the situation and most the problems cited by the GIIs cleared up. By this time, Region 5 had done a fine job contacting important groups such as labor and women's organizations, which were added to the "tickler list" of important organizations and clientele. This list included irrigation districts for water issues, the Sierra Club and outdoor clubs for recreation, the Western Lumber Manufacturers, Inc., for timber, the California Cattlemen's Association for grazing, the Western Mining Association for mining, and the Izaak Walton League and the Audubon Society for wildlife. Under Regional Forester Hendee, Region 5 had also greatly increased congressional "show-me trips," and top regional office staff and field staff had assumed more of their share of I & E responsibilities (Parkinson 1954: passim). Moreover, Region 5 had developed and printed several informative pamphlets such as "Public Campgrounds on the National Forests" (USDA Forest Service 1950) and fact sheets for public distribution and consumption such as "A Few Highlights on California National Forests" (USDA Forest Service 1953a) and "Facts About the Resources and Management of the National Forests of the California-R-5-Region" (USDA Forest Service 1954). In conjunction with these publications, I & E staff and various national forests continued their normal program of visiting schools, participating in seminars at forest schools, releasing press and magazine articles, and participating in radio interviews. They also experimented with television programming for the first time (Parkinson 1954: passim).

In the meantime, in 1951, a second round of GIIs for individual forests began, an inspection cycle that was not completed until 1955. The Stanislaus National Forest was the first forest to undergo the second inspection round (1949), followed by the Inyo, Klamath and Los Padres National Forests (1950), the Angeles, Cleveland, Eldorado and San Bernardino National Forests (1951), the Lassen, Sequoia, Shasta and Tahoe National Forests (1952), the Trinity National Forest (1953), the Modoc and Plumas National Forests (1954) and finally the Mendocino and Sierra National Forests (1955). However, by 1955, inspections continued to fall far short of Forest Service standards and frequency for GIIs (Hendee 1955b). At this time, there was talk of stopping all internal GIIs, and having it done by an outside organization, but nothing came of it (Hendee 1955c with Attachment).

In addition to scheduled WO inspections of Region 5, there were also the periodic "unscheduled" inspections, which caused much consternation among Region 5 officials. The first came in 1951, when Leslie S. Bean was sent by the WO to Region 5 to report on the conditions of various physical improvements such as ranger stations, roads, telephone lines and recreational facilities. Bean's findings and recommendations were less than flattering, citing poor maintenance and sanitation around various stations, guard cabins,

Heavy recreation use on Pope Beach Picnic Area on the south shore of Lake Tahoe, Eldorado National Forest.

lookouts and campgrounds. Bean's report prompted Chief Watts to write to Region 5 with concern, stating, "Is there something wrong with the inspection system, or the follow-up on inspections, that permits these conditions to exist around our stations? Have they been reported before but corrective action not taken?" Caught off guard by the Bean Report and Chief Watts' comment, it inexplicably took several years for Region 5 to respond to them, perhaps because the regional forester did not wish to justify the Bean report. However, after the regional office received a direct request from the chief for a follow-up, Regional Forester Hendee finally responded. In his reply, Hendee reacted sharply to Bean's report and its "shock treatment" or "exposé" approach, stating that Bean's report clearly did not reflect a true, overall summary of conditions as they actually existed. Furthermore, Hendee wrote, "While we do

Using field telephone on the Plumas National Forest. 1945

need to know our shortcomings, so we can focus our efforts to improve them, we feel that such an inspection should be constructive by suggesting ways and means of improvement, or outstanding accomplishment, as demonstrated elsewhere." Privately, Region 5 staff felt Bean, who knew full well that the growing lack of funding caused maintenance problems on Region 5 and elsewhere, had ambushed them. Thereafter, several supervisors were cautioned to be careful when speaking to "official" visitors because they might take considerable advantage of the rather frank discussions during visits and use Region 5's self-criticism to tear down the region (Bean 1951: passim). One can only speculate how S. B. Show would have handled this untoward WO criticism, but it is certain he would have taken a more aggressive approach.

A more "friendly" assessment of Region 5's performance and progress came in 1954, when two additional impromptu inspections took place. The first visit came from Assistant Chief Edward P. Cliff in the fall of 1954, years before Cliff became chief in 1962. The objective of Cliff's visit was to study several problems, including recreational problems on California's national forests (particularly on the Tahoe National Forest), the fire-watershed relationship on Southern California national forests, timber management problems in general and land-use planning (Cliff 1954: passim). Assistant Chief Edward C. Crafts, who also delved into many of the same problems, followed up Cliff's visit with a visit to Region 5. Both memoranda-reports offered constructive observations and were taken that way by the Region 5 staff, although Craft's report was considered highly impressionistic and therefore suspect (Craft 1954: passim).

While worrying about inspections in the early 1950s, Region 5 staff also began seriously considering long-range, or area planning. Area planning was that part of national forest planning that was concerned with integrated management controls of all land use within a specified national forest area.

The roots of national forest planning stem from the 1897 Organic Act, which provided two directions for management for the newly-established forest reserves: one, multiple use of land (timber, forage, water, minerals), and two, the greatest good to the greatest number in the long run. Over the years, the concept of area planning developed along the line of these two management premises. Out of necessity to meet growing pressures for use of public lands and resources, a refinement of this policy eventually included recreation and wildlife resources. By the 1950s, the three basic objectives of area planning were formulated: (1) to direct administration in event of conflicting demands for a particular land area; (2) to provide a guide for making the best use of land; and (3) to provide for continuity in administration of use within specific land units (USDA Forest Service 1952: 1-5).

Once area planning was considered for a project, several steps were required to implement it. The first step was to determine the limits or borders for the area planning unit. Boundaries were determined by a checklist of use and occupancy factors such as key communities, source of users, nature and acuteness of problems, land ownership, political units, road systems, reservoir sites and existing resource management units; and natural factors such as soils, vegetation, topography and rivers. Region 5 gave itself great flexibility in the definition of the area planning unit in order to provide considerable latitude in the size of the unit, thereby making it possible to plan at a forest or sub-regional level. After the boundaries of an area planning unit were established, the next step was to divide the unit into management zones. Management zones represented the principle of highest use. Zone definition simplified the job of analysis of problems and was used to set up smaller tracts for which more specific controls could be written (USDA Forest Service 1952: 1-5).

Once an area planning unit's management zones were established, an inventory of resources within the unit, based upon the principle of multiple uses, was required. There were six major classes of resources or land uses to be considered: soil, water, recreation, timber, wildlife and forage. Where a choice had to made between conflicting claims, and assuming that possibilities for compromise had been exhausted, in theory Region 5 policy and decisions safeguarded resources or land use in the following order, under certain conservation, legal or administrative, and resource management classes:

- **Soil** was Region 5's first consideration – conserved to prevent erosion and to maintain productivity and permeability, administered as a priority for agriculture, and managed to select the most desirable cover types commensurate with soil capabilities.

- **Water** ranked second – conserved by flood control, administrated through cooperation with water-owning agencies to prevent contamination or pollution, and managed through the regulation of yield.

- **Recreation** placed number three in importance – conserved to preserve scenic and use values, administered to set priorities of use and to establish wilderness areas, and managed through special-use permits and layout plans.

- **Timber** held the fourth position – conserved and protected from fire, insects, and disease for sustained yields, administered under working circles and for community stabilization, and managed through marking standards and sale contracts.

- **Wildlife** held the fifth position – conserved to protect habitat and species from extinction, administered by cooperation with the State of California in maintenance of stream flows and public hunting areas, and managed through habitat improvement and fish and game seasons.

- And, finally, **Forage** – conserved for sustained yield, administered by term permits, commensurable standards and a stabilized livestock industry, and managed by grazing seasons and utilization standards.

Marking with paint gun on Plumas National Forest. 1945

Source: Forest History Society

Most of the above uses had a considerable degree of tolerance for each other, and competition could be kept within bounds by the exercise of ingenuity and foresight. But occasionally, as the history of the next few decades proved, competition occurred between two or more incompatible land uses, in which case competition became conflict. Conflicts could occur between man and nature, resulting, for example, in erosion or wildlife extinction, or they could occur between classes of users with special interests. When this happened, the principle of "multiple use" sometimes gave way to the principle of "highest use" (Ibid.: 20-24).

By the early 1950s, the phenomenal growth of California had changed many conditions that affected the administration of national forest lands in the state. Pressures for various resource uses and watershed protection after World War II, especially in Southern California, were far greater than Region 5 ever really anticipated, and therefore regional officers reviewed the region's land management policies. To meet the challenge, in 1950 Region 5 turned to area planning as a scientific management tool for comprehensive land- use planning. To help forest supervisors understand the concept, Region 5 prepared an "Area Planning Guide" to explain it. Furthermore, because of the far-reaching consequences of this concept, Region 5 pre-tested the idea on several Southern California forests prior to embarking on a systematic planning program, which resulted in the *Management Direction for the National Forests of Southern California* (1953). By 1955, Region 5 was so satisfied with the results of this plan, that Regional Forester Hendee recommended that the Southern California type of plan be extended to the remainder of the region (USDA Forest Service 1951a: 28; 1961: Foreword; Hendee 1955a).

Conservation and Multiple Use of Region 5, 1946-1955

Even though timber placed fourth in the classification scheme for Region 5's area planning guide, during the period 1946 to 1955, it dominated much of Region 5's activities, yearly supervisor meetings, inspection reports and annual planning program. Up until World War II, lumber had been the "stepchild" of the Golden State. Prior to the 1950s, wood played a very important part in the state's economy but was sometimes so commonplace that it was taken for granted. For instance, wood provided necessary material for the boxes used yearly for shipping fruits and vegetables. The citrus crop alone required 15 million boxes annually. Other industries were also dependent on wood. Countless fence posts and pole corrals were an accepted part of the far-flung livestock industry, an annual usage that was little considered. The importance of wood products in California's economy was finally highlighted during World War II, as nearly 400 pine and redwood sawmills produced record quantities of wood for war needs. By war's end, California ranked first in the nation as a consumer and third as a producer of lumber. On the basis of these contributions, both in war and peace, timber deserved to be more than a stepchild in California's family of basic commodities (Show 1944: 4-5).

Insatiable demand for timber during World War II required regulation to protect the land from the greed of exploitation and the expediency of speculation. Region 5 tried to encourage private timberland owners and operators to manage their land for the best interests of all through sustained-yield management techniques. In 1944, Congress passed the Sustained-Yield Forest Management Act for this very purpose. This legislation authorized the

Forest Service to enter into long-term, noncompetitive contracts with local lumber mills in timber-dependent communities to assure a continuous supply of wood products. In 1946, Region 5 expected that demands for lumber and other wood products would increase within the state, but with the continued upward trend in population and unanticipated industrial development, demand far exceeded expectations both in scope and quantity. To counter this demand and conserve timber, Regional Forester Thompson applied the Sustained-Yield Forest Management Act to the situation. In 1946, he announced a 1.7 billion board feet, fifty-year agreement with privately-owned timber operators on the Plumas National Forest under the Sustained-Yield Act to assure a sustained yield of forest products from 106,000 acres of timberland near Woodleaf, California. Logging on the unit was to be done in accordance with Forest Service cutting practices, which provided for leaving a reserve stand of timber in the cut-over area as a source of future crops of timber, with the land maintained in productive condition perpetually (*Administrative Digest- California Region* 1946: No. 6).

However, most lumbermen, represented by groups such as the National Lumber Manufacturers Association (NLMA), American Forest Products Industries (AFPI), and the American Paper and Pulp Association (APPA) rejected Forest Service efforts under the Sustained-Yield Act as "collectivism or socialism," (Steen 2004: 259-269). Furthermore, county chambers of commerce openly stood in the way of any federal regulation of private forestlands. They also blocked Region 5's efforts to acquire timbered lands through purchase and exchange, openly opposing Region 5's purchases of redwood tracts in Northern California to create a Franklin Delano Roosevelt Memorial Redwood Forest, and pine tracts in the Sierra foothills. Most California lumbermen in the post-World War II period, with the exception of a few progressive lumber companies, continued to operate on a liquidation basis, with no conscious regard for the condition or the future of forests. In fact, the 1946 Loveridge-Dutton GII predicted that 90 percent of the pine on private lands would be liquidated in fifteen years' time if the fifty-two operators continued production at their present levels, and only twelve of them would survive past twenty years (*California Region-Administrative Digest* 1946: Nos. 18, 46; Loveridge and Dutton 1946: PF 1-4, TM-1). The 1951 Loveridge-Nelson GII reiterated this problem, pointing out that the timber drain on private forestry lands had doubled from 1944 to 1950 from roughly two billion to four billion board feet. Moreover, in the interim, a "mass migration" of lumbermen and plywood manufacturers from the Pacific Northwest set up businesses in California, tending to exacerbate private forest conditions. Region 5 officials hoped that the recently passed California Forest Practices Act would provide the needed regulation of cutting practices to obtain better forestry on state and privately-owned lands (Loveridge and Nelson 1951: PF-1, TM-1). In the intervening time, the *San Francisco News* wondered out loud if the wholesale harvest of timber by logging companies, which the Forest Service estimated would reach an estimated 364 million board feet by 1948 – or enough for 36,000 ordinary six-room houses – would ever be replaced (*California Region-Administrative Digest* 1946: No. 38).

In the meantime, Region 5 realized that timber demand could no longer be met through private forestry, especially as mills and lumber-oriented communities increasingly looked to California's national forests for their future supplies. Considerable pressure was put on the Forest Service to open up and provide more logging opportunities. To meet this reliance on national forest timber and avert the impending timber crisis, Region 5's timber management policies gradually changed in one important respect – there was a push for new forest roads to access fresh timber stands and to speed the flow of timber to market from California's national forests.

The initial push for timber access roads began as early as 1946, when California Governor Earl Warren supported the effort (*California Region-*

Forest Service scaler scaling logs near a landing on the Cone Mountain Sale. Lassen National Forest. 1961

Administrative Digest 1946: No. 19). But the real push did not come until 1951, following an upswing in California timber sales. In 1945, Region 5 timber sales amounted to approximately $750,000. Two years later, regional sales topped $1.5 million (*California Region-Administrative Digest* 1946: Nos. 5, 30; 1947: No. 29). But in 1950, demand for all western species came from every direction, including the outbreak of the Korean War on June 30, 1950. Demand dwarfed all available supplies, leading the West Coast into the wildest lumber market in its history. Frantic buyers placed orders on carloads of lumber, with the price left open. By year's end, California forest industries set a new all-time production record of close to 700 million board feet (*Administrative Digest-California* Region 1950: No. 36). Timber sales on national forests nationwide became so heated that in January 1951, the Office of Price Stabilization halted them until the Forest Service could devise a valid procedure for setting ceiling prices for saw logs, posts, pulpwood and other national forest wood products. Timber sales were halted again from March until April because permissible prices determined earlier had been inadequate (*Administrative Digest-California* Region 1951: Nos. 7, 11, 14). Meanwhile, to meet this domestic and defense lumber demand, in early February 1951, timber management chiefs of the western regions met in San Francisco with Ira J. Mason, chief of timber management for the Forest Service, and with industry groups such as the Western Pine Association (WPA). In San Francisco, they discussed timber policies, and Mason concluded that additional production of national forest timber for defense needs should come from the development of the more inaccessible working circles by means of access roads rather than from rapid liquidation of timber in the more accessible areas, where inventories should be retained for sudden and greater emergencies (*Administrative Digest-California* Region 1951: No. 8).

In response to this decision, the 1951 Region 5 supervisors' meeting gave special attention to timber access roads and other phases of management that were closely tied with the production of materials for national defense (*Administrative Digest-California* Region 1951: No. 15). A month later, Regional Forester Hendee presented to the Region 5 advisory council a cooperative industry-government plan to obtain more timber for defense needs by building access roads into a vast area of Douglas-fir in the national forests of northwestern California. "This is a big step," according to Hendee, "toward our goal of increasing the production of timber from California's national forests, from 560 million board feet per year to 1.3 billion board feet per year in the next ten years" (*Administrative Digest-California Region* 1951: No. 19). Chief Watts reiterated these numbers during a visit to the California Region in July of 1951 (Watts 1951: 1-2). Subsequently, the advisory council endorsed Region 5's plan, believing that the timber access road program would be a "sound investment by the lumber industry and the various governmental agencies taking part" (*Administrative Digest-California Region* 1951: No. 19). Thereafter, the financial aspects of the timber access road program were worked out by the regional office and presented by the end of the year in a publication entitled "The Need for Timber Access Road Development and Responsibility for Financing Construction." Region 5 justified financing the roads through an expected increase in national forest receipts, along with expenditures from the State of California, counties, and timber operators (Loveridge and Nelson 1951: E-2). By the end of the year, "maximum material production had become a national duty and a moral imperative," which President Harry Truman's Materiel Policy Commission expressed in their voluminous publication, *Resources for Freedom* (1952), in which they advocated maximum production of lumber and pulpwood in the interests of national security. Before long, the foresters began to believe that their "overriding purpose was not so much to protect the national forests but rather to develop their resources to meet the material needs of the American people." In other words, "they saw their mission as one of overcoming limits, not establishing them" (Hirt 1994: xxii).

In 1953, this "can-do" optimism permeated Region 5's timber policy as foresters promoted the idea of developing access roads to inaccessible timber stands. Using a ten-county area in northwestern California as an example of the situation, Region 5 estimated that this area contained some seven million acres of timberland; upon which grew 153 billion board feet of saw timber – 59 billion board feet of which was in national forests. Region 5 officials argued that this timber was needed not just for the country's expanding defense effort but also for satisfying the country's peacetime requirements. They argued that it could be harvested on a continually productive program if strategically planned timber access roads were constructed in order to transport timber from the forest to points of manufacture. Existing roads within this targeted area – the Six Rivers, Klamath, Shasta, Mendocino and Trinity National Forests – meant that an additional 344 million board from this area could be expected (USDA Forest Service 1953b: 1-4). With this optimistic assessment regarding Region 5's annual allowable cut (AAC, now called allowable sale quantity or ASQ) and the support of timber industry groups such as the NLMA, the mileage of roads in California's national forests steadily rose. As roads reached into these rugged, inaccessible and undeveloped areas of timber stands, few raised the question of conducting too much logging (McKinsey & Company 1955: 2-3-2-6).

By 1955, public timber represented the major remaining source of supply in the areas of heaviest lumber production. By this date, the national forests of Region 5 had become an active source of supply for current timber needs, instead of serving as static reserves. In hearings before Congress a year before, WO Chief of Timber Management Mason explained to congressmen that efforts to build up the AAC to full sustained-yield cutting capacity had placed timber sales in the category of "big business." Forest Service personnel, Mason explained to them, had become "resource businessmen" rather than "resource managers." Foresters, who had grown up in an era when selling timber was a very minor part of the job, now had new responsibilities and needed to recognize that business management was now as important as technical forestry (McKinsey & Company 1955: 1-5). The idea of timber sales as "big business" sowed the seeds of timber mismanagement for the following twenty years and resulted in public distrust (Cermak 2005).

By the end of 1955, timber had become big business, although there were several problems that needed working out, such as managing and paying for the huge right-of-way and road construction costs. One solution was to have the company furnish the right-of-way as part of the sale. This forced the companies to cooperate and jointly participate in road construction. However, the Forest Service wished to have multi-purpose roads, and that burden could not be placed on companies as part of a standard sale contract. That problem was not solved until the passage of the 1964 Roads and Trails Act, which allowed the Forest Service to use public funds to build roads for all uses and then collect fees from commercial haulers who wanted to use them (Leisz 2004: 33, 113).

If timber had become big business in Region 5 during the mid-1950s, then so had fire control, for it was needed to protect this sizeable and valuable resource. Starting in 1945, Region 5 made several changes in its fire control and protection management program. First, in April 1945, Region 5 acknowledged the growing importance of it when S. B. Show promoted fire control from a section in the Division of Operation to a division, itself, a status it held prior to 1939 (*California Region-Administrative Digest* 1945: No. 16). The Loveridge-Dutton GII approved this action because of the enormity of the fire control duties on Region 5 (Loveridge and Dutton 1946: FC-2). Fire losses during the war years were generally low, but during the late forties, they sharply increased. Region 5 embraced new technology and experience learned from World War II – first smokejumpers, then helicopters.

Air drop of supplies over Angeles National Forest. 1947

During World War II, Region 5 heavily relied upon the military for backup power to fight fires, including providing smokejumpers who worked wilderness fires. Actually, the parachute program had developed out of other phases of aerial warfare against fire that began shortly after World War I. At that time, the Forest Service and the Army Air Corps cooperated in organizing an aerial patrol, first in California and later in Oregon, Washington, Idaho and Montana. By 1934, it was suggested to drop firefighters by parachute, an idea many laughed at, but when a parachute was developed to safely drop men into timbered country, the idea came of age. By 1944 the experimental period was over and parachutists were reportedly used successfully on one fire in Region 5 that year, so when thousands of veterans with parachute

Smoke jumper in mid-air. 1940

The Ever-Changing View ■ 350

experience returned to California after World War II, they bolstered Region 5's pool of smokejumpers. Following the war, Region 5 worked out a program to use them. By 1951, smokejumpers were seen as the quickest way to the fires in the backcountry and had become commonplace on Region 5 (Loveridge and Dutton 1946: FC-7; Clark 1951: passim; Cermak n.d.: 526-532).

The real innovation in the war on fire, however, came with the incorporation of the helicopter into Region 5's fire control arsenal. In 1944, during the height of the war, Region 5 proposed adapting helicopters to forest fire control work, an idea that the Loveridge and Dutton GII found to

Cowboys looking at helicopter, Big Bear Lake, San Bernardino National Forest. 1946

be an original and excellent idea that offered something "real and substantial to fire control of this Region" (Loveridge and Dutton 1946: FC-7). In late 1945, Region 5 pushed forward on the idea. Region 5 initiated testing of their idea when the Fourth Air Force assigned six R-6 Vought Sikorsky helicopters from March Field to the Forest Service in December of that year. Working on the Angeles National Forest, the Sikorsky helicopters were first used to conduct fire detection patrols and to determine the practicability of discharging and picking up firefighters. In the latter task, rope ladders were raised and lowered on hoist lines. Thereafter, they were employed in a variety of tasks unrelated to firefighting, such as "making game and stock counts, sowing seed over burned over areas, photographic and map work, and even stocking remote mountain lakes with fish" (*California Region-Administrative Digest* 1945: No. 50; USDA Forest Service 1945a). The firefighting flights, and others made by the newer R-5 Sikorsky helicopter, were so successful that Region 5 triumphantly demonstrated this firefighting method to the public on the Angeles National Forest by mid-1946 (USDA Forest Service 1946d). Another chapter was added to Forest Service fire history when in September a Sikorsky R-5a helicopter was used for the first time to scout a hunter-caused fire that started on the Angeles National Forest – the Castaic Fire (*California Region-Administrative Digest* 1946: Nos. 25, 37; USDA Forest Service 1946a). Naturally, Region 5 provided the press with many catchy-titled articles such as "Helicopter Hopes

Unloading injured person on the Angeles National Forest.

for Forest Fire Control," "Flying fire-Engines" and "Egg-beating A fire – Southern California Style" in magazines and journals such as *American Helicopter*, *American Forests*, and *Journal of Forestry* (USDA Forest Service 1946d; USDA Forest Service 1947b). As more tests took place, production of newer helicopters was already underway by the Bell Aircraft Company (Cermak n.d.: 534). One firefighter remembered these old "no-door" Bell helicopters well, especially not knowing if that thing was going to take off or not. Firefighters appreciated the ride to the fire, but it was a one-way ride. They were expected to walk out, sometimes many miles to a nearby road (Beardsley 2004: 14-15).

The helicopter became a great tool for fire control, but it was no match for the bad weather conditions that occurred during the 1950 fire season. In that year, fire conditions in California were at their worst since the early 1920s – years of subnormal precipitation, a warm and dry spring followed by a one of the hottest summers on record. Major fires occurred on the Modoc, Los Padres and Plumas National Forests, with the largest fire of the 1950 season occurring on the Cleveland National Forest. By season's end, more than 220,000 acres of California national forest land had been blackened. The 1950 season, however, could have been much worse if not

Fighting the Wheeler Springs Fire on the Los Padres National Forest. 1948

for the help provided to the exhausted firefighters by off-forest groups, which included ten regional "Hot Shot" crews from the Cleveland, San Bernardino, Sierra, Eldorado, Plumas and Shasta National Forests, and the importation of out-of-region help such as the Mescalero and Hopi Indian crews – the first time Region 5 used Native American crews from the Southwest. They worked side by side with the regular firefighters cutting fire lines and established such a good record during that year that Region 5 continued to use them as a major source of reinforcements thereafter (Cermak n.d.: 571-578; Beardsley 2004: 13).

Using water trucks to fight the Indian Canyon Fire on the Santa Barbara National Forest. 1933

The losses of the 1950 season shocked Region 5. The white glare of urgency for fire control became the most critical internal problem facing Region 5, much as it had after the devastating 1924 fire season (Loveridge and Nelson 1951: FC-1). Subsequent fire seasons were more tragic, such as the 1953 fire season still remembered by many in Region 5 because of the terrible tragedy of the Rattlesnake Fire on the Mendocino National Forest (Asplund 2004: 18-19; Peterson 2004: 27; Radel 1991: 11-13; Leisz 2004: 21-22).

The 1953 fire season started out badly, when the Monrovia Peak Fire started in late December 1952. Santa Ana winds picked up and threatened the observatory at Mount Wilson and the television stations located nearby. There was even talk of canceling the Rose Bowl parade and game. The fire was contained and miraculously no lives were lost, but many cabins were destroyed. Public concern over potential structural damages along the wildland interface with urban communities led to a new fire priority that in some cases sacrificed "perimeter control" of a fire for protection of structures. The Monrovia Peak Fire also eventually helped create the Los Angeles County Watershed Commission, as well as an organization later known as the Southern California Watershed Fire Council (Wilson 1991: 23-27).

It was the Rattlesnake Fire, however, that later really changed Region 5's firefighting policy. The Mendocino National Forest had an incendiary problem, people purposely setting fires to clear brush or to get a job, and the Rattlesnake Fire was purposely started by a young man who needed work. When the fire unexpectedly changed fronts during the night, the intense heat and fire overtook fifteen men who tried to outrun the fire by going downhill. All were consumed in the flames, including two Forest Service officers. The majority of the dead were well-disciplined firefighters from a Christian overseas missionary camp who called themselves the New Tribes Mission. The arsonist was charged with murder, but he was convicted of only two counts of arson and sentenced to a long prison term. The senseless tragedy left Region 5 shaken once again. The 1953 fire season, which left more than 130,000 acres burned in three California fires and left fifteen dead from the Rattlesnake Fire, led to many Region 5 changes and marked the closing of an era (Radel 1991: 11-13; Cermak n.d.: 606-610, 620; Leisz 2004: 22-24).

After the Rattlesnake Fire, fire control became more and more a separate field, developing into a new, larger, cooperative fire organization and, indeed, a fire department. Increasingly thereafter, Region 5 placed fire suppression as the top priority responsibility in the region. After all, it was argued, "if resources are destroyed, management becomes futile" (Loveridge and Nelson 1951: P-1). To meet this priority, Region 5 progressively employed more aircraft to attack fires on the frontlines instead of men. Though helicopters had been used for several years by this date, this was the first time they were used for a direct attack on fire with tanks and buckets of water – or helitack – now a common Forest Service term. Region 5 became the most aggressive region in organizing helitack crews (Wilson 1991: 33-36; Leisz 2004: 68-69). The Forest Fire Laboratory at Riverside contributed to this step through Operation FIRESTOP. This program was the genesis of using air tankers to spread fire retardant (the first retardant was borate), demonstrating that the method was both feasible and effective in combating wildland fires. These air attack techniques and tools, inspired by FIRESTOP, revolutionized fire suppression. They also made firefighting exponentially more expensive (Wilson and Davis 1988: 6; Cermak n.d. 621).

As the war on fire heated up, another war was being fought – a range war. As will be remembered, in 1945, a range allotment analysis resulted in the conclusion that Region 5 completely lacked control of range use (USDA Forest Service 1945b). In the words of one report, Region 5 had "arrested depletion and permitted recovery of some areas; yet, because of the lack of full-scale facilities to control livestock and to aid nature in the job of rehabilitation, these forest ranges were far from being fully productive."

Dry Lake Allotment at head of Deadcow Creek. Cattle in poor condition on severely depleted upland range, Klamath National Forest. 1945

While Region 5 acknowledged that an intensive range management program was desirable, their initial approach to the problem was additional livestock reductions (USDA Forest Service 1947a: 7-8). Reductions were especially needed in the numbers of livestock in the High Sierra and in the high mountain country of the Klamath and Trinity National Forests. Now that the war emergency was over, recreation demand had both returned to and markedly increased in these areas. In light of this recreation pressure and the opening of new recreational developments, both the Loveridge-Dutton and the Loveridge-Nelson GIIs pointed out that Region 5 needed a fresh look at grazing use in the context of conflicts with recreational use (Loveridge and Dutton 1946: G-1-G-6; Loveridge and Nelson 1951: RM-3).

Region 5 expected resistance from the livestock industry to reductions in livestock numbers, and it came in early 1947 from two organizations, the National Wool Growers Association (NWGA) and the American National Livestock Association (ANLA). Both organizations contended that much of national forest land was primarily valuable for grazing, and both adopted resolutions inviting the federal government to make this land available for purchase at low prices by the permittees who grazed their cattle and sheep upon these ranges. In fairness to the California livestock industry, while not all stockmen and livestock organizations in the state supported the range-transfer program espoused by the NWGA and the ANLA, it seemed the "ultimate" goal for cattlemen and woolgrowers (Thompson 1947:4-5; Steen 2004: 272; Clawson 1950: 109-110). In California, Regional Forester Thompson naturally lashed out against this proposal. Using words reminiscent of Gifford Pinchot and Teddy Roosevelt's Progressive-era rhetoric, Regional Forester Thompson labeled these organizations "special interests" that threatened the philosophy behind the national forest management under multiple-use principles. During the fortieth anniversary celebration of designating the forest reserves "national forests," Thompson declared that it was time for the people of California to decide whether their national forests should be managed for the "greatest good of the greatest number in the long run, or should they be preempted by a single class of forest user for a single purpose" (Thompson 1947:4-5).

Meanwhile, Region 5's efforts to reduce livestock numbers on the Modoc National Forest after World War II drew national attention when livestock owners who grazed stock in the North Warner ranges of the forest rebelled. They felt the Forest Service reductions were ruining the main industry of the county and denying them the right to a living. Instead of reductions, they preferred greater Forest Service investments in range improvements (fencing, water developments and reseeding), which together with better stock handling would build up the carrying capacity of ranges. Consequently, in 1949, they banded together and hired an attorney to lobby their cause to various groups such as the Chamber of Commerce, Rotary Club and the Bank of America, a major investor in loans to California's livestock industry. Furthermore, they collectively closed 10,000 acres of their land to urban hunters by posting signs that read in bold letters "NO TRESPASSING: Due to Forest Service Regulation, The Owner of This Property Feels Compelled to Post This Notice." This tactic was clearly devised to drive a wedge between the Forest Service and one of its more important constituents, hunters (U.S. House of Representatives 1949: passim; Rowley 1985: 198).

The ranchers' lobbying and campaign resulted in a congressional hearing on the matter. Held on the Modoc National Forest, the hearing was headed by six-term Congressman Clair Engle (D-California). At the hearing, Regional Forester Thompson defended Region 5's reduction program. Thompson admitted that Region 5 could be criticized; forest officials had failed to make adjustments on some of the overstocked allotments because they were reluctant to face opposition or to cause financial loss to permittees. Region 5 also hoped that wet years might improve grass and soil conditions, or that changing economic conditions would automatically bring about reduced demands for forest range. Thompson's statement was given not as an excuse but as an explanation. Nonetheless, Thompson placed part of the blame for overgrazing conditions on the stockmen. To Thompson, they refused to recognize the facts of depleted ranges; they temporized, resisted adjustments, and also hoped that favorable conditions would develop and remove the need for changes. Thompson agreed with the ranchers that range improvements were needed, but he informed the congressmen that they would take time to work. In the meantime, Thompson felt that a start must be made somewhere and that stock reductions were in order. Despite Thompson's pleadings, Congressman Engle sided with the cattlemen. At the end of the hearing, Engle, according to Thompson, summed up the hearing stating, "As long as fat cows came off the ranges, he was convinced the ranges could not be over grazed." Then Congressmen Engle made a big play for the cattlemen's approval by stating he intended to work to have all grazing lands transferred to and administered by the newly-created Bureau of Land Management (BLM) (U.S. House of Representatives 1949: passim).

In the next election, Congressman Engle was defeated, but the divergence of opinion between the Forest Service and ranchers over the rate of stocking, as well as the rancher's quest for security in his use of federal range, continued. Even though relationships elsewhere in the region with the livestock industry were noted as quite satisfactory, the 1951 Loveridge-Nelson GII put Region 5 at the bottom of the national list in management of range and related resources. They ranked the region low because of Region 5's tumultuous relations and deteriorating range conditions in the Modoc country. To remedy the state of affairs, in the years to come Region 5 promulgated a five-year range management program, which placed greater emphasis on range improvements and research and less importance on stock reduction (Clawson 1950: 117-118; Chapline 1951: 635-636; Loveridge and Nelson 1951: PR-2, RM-1-RM-3).

If the 1949 Modoc "range war" were not explosive enough, another range issue – brushland burning or controlled burning – became the most explosive situation facing the Forest Service in California in some time. The issue of brushland burning flared up when the California State legislature authorized the California Division of Forestry (CDF) to experiment with controlled burns (later called prescribed burns) for range improvement. In many ways, this practice harkened back to the "light burning" controversy of the early 1920s. With increased population pressure in California and high returns from livestock production, farmers and ranchers passionately desired to convert seemingly worthless brush fields into useful grass-producing range. Some claimed that up to one million acres could be converted to grass through prescribed burns (Loveridge and Nelson 1951: BB-1-BB-2, Cermak n.d.: 565).

Unlike his predecessor, who had such a experience with the Modoc range war, Regional Forester Clare Hendee avoided public confrontation by initiating positive action. Though some Region 5 forest supervisors, whose forests adjoined the burns, naturally feared the program, Regional Forester Hendee decided to take a constructive approach to the matter. In 1951, he authorized cooperation with the San Dimas Experimental Station to study the subject, and he instructed Region 5 forest supervisors to meet with local burn committees so they would better understand Forest Service policy and problems with the practice. Region 5 conducted only a few tests of prescribed burning before Arthur W. Sampson and L. T. Burcham published their report, *Costs and Returns of Controlled Brush Burning for Range Improvement in Northern California* (1954). This report indicated that only a small portion of California's nine million acres of brushland were suitable for conversion to grass, that chamise-covered lands were usually unsuitable for conversion and that brushland burning costs were prohibitive. With this information,

Region 5 de-emphasized the prescribed brushland burning program until the 1970s, when renewed interest arose in relation to fuels material management (Cermak n.d.: 566, 569-571). At that time, the program was expanded, often using habitat funds from the California Department of Fish and Game. What developed was an excellent fawning range, a more friendly mosaic of brush and a reduced fire hazard (Leisz 2005).

Controversies other than those associated with range management also plagued Region 5 in the decade following World War II. The most public and important one involved managing California's phenomenal recreation growth after the war, which put the Forest Service's multiple-use policy to the test.

With the end of World War II, Californians once again visited and traveled to national parks and monuments. As recreation picked up in peacetime, California's national forests received a tremendous load of recreational use, largely because of their extensive area and geographical distribution and because of their beauty, accessibility and low cost. In contrast to other forest resources, forest recreation was one product from forestlands that served the state's population directly. For this reason, California's astonishing population influx immediately following the war stimulated an upward trend in demand for recreational facilities that needed to be met. The changing nature of recreational demands was another problem confronting the region. As California's population increased, prospects for outdoor experiences in more natural settings decreased, especially for children. Camps for groups such as the Boy Scouts were needed to provide the camping experience for youngsters ten to seventeen years old, and California's national forests were inundated with applications in the postwar years. For instance, Barton Flats on the San Gorgonio Wilderness, one of the original wildernesses established by administrative order, at one time had more than 300 people camped around a single meadow. "Everybody and his brother," according to former Chief Max Peterson, "wanted an organization camp in Barton Flats, and we were just running out of space, and then people would build an organization camp, they wouldn't have the money to maintain it" (USDA Forest Service 1947a: 9; Peterson 2004: 42-45).

There was also a dramatic rise in winter use of the mountains. Skiing in California goes back to the gold mining days, when many communities held cross-country ski races. In the late 1920s and early 1930s, interest in skiing picked up once again, especially in ski jumping, but the Forest Service did not get involved in winter sports until the late 1930s. By America's entry into World War II, winter sports on national forests had become popular in the West (Bachman 1990: 36-37). By 1945, many of the national forests' winter playgrounds had become known all over the country, including Timberline Lodge on Oregon's Mount Hood National Forest, Alta at the backdoor of Salt Lake City in the Wasatch National Forest and Sun Valley in central Idaho, surrounded by the Sawtooth National Forest. After the war, interest in the packed slopes of winter ski centers expanded dramatically in California and elsewhere. Snow in the mountains, viewed for more than a century after the California Gold Rush as stored water for irrigation ditches, metropolitan viaducts and hydroelectric penstocks – the very lifeblood of the American West – now came to be viewed as "white gold" lying on mountain slopes to create wealth for the people of the valleys below. This winter demand put additional pressure on the backcountry as young men from the mountain regiments, who had learned the joys of forest and mountain travel during the war, returned home to California and now wished the Forest Service to build ski trails and shelters for their enjoyment (Sieker 1945: 1-2; USDA Forest Service 1947a: 9). In time, ski fever during the 1950s led to the development of many winter resorts on national forest land, such as Mammoth Mountain Ski Area on the Inyo National Forest. Entrepreneur David McCoy, who started this popular winter spot with a few rope tows in tandem (Feuchter 2004: 6-7), was enthusiastically encouraged by Wilfred "Slim" Davis, forest supervisor for the Inyo National Forest at the time, who had been in the Tenth Mountain Division during World War II. Skiing was one way to sell recreation to the Forest Service, which did not need much encouragement because of the revenue generated by winter sports. By encouraging developers to provide the facilities for the sport on national forest land and to operate and share the revenues through special-use permits, Region 5 demonstrated that it could do something besides cut timber and put out a fire (Radel 1991; 27, 43-46, 50; Rice 2004: 21-22).

As noted earlier, recreation placed number three in importance in Region 5's area planning scheme. With the predicted rise in future usage, in 1946 the major problem faced by Region 5 was how to develop ways and means to meet these growing recreational demands. For an orderly expansion and development in terms of roads, infrastructure and protection of scenic values, correlated with other uses of forest lands such as timber, grazing, watershed, wildlife and wilderness, Region 5 had to create an overall, broad-scale recreational management plan in advance. Unfortunately, Region 5 did not have the luxury of adequate funding or time (USDA Forest Service 1947a).

In 1949, recreationists made 3.8 million visits to California's national forests, which included 1.1 million visits to campgrounds, an 8 percent gain over the previous year. Of these 3.8 million visitors, 531,000 visitors were picnickers, a 16 percent increase over the previous year, and 145,000 youngsters were organization group campers, a step-up of 13 percent (*Administrative Digest-California Region* 1951: No. 20).

Congestion in front of Pinecrest P.O. Stanislaus National Forest. 1946

The inability to meet this kind of exceptional forest recreation demand with new development caused many of Region 5's recreation problems at this time. For instance, Region 5 recognized the need to expand recreational facilities on the Pinecrest recreation unit on the Stanislaus National Forest. At that time only a four-and-one-half-hour drive from the San Francisco Bay Region (now it's only two hours), Pinecrest's summer population soared to 6,000, three times its desired capacity, and it got worse. Sewer and water systems, beaches and campgrounds, and road and nearby trails were overtaxed. To relieve the congestion, Region 5 planned several new organization camps and the development of new public campgrounds. But despite these and similar facility expansions, Region 5 was unable to keep pace with Californians' pent-up demand for outdoor recreation after the war. The question was not the will to manage recreation on Region 5's national forests; the question was where to find the funds (Loveridge and Dutton 1946: U-2).

In the postwar period, Region 5 funding for recreation was woefully lacking, and facilities built by the Civilian Conservation Corps (CCC) had deteriorated in some cases to the point of being "campground slums" (Loveridge and Dutton 1946: U-3). Region 5 leaders during the war years certainly were more concerned with timber production and fire control than recreation. Immediately after the war, little attention was paid to recreation except for a Forest Service special request and appropriation in fiscal year 1947 that brought $200,000 into Region 5 to restore several national forest recreation areas to a safe and sanitary condition (Bachman 1967: 7; Coutant 1990: 4). Nonetheless, campground sanitation problems persisted. On Labor Day weekend 1951, GII Inspector M. M. Nelson witnessed the predicament first hand on the Stanislaus, Inyo and Sierra National Forests. During this holiday, all campgrounds on these forests were filled to capacity, and many had crowds two and three times more than what the camps were designed to handle. A great many campers used areas with no sanitation facilities. In fact, the Bass Lake campgrounds on the Sierra National Forest were so overcrowded and unsanitary that the Madera County health officials warned the Forest Service that it would publicize these conditions if the Service did not immediately bring them up to standard (Loveridge and Nelson 1951: U-2). The underside of this health warning was that one Forest Service official believed that the Madera County officials took this action to threaten the Forest Service regarding a future move of the forest headquarters from North Fork to Fresno (Kirchner 2004: 29-30).

The Bass Lake incident was one example of the seriousness of the problem. There were simply insufficient funds for sanitation and maintenance to do a satisfactory job, let alone monies to get on top of a badly needed betterment program to replace the aging CCC and pre-CCC constructed recreational facilities (Loveridge and Nelson 1951: U-2). Even so, the Bass Lake incident had a silver lining. Though the event caused a great deal of bad publicity for the Forest Service when this news reached the governor's office, the State Recreation Commission, leaders of the state legislature and several California congressmen, it brought action. Most likely in response to the situation, the chief requested that Regional Forester Hendee prepare a report on the California Region's recreation needs for its operation and maintenance, and any new construction. In his report, Hendee asked for a 50 percent increase in appropriations to meet Region 5's annual needs. The Hendee report was eventually presented to Congress, which took almost immediate action. By fiscal year 1954, Region 5's appropriation increased from $215,000 to $605,000. This amount increased in fiscal year 1956 to $1,106,200 and doubled in fiscal year 1957 (Bachman 1967: 7).

Aside from recreation management funding problems, Region 5 experienced conflicts between recreation management and other types of resources management. Recreation/grazing conflicts have already been noted, but there were also multiple-use correlation battles between recreation usage and timber production. For instance, in 1946, Region 5 had not given any special consideration to curtailment of clear-cutting logging in potential scenic areas (Loveridge and Dutton 1946: U-7). This logging practice was just beginning to take root in the California Region in the postwar period. To meet the "impending timber crisis," Region 5's timber management policies gradually turned to clear-cutting over selective cutting. Many Region 5 leaders believed that selective cutting led to "high-grading," leaving poor quality trees and stands. Historically, conservationists had always claimed that selective logging was preferable to clear-cutting – largely as a reaction to the "cut-and-run" operations from the previous century. When recreationists such as Sierra Club members happened upon these early clear-cut patches in the midst of California's national forests, they were disturbed at the destruction and had difficulty distinguishing the new clear-cutting method from the past rape of the forests. In 1953, Sierra Club members especially fulminated over Region 5 plans to harvest a 3,000-acre virgin Jeffrey pine stand at Deadman Creek on the Inyo National Forest, and two years later, they further opposed planned harvests on the Kern Plateau in the Inyo and Sequoia National Forests. which will be discussed in detail in the next chapter (*Pacific/Southwest Log* 1985: July-6).

The episode at Deadman Creek was particularly significant and caused David R. Brower of the Sierra Club to begin to doubt the Forest Service's judgment. At the time of the sale, a museum of natural history planned to make a diorama of this unique area, which was one of the few remaining virgin Jeffrey pine forests in the country. Region 5 saw the Deadman Creek sale as a way to provide logs for a financially struggling local sawmill rather than as a way to protect and enhance the area's unique recreational benefits. Similar situations throughout the West, which required the Forest Service to make a choice among several uses, caused Brower and the Sierra Club to advocate protecting wilderness by law from the dangers of such administrative decisions. Brower and the Sierra Club moved on to fight other battles, such as the Echo Park dam controversy in Dinosaur National Monument, Utah, where they gained tremendous strength and support from the general public (Steen 2004; 302; Roth 1984: 120). Region 5 learned from the Deadman Creek decision as well. Several years later, when a 380-acre stand of some of the last virgin sugar pines in Tuolumne County became available to the Forest Service through a lumber company land exchange, it administratively set them aside as the Calaveras Big Tree National Forest, which was administered by the Stanislaus National Forest. This was the first such dedication of national forest lands in California to be administered for the exclusive purpose of preserving scenic recreational values (USDA Forest Service 1955).

The complexity involved in multiple-use management became evident again when a rich oil strike was made in the Cuyama area of the Los Padres National Forest in the late 1940s. Once the strike was made, oil and gas speculators on the Ojai, Santa Barbara, Santa Maria, Mount Pinos, San Luis and Monterey ranger districts received thousands of lease applications, totaling in excess of 700,000 acres. The filing and lease fees alone generated more than $350,000 for the forest. Though revenue from this oil boom may have tempted some Forest Service leaders, in time Region 5 decided that oil and gas leasing would adversely impact the Los Padres National Forest watershed (*Administrative Digest-California Region* 1949: No. 40; Loveridge and Nelson 1951: U-5-U-6). Therefore, the Forest Service influenced the BLM to reject some seventy applications for oil and gas leases in the watershed of the Santa Ynez River in the Los Padres National Forest (Santa Barbara County). According to a Region 5 press release, the Forest Service based its recommendation on the "life and death importance of the Santa Ynez Watershed as the source of water for domestic use in Santa Barbara and surrounding communities, where the water supply for about 50,000 people has been growing scarce; and on the importance of that watershed as the source of water for irrigating crop and pasture lands." Additionally, "the two agencies felt that any major activity in the watershed, such as oil and gas development, would increase the danger of forest fire, erosion, and pollution of streams or underground water" (Loveridge and Dutton 1951: U5-U6; USDA Forest Service 1951c).

In conjunction with this action, in 1948, Region 5 withdrew as a sanctuary a 35,200-acre condor nesting area on the Los Padres National Forest in Ventura County. The withdrawal eliminated disturbances from most forms of use, except mineral and oil activity. For that reason, in 1951, with strong urging from groups such as the National Audubon Society, the secretary of the interior ordered that an area identical to the Forest Service refuge be withdrawn from oil leasing in order to protect the rare but vanishing California condor (USDA Forest Service 1947f; 1951b; National Audubon Society 1950: 1-7). This controversial move was fully captured for the American public by the *Saturday Evening Post* in an article titled: "The Fabulous Condors' Last Stand," which described how the largest winged creature in North America – all sixty of them – were saved from oil-hungry prospectors drilling at the edge of their stronghold (*Saturday Evening Post* 1951: 7 April). In the end, the rejection of oil and gas leases on the Los Padres National Forest exemplified the two Forest Service management principles inherent in the Organic Act of 1897 – one, multiple use of land (minerals versus watershed and wildlife), and two, the greatest good to the greatest number in the long run (oil speculators versus people of Santa Barbara County).

Region 5 also faced a similar, but perhaps more difficult decision, in the case of a proposed winter ski center development on San Gorgonio Mountain on the San Bernardino National Forest. In 1931, the Forest Service set aside the San Gorgonio Primitive Area to be preserved as a natural wilderness under the L-20 Regulations that covered all such national forest primitive areas. "These regulations," as one source noted, were not binding but "were simply strong recommendations to Forest Service field personnel suggesting limitations on unplanned development in untouched areas" (Roth 1988: 2-3). At the time Regional Forester S. B. Show approved this primitive area in the early 1930s, there was little or no recreational skiing in California. However, within a few years thereafter, as noted previously, skiing rapidly increased in popularity in the nation. This interest led many Southern Californians and business groups associated with the ski industry in 1936 to search for sites near to the Los Angeles area. The San Gorgonio Primitive Area had the slopes and snow conditions they sought, and in 1937, they made a proposal to modify the area to allow ski development. Initially, Regional Forester Show rejected their proposal. However, as Show became more cognizant of the rapid increase in popularity of skiing, he opened the door to possible development of San Gorgonio Mountain, the highest peak south of the Tehachapis. Increasing demand for skiing facilities at San Gorgonio culminated in a public meeting in March 1942, but because of the war, Region 5 withheld its decision until the summer of 1946. At that time, the Forest Service made a cautious decision to change the boundaries of the 35,635-acre San Gorgonio Primitive Area and eliminate approximately 3,000 acres in the heart of the primitive area, to be tentatively developed as a ski center (USDA Forest Service 1947d; *Daily Redlands Facts* 1947: 3-5 February).

Prior to making this decision, Regional Forester Thompson contacted the Sierra Club's board of directors in the summer of 1946 regarding the ski center proposal. Relations between the two organizations were amiable at this point, and it was almost policy to discuss land-use issues with the Club. After hearing from Thompson, the Sierra Club decided against the project. David R. Brower, editor of the *Sierra Club Bulletin*, analyzed the proposal and concluded that the issue was not just about a particular land- use change, but also a question of nationwide importance – wilderness versus ski resorts. His analysis concluded that if pressure groups such as the California Chamber of Commerce or the California Ski Association could effect such changes in the Forest Service wilderness policy, then it would destroy the value of the Forest Service policy because the basis for denying changes to other groups would be gone. The Sierra Club's answer to the Forest Service was a firm "Hands Off!" – and they meant it (*Sierra Club Bulletin* 1947: No. 4; Anonymous 1947a: 35-36).

Meanwhile, in the fall of 1946, Regional Forester Thompson made a thorough presentation of the idea to Chief Lyle Watts. After the presentation, Watts concluded that "the public value of the ski area seems to be so much greater than the value of the area as a wilderness that modification of the area seems to be a public necessity" (USDA Forest Service 1946b). The commercial ski industry

San Gorgonio Wilderness, San Bernardino National Forest. 1963

naturally supported Chief Watts' conclusion and thereafter clamored for the development. San Gorgonio, in their estimation, had what Southern California skiers wanted: good terrain, dependable snows, long seasons and proximity to the homes of some fifteen million people. San Gorgonio was "convenient and economical for the mass of skiers" who wanted a resort they could reach easily by automobile (*Redlands Daily Facts* 1947: 3-5 February). With the chief's approval and Region 5's support, it seemed almost certain that the ski resort would be developed. They just needed to hear from the public.

In early December 1946, Regional Forester Thompson announced that in February 1947, a public hearing would be held regarding Region 5's proposal to modify the San Gorgonio Primitive Area's boundaries so that its snowy slopes could be developed for winter recreation. Thompson assured the public that "all developments would be strictly in accordance with a Forest Service plan [not even drafted at the time], and operated under Forest Service supervision" (USDA Forest Service 1946c; *California Region-Administrative Digest* 1947: No. 7). These reassurances notwithstanding, opposition to the San Gorgonio proposal from diverse groups snowballed prior to the hearing date of February 19, 1947. On the one hand, there was the issue of watershed. From the practical viewpoint of agriculturalists, the opening of the primitive area was contrary to every instinct of water conservation. Citrus growers worried the most and openly opposed the project because their land was dependent upon the flow of water that came from the San Gorgonio snowpack. They realized that this kind of development would bring added population to the area and thereby increase area water consumption. Furthermore, they understood that as the population increased in the area, the danger of fire and harm to the watershed would increase as well. On the other hand, there was the issue of wilderness. The San Gorgonio Primitive Area had been created to preserve the area from the works of man – roads, machines and buildings for the enjoyment of "summer wilderness use," which at that time included some 15,000 individuals from seventy-two different organizations ranging from various wilderness groups to the Boy Scouts of America. To them, the San Gorgonio Primitive Area offered Southern Californians a unique wilderness opportunity. They contended that the demands for skiing could be reasonably met elsewhere without harming San Gorgonio. Finally, there were those who objected to the proposal from a conservationist perspective. For them, the terms were unconditional surrender and blind trust of a Forest Service that had neither a viable plan nor control of all the land needed for the ski center, experience in winter sports development or even enough recreation funds to "keep the garbage cans emptied in the summer camp grounds" (*Redlands Daily Facts* 1947: 3-5 February).

On the first day of the hearing, more than 500 citizens attended. By the second day, eighty individuals had delivered 90,000 words of testimony. This oral presentation was in addition to the briefs various groups had filed earlier with Regional Forester Thompson. The hearing was a remarkable event because it allowed a full discussion of the various viewpoints without any legislative mandate to ensure that all were heard. Skiers and ski organizations, the California Chamber of Commerce and some local chambers, and youth groups interested in winter sports testified in support of the Forest Service proposal. The opposition came from wildlife organizations, water users, youth organizations with camps within the area, fish and game organizations and scientific groups. Their testimony indicated to the Forest Service the intense interest in national forest recreation as well watershed protection (*California Region-Administrative Digest* 1947: No. 9).

What started as a local debate about a specific ski area proposal blossomed into a broad, national debate on Forest Service wilderness policy, a portent of the future environmental movement. While the Forest Service took ninety days to make its decision, wilderness groups conducted an organized campaign against the project. Writing in national magazines such as *Nature Magazine*, *The Living Wilderness*, and the *Sierra Club Bulletin*, wilderness advocates took the offensive. In articles such as "Why We Cherish San Gorgonio *Primitive*," they contended that the San Gorgonio decision was not a local one but that the boundary decision would jeopardize the entire system of wilderness and wild areas in the national forests. They reminded the Forest Service that its mission was to protect wilderness for the *entire* population of the nation – *the greatest good for the greatest number in the long run*. Organizations like the Sierra Club and Wilderness Society passed resolutions to take a firm stand against any change in the boundaries. Before long, Chief Watts' and the regional office's daily mail were flooded with several thousand protest letters from around the nation (Wilderness Society 1947: 1-7; Anonymous 1947b: n.p.). After careful appraisal of the criticism by forest officers, Chief Watts was persuaded by the opposition, and he announced that the San Gorgonio Primitive Area had "higher public value as a wilderness and a watershed than as a downhill skiing area" (USDA Forest Service 1947e; *California Region-Administrative Digest* 1947: No. 26). The local press celebrated and saluted the Forest Service for its courage and integrity in resisting the pressure of the ski industry (*Redlands Daily Facts* 1947: 18 June). Though ski developments became important to Region 5 in the decade following the war, the San Gorgonio event highlighted the fact that water was still the most valuable "crop" of California's mountains. Clearly, in Southern California other resources, no matter how valuable, were to be managed so that they did not interfere with or jeopardize vital watershed (*California Region-Administrative Digest* 1943: No.16). With the rejection of San Gorgonio as a winter sports center, Region 5 turned its development attention to the small valley of Mineral King in the High Sierra, equidistant from Los Angeles and San Francisco. Covered in the next chapter, the Mineral King winter sports and recreation development would become one of the most controversial conservationist conflicts in California history (Robinson 1975: 131).

In the late 1940s and early 1950s, avalanche control in Region 5 shared the spotlight with flood control and watershed management, but the latter two were deemed far more important. Following the passage of the Omnibus Flood Control Act of 1936, several projects to halt periodic flooding throughout Region 5 were successful, such as the Arroyo Seco channel stabilization program on the Angeles National Forest, which effectively handled flood waters during World War II (*California Region-Administrative Digest* 1943: No. 5). But despite progress made to date in meeting this problem, flood damage continued to offset flood control almost every year (USDA Forest Service 1947a: 3).

However, disastrous floods, causing loss of life and destruction of property, were only one problem confronting Region 5's watershed managers in the postwar period. Supplying and meeting water demand storage was another. After World War II, California's industrial and domestic requirements exceeded a billion and half gallons of water a day and were increasing rapidly (USDA Forest Service 1947a: 3). Water storage issues were most serious in Southern California. Though more than 50 percent of the state lived here, the region had somewhat less than 2 percent of the state's water supply. By 1955, many already realized that a shortage of water was the limiting factor in the future development of this area (Kraebel and Sinclair 1947: 1-2). Besides for residential use, Southern California also required adequate supplies to meet the essential needs of single-purpose water users: agriculturalists and power companies. The mountains and forests of Region 5 provided irrigation water and also produced the hydroelectric power needed by industry and cities. As California's population and industry grew, Region 5 realized that a comprehensive, fully-integrated plan was required to control the catch of rain and snow on the 73,000 square miles of its forested watershed lands. Out of this realization came a new watershed protection concept. At this point, Region 5 began to think of water as a "forest crop." Like other crops, Region 5 intended to manage this downstream with the same intensive methods practiced in handling other forestland crops. Ultimately, Region 5 sought to capitalize on the 60 to 75 percent of California's precipitation caught on forestlands through intensive management. Region 5 began to reduce waste due to uncontrolled run-off and evaporation. To obtain the maximum possible upstream use, Region 5 developed headwater and intermediate reservoir storage facilities and spreading grounds, and turned to cover improvement planting, channel stabilization and other control measures, erosion survey and rehabilitation of damaged lands (USDA Forest Service 1947a: 3-4). Region 5's long-range multiple-use objective was "to develop principles and methods of managing mountain watershed lands to produce the maximum yield of <u>usable</u> water compatible with adequate control of soil erosion and floods, and with other legitimate uses of those lands" (Kraebel and Sinclair 1947: 1-2). To implement this intensive management, Region 5 first inventoried all resource values on the watersheds of each of the eighteen national forests (USDA Forest Service 1947a: 4). Region 5 also began to inventory the amount of water which each watershed produced, the quality of the water, what it was used for and where it was used. This information was critical, for without it, Region 5 could not begin to regulate and produce the maximum usable water from California's national forests (Littlefield 1947: 3-4). The result of these efforts was the construction on California's national forests of more than 300 storage reservoirs with more than 100 acre-feet of capacity each. By 1954, 8.5 million Californians received all or part of their domestic water from them (USDA Forest Service 1954: 3).

Beyond domestic usage, water was also necessary for developing and managing fish and wildlife. As with other resources, California's wildlife had severely suffered from exploitation before and after the creation of forest reserves, but up until the postwar period, it would be fair to state that wildlife conservation efforts by the California Region had been misdirected. Essentially, its policy regulated sportsman rather than restored wildlife habitat. Region 5 wildlife protective measures focused on setting hunting and fishing seasons and bag and creel limits. Wildlife refuges were established, and certain species were removed from takeable game. When these limitations failed, Region 5 began transplanting native species, introducing exotics into the forests and artificially propagating upland game birds and fish. Contributing to the complexity of the situation was the regression of forested lands to brush fields, the overgrazing of range and the misuse of fire, all of which had the effect of crippling the native ecology's capacity to sustain wildlife. An outstanding example of Region 5's failure to recognize wildlife values was the destruction of fish life caused by its management of water. The diversion of water for agriculture, mining, power companies and municipal use severely impaired fish habitat. The construction of dams and other diversions blocked off streams, and without the installation of adequate fish ladders, fish could not get around these man-made barriers in order to spawn in upland streams (USDA Forest Service 1947a: 8). Of course, some of these diversions were established before the forests were created, and the Forest Service could not ignore state-granted water rights or rights established before a forest reserve was created (Leisz 2005).

Beginning in 1946, a newly-created division of wildlife management in Region 5 considered the seriousness of California's wildlife situation. At this time, Region 5 began a program of natural habitat improvement – a program to attain optimum conditions for game animals, upland game birds, water fowl, and fish life; to integrate wildlife production with other necessary and desirable uses of the land and water; and to obtain fundamental scientific facts founded on biological principles to build a progressive plan of wildlife resource management. This program included several remedial measures. There were technical measures to protect wildlife values during construction, timber, range and other resource programs in the administration of these Forest Service functions. There was the installation of facilities needed to remedy imperfections in habitat. For fish, Region 5 constructed small dams for regulating water flows, installed fish ladders over barriers and fish screens, and developed new spawning areas. This effort included the removal of rough fish and the construction of holding ponds for better fish distribution. For game and birds, Region 5 developed watering places, food and cover planting, and provided an interspersion of food and cover types for wildlife. Finally, Region 5 wildlife management programs sought to reestablish desirable species on California's national forests, including reintroducing beaver and antelope and building holding pens for better game bird distribution, while controlling or removing undesirable predatory species (USDA Forest Service 1947a: 8).

By 1951, the Loveridge-Nelson GII recognized Region 5's progress in wildlife management and cited several examples, such as greater cooperation between Region 5 and the California State Game Department on overpopulated deer areas (USDA Forest Service 1947a: 8). However, the estimated number of wildlife on California's national forests presents a clearer picture of the scope of the problem. In 1953, there were only an estimated 418,000 deer, 3,100 antelope, 16,000 bear, 520 elk and 490 bighorn sheep for all of California's eighteen national forests (USDA Forest Service 1954: 5). Numbers for deer, antelope, bear and elk had nearly doubled from those reported by Region 5 in the 1920s, a marginal improvement under Region 5's custodial care, but numbers of bighorn sheep had dropped significantly from the 600 reported thirty years earlier.

California: Bellwether for Conservation

By the mid-1950s, Region 5's active custodial, multiple-use management of the pre-war period had evolved into integrated, planned, multiple-use utilization of California's national forests. Forest Service rangers had not only regained their status as protectors of the forests but were now seen by many as "wise" managers of the many resources found on these forests. A growing California recreational population entering the woods saw rangers out in the forests engaged in timber sales, building forest roads and campgrounds, maintaining trails, protecting and restoring watersheds and, of course, fighting fires. In rural communities, Forest Service personnel and their wives still had the time and inclination to participate in service clubs, school boards, parades and the like. But times were changing, and in many ways, the history of Region 5 in the postwar period was a bellwether for the many problems and changes the Forest Service would face in the future.

Housing demand in California and elsewhere in the nation led to increased timber sales.

First, by the end of World War II, Regional Forester Show and several other Region 5 officers had ingrained in themselves the concept of multiple-use management, but it was not until 1949, after Show's retirement, that Regional Forester Thompson openly declared that each forest would thenceforth be managed on a "multiple use" basis. Thereafter, Regional Forester Hendee coupled Show's concept of multiple use with long-range or area planning methodology to create a system of integrated management controls of all land use within a specified regional or national forest area. Though they did not realize it at the time, between 1952 and 1955, Region 5 had all but laid out the groundwork for national forest planning for decades to come, starting with the passage of the Multiple- Use and Sustained-Yield Act of 1960. This act would authorize and direct the secretary of agriculture to develop and administer the renewable resources on the nation's national forests – outdoor recreation, range, watershed, timber, wildlife and fish – for multiple use and sustained yield of the many products and services obtained from the forests. Essentially, the 1960 Multiple-Use and Sustained-Yield Act formalized what Region 5 officials had understood and practiced since the end of World War II.

Second, conflicts between the timber industry and the Forest Service, which had been the norm during the New Deal and World War II because the Forest Service wanted to regulate logging on private lands, had subsided by the early 1950s. At this time, great public pressure was put on Region 5 to produce timber to meet the housing needs of California's exploding population and the defense needs of the nation for the Korean War, so Region 5 changed its timber policies to meet the demand that could no longer be met through private forestry. The active custodial role of Region 5, which handled only limited timber sales prior to World War II, changed into a timber extraction business. Timber harvesting practices changed in the early 1950s as, more often than not, the key phrase became "Are you meeting the cut?" This change from active custodial care to timber industry business partnership started out moderately, with guided cuts that involved salvaging dead, dying and high-risk trees on each national forest. These sales, however, were a harbinger of the future. In time, more and more personnel were thrown into the mix to meet ever-optimistic annual allowable cuts. Eventually, timber dominated Region 5's priorities in the period after 1955. Pressured by private industry, the Eisenhower administration and Congress to meet this demand, Region 5 began a program of building timber access roads to reach inaccessible stands. Timber access roads allowed timber sale offerings to double and then triple in the future, thus making it harder and harder to balance timber harvesting targets with other forest uses.

Other events in the early 1950s also foretold the future. In fire control, the Rattlesnake Fire tragedy on the Mendocino National Forest prompted Region 5 to increasingly turn more to technical firefighting strategies such as helicopters and air tankers, rather than risk the lives of men. As with so many previous fire disasters, Region 5 learned from it, and in time the California Region became more and more sophisticated in the use of technical improvements. Another portent for Region 5 was the controversy over developing a portion of the San Gorgonio Primitive Area into a winter resort, which foreshadowed future controversies between utilitarian conservationists and preservationists. The wilderness boundary issue at the heart of the San Gorgonio dispute in many ways led to the introduction of a key wilderness bill by preservationists in 1957, which evolved into the Wilderness Act of 1964, which required that the status of primitive areas like San Gorgonio be reviewed and submitted to Congress for permanent wilderness designation, in order to preserve their primeval character.

Finally, regarding wildlife, by the early 1950s Region 5's programs had become more sensitive to protecting streams, riparian vegetation and wildlife habitat, while mitigating impacts from extractive resource activities such as mining and logging. The establishment of a refuge for the endangered California condor on the Los Padres National Forest predated the passage of the Endangered Species Act of 1973, which aimed at insuring the survival of all native species of fish and wildlife, even non-game animals, and authorized survival programs for those species threatened with extinction such as the magnificent condor.

The environmental issues with which Region 5 wrestled during the early 1950s were a bellwether for the environmental movement of the 1960s. Unlike previous reform impulses, the 1960s environmental movement found support among all ages, across a broad spectrum of economic classes and across all political lines.

Two unnumbered pages have been inserted here to provide the map of the National Forests in California as of 1955.

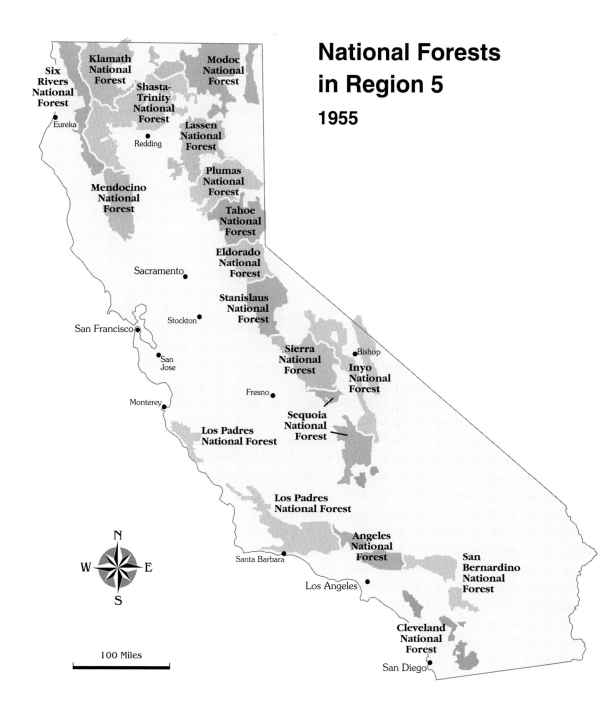

Chapter IX: **1955-1967**

Programmed Multiple Use *Maximus*

California: Conformity to Conflict

Life in California in the late 1950s was a world of growing suburbanization, middle-class affluence and growing leisure time, based on triumphant capitalism. Economic trends for the state in the last half of the 1950s decade could not have been better.

In 1955, agriculture was still the state's largest single industry and would remain so until at least 1965. While the state's economy grew to become remarkably diversified, based upon manufacturing and the defense and aerospace industries, agriculture – California's very own "green gold" – still led the way. Large mechanized and highly capitalized industrial farms developed and concentrated their energies on valuable income-per-acre crops such as fruits, vegetables, dairy products, poultry and eggs – all California specialties that the average American could now afford, thanks to America's continuing growth of real income and purchasing power. California's agribusiness success in the decade from 1955 to 1965 was in part attributed to cheap farm labor. From World War II to 1965, California's fields were the province of Hispanics. After the war, the *bracero* program continued in full measure after Congress adopted the agreement with Mexico, when it passed Public Law 78 (1951). In 1957, the importation of these "strong-armed ones" into California reached its peak number of close to 200,000 workers, making up about one quarter of California's seasonal farm labor workforce. Most of them worked on large corporate farms, and to their employers, they represented a "supply of cheap, docile, and dependable labor." If the success of California's agriculture depended on this non-unionized workforce, the future of the industry was linked to subsidized irrigation water, which was indispensable for not just agricultural growth but the future growth of the state. That future was assured when in 1960 California voters passed a $1.75 billion bond to begin construction of the California Water Plan (CWP). The CWP envisioned 376 new reservoirs and a vast network of aqueducts throughout the state to bring water from the northern half of the state to where it was most needed, Southern California farms and municipalities. The Feather River Project, the first phase, began in 1961 with the construction of the Oroville Dam, assuring controlled flow of water to the San Joaquin delta region (Bean 1968: 483-501).

Meanwhile, the technological explosion in aircraft, weapons systems, nuclear energy, aerospace and electronics challenged agriculture's central role in California's economic growth. After 1957, the Cold War intensified when the Russians launched *Sputnik* and the United States entered the "space race." Thereafter, the National Aeronautics and Space Administration (NASA) poured billions more into California's economy for various satellite and space-exploration projects, including the Mercury, Gemini and Apollo manned flights. As the Cold War intensified, the Dwight D. Eisenhower administration poured even more money into defense. By 1960, the Defense Department was dispensing more than $25 billion each year in California for the development of military aircraft such as the B-58 Hustler bomber, intercontinental ballistic missiles (ICBMs) and a variety of other military products to fight the Cold War. After *Sputnik*, the educational community raced to catch up with the Russians in the areas of science and math. Soon entire scientific communities supporting advancements in fields related to the new military-industrial complex grew up around key schools and universities such as the University of California at Berkeley, Stanford University and the California Institute of Technology in Pasadena. The California education system also enjoyed a huge building expansion in order to school the postwar "baby boom" (Watkins 1973: 463-465; Rice, Bullough and Orsi 1996: 493).

The agricultural and industrial achievements were integral to the continued development of California as part of the "New American West." In the late 1950s, California continued to grow. By 1960, there were 2.3 million people living in California who had not been in California five years earlier, most of whom settled in Southern California. Very soon thereafter, California overtook New York as the nation's most populous state. Housing, school construction and transportation improvements matched this phenomenal growth, and many portions of counties changed virtually overnight from agricultural to residential in character. While they varied in size, design and geographic location, each new community that contributed to the growth of cities like San Diego, Los Angeles and San Francisco had one thing in common, the "California lifestyle," which wedded affordable tract-housing with "out-of-doors" living on covered patios bedecked with redwood furniture and with swimming pools, clipped shrubbery, sweet-smelling, freshly cut lawns and grape-stake fences. This way of life was centered on creating a small piece of paradise in a troubled world and became a religion of sorts for many. Exhorted by magazines such as *The Magazine of Western Living* and *Sunset*, which defined, informed and reflected upon this uniquely California lifestyle for their readership, Californians learned about food, fashions, gardening, home improvements and the travel delights of the Golden State. "Its finest moment," as one source commented, was illustrated by the "picture of a man standing on his patio on a Saturday evening, a drink of good Scotch in one hand and a spatula in the other, barbequing a few two-dollar steaks while his wife prepares a cool green California salad, his children cavort on the lawn with the dog, and the patio lights dance on the chlorine blue surface of his kidney-shaped swimming pool." California became the quintessential example of the "barbeque" lifestyle of the American Southwest, as the "Grapes of Wrath" folk of a decade before reversed their fortunes and enjoyed the suburban life and the Californian style (Watkins 1973: 450-458; Rice, Bullough and Orsi 1996: 489-491, 495).

Of course, California living, and all of its ramifications, would not have been possible without the automobile and an improved freeway network. The automobile spawned this new economic and social culture, which took off after 1957, when the California State Division of Highways ambitiously committed itself to completing 12,500 miles of new freeways by the year 1980. The shopping center – 90 percent parking lot and 10 percent stores, along with gas stations, drive-in banks, restaurants, movies and even churches at each intersection – reflected California's dependence on the automobile. The ultimate automobile destination was Disneyland, a revolutionary amusement park that opened in Anaheim in 1955. It was designed to promote the Disney organization's latest movie and television ventures, such as the Mickey Mouse Club, which began airing the year before (Watkins 1973: 450-458; Rice, Bullough and Orsi 1996: 489-491, 495). At mid-century, according to the title of eminent California historian Robert Glass Cleland's history of California, the state had gone *From Wilderness to Empire*. Five years later, California had become the "Magic Kingdom" thanks to Walt Disney (Cleland 1962). But the middle-class magic of California's kingdom and lifestyle, based on big business, science and imagination, had a darker, undesirable underside – traffic congestion, the deterioration of the quality of life for some in California, middle-class silence and indifference to national social and political problems, the exploitation and discrimination of various ethnic groups, and urban blight and the ghettoization of its inner-cities.

California's "age of conformity" gave way to years of confrontation starting in the mid-1950s, when California's "Beat Generation" first began to find fault with the California lifestyle. Led by San Francisco authors and poets such as Jack Kerouac and Allen Ginsberg, they rejected this conventional, middle-class, business-oriented world and reacted against the accumulation of material goods, suburban life and stultifying, nine-to-five workdays of their fathers.

In novels such as *On the Road* (1957) or the *Dharma Bums* (1958), Kerouac and other authors romantically and innocently sought a more vibrant and spontaneous world divorced from modern-day realities such as the Cold War and the potential for atomic war and world annihilation (Rice, Bullough and Orsi 1996: 502-503). Interestingly, Kerouac and other Beat writers, such as Gary Snyder and Philip Whalen, once worked as fire lookouts (Suiter 2002).

However, others reacted more aggressively to these modern day problems with a variety of movements. Many cities especially witnessed the social cost of building freeways through viable communities, wiping out large sections of housing, razing historic landmarks and destroying community cohesion. For example, the people of San Francisco rose up against a proposed freeway through Golden Gate Park, and the city of Laguna Beach openly rebelled against a proposed coastal freeway that threatened to destroy the attractive qualities of this artistic beach community. Events in the late 1950s and early 1960s led to a student revolt on California's campuses, and Berkeley became the flash point for student unrest over civil rights, atomic warfare and United States involvement in Southeast Asia. In 1964, when the administration and the Board of Regents restricted on-campus political activity, Berkeley students united in the Free Speech Movement. They occupied one building until police were ordered in, resulting in the arrest of more than 600 students for trespass and resisting arrest. By the end of 1965, this first major academic rebellion in United States history led to other campus protests, giving momentum to politically radical groups such as the Students for a Democratic Society (SDS), as well to the so-called "hippies." These children of California's affluent society rejected established institutions and values. Instead, they sought spontaneity, direct personal relations expressing love, and expanded consciousness by means of psychedelic drugs.

On the other hand, Hispanics, California's largest ethnic minority, agitated against the low wages and poor working conditions in the fields. The 1965 agricultural workers' strike at Delano, mobilized by Cesar Chavez and his National Farm Workers Union, was one example of such a movement. The strike led to a nationwide boycott of table grapes from the Imperial Valley, and eventually to the acceptance of Chavez's union. Finally, racially segregated "palm-tree ghettos" that differed from the social and economic conditions of the "typical" California neighborhood erupted from the frustration of being left out of the California dream. One night in 1965, the Watts community of Los Angeles exploded into violence and gunfire. In the aftermath of the looting and burning of Watts, militant and radical political action groups rapidly emerged around the state and the country, most notably the Black Panther Party, founded in Oakland in 1966 (Rice, Bullough and Orsi 1996: 520, 537-542; Watkins 1973: 486-495).

Naturalist and visitors on nature trail, Angeles National Forest. 1962

Activism in the period 1955 to 1967 also included discontent over the declining quality of California's environment. Part of the leisurely California lifestyle included outdoor recreation, and the state's recreation industry blossomed with a whole range of activities including skiing, surfing, boating, hiking and camping. But as people got outdoors, increasingly they witnessed a growing degradation of California's air, water and natural resources. All of the causes that would drive the conservation movement of the early 1960s and 1970s – the pollution of air and water, the mistreatment of forests and the attrition of undeveloped areas and parklands by uncontrolled urban development – appeared as early as 1955. As Californians began to experience smog that irritated eyes, noses and throats, and aggravated respiratory conditions; as Californians witnessed factories, cities, towns and farms pumping toxic effluents, deadly poisons and raw sewage into the state's rivers, streams and bays and into the Pacific Ocean; as Californians watched incredulously at the rate that open space, coastline and farmland disappeared to urban sprawl; and as Californians observed freeways bisect and then trisect neighborhoods into bits and pieces, they became overwhelmed with worry. In 1962, Rachel Carson's eye-opening and enormously popular book, *Silent Spring*, warned of the impact of chemicals on the environment and for the first time alerted Americans to the threats to their environment. Public anxiety about the state of California's environment was

Backpacking was one part of the California outdoor recreation lifestyle. Minarets Wilderness, Inyo National Forest. 1967

further amplified by Samuel E. Wood and Alfred Heller's *California, Going, Going...* (1962), Raymond F. Dasmann's *The Destruction of California* (1965) and Richard G. Lillard's *Eden in Jeopardy* (1967). These books and others made Californians acutely aware of the past, present and dire ramifications of failing to safeguard the state's environment (Watkins 1973: 505, 511; Clepper 1971: 60, 89).

Programmed Multiple Use – The Eisenhower Years, 1955-1959

Against the economic and social backdrop of the period described above, Charles Arthur Connaughton became Region 5's seventh regional forester – the region's third since S. B. Show retired in 1946. Chief Richard E. McArdle (1952-1962) appointed Charles, or "Charlie," Connaughton as regional forester in November 1955 after Clare Hendee moved on to the Washington office (WO) as assistant chief (*California Log* 1955: No. 12). Connaughton was a

Charles A. Connaughton, Region 5's seventh regional forester (1955-1967)

strong regional forester, very much in the tradition of S. B. Show, but unlike Show, he was less independent of the WO, and he fit well into McArdle's national forest policies. Under Chief McArdle, the first chief to hold a Ph.D. and to be a researcher, the Forest Service sought to increase intensive management aimed at increasing national forest commodity outputs, provide for reforestation of logged and other lands, curb mining and grazing abuses, and accelerate various recreation projects. McArdle also sought to improve relations with the timber industry by backing away from regulation of timber harvesting practices on private lands (Williams 2000: 94).

Connaughton was born and raised in the mountain community of Placerville, Idaho; attended the University of Idaho, where he was awarded a B.S. in forestry; was a Yale forestry graduate in 1934; worked for a short time in the WO in the Civilian Conservation Corps (CCC) program; and then became the director of the Rocky Mountain Forest and Range Experiment Station (1936). In 1944, Connaughton went to the southern United States, where he served as the director of the Southern Forest Experiment Station in New Orleans, Louisiana. Connaughton was appointed regional forester for the Southern Region with its headquarters in Atlanta, Georgia, in April 1951. His responsibilities there included national forest management and oversight of state and private forestry programs. During his time in the South, Connaughton enjoyed what he later termed a great American revolution in technical forestry. This development was based on early research coming to fruition and on new protection programs being implemented, and this revolution was stimulated by new markets that made it economical to grow trees there (*Administrative Digest* 1951: No. 17; Connaughton 1976: 116-117; Clepper 1971: 77-78).

Connaughton was articulate and conservative, and an indefatigable public servant. To those who knew him intimately, he had a reputation for living "by the book" and expecting the same from others. A note signed "C.A.C." was not to be ignored. Essentially, Connaughton ran Region 5 as if he were a general in the army. Each year, he had a week's meeting with beginning professional employees. For several hours, he discussed with them the history, aims, goals and aspirations of the Forest Service. He emphasized to them loyalty to the Forest Service and recommended to them that, if they couldn't be part of the organization and subordinate themselves to its policies, they ought to consider a new job. In Connaughton's opinion, the Forest Service was no place for "individual stars." He invited discussion and disagreement prior to a decision, and he detested ambiguous policy statements. As far as public expression of opinion, to him there was only one position, and that was the position of the Forest Service. In other words, once the Forest Service decided on a direction to go, employees were expected to wholeheartedly support it or quit (Connaughton 1976: iv-v, 141-142; Leisz 2005).

As a master of the English language, Connaughton reviewed letters sent out to the public by the division chiefs to ensure there were no misspellings (James 2004: 25). A typical "Charlie" Connaughton "story," told by one retiree, related how Connaughton once took the California advisory council on a routine trip to one forest. While there, he took them over to see some signage being prepared by a fire crew, only to find the word "forester" misspelled on almost every sign (Peterson 2004: 32). Though he could be completely flustered by such moments, they were rare. Some retirees recalled Connaughton's marvelous memory, especially for people's names (Leisz 2004: 96; Leonard 2004: 23). Another retiree also remembered that he was a gifted speaker and did an excellent job of advocating the concept of multiple use, especially making the point that uses would change over time and that the use of today might not be the use of tomorrow (Leisz 2004: 64). Ultimately, Connaughton provided leadership in harmonizing and reconciling competitive uses of forested lands and facilitated understanding among competing user interests (Clepper 1971: 77).

Connaughton was at heart a timber man, which undoubtedly earned him his position in Chief McArdle's program. Regional Forester Connaughton worked very closely with the timber industry and with industry-leaning organizations, and won a seat on many types of councils, including the Western Forestry and Conservation Association (WFCA), Western Wood Products Association (WWPA), Society of American Foresters (SAF) and the American Forestry Association (AFA) (Connaughton 1976: iv). In fact, within months of assuming the position of regional forester, Connaughton was elected director of the AFA for a three-year term (*California Log* 1956: No 2), a position he held until he became president of the organization in 1971 (Clepper 1971: 77). To Connaughton, foresters were the accepted stewards of the country's lands, and to him, the greatest challenge facing him and other foresters, which he told the SAF in his year as their president, was to acquaint the public with the fact that foresters were competent land managers and that they practiced "land-use and not land-abuse" (*California Log* 1963: No. 1). Though he was a timber man through and through, he also had a broad

Camper driving up Mammoth Road, which was constructed for a logging access road on the Sierra National Forest. 1963

view of forestry stemming from his experience in research. For instance, very early on, he was sold on the idea of using prescribed burning to reduce the possibility of major fires (Peterson 2004: 40). However, in the short term, he was also skeptical about issues such as skiing development at Mammoth Mountain on the Inyo National Forest and other recreational opportunities for California national forests (Radel 1991: 46).

When Regional Forester Connaughton assumed office, he provided leadership in administrative matters, which had been largely lacking under

Arch Rock in John Muir Wilderness. (L to R) Sierra Forest Supervisor Walter Puhn, R5 Regional Forester Charles A. Connaughton, R6 Regional Forester J. Herbert Stone, and packer Arch Mann. Sierra National Forest. 1964

the previous two administrations. He faced many of the same problems confronting his predecessors, such as aging facilities and inadequate housing for employees that dated to the CCC era, ranger district workloads, and replacement and recruitment problems. Fortunately, Chief McArdle's policies addressed many of these problems, including upgrading Forest Service personnel by hiring new specialists to bring about more intensive management and to enhance the professionalism of the Forest Service. During the 1950s, forest engineers, landscape architects and silviculturists became common (Williams 2000: 94).

Connaughton also worked on completion of the sub-regional area plans laid out by Regional Forester Hendee. In 1955, Chief McArdle stressed that Region 5 must do a better job of coordinating work planning at all levels, and sub-regional area plans assisted this policy goal. Regional Forester Hendee had hoped to have all subregional plans completed by 1956 (USDA Forest Service 1955b), and Connaughton worked toward this goal. Area plans were not an end in themselves, but their value was derived from their use in annual and long-range work programs in research and administration, in the preparation of new land-use proposals, in providing continuity in administration, in furnishing clues to improvement of the organization and in pointing out the need for close working relationships with other agencies and groups on mutual problems (USDA Forest Service 1955a). In addition to this goal, Connaughton sought to meet the WO's goal of providing priorities to make multiple-use management work more effective (USDA Forest Service 1956b).

Stating that the past record was disappointing, Connaughton also tightened the expectations and schedule for general integrated inspections (GIIs). He expected his forest supervisors to make sure they were conducted on time and by the book. He also expected that gaps between objectives and actual accomplishments be eliminated (USDA Forest Service 1956a), an action that was in line with Chief McArdle's program for 1955 (USDA Forest Service 1955b). At mid-term, the quality of the GIIs appeared to have improved, but meeting the schedule continued to elude the staff (USDA Forest Service 1957a: 39).

During this inspection cycle, Edward C. Crafts and Russell B. McKennan completed the fifth GII for Region 5 in 1957. Unlike previous GIIs, which were little more than a collection of sketchy functional inspections, the Crafts-McKennan GII focused on the major problems facing the Forest Service in the California Region: policy leadership, effectiveness of controls and constructive critical evaluation of broad program direction. First, the Crafts-McKennan GII commended Region 5 in many areas. It praised Region 5 for its progressive thinking in watershed management and in the problems and needs of reforestation programs and for efforts to consolidate land ownership for better resource management. The Crafts-McKennan GII also praised Region 5 for the high esteem in which it was held by conservation leaders in the state. Crafts and McKennan were also impressed with the cooperation that existed between federal, state and county authorities in Southern California in fire prevention and suppression, and commended the region for its FIRESTOP program, started in 1954. Finally, the Crafts-McKennan GII commended the California Region for its responsiveness to the needs of the timber industry. Under Hendee's and Connaughton's direction, Region 5 had increased timber harvest from 650 million to one billion board feet annually in a period of just three years (USDA Forest Service 1957a: 1-3; Wilson 1991: 31).

However, the Crafts-McKennan GII also criticized the region in one major area: losing, or being in danger of losing, its conservation leadership role in certain fields to the State of California and to Los Angeles County. For instance, in recreation, the Crafts-McKennan GII felt that the State Division of Beaches and Parks was doing a good job, had expansionist aims and was outstripping the Forest Service. In addition, fire leadership in Southern California seemed to rest in the Los Angeles metropolitan area and not Region 5. The Crafts-McKennan GII argued that in a very real sense, the situation now confronting the Forest Service in California was a foretaste of what might come later in varying degrees and at varying times in other western states. "If the Forest Service failed to meet this competition in conservation," according to the Crafts-McKennan GII, "it is to be expected that recreation areas will be sought for transfer to States or counties, certain timberlands to States or private owners, reservoir areas perhaps to the State, watershed protection forests perhaps to the State or counties." Essentially, the Crafts-McKennan GII saw California as a testing ground for the ability of the Forest Service to adjust its activities to the times and to the needs of a rapidly growing West. Chief McArdle seemed to agree with this point (USDA Forest Service 1957a: 12-14; 1958b: 1-3). To counter this trend, the Crafts-McKennan GII thought Region 5 ought to emphasize additional public relations work and recommended that the region seriously consider the need for an additional information and education (I&E) specialist to be stationed in Southern California. The primary purpose of such a move was to maintain more effective contact with the radio, television and motion picture industries and to take advantage of the mass educational opportunities afforded by the concentration of population in that area (USDA Forest Service 1957a: 11). Interestingly, at this time, some Region 5 conservation projects in Southern California were cancelled because the State of California issued new textbooks covering the topic (USDA Forest Service 1955c: 44).

Up until the Crafts-McKennan GII comment, Region 5's I&E had published rather general, mundane pamphlets on the California Region, such as *Facts about the California Region* (1956) (USDA Forest Service 1956c). Following the Crafts-McKennan GII, Grant A. Morse, head of the I&E, began preparing the photo-enriched *Regional Forester's Report* for public consumption. The *Regional Forester's Report* was peppered with photographs of scenic and resource values pertaining to timber, watershed, range, recreation and wildlife, as well as dramatic action photographs of the Forest Service fighting fires with tankers and crews, and helicopters and air tankers. At the

Grant A. Morse, head of the Region 5's Information and Education Division on a fishing trip. Lake Schmidell, Eldorado National Forest. 1965

same time, the I&E initiated a process to collect photographs that documented national forest activities for future annual reports of this nature (USDA Forest Service 1957c; 1958b: 38). The I&E did not just limit itself to publications. Thereafter, the I&E actively began to explore opportunities with television and motion picture producers (USDA Forest Service 1957b: 43; 1958b: 37).

Another Crafts-McKennan GII criticism of the region centered on the lack of understanding among personnel regarding the long-range objectives for multiple-use policy. To meet this deficit, in 1958, in cooperation with the WO, Region 5 developed a program to effectively present multiple use on California's national forests to its field personnel and the public (USDA Forest Service 1958b: 38). This public relations effort culminated in 1959, when the *Regional Forester's Report for 1959* opened with a discussion of the general principle of multiple use and how the concept was being tested on California's national forests at an intensity beyond anything experienced in times past. Intensive multiple-use management was promulgated by Region 5 as the solution to the dilemma between present consumption and future use of resources (USDA Forest Service 1959a: 1-2).

To accentuate this point, Regional Forester Connaughton gave a speech to the WFCA that year, entitled "How Can We Resolve Conflicts in Forest Land Use?" In his talk, Connaughton defined the multiple-use management approach to industry leaders, stating that he hoped to lessen confusion and conflict with the timber industry regarding multiple- use policy. He spotlighted how multiple-use management planning worked and assured them that if the proper steps were taken, gathering all possible facts, analyzing them realistically and then drawing a balance between facts and management objectives, then most potential conflicts would be resolved early in the process, if not avoided altogether. If for some reason conflicts continued, Connaughton suggested open arbitration involving public meetings or hearings in order to shed further light on the problem and to garner majority support for a decision (USDA Forest Service 1959b: 1-18). This statement, and his own inability to recognize such a situation in a future decision regarding the Mineral King Valley, would come back to haunt the Forest Service later.

Programmed Management of Multiple Use, 1955-1960

At the time Regional Forester Connaughton gave his talk to the WFCA, Region 5 was embroiled in a conflict over the application of multiple use on the Kern Plateau, an area of about 500,000 acres in the southern Sierra Nevada on the Sequoia and Inyo National Forests. This large, undeveloped expanse was not cut by roads but held 200,000 acres of commercial timberland with a sustained-yield cut estimated in 1959 at 30 million board feet. The Kern Plateau also had a potential for recreational overnight camping estimated at

55,000 people and would eventually become popular for hiking, hunting, fishing and off-road recreation. Nonetheless, up until 1947, there was little pressure to develop the area. At that time, many people wished to designate the Kern Plateau formally as a wilderness area because of its pristine beauty, but because of its proximity to the population centers of Southern California, the secretary of agriculture concluded that it should be developed according to the concept of multiple use (USDA Forest Service 1959b: 14-15).

Camping, Sierra National Forest. 1963

Source: Forest History Society

Despite the agriculture secretary's supposition, no action on developing the Kern Plateau was taken until 1956. Meanwhile, an epidemic of insect damage had built up in the commercial timber, and there was a clear and growing market for sawlogs in Region 5. In fact, when Connaughton assumed office, timber management had become a priority in Region 5. The annual allowable cut (AAC) from all of the region's seventy-seven working circles stood at 1.3 billion feet, and as a result of an increased sales program, Region 5 closed in on this goal (*California Log* 1955: No. 12). Concomitant with the demand for logs was a demand for new recreation opportunities accentuated by Southern California's tremendous population boom.

The above two circumstances and others brought pressure on the Forest Service to open this area for timber harvesting and recreation through the construction of a timber access road into the Kern Plateau, so, in 1956, Region 5 began to prepare a multiple-use management plan for the Kern

Plateau to see what effect timber sales would have on the area and to safeguard the recreational values of the area. To gather recreational data for the report, Region 5 spent the entire summer of 1957 surveying the recreation potential for the Kern Plateau. With this multiple-use plan drafted, in 1957, Region 5 proposed a timber sale for the plateau. This action drew vociferous opposition from wilderness supporters in public meetings, but because the opponents presented no new facts, the Forest Service made the sale and built an access road onto the Kern Plateau. To protect recreational values, special provisions were inserted in the contract to assure that timber sale operators recognized them as they logged.

However, when a second timber access road was proposed on the opposite side of Kern Plateau shortly thereafter, organized opposition again naturally followed. This time, using the techniques and tactics they learned from the San Gorgonio Primitive Area dispute of a few years earlier, wilderness advocates solicited support from leading California citizens, released a special pamphlet in opposition, garnered publicity in magazines and newspapers, and called for a congressional ground review of the situation. However, in spite of this strong protest by the Kern Plateau Association, the Sierra Club and the Wilderness Society, Region 5 won the argument. Subsequently, Region 5 advertised the second road, which was also built. The next year, a multiple-use coordination plan for the Kern Plateau, entitled *Multiple Use Management on the Kern Plateau*, originally prepared in tentative form in 1956, was reviewed, revised and, in 1960, placed into use. Eventually, the Kern Plateau would become a Region 5 showplace for multiple-use management, but in the late 1950s, the Kern Plateau controversy was one of the sharpest points of environmental contention in the nation. It certainly drove the wedge between the Forest Service and wilderness-oriented people deeper. According to former Regional Forester Douglas "Doug" Leisz, "The Kern Plateau land use controversy was the beginning of the preservation versus use fight which has since touched public lands over the entire country" (USDA Forest Service 1959b: 15-16; Feuchter 2004: 23-24; Leisz 2004: 54; Smith 1990: 23; Connaughton 1976: 119; USDA Forest Service 1959a: 17; 1960a; 1960b: 5; Leisz 1990: 117). While Leisz' appraisal may be an exaggeration – for example, some would point to the Pinchot and Muir fight over the Hetch Hetchy Valley as the beginning of the "preservation versus use fight" – it definitely indicates the importance of the Kern Plateau controversy in framing future land-use disputes (Conners 2005).

During the second half of the 1950s, Region 5's timber management activities had three general thrusts: increasing the timber harvest, reforestation and protection. In 1952, at a time when many feared a timber famine, the WO announced that it was undertaking a massive reappraisal of forestry in

Machine planting for reforestation on California national forests. Shasta National Forest. 1961

America. Six years later, the Forest Service published the final product of this effort as *Timber Resources for America's Future* (1958), known also as the *Timber Resources Review*. While this 715-page report toned down the gloom over a timber famine, it did indicate that nationwide there was a significant problem in timber supply and demand. The *Timber Resources Review* also implied that harvests in some parts of the country exceeded sustained yield – in other words, the nation's forests needed more intensive management, including greater reforestation efforts and stand improvement. However, the assumption made by the *Timber Resources Review* that demand could be met did not take into account timber losses from many variables, such as fire, insects, disease and weather, which could affect the demand-supply equation (Hirt 1994: 139-140; Steen 2004: 286-290).

In conformity with the *Timber Resources Review*, Region 5 set out on a program of timber stand improvement (TSI), renewed reforestation and a better timber protection program. First, in the years 1957-1960, Region 5 conducted many TSI projects throughout the region. By this date, successful pine plantations in California from the CCC days had reached pole size and required thinning and pruning in order to increase rates of growth and the quality of lumber being grown. Thinning and release work also consisted of killing what were then viewed as undesirable hardwoods in coniferous stands to promote better growth of the conifers. These intensive cultural treatments were conducted on hundreds of thousands of acres in Region 5. Along with TSI projects, Region 5 next intensified its reforestation program. Nineteen fifty-seven was a record planting year on California's national forests, in large part due to the success of the Mount Shasta and Oakdale nurseries in producing conifer seedlings. A year later, the Oakdale site was terminated in favor of a new nursery site at Placerville, and Region 5 planned to select 200,000 acres of new sites throughout Region 5 for reforestation. One objective of the newly established Placerville Nursery was to produce hybrid forest trees. Along with the Institute of Forest Genetics, the Placerville Nursery hoped to grow trees with superior qualities of growth and resistance to diseases and pests. Some of the larger reforestation projects in Region 5 were in worked-over brush fields located on the Lassen, Shasta and Klamath National Forests. In conjunction with the latter effort, Region 5 increased the tempo on protection of forests from diseases, insects and animals. By the 1950s, diseases were taking a heavy toll in California's forests, and by 1959, Region 5 estimated that losses from disease alone were more than 200 million board feet a year.

Ranger checking growth of pine. Summit Creek Plantation, Sierra National Forest. 1960

The principal diseases threatening California's forests were heart rots, rusts, root diseases, mistletoes and needle diseases. To meet these threats, disease control work included the elimination of currants and gooseberries to control blister rust, which destroyed sugar pine and other white pines. Insect control work also included a statewide systematic survey and field reporting on timber killing insects such as bark beetles. Thereafter, direct treatment of infested trees included disposing of infested trees through sanitation-salvage sales or by aerial spraying with powerful chemicals such as DDT. On the other hand, animal control work consisted of working closely with the U.S. Fish and Wildlife Service to reduce porcupine populations, as well as mice and other seed-eating rodents. Finally, in conjunction with TSI, reforestation, and pest and disease control, Region 5 stepped up its timber management plans and inventories. Increased demands on the timber resources of the region dictated speeding up the preparation of detailed plans, and each year between 1957 and 1960, funds and manpower were directed toward this end. By 1960 the work was all but completed (USDA Forest Service 1957c; 1958b; 1959a; 1960a; California Forest Industries 1959).

In the interim, timber sales rose dramatically. In 1957, one hundred and ten years after John Sutter had built his sawmill on the American River, California still had 360 billion board feet of standing timber awaiting use. According to one publication, this amount was enough to build 35 million new homes, which, at the time, was enough to rebuild every dwelling in the United States. The majority holder of California's timber was the Forest Service, with 49.7 percent of the state's 8,573,000 acres of commercial forest land. This commercial forested land had a potential volume of close to 179 billion board feet (California Forest Industries 1959). In 1957, the timber harvest from Region 5 was just over 969 million board feet. By 1960, in spite of a slump in the lumber market, the "cut" had reached an all-time high of 1.4 billion board feet. This amount had been reached by an aggressive program aimed at controlling unusually heavy insect activities and by salvaging large volumes of timber killed by fires in 1959 and 1960 (USDA Forest Service 1957c; 1960a).

During the late 1950s, the focal point for fire control of Region 5 policy was to support ground forces through improved air attacks using fixed-wing aircraft, helicopters, air tankers, and smokejumpers. Fixed-wing aircraft and small helicopters continued to be effective fire control tools, and they continued to be used to drop cargo or transport personnel to the fire lines. Air tankers, usually converted World War II torpedo bombers, continued to drop their payloads of hundreds of gallons of water and borate slurry retardant, and, by 1957, Region 5 had even established its own smokejumper base at Redding, California, which had twenty-six jumpers. Moreover, in that year, Region 5 began to experiment with new, larger fire organizations on all of its forests. For example, the Southern California air attack (SCAA) project experimented with using a single command post to effectively coordinate air tankers and small helicopters with ground forces – in other words, an integrated centralized attack on forest fires. By 1959, SCAA could put fifteen air tankers, six lead planes and six helicopters on any fire in Southern California within thirty minutes. Notwithstanding, human-caused fires continued to be the bane of Southern California. For instance, in 1957, 80 percent of the entire acreage burned were caused by two such fires: one by children playing with matches that caused a 11,000-acre fire on the Angeles National Forest, and the other by a young man firing a tracer bullet that started a fire in the parched watersheds of the Cleveland National Forest, which burned more than 66,000 acres, 42,000 within the national forest (USDA Forest Service 1957c).

Though there were no fatalities associated with these two 1957 fires, the subsequent fire season once again demonstrated how dangerous a job forest fire fighting was. That year, eleven men lost their lives fighting fires on the Inaja Fire on the Cleveland National Forest, and in 1960, nine men lost their lives while on fire duty. Four of these men in the 1960 fire season were pilots of air tankers. Because of this mounting death toll, at the end of 1960

Inaja Fire on the Cleveland National Forest. 1956

Region 5 reorganized and enlarged its Division of Fire Control. By doing so, Region 5 hoped to give improved direction and coordination in handling the tremendous fire control job on California's national forests. Additionally, greater emphasis was placed on fire safety and training in fire behavior at all operating levels, and plans were initiated to fund the Fire Laboratory in Riverside, California. This new direction included several fire control experiments in Northern and Southern California. One such experiment was the fuelbreak program, started in mid-1957 in Southern California. The idea behind the fuelbreak program was to remove the brush on key ridges and replace it with perennial grass that burned less intensely than brush because perennial grass was green later in the year than annuals. Everybody in Southern California was excited about the initial success of the fuelbreak program, and in 1962, a similar program, called the duckwall project, was developed to create similar fuelbreaks in timbered areas of Northern California (USDA Forest Service 1957c; 1959a; 1960a; Wilson 1991: 39, 42, 54-55).

While the Division of Fire Control coped with its problems, a new Division of Range and Wildlife Management was created in 1958. This division's main objective was to systematically improve range forage in order to benefit domestic livestock and to create habitat for wildlife. It was first headed by William Dasmann. (Fred P. Cronemiller, who headed the old division, retired in January of that year.) By the end of the Eisenhower years, domestic livestock use of national forest ranges in the state appeared in balance with forage supplies, with about 125,000 head of cattle and 120,000 head of sheep grazing under permit. However, a number of problem spots still existed. For instance, what were deemed excessive numbers of deer caused a steady decline

District Ranger talks "sheep business" with range permittees. Plumas National Forest. 1958

of forage plants on some critical ranges, but Region 5 attempted to correct these "sore spots" through the reduction of game animals. Furthermore, slow but steady progress was also made by Region 5 in the construction and maintenance of fences, water developments and other improvements to better manage and control livestock use. Along with these improvements, rangeland was treated with measures such as seeding, water spreading, fertilization and chemical control to increase livestock and wildlife forage supplies. Finally, to give rangers realistic figures on allotment capacities for livestock and game, in 1958, Region 5 began a new range analysis program (USDA Forest Service 1957c; 1958b; 1959a; *California Log* 1958: January).

Regarding wildlife habitat, the new Division of Range and Wildlife Management recognized that wild animals and fish were products of suitable habitat, and therefore each national forest needed to coordinate any major activity with wildlife habitat maintenance. This practice received increasing attention on California's national forests as part of the region's multiple-use management creed. For instance, on major water projects that impounded water for the generation of electricity, irrigation, manufacturing and household use, Region 5's considerations now included how such projects might impact wildlife habitat. Toward this objective, the Forest Service worked closely with the U.S. Fish and Wildlife Service, the Army Corps of Engineers and especially with the California State Department of Fish and Game. Lastly, in 1959, Region 5 initiated a survey program aimed at evaluating all big game habitats located in the national forests of California. This program looked at browse condition, improvement possibilities and the degree of stocking by game animals, and included the development of habitat management plans for each important big game area (USDA Forest Service 1957c; 1958b; 1959a).

Another new division within the region was the Division of Watershed Management, which was created in 1958 and first headed by Lloyd A. Rickel. The 1957 Crafts-McKennan GII pointed out the growing importance of watershed management on California's national forests, from which half the surface run-off water of the state originated, much of it falling as snow. With the start of the CWP, one of the most important projects started that year was cooperative work with many agencies in the water and power development program of the Sacramento Municipal Utility District (SMUD). This program provided for power developments on the American River within the Eldorado National Forest. Construction started in 1958 included increasing the capacity of several reservoirs, as well as building new ones in the American River drainage. Other major water power projects that had an impact on national forest management included Ruth Dam on the Six Rivers National Forest and the Oroville-Wyandotte Project on the Plumas National Forest. These developments, which involved multiple-use coordination, were only the tip of the iceberg for water development projects to come once the California Department of Water Resources sired the California water plan (USDA Forest Service 1957c; 1958b; 1960a).

Managing mining development was not left out of the picture either. In the early 1950s, the Forest Service, along with conservationists, launched a campaign to expose abuses of various mining laws. Investigations followed that uncovered widespread problems such as mining claims being used to illegally cut timber and to build home and recreation sites and hunting and fishing camps on national forests. This campaign resulted in the passage of two important laws. First, in 1954, Congress passed the Multiple Mineral Development Act (1954), which reserved the use of the surface area of mining claims to the public land agency (in this case the Forest Service) for the purpose of managing non-mineral surface resources so long as such management did not materially interfere with legitimate mining operations. Next, Congress passed the Multiple Use Mining Act (1955), which established a program to examine all national forest lands for mining claims and to resolve occupancies on invalid mining claims. Under the latter law and with proper notice, claimants now had to prove the validity of their mining claims. In California, under a ten-year surface rights determination program, this procedure, thanks in part to the cooperation of the mining industry, quickly resulted in the elimination of thousands of abandoned mining claims. By 1960, Region 5 had given notice on 4.5 million acres of disputed mining claims. Moreover, multiple-use management was re-established on 750,000 acres of invalid claims, many of which were located in prime recreational areas (Williams 2000: 97, 100; USDA Forest Service 1957c; 1959a; 1960a).

Recreation management also benefited at this time from several events. First, in January 1957, the Forest Service presented "Operation Outdoors" to Congress for funding. Operation Outdoors, a five-year expansion and renovation plan for Forest Service recreation facilities, was patterned after the National Park Service's (NPS) Mission 66 program, which was developed to accommodate rising demand on national parks. One important element of the NPS Mission 66 program proposed transferring national forest recreation

Overcrowded campgrounds and lack of adequate facilities led to "Operation Outdoors." Inyo National Forest. 1958

lands to NPS management, and Operation Outdoors counteracted this element. Operation Outdoors was indicative of the long competitive history between the two organizations, but it was also the culmination of efforts by chambers of commerce, recreation and conservation organizations and county and state officials to get Congress to address the many problems faced by recreationists and Forest Service officers. Congress responded to the proposed Operation Outdoors by appropriating close to $6.5 million for fiscal year 1958. Six months later, Congress favored recreation in its budget again, this time by establishing the Outdoor Recreation Resource Review Commission (ORRRC) with a $2.5 million budget. ORRRC's main task was to inventory and evaluate outdoor recreation in the nation to meet the public's demands up until the year 1976. By 1961, government agencies needed to report back to Congress with data and recommendations for outdoor recreation on a state-by-state basis. The Forest Service quickly moved to participate in the project and began studying the recreational potential of the forests (Steen 2004: 311-312; Hirt 1994: 157-159; Bachman 1967: 70).

The California Region was naturally very important to both Operation Outdoors and the ORRRC. As a consequence, Region 5 hired several recreation professionals and landscape architects over the next five years to help inventory lands suitable for outdoor recreation use and to prepare plans to meet the recreational demands of the future. In general, the period from 1957 to 1965 was one of rehabilitation of old recreation sites and facilities throughout Region 5. Rehabilitation consisted of new layout plans and the installation of new standard facilities. New sites were developed where recreational pressures were the greatest, but unlike the NPS, the Forest Service experimented little with their designs, and innovations were limited to construction for permanence and ease of maintenance. By 1965, there were close to 1,500 developed picnic and campgrounds in the California Region. Nonetheless, heavy recreational demand continued to outstrip available facilities. One bright light in Region 5's recreational program came in 1957 with an exchange with the Army of lands adjacent to the Hunter-Liggett military reservation. This land exchange resulted in the Forest Service acquiring more than 26,000 acres of Pacific coastline with a high degree of recreational potential under Operation Outdoors (Bachman 1967: 70; 1957c). Another important outcome of Operation Outdoors was the discovery during survey work that ancient bristlecone pines found near the summit of the White Mountains on the Inyo National Forest were the oldest living plant life known. After the discovery of one pine nicknamed "Methuselah" because it was 5,300 years old, special protective measures were immediately taken by the Forest Service to protect and preserve this unique alpine forest area and these "patriarchs" of the plant world (USDA Forest Service 1957c; Rice 2004: 6; Radel 1991: 28-38).

In 1957, the State of California also adopted a law calling for a statewide study to document what the state had in outdoor recreation resources and what they would need in the future and to coordinate recreation development on a

Picnicking on the San Bernardino National Forest. 1952

Ancient Bristlecone Pine flings its bony arms into the wind high in the White Mountains, Inyo National Forest. 1958

Source: Forest History Society

statewide basis. To meet the federal mandate and the state programs, Region 5 immediately began a complete inventory and study of the recreational resources of California's national forests. Forest Service regional and forest level personnel conducted field work, compiled data, prepared maps and integrated recreation plans with other forest resource uses to provide the required material for both the WO's and the State of California's planning purposes (USDA Forest Service 1958b). In 1959, Region 5 turned over its recreational data to the California outdoor recreation planning committee. Region 5's data formed an important part of the state's survey, since California's national forests included one-fifth the area of the state and a large proportion of the state's recreation attractions (USDA Forest Service 1958b; 1959a).

However, before the door closed on the Eisenhower administration, on June 12, 1960, Congress passed a significant piece of legislation that affected not only recreational resources but all resources located on California's national forests. That legislation was the Multiple-Use Sustained-Yield Act (MUSYA) of 1960, which authorized and directed the secretary of agriculture to develop and administer renewable resources (timber, range, water, recreation and wildlife) on the national forests for multiple use and sustained yield of their products and services. For the first time, these five major uses were contained in one law, with no single use having priority over another (Steen 2004: 307).

The movement toward MUSYA started in 1956, when Senator Hubert H. Humphrey (D-Minnesota) introduced a bill to protect and preserve national forest resources. This first multiple-use bill was introduced with the backing of dissatisfied recreation and wildlife advocates. Initially, the Forest Service had mixed feelings regarding the bill, but by 1958, with the support of the Eisenhower administration, the Forest Service aggressively promoted it before Congress, after references to sustained yield were added to a redrafted bill by the service. In doing so, the Forest Service optimistically expected that the bill would reduce conflicts between user and interest groups by clearly defining management priorities on the national forests. By this date, Chief McArdle had decided that legislative sanction of the Forest Service's multiple-use and sustained forestry policy, which it claimed stemmed from its organization in 1905, might be to the service's "advantage in the contentious climate of demands." With this objective in mind, the Forest Service began to court congressional and interest group allies. The Society of American Foresters immediately jumped into the policy debate in 1958, when it dedicated its annual meeting to the subject "Multiple Use Forestry in the Changing West." At the SAF meeting, various groups voiced their opinions. "In a nutshell," according to one source, "multiple use for the timber industry meant maximum economic stimulation through more wood production. [But] for recreation and wildlife interests, it meant greater protection for the non-market multiple uses and values that had been generally neglected in the drive for wood and forage production." To the Forest Service, its definition of multiple use was to "maintain the status quo that allowed it wide discretion, while commodity interests sought to firmly establish their priorities and non-commodity interests sought to alter the priorities" (Steen 2004: 303; Hirt 1994: 171-173, 176-181). As pointed out earlier, Regional Forester Connaughton was an avid advocate of multiple use, and in a widely distributed pamphlet entitled "Multiple Use of Forest Land" (1959), he held that full utilization was the ultimate goal of multiple-use management. Connaughton characterized the Forest Service's job to be that of negotiator of public demands, and when disharmony occurred, minor uses were to be adjusted to exclude conflict with major or dominant uses. This opinion reflected Connaughton's and the timber industry's hierarchical notion of multiple-use priorities (Hirt 1994: 181).

The final debate on MUSYA took place in March 1960. During the debate, virtually every conceivable interest group testified at the congressional hearings. Timber industry lobbyists such as the AFA and the NLMA pushed hard to make timber production a priority among the multiple uses and came on board only when they were assured that the law would be "supplemental to, and not in derogation of" the 1897 Organic Administration Act. The livestock industry eventually supported the measure when reference to "range" in the enumeration of uses was changed to "grazing," which they felt better reflected livestock use. In their opinion, "range" was a resource that populations of deer, antelope and elk used as well. Finally, a half dozen conservationist groups such as the National Wildlife Federation and the Izaak Walton League expressed support for the bill once the Forest Service agreed to include a clear statement that recreation should be equal to other uses and that wilderness would be considered a legitimate multiple use. Explicitly recognizing recreation as one of the multiple uses for which national forests should be managed assured them that the Forest Service was in the business of recreation to stay. However, David Brower and the Sierra Club did not attend the hearings. Fresh from victory over the proposed Echo Park Dam in Dinosaur National Monument, they refused to endorse MUSYA because they believed the law to be unimportant and patently ambiguous. During this debate, the Forest Service gave in to user demands when necessary and reasonable. In the final analysis, the Forest Service thought it retained its full discretionary control over national forest management and got its statutory endorsement of its policies. The Forest Service expected that MUSYA would strengthen the service's ability to prevent overuse by the timber and livestock industries and resist pressure from preservationists seeking recreation and wilderness preservation and protection. (Steen 2004: 303-307; Hirt 1994: 182-190).

Maturation of Multiple Use – The Kennedy Years

In January 1961, John F. Kennedy took office, and for millions of Americans there was a mood of optimism as his "New Frontiersmen" set about governing. Fresh policies and programs were instituted as part of President Kennedy's New Frontier program. But regarding the Forest Service program, multiple-use management planning continued to be the main thrust of the agency.

For instance, in early 1961, President Kennedy presented to Congress "A Development Program for the National Forests," which the legislative body adopted. However, this "new" program was in fact only a dusted-off version of a 1959 Forest Service plan called Operation Multiple Use, which was the Forest Service's first fully-integrated, multiple- use management plan to manage *all* resources and activities. Operation Multiple Use was a long-range program, which was very similar to other Forest Service long-range programs of the 1950s such as the *Timber Resources Review* and Operation Outdoors. Over the long term, through intensive management planning, Operation Multiple Use anticipated raising timber harvests through additional road building, increasing the quantity of water runoff from forests to enhance the supply of water to consumers and commercial users downstream, building forage production through range improvement and rehabilitation of existing livestock developments and meeting the demand for recreation through development. When Operation Multiple Use was reissued under the Kennedy administration, the only significant changes were a 20 percent increase in the timber harvest target by 1972, an acceleration of the roads construction program to achieve these development goals and a one-third increase in estimated recreation use, also by 1972. So in essence, planning trends in national forest management that developed in the 1950s continued into the 1960s uninterrupted. The Forest Service continued to favor commodity development without adequate environmental protection, which would prove to be a breaking point between the Forest Service and leaders of a growing environmental movement in 1960s (Hirt 1994: 193-203, 216, 222).

In the early 1960s, the California Region toed the Forest Service national party line. In 1960, in speaking about Operation Multiple Use, Regional Forester Connaughton felt that there was no question that skillful application of multiple-use principles was the best answer to pressures from an expanding California population, which had soared from nine million to 19 million people from 1946 to 1966, with 85 percent of the population living in urban centers. When President Kennedy's "Development Program for the National Forests" was brought forward, Regional Forester Connaughton enthusiastically publicized it in each issue of the regional forester's report, stating that the "program recognized the increasing importance of the National Forests to the Nation's growing population and sought to ensure scientific management of their natural resources for public benefit…" Thereafter, the tempo increased in the development of multiple resources on Region 5. Multiple-use planning in the California Region blossomed, and its national forests were seen as "lands of many uses" (USDA Forest Service 1960a; 1960b; 1961a, 1962; 1963; 1966; 2).

In 1960, Region 5 revised and reissued *Multiple Use Management on the Kern Plateau* (USDA Forest Service 1960b) and in the same year issued *Multiple Use Management for the Northwest Subregion*, which included the

Mendocino, Shasta, Trinity and Six Rivers National Forests, as well as all of the Klamath National Forest west of Highway 99 (USDA Forest Service 1960c). A year later, Region 5 issued *Multiple Use Management Guides for the National Forests of Southern California*, which revised a 1953 plan and updated it to reflect the changing times. The original plan had demonstrated its worth, but with the passage of MUSYA, Region 5 took the opportunity to review and revise it. The Southern California *Guide* served as the basis for direction in the preparation and use of functional resource and development plans, for management decisions involving the integration and coordination of use and resources and for guidance in the preparation and revision of ranger district multiple-use plans in California's four Southern California national forests (USDA Forest Service 1961b). By 1962, Region 5 completed the job of developing multiple-use management plans for each ranger district throughout the region. These ranger district multiple-use plans proved useful to rangers and supervisors in explaining how they were carrying out Congress' wishes under Kennedy's "A Development Program for the National Forests" (USDA Forest Service 1962).

At the same time that Region 5's forests were preparing their ranger district multiple-use plans, many were also busy preparing formal, multiple-use impact surveys designed to mitigate the negative effects of construction projects for which they had a role in permitting or licensing. These types of projects ranged from high-voltage power transmission lines and reservoirs to major construction projects such as the state's highway construction program or the California Water Plan (*California Log* 1963: Nos. 4, 5). A prime example was the Kern River Project. This multiple-use and multiple cooperation project between the Forest Service and Southern California Edison, the U.S. Bureau of Sport Fisheries and Wildlife, and the California Department of Fish and Game guaranteed water releases and other protection measures for fish and wildlife on the Kern River. Cooperation among these groups improved conditions for trout, wildlife and recreation along a sixteen-mile stretch of the Kern River that flowed through the Sequoia National Forest, which before these negotiations flowed as slowly as two cubic feet per second (*California Log* 1963: No. 8).

In the Kennedy years, a special management plan was also developed for the national forest lands in the Lake Tahoe Basin, which were being threatened with overdevelopment on private lands. Starting in the 1930s, communities, some with spectacular vacation estates, grew up along the geographically accessible shores of Lake Tahoe, which was heavily promoted as a recreation and vacation paradise. Post-World War II prosperity, the construction of the modern trans-Sierra and interstate highway system and the construction of gambling casinos on the Nevada side in the 1950s spawned more visitors to Lake Tahoe's shores. The 1960 Squaw Valley Olympics, during which the Forest Service provided avalanche forecasting for the safety of the skiers because a portion of Squaw Valley resort was on Tahoe National Forest land, brought winter tourists as well. By the early 1960s, a development boom at Lake Tahoe had moved into high gear. To meet this situation, the Forest Service took two actions. First, it pursued land acquisition of key tracts to prevent development and provide for public use, including two miles of shoreline on the south end of the lake. This was the beginning of a major acquisition program for the Lake Tahoe Basin, including the Eldorado, Tahoe and Toiyabe National Forests. Second, Region 5 developed a multiple-use management plan for the region to halt overdevelopment (USDA Forest Service 1962; Norman, Lane, and McDowell 2004: 4-6; Bachman 1990: 40-42; Feuchter 2004: 2-3; Lessel 2004: 9; Rice 2004: 10, Leisz 2005). Multiple-use management on Region 5 had matured into a proscription, as well as a prescription, for land and resource development.

Meanwhile, other significant activities happened during the Kennedy years, indicating that the decade of the 1960s would be remembered for growing national and world strife. On a national level, the culture of affluence of the 1950s gave way to a decade of anxiety over poverty and unemployment, troubled international leadership over the Cold War, the civil rights movement and the women's liberation movement.

In response to Michael Harrington's popular book, *The Other America: Poverty in the United States* (1962), in November 1962, Congress passed the Accelerated Public Works (APW) Program, which was designed to help alleviate unemployment in critical areas of the country through government-funded employment. The Forest Service participated, and Region 5 received fifteen million dollars of program funds, which it used to hire the unemployed to help complete the construction and maintenance of administrative buildings, lookouts, erosion control structures, heliports, and roads and trails. For instance, APW funding was used to begin work on a Lake Tahoe visitors' center, and funds were used to build much-needed modern housing for its employees and their families at certain stations. As one retiree put it, "We had some conditions that, when I look back on it now, weren't the best, but

Sierra Buttes Lookout Station, Tahoe National Forest. 1964

both my wife and I were enthused about working for the Forest Service, and we were willing to put up with some of these kind of things." APW funding was also used for measures such as range revegetation, insect control and soil stabilization for erosion control (USDA Forest Service 1963; Church and Church 2004: 27-28; Dresser 2004: 9; Leisz 2005). Many forests used large portions of their APW funds on recreation facilities – some new, some to be rehabilitated. Thanks to good management and fast appraisals of needs, the APW program in the California Region rapidly employed about 900 civilians. These men, who were from areas suffering severe unemployment, benefited greatly from the program, while the Forest Service was also able to get much needed work accomplished (*California Log* 1963: No. 1).

Region 5 also moved quickly in response to the Cold War with the Soviet Union. In the early 1960s, America perceived a widening "missile gap" with the Soviet Union and reacted by stepping up its deployment of the Nike missile defense system as a deterrent. Starting circa 1962, under a special permit with the Forest Service, missile tracking devices, radio telescopes, Nike launchers, radar warning stations, aircraft control stations and similar installations occupied sites on California's national forests (USDA Forest Service 1962). For instance, in the late 1950s, battery units for the Nike missile anti-aircraft defense system were located on forests that were near major urban areas such as the Angeles National Forest. In fact, the Mount Gleason Nike base on the Angeles was the highest Nike base in the world. Remains of many of these sites can still be seen on the forests and in some cases are being used for other purposes (Lux 2005).

Conversely, the Forest Service was slow to respond to the growing women's liberation movement, which called for equal opportunity in jobs and education, sparked in part by Betty Friedan's *The Feminine Mystique* (1962). For instance, Region 5's publication, *The National Forests of the California Region (R-5) What, Where & Why* (1961), explicitly stated under job qualifications that "women are employed by the Forest Service in clerical,

Clerks in supervisor's office. Standing left to right, Dorothy Minton, Hope Glatley, Jessie Holt, Lois Hoffman; sitting from left to right, Ruby Engstrom, Jiame Montgomery. Plumas National Forest. 1953

technical or lookout capacities. They are not appointed as district rangers or to other field officer positions." Women in particular were not accepted in wildland firefighting service at this time (USDA Forest Service 1961c: 8; Macebo 2004: passim; Gray 2004: 13-14; Grosch 2004: 34; Peterson 2004: 81-83). The Forest Service, however, did provide them with their own uniform the next year, which was worn by women employees who regularly had contact with the public. However, uniform components, such as the Forest Service badge, were mini-versions of the real thing (USDA Forest Service 1963).

The upsurge of programs such as the APW, along with multiple-use planning, kept the regional office very busy during the Kennedy years. For all that, the region's regular responsibilities continued. The inspection program routinely plodded onward, which appeared less and less important given the multiple-use planning effort. In 1963, Associate Deputy Chief Gordon D. Fox and General Inspector R. B. McKennan visited the California Region and conducted the region's sixth GII without incident. The Fox-McKennan GII pointed out many of the continuing and persistent problems that had plagued the California Region during the past decade, namely the demands for intensive silviculture to meet housing needs for an increasing population, the growing cost of protecting California forest resources from fire and the problems associated with overcrowding of recreational facilities.

One significant event related to the individual GIIs occurred at the end of this cycle. In 1964, Region 5 completed an individual GII for Hawaii, which included the Pacific Territories. At the end of World War II, the United States, under the aegis of the United Nations, took administrative responsibility for groups of widely-scattered islands in the Western Pacific, such as Samoa, Guam and the Trust Territories of the Pacific Islands (TTPI). The United States chose not to treat them as newly-won possessions but rather to take responsibility to promote their economic, social, educational and political advancement, with the goal of helping them become self-determining, self-governing states. As the lead agency for forestry, the Forest Service provided assistance to develop and protect island resources. The major resource problems throughout these islands were inadequate water supplies, declining forest and agro-forestry plantations and loss of fisheries and wildlife (*Pacific/Southwest Log* 1982: August). In early 1959, the Hawaiian Islands became the fiftieth state in the Union, and the United States Territories in the Western Pacific also became more important. Following Hawaii's admission, the California Forest and Range Experiment Station changed its name to the Pacific Southwest Forest and Range Experiment Station to reflect the Station's growing Pacific responsibilities (*California Log* 1959: Nos. 3,4,5). Region 5 also recognized its growing responsibilities to this area as well. The state and private forestry division provided increasing amounts of cooperative guidance and financial assistance to Hawaii and the Pacific Islands, while the islanders carried out cooperative fire control, tree planting, forest management and agricultural conservation work. At this time, the State of Hawaii was making a big effort to grow timber for its own needs so that it would not have to import timber from overseas. In 1961, the Forest Service helped Hawaii develop a multiple-use program for its state forests, and in 1962, Region 5 signed an agreement with the Hawaii state forester to provide for cooperative handling of forestry work on small watershed projects (USDA Forest Service 1960a; 1961a; 1962). For many years, Region 5's Division of State and Private Forestry had provided cooperative and assistance programs to the State of

California. The California programs were related mostly to fire protection and control programs and to forest management advice regarding tree planting, timber stand improvement, nurseries and insect control. By 1963, one-third of all trees planted on private lands in California were planted under the Agricultural Conservation Program (ACP), which was administered by Region 5 (USDA Forest Service 1963).

Multiple-Use Management Guidance, 1960-1963

Under intensive multiple-use management guidance, timber continued to be sold and harvested in increasing volumes, which in hindsight to some overshadowed the management of other resources (Dresser 2004: 8). During the 1950s and into the early 1960s, Forest Service officials believed there was a political consensus in the country that the national forests were to be used and managed for timber (Leonard 2004: 28). In one retiree's words, at this time "timber was God," and when a forest supervisor was behind in the cut, he would receive a telephone call or a note from Connaughton asking why and imploring him to "get that cut out." (Millar 2004: 14, 18). If a person were good at making the allowable cut, he would likely to be promoted (Radel 1991: 18).

There are many reasons why Region 5 and Regional Forester Connaughton tried to keep timber harvest levels up. The "heat" came from a well-organized timber industry, led by groups such as the Western Timber Association (WTA), later renamed the California Forestry Association (CFA). They hired economists to monitor the planning process to make sure that harvest levels were kept up because accessible private timber in California was just about cut over and established mills looked to the national forests as their only source of timber. Local communities such as those surrounding the Plumas and Stanislaus National Forests also provided strong support for rising timber sales. Many people in these communities depended on the timber industry for employment. Furthermore, county governments and local school boards wanted the Forest Service to keep the cut going, because their institutions were dependent on the income they derived from Forest Service timber sales (Leonard 2004: 11, 13-14).

By 1962, a record 1.5 billion board feet of timber was sold, to be harvested from California's national forests. After this remarkable record was achieved, Region 5 raised the allowable annual cut (AAC) of timber to a new high of 1.975 billion board feet. Forest budgets depended on offering and harvesting target volumes within the calculated AAC. Region 5 took this action based on new timber management plans. Reportedly, better growth and inventory data, intensified forestry practices and shorter rotation periods for the timber crops also contributed to this change. A year later, another new record was reached for timber sold and harvested. In 1963, 1.9 billion board feet was sold and 1.65 billion board feet actually harvested. Naturally, as timber sales rose, Region 5's timber staff gained power in terms of controlling the dispensation of appropriations within the region (USDA Forest Service 1961a; 1962; 1963).

As timber sales dramatically climbed, many believed that sustained-yield practices could be met through a variety of actions. First, Region 5 timber management optimistically believed that reforestation efforts would replace these stands. By 1961, there was a breakthrough in getting better survival of planted trees on the region's reforestation projects, and the growing stock of seedlings at the Mount Shasta, Placerville and recently- planted Humboldt Nursery (1961) were deemed adequate to meet reforestation demands. Moreover, Region 5 timber management gambled that scientific achievements and advancements in the suppression of disease and pests through spraying great amounts of insecticide would keep troublesome timber losses to a minimum. In the early 1960s, one knotty timber management problem was how to get sufficient well-designed, well-located roads built fast enough to keep pace with the demand for timber resource development. To solve this problem, Region 5 added several professional geologists to its staff and purchased new equipment to improve its planning and construction techniques. New staff and equipment was needed because many of the desired roads were in remote and steeply sloping areas with unstable soil conditions. As better access was developed through new road construction and right-of-way acquisitions, Region 5 moved closer and closer to harvesting the AAC of 1.975 billion board feet (USDA Forest Service 1961a; 1962; 1963).

Another difficult problem confronting timber management was preventing human-caused fires from burning up California's national forests. Region 5 was very fortunate in the period 1961-1963: Despite some adverse weather conditions, major fires and burned acreage were kept to a minimum. In 1962, only 25,000 acres burned on the national forests, the second-lowest burned acreage on record. In 1963, despite its being a year of greater than normal fire danger, just over 9,000 acres burned, the fewest in Region 5 history. Furthermore, for the first time since 1938, there were no Forest

Service fatalities. While weather conditions had much to do with holding down the number and average acreage of the fires, other factors entered into the picture. Chief among them was the completion in 1961 and initial implementation of a new overall fire plan for District 5, which showed a need for larger striking forces and more land treatment work on the national forests. An increased strike force would enable Region 5 to attack fires with more people at an early stage. Furthermore, Region 5 hoped to achieve greater efficiency in dispatching through the use of preplanned aerial attacks that would ultimately boost helicopter and helitack crews' responsiveness (USDA Forest Service 1961a; 1962; 1963; Leonard 2004: 18-20).

Steady achievements were also made in range and wildlife management, thanks to multiple-use management guidance. Range improvements and range revegetation continued to build healthy range conditions, while the 1,500 California ranches that used national forest ranges under Forest Service permit kept livestock numbers steady. Moreover, increased congressional appropriations and cooperative multiple-use programs with the California Department of Fish and Game and other cooperating agencies allowed for accomplishments in habitat management that were greater than before. These measures included clearing brush-covered lands through prescribed burning to create openings, seeding browseways, creating big game management units, managing deer habitat and improving fish habitat through various water and stream improvement projects such as the Kern River Project (*California Log* 1963: No. 3; USDA Forest Service 1961a; 1962; 1963).

Multiple-use coordination with the State of California also extended to new water developments all over the state of California and greatly affected national forest watershed management in Region 5. Cooperative watershed management and improvements occurred throughout the state as impact surveys were made in advance of water developments on the Yuba and Bear rivers for Nevada County's proposed developments, on the state's Feather River developments, on Placer County's American River proposal, on the state's Cedar Springs reservoir on the San Bernardino National Forest and on the Santa Ynez River Flood Prevention Project on the Los Padres National Forest. Others followed. Then, as part of a nationwide program, in 1963 Region 5 completed an inventory of all municipal watersheds tied to California's national forests. According to the inventory, there were a remarkable 297 national forest watersheds supplying an estimated 787,000 acre-feet of water annually to fulfill the municipal water needs, in whole or in part, of some 10.5 million Californians. This figure was in addition to the millions of acre-feet California's national forests provided for agriculture, industrial and recreational purposes (USDA Forest Service 1961a; 1962; 1963).

Not surprisingly, California's national forests continued to grow in popularity with the public during the Kennedy years, and multiple-use management guidance affected how Region 5 treated recreation. In the early 1960s, visitation by people from around the nation topped 20 million in California. This number increased as winter sports grew even more popular after the 1960 Squaw Valley Olympics. As a result of these Olympics, a new facility at Alpine Meadow near Squaw Valley on the Tahoe National Forest and the state-of-the-art aerial tramway at Heavenly Valley on the Eldorado National Forest became immediate successes. Recreational activities, however, varied by national forest. They ranged from enjoying the solitude and beauty of the High Sierra Primitive Area, to white water kayak racing on the Feather River of the Plumas National Forest, to leisurely enjoyment of Gallatin Beach on Eagle Lake on the Lassen National Forest. By this time, seventeen of eighteen national forests had completed general recreation plans, and more detailed future plans were scheduled for lakes, basins, reservoirs, rivers and other key areas. At the same time, Region 5 introduced a new self-service charge system for collecting fees from the public for using national forest recreational facilities such as campgrounds, picnic grounds and beaches, and, thanks to the APW funding, new campgrounds were under construction and others were being rebuilt (USDA Forest Service 1961a; 1962; 1963).

At this same time, Region 5 began a program to reclassify existing primitive areas and, by 1963, had established several new wilderness areas. In 1963, twenty wild, wilderness and primitive areas, covering 1.7 million acres, existed in the national forests of California. In that year, lands rich in wilderness values were added to them, while others were reclassified, a process that often involved simply adjusting boundaries to arrive at workable administrative units. New areas in Region 5 included the 62,561-acre Dome Lands Wild Area, which was located in Tulare and Kern counties, at the south end of the Kern Plateau in the Sequoia National Forest; the 50,540-acre Mokelumne Wild Area of rugged lands along the Upper Mokelumne River drainage on the Eldorado and Stanislaus National Forests; and the 109,500-acre Minarets Wilderness, which was formed by enlarging, reclassifying and renaming the Mount Dana-Minarets Primitive Area. The latter area was located in the Sierra and Inyo National Forests in Mono and Madera counties. Finally, Region 5 proposed that 122,000 acres be added to the High Sierra Primitive Area to form the High Sierra Wilderness – still the largest wilderness in California (USDA Forest Service 1961a; 1962; 1963, 1976).

Multiple Use *Maximus* – The Johnson Years

On November 22, 1963, President John F. Kennedy was assassinated in Dallas, Texas. Lyndon B. Johnson entered the Oval Office and was then returned to the presidency in a Democratic victory at the polls in 1964, elected on a platform of working toward a Great Society. President Johnson's domestic policy aimed for equality in the form of the civil rights movement, for a war on poverty, and for breakthroughs in education and health, urban renewal and environmental protection. President Kennedy had endorsed many of these ideas prior to his untimely death, including environmental concerns. Influenced by writers such as Rachel Carson and Barry Commoner, who warned of increasing ecological dangers, and by the growing public awareness of the implications of the postindustrial pattern of metropolitan growth, President Kennedy had wished to spend more federal funds on conservation and environmental controls. In his Great Society programs, President Johnson, the consummate politician, pushed forward Kennedy's agenda with skill, energy and resourcefulness (Grantham 1976; 226; Bachman 1990: 47-48; Stewart 2004: 11).

At the start of the Johnson years, the Forest Service's multiple-use philosophy reigned supreme but soon thereafter ran into deep trouble. Coincidentally, former Regional Forester S. B. Show, after a lifetime in the cause of forestry, conservation and advocating multiple use, passed away shortly after troubles began for his beloved Forest Service (*California Log* 1963: No. 11). The Forest Service thought that the passage of the Multiple-Use Sustained-Yield Act would stave off "unacceptable demands from any one sector, to resolve conflicts between user groups, and to prevent overuse of any of the national forest resources," but the Sierra Club and the Wilderness Society did an end run. In 1964, thanks to the hard work of Howard Zahniser of the Wilderness Society, these environmental groups successfully sought for and passed the Wilderness Act (Williams 2001). Neither the Wilderness Society nor the Sierra Club had opposed MUSYA, largely because it contained a key phrase stating that the establishment and maintenance of areas of wilderness was consistent with multiple use. However, time and previous battles with the Forest Service in the 1950s, such as over the San Gorgonio Primitive Area and the Kern Plateau, had convinced Sierra Club and Wilderness Society members they could no longer trust the Forest Service to make the right administrative decisions favorable to their interest in protecting and enlarging wilderness. Simply put, they wanted wilderness protected by specific legislation, and the passage of the Wilderness Act demonstrated the growing political power gained by preservationists in the 1950s and early 1960s (Steen 2004: 309, 313-314).

Prior to the passage of the Wilderness Act, Regional Forester Connaughton and the California Region tried to hold back the rising tide of the wilderness movement because it conflicted with Forest Service multiple-use policy. When Connaughton reviewed the first draft of the wilderness bill in 1957, he believed it to be too expansive and, furthermore, did not see much point in a "legal" wilderness system. While the 1957 bill and later drafts were stalled in Congress in part because the mining industry opposed them, for many years regional foresters discussed the various drafts of the bill with Chief McArdle and later with Edward P. Cliff, who replaced McArdle in 1962. Eventually, an acceptable bill was drafted that could be supported by the Johnson administration and, reluctantly, by the Forest Service. In its final version, the Wilderness Act established a National Wilderness Preservation System, including those areas that had been previously classified by secretarial order as "wild" or "wilderness," and it set up the remaining "primitive areas" for study over a ten-year period to be proposed for inclusion in this system. One key provision of the Wilderness Act was that mining activities could continue in any new wilderness areas for another twenty years but that after 1983, no new claims could be patented in wilderness areas. Another key provision called for the evaluation of any roadless areas on national forests for inclusion for future wilderness status, a process that began in 1967 as the Roadless Area Review and Evaluation (RARE), which will be discussed in the next chapter (Connaughton 1976: 100-101; Williams 2000: 105, 108-110; Steen 2004: 313-314).

The Wilderness Act of 1964 greatly affected the Forest Service, and the California Region immediately started the process of re-evaluating primitive areas for the new National Wilderness Preservation System by reexamining the Salmon-Trinity Alps Primitive Area (1964) for inclusion in it, which also included thousands of acres of private lands owned by the Southern Pacific Land Company (Leisz 2005). Regional Forester Connaughton's support for wilderness designation can be best be described as lukewarm, cautious and utilitarian. In his own words, Connaughton stated, "I have always been exceedingly interested in and supportive of the wilderness classification. I'm antagonistic, however, toward promiscuous proliferation of wilderness designations.... I don't believe in continuing to add wilderness land, involving controversial decisions and resources that can be used for other purposes." Once the Wilderness Bill was signed, however, Connaughton went "by the book." Thereafter, Region 5, as well other regions, set out to develop the best set of management regulations for the system by law, and often consulted with groups like the Sierra Club and the Wilderness Society on the subject (Connaughton 1976: 102-106). By July 1966, the Forest Service developed a set of regulations that were approved by Secretary of Agriculture Orville Freeman (*The Log* 1966: July). However, Connaughton's inability to see beyond the law and his own utilitarian perspective limited his ability to understand the full impact of the impending environmental movement, and put him out of step with the times.

On the home front, Regional Forester Connaughton continued to manage Region 5 with diligence under MUSYA. By 1965, the WO had developed an elaborate flow chart describing how to implement the Multiple-Use Sustained-Yield Management Act. This chart explicitly diagrammed, through boxes and arrows, how the "authoritarian" Forest Service, at the top of the chart, would use this law to work for the public good at the bottom of the chart – very much like Moses handing down the Ten Commandments to his people. Forest Service administrative "know how" and research "findings" would lead to "inventories" of people's needs and resources. Once these inventories were completed, they would be analyzed using regional multiple-use management guides and district ranger multiple-use plans. Once analysis was complete, then additional "planning," "action plans" or "multiple use surveys," would follow up on the analysis, leading to "action programs." These action programs would then be executed in the form of contracts, agreements, permits, licenses and cooperative projects. Interestingly, there was no room for public input in this entire process until the very end. There, at the bottom of the chart, appeared the phrase "public acceptance," with an arrow and small type for "feedback" leading upward (USDA Forest Service 1964).

Along these lines, in September 1965, Region 5 prepared *A Guide for Multiple Use Management of National Forest Land and Resources: Northern California Subregion.* The Northern California subregion, according to this

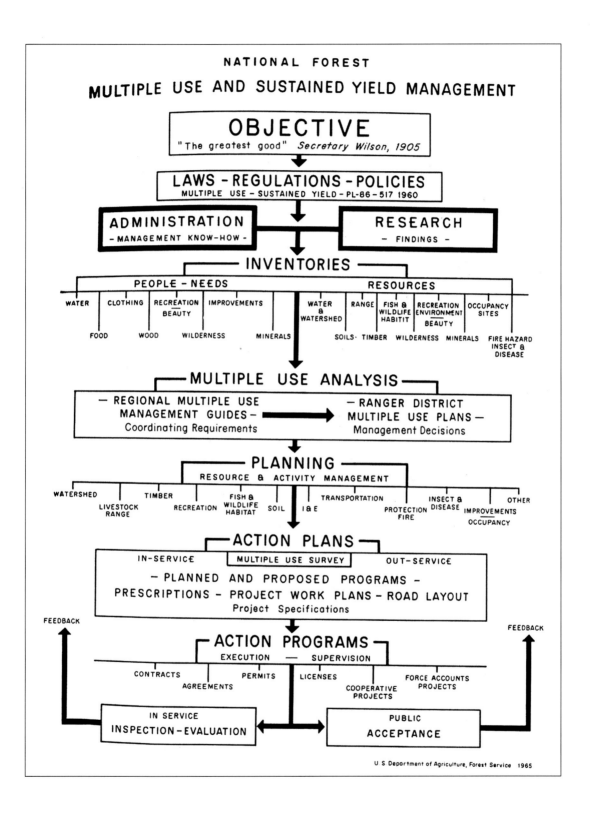

Flow Chart of Program Implementation. 1965

multiple-use document, embraced the Sierra Nevada Mountain Range, much of North Coast and Klamath mountain ranges, and the high plateaus and mountain ranges in the eastern and northeastern portions of California – in other words, all of California's national forests from the Klamath National Forest in the north down to the Sequoia National Forest in the south. The Northern California subregional *Guide* was designed to provide direction for Forest Service administrators in their multiple-use management of these national forests. The *Guide* replaced any and all previous subregional multiple-use plans, except for those already in place for the Lake Tahoe Basin and Kern Plateau (USDA Forest Service 1964).

The Northern California subregional *Guide* was an interpretation of the WO's most recent instructions as they related to subregional multiple-use management planning. First, the document defined and explained the various functional uses (timber, watershed, livestock forage, minerals, lands, wildlife and recreation), including a list of assumptions pertinent to each. This first section of the document indicated the extent and significance of each resource, demands upon that resource, how anticipated future impacts would effect management and what was needed to meet present and foreseeable needs. Next, the document provided a listing of general coordinating instructions that applied throughout the subregion. This section was followed by a list of coordinating instructions that applied within the zones (crest, general forest, front, travel influence, water influence and special). In 1970, and at not greater than five-year intervals thereafter, management situations and assumptions were to be reviewed for validity as the basis for this *Guide* (USDA Forest Service 1964). The Northern California subregional *Guide* signified multiple-use management planning to the maximum, and it is instructive to view the management situations and assumptions of Region 5 for the various functional uses as depicted by the Northern California subregional *Guide*.

Regarding timber, Region 5 firmly assumed that there would be continued demand for the full sustained-yield allowable cut of each national forest working circle. Lack of roads and satisfactory rights-of-ways would persist as a deterrent to full production of the allowable cut of timber. Despite these obstacles, according to the *Guide*, the allowable cut was necessary to support dependent industry and local community growth. The national forests of the Northern California subregion provided 33 percent of the mill capacity for the sawmills of the subregion. Region 5 optimistically hoped to increase the AAC through reforesting the backlog of land deforested by early-day fires, logging, and insect and disease infestations. The AAC could also be increased through timber stand improvement, including treatments such as pre-commercial thinning and weeding, as well as increased timber utilization. Utilization could be improved through the timber industry's innovations in harvesting and manufacturing and through the development of new products that took advantage of what had earlier been considered waste. Plywood and particleboard products are just two examples. Region 5 also acknowledged the need to inform the general public of work being done in the conduct of timber harvesting in order to address their concerns about the growing timber management practice of patch and group clear-cutting (USDA Forest Service 1964: 13-15).

Typical patch cut of timber. Lower Trinity District, Six Rivers National Forest. 1964

On the subject of watersheds, the *Guide* described how vitally important water originating from the Northern California subregion was to the California Water Plan, as well as innumerable existing and proposed water uses not mentioned in the CWP. Region 5 assumed that California's population would double by the year 2000 and that this expanding population would continue to be dependent upon water from national forest watersheds. Therefore, Region 5 felt it had to carefully manage practices in some areas to influence quantity and timing of water yield and to improve water quality. Region 5 also had to restore eroded or damaged soil areas to their full sustained-yield production. At the same time, Region 5 was cognizant of the growing development of hydroelectric power in California generated by Federal Power Commission (FPC) power sites. The FPC was a public utilities regulatory body that, under the Federal Water Power Act passed in the early 1920s, gave preference to proposed developments that offered the most public benefit; such as hydroelectric power, irrigation, flood control, fisheries protection, recreation and other uses. In the Northern California subregion, relicensing procedures for FPC power sites involved not just watershed regulation but also recreational management. Sites involved recreational facility constructions and recreation management and water bypass for fish life. Power company plans were now required to provide the optimum flow for recreation (USDA Forest Service 1964: 8-10; Feuchter 2004: 24-25).

In relation to livestock forage, Region 5 believed that the demand for meat, wool and leather would increase and that livestock production would continue to be an important industry in this subregion. However, the growing trend among livestock owners was to turn to intensively managed private pastures or to feed lots. In fact, the number of livestock permitted to graze on national forest land had steadily decreased over the decades from approximately 190,000 cattle and 425,000 sheep in 1920 to just 90,000 cattle and 83,000 sheep in 1960. That notwithstanding, Region 5 felt that national forest lands suitable for grazing of livestock would be used indefinitely for that purpose. Therefore, Region 5 held that forage production on national forest land could be increased by more intensive management, cultural and structural range improvements, irrigation and water spreading, reseeding and more use of fertilizers (USDA Forest Service 1964: 19-20).

With respect to the mineral and land adjustment plans, the *Guide* had little to say. Under the *Guide*, national forest managers would cooperate with legitimate mineral operators in the development of mineral resources so that results would be compatible with other national forest resource values and uses. The *Guide* also stated that abuses of mining laws would require continued administrative action to eliminate unauthorized occupancies. Under MUSYA, Region 5's land allocation assumption was that surveying and classifying national forest land for special purposes (wilderness, geologic, botanical, scenic, etc.) was no longer justified. Changing conditions and the passage of MUSYA in 1960 had reduced the need for future formal classifications of this kind. Formal classifications were found to be less satisfactory than special zoning under multiple-use planning (USDA Forest Service 1964: 21-24).

If mineral and land considerations ranked low in multiple-use management priority, wildlife received more attention. The Northern California subregional *Guide* recognized that the subregion was well known to sportsmen for big game hunting, fishing and bird hunting and that the California national forests were facing public pressures that demanded accelerated wildlife management programs. Hunter visits had more than doubled from 1951 to 1964, while fisherman visits had more than tripled in the same period. Past actions showed a need for better coordination of programs and planning between federal and state agencies, private groups and private landowners. Problems confronting them included managing big game and overcrowded ranges, developing programs for catchable fish, providing more public hunting grounds on principal flyway routes, improving hunting and fishing access to national forest lands and accelerating programs for the improvement of fish and wildlife habitat. The *Guide* judged that fishing would continue to be the number one outdoor sport in California, that hunter numbers would double by the year 2000 and that rapidly-increasing recreation use of the subregion would focus more attention on the significant and increasing social and economical values of fish and wildlife resources of the subregion (USDA Forest Service 1964: 17-18).

Recreation ranked as the most urgent subject in the Northern California subregional *Guide*. Increased demand for outdoor recreation had been so rapid and had involved such a great number of people that competition among recreational users for the same areas had become commonplace. Public campground needs competed with recreation residences. Motorcycles vied for trails used by pack stock, equestrians and hikers. Demand for more camping and skiing facilities conflicted with wilderness usage. The *Guide* admitted that in the past, while conducting timber sales and clear-cuts, Region 5 had sometimes given inadequate attention to managing for an aesthetically pleasing environment within view of recreational hikers. The

Private summer cabin under special use permit at Whitney Portal, Inyo National Forest. 1958

region's timber managers had inadequately understood what kinds of views would be desirable or acceptable, and Region 5 had insufficiently recognized tomorrow's need for aesthetic landscapes. Region 5 now sought to understand how multiple-use management could be practiced so that commodity uses were compatible with aesthetic values. Reflecting at least a modest level of sensitivity to the subject, Region 5 cited its multiple-use plans for the Kern Plateau and Lake Tahoe as examples of coordinated and more balanced resource management. Timber management had been carried out in these areas in a manner that was compatible with recreation and scenic values (USDA Forest Service 1964: 11-12).

Regardless of the past, the Northern California subregional *Guide* presupposed that recreation use in this subregion would quadruple by the year 2000, that increased demands for public recreation would require improved procedures for coordination with other resource uses in areas of significant recreational use, and that in the interests of sanitation, it would be necessary in some areas to limit camping and picnicking to developed sites. Furthermore, Region 5 realized that there was a need to recognize important view areas, where special treatment was needed in resource management to assure a pleasing landscape. According to the *Guide*, Forest Service officers needed to better understand what kinds of modifications would be aesthetically acceptable to the majority of the people. They then had to develop standards and procedures for resource management of areas important for recreation aesthetics. Finally, the *Guide* also realized the urgent need to step up construction of recreational facilities, as shown by the Outdoor Recreation Resource Review Commission or ORRRC (USDA Forest Service 1964: 11-12).

By the mid-1960s, recreation clearly was the focus of public attention on Region 5 national forests. Regional Forester Connaughton belatedly realized this, for thereafter, the California Region began to capitalize on this interest by combining national forest protection and multiple-use activities, forestry research and cooperative state and private forestry with the lure of recreation as a way to achieve greater support for the total Forest Service program (USDA Forest Service 1966: 7). A prime example of this direction happened in 1965, when Region 5 made the decision to promote development of the Sequoia National Forest's Mineral King Valley as a major ski resort. The "Mineral King" decision, however quickly became a *cause célèbre* for preservationist groups such as the Sierra Club and rivaled the battle over the

Hetch Hetchy valley that had occurred earlier in the century. This disputed development once again raised the question of who was a conservationist and who was not, and who best represented the public interest.

The Mineral King Valley was an area of approximately twenty square miles, and its name derived from a 1880s mining boom. The valley was located in the High Sierra within Sequoia National Forest, but the area was

Mineral King Valley overlooking old mining camp. 1900

Source: Forest History Society

also bounded on three sides by Sequoia and Kings Canyon national parks. When the search for gold and other valuable minerals went bust shortly after it began, some people stayed on in this picturesque valley. By the 1920s, Mineral King's scenic qualities had long been recognized, but it had not been considered as pristine wilderness because of the impacts of past mining and settlement. In 1926, Congress transferred Mineral King Valley to the Forest Service and designated it the "Sequoia National Game Refuge" to protect the wildlife in the area. This designation allowed other uses, such as mineral exploration and recreation. By the 1940s, developers were interested in establishing a ski resort in the area, and Region 5 initiated an investigation of the valley's winter recreation potential. The Forest Service surveyed the Mineral King basin during the winter of 1947-1948 and concluded that it had superlative ski terrain. Even the Sierra Club seemed to agree. At that time, the Sierra Club was very concerned over the reclassification of the San Gorgonio Primitive Area near Los Angeles to make it eligible for a winter resort. In its fight over the San Gorgonio Primitive Area, the Sierra Club argued that alternate sites, such as Mineral King, were available. In the midst of the San Gorgonio dispute, the Forest Service issued a prospectus inviting proposals to develop Mineral King for skiing. However, no proposals were

Sign and leaflet dispenser for San Gorgonio Multiple Use Tour, Mill Creek Ranger Station, San Bernardino National Forest. 1962

forthcoming from private developers, partly because of an access road problem and partly because at the time, it was considered too far from population centers. In 1953, the Visalia Chamber of Commerce tried to revive interest in developing the Mineral King Valley. They held a meeting to stimulate interest in the project, but had no luck. Use of the Mineral King Valley continued to be limited to summer visitation, which included continued use of recreational residences on Forest Service land under permit (Nienaber 1973: 31-36; Pendergrass 1985: 167-169; Feuchter 2004: 32-34).

Interest in Mineral King surfaced again in the late 1950s, when Walt Disney contacted the Forest Service and indicated that he was interested in developing a year-round recreation area in Southern California on national forest land. Disney, an avid skier, had a commercial interest in skiing dating to the late 1930s. He was most interested in the San Gorgonio area, but the Forest Service discouraged him since the San Gorgonios were under wilderness protection. Instead, the Forest Service suggested the possibility of Mineral King. Reportedly, Disney was reluctant, stating that the Mineral King area was too far from Los Angeles, but in the early 1960s, his company began systematically purchasing the developable private land in the Valley (Nienaber 1973: 61-62).

Following the report of the ORRRC in the early 1960s, Region 5 actively sought to develop new recreational sites for California's relatively affluent, growing middle-class population. Region 5 wanted to meet their need for more leisure-time activities and opportunities, and focus turned toward the Mineral King area again. In March 1965, Region 5 decided to issue a second prospectus seeking a developer for a year-round recreational area with skiing as the principal winter-sports attraction. Under this prospectus, the developer would also have to continue summer use of the area, but on a more highly developed and supervised level. According to one retiree, "big money" was a motivator for the Forest Service. Some people wished to see "what a person with good business sense and all…would be willing to bid for the right to develop winter sports on the National Forest." The second prospectus was publicized by the larger California newspapers and via radio and television and immediately drew a lot of attention. Two proponents immediately were considered in the forefront – Walt Disney Productions, and Robert Brandt, a Los Angeles businessman, who was married to actress Janet Leigh and had an interest in recreational developments. To study the area, Brandt hired Ed LaChappelle, a Forest Service avalanche expert from Alta, Utah. Prior to acceptance of either offer, both parties lobbied the state legislature hard and got it to appropriate funds for a state-owned paved road of at least two lanes. The hurdle of financing an access road to the area had been overcome, leaving one less obstacle in the path of development (Nienaber 1973: 48-58; Connaughton 1976: 111; Feuchter 2004: 34-35).

With the difficult problem of financing the road resolved, the Forest Service now had to select the developer. Normally, the decision would have been left to the regional forester, Connaughton. However, prior to the decision, the Sierra Club entered the fray, requesting a deferral of the award and a thorough review of the decision. The Sierra Club and other environmentalists believed that Mineral King had potential as wilderness, which the Forest Service had not considered. The Sierra Club also argued that development of Mineral King would seriously endanger Sequoia National Park. Needless to say, the Forest Service did not expect this opposition, since the Sierra Club had stated in 1949 that it had no objection to any winter sports development there. However, the Sierra Club countered that the development they approved in 1949 was not in violation of wilderness values because of its size, but that the newly envisioned development, which included millions of dollars of investment for accommodations, parking and lifts, certainly did conflict with these values. Nonetheless, the Forest Service simply dismissed the Sierra Club's request on the grounds that the situation had already been discussed for more than a decade and half and needed no further public comment (Nienaber 1973: 58-70). Though the Forest Service was not legally required to hold public hearings on such important decisions, Connaughton's decision was a big mistake, and it clearly went against the grain of his multiple-use pronouncements to the WFCA in his 1959 speech entitled "How Can We Resolve Conflicts in Forest Land Use?" In that talk, Connaughton stated that if for some reason conflicts continued over any land-use, open arbitration involving public meetings or hearings would be needed in order to shed further light on the problem and to garner majority support for any decision.

Meanwhile, Agricultural Secretary Freeman short-circuited the ordinary administrative decision-making mechanism by suddenly announcing that Washington would take the responsibility of making the final decision on which developer to select. Regional Forester Connaughton should have been chagrined over this loss of autonomy, but instead he was probably relieved not to have to make the decision. According to Connaughton, he sized up the situation and said to himself, "Why should I make a decision …it's going to have to go to Freeman ultimately on appeal anyway, so why doesn't Freeman make it in the first place?" Both Brandt and Disney traveled to Washington, where they were allowed to present elaborate displays and proposals to Freeman and a selection committee. Both bids were considerably over the minimum required, and Brandt's bid was higher than Disney's. Notwithstanding, on December 17, 1965, Freeman announced that Walt Disney Productions would be granted a preliminary three-year planning permit, partly because of Disney's reputation and his investments in the valley. The winning proposal called for the $35 million development to be completed by 1975, just in time to meet the public's demands, as outlined by ORRRC. The Mineral King Valley project was the biggest special-use development ever to occur under national forest management up until that time. The Disney Production Company immediately went to work on the project, gathering snow information and meeting with California Governor Edmund G. "Pat" Brown to ensure construction of the needed access road (Nienaber 1973: 58-70; Pendergrass 1985: 169-170; *The Log* 1965: December; Connaughton 1976: 112-113; Feuchter 2004: 35). Once this difficult decision of selecting Disney over Brandt was reached, the Forest

Service and the Department of Agriculture probably felt they were out of the woods. No one realized that the Mineral King controversy was just beginning and that opposition to the project would come from many quarters. What Connaughton and the Forest Service failed to recognize was that times had changed, and the Sierra Club had changed with them. It was no longer just a hiking club, but a major power in the growing environmental movement.

The year 1966 was a triumphant time to Regional Forester Connaughton for many reasons. Nearing the AAC goals for Region 5 during each year of his administration should have made Connaughton, as a timber man, very proud of his accomplishments. He certainly could be proud of his multiple-use management planning efforts. Multiple-use planning was in full stride throughout Region 5 based on two guides: *Multiple Use Management Guides for the National Forests of Southern California* and *A Guide for Multiple Use Management of National Forest Land and Resources: Northern California Subregion*. Additionally, the Kern Plateau and Lake Tahoe Multiple Use plans served as "models" for the rest of the region. The Kern Plateau and Mineral King controversies could not be forgotten, but they were mere bumps in the road to multiple-use planning *maximus*. Certainly, they were difficult and aggravating, but his multiple-use policies had seemingly triumphed over the nascent single-interest environmental movement (Connaughton 1976: 119-120). Notwithstanding, the Mineral King decision had been successfully made, and shortly thereafter, the AFA honored Walt Disney for instilling the "love of animals and forests and all wild things in the hearts of generations of young people and adults." Connaughton and other Region 5 forest officers felt that Walt Disney Productions would give the Mineral King Valley project wide national, and perhaps international, publicity. As this destination resort project moved along, the regional office undoubtedly believed they would find opportunities to maintain and expand Forest Service identification in the project (*The Log* 1966: December; USDA Forest Service 1966: 12). Besides, the Forest Service's popularity could not have been greater, thanks to the television program *Lassie*, a series about a dog owned by a forest ranger named Corey Stuart, who inherited Lassie after a child named Timmy and his non-Forest Service family moved to Australia. By the mid-1960s, this show, which ran until 1972, was among the golden top ten for percentage of television viewers. Thanks to *Lassie*, Hollywood had gone to the woods and the Forest Service's reputation radiated positively in the public's eye. The television show, which rang true to many Region 5 personnel, was filmed on compounds located on several Region 5 national forests, and Forest Service personnel were sometimes even used as extras (*The Log* November 1965; Beardsley 2004: 6-7; Feuchter 2004: 31-32; Macebo 2004: 13-14; Rice 2004: 10-11; Peterson 2004: 50-53; LaLande 2003: 18-19).

The Forest Service and Lassie support the Johnson administration's "Keep America Beautiful" campaign. 1967

By 1967, Regional Forester Connaughton was at the top of his game in Region 5, but it was a time to move on to new challenges. In mid-1967, when the WO offered him a transfer to the Pacific Northwest Region (Region 6), this intensely loyal public servant accepted without question.

Conservation Broadens to Environmentalism

During his twelve-year term as the California regional forester, Charles Connaughton kept the lid on many conservation/preservationist disputes, such as the Kern Plateau and Mineral King controversies. He did so through his faith in multiple-use management and by keeping a firm hand on policy matters, both with the public and within the agency. But no matter how well planned his conservation efforts were to control resource development through multiple-use planning, during the Connaughton years, the timber industry dominated resource and harvesting programs. Furthermore, even multiple-use management for recreation values clashed with wilderness preservation. The proliferating roads, campgrounds, ski resorts and visitors' centers designed to accommodate the recreating public – along with continued logging, grazing and mining under Forest Service multiple-use planning – jeopardized precisely the areas of pristine nature that the Forest Service hoped to save.

Soon after Connaughton moved on to Region 6, mainstream California began to take notice that rampant and unplanned population growth and overdevelopment had caused enormous strains on California's finite resources such as air, water, forests and wilderness. "After several decades of post-World War II boom," according to one source, "most of California's remaining resources – forests, petroleum, and natural gas, farmland, scenery, open space, water, and even the air itself – approached exhaustion. Like the overused older states to the east, much of the state was congested, littered, polluted, and disfigured. Californians were destroying the unique natural heritage that had contributed so much of their success. Smog, traffic jams, urban sprawl, overcrowded parks, and polluted water were only the most obvious manifestations of profound environmental transformation." In his book *The Destruction of California*, biologist Raymond F. Dasmann lamented that greed and ignorance had turned California into the "not so golden state" (Rice, Bullough and Orsi 1996: 576).

Shadow Lake with Banner Peak in the background within the Ansel Adams Wilderness. Inyo National Forest. 1967

Dasmann's comment was part and parcel of a new science called ecology that "emphasized the interrelationships of nature's parts, the importance of protecting all species to maintain genetic diversity, and the capacity for human-induced environmental changes to spread dangerously in many directions." Many believed that civilization could not survive unless harmony was reestablished between people and the environment. This reassessment called into "question conventional forms of technological progress, as well as practices in water development, land use, and wildlife management that conservationists once condoned." Conservation was evolving into modern environmentalism. which emphasized "preservation of the environment as a whole and criticized unrestrained economic and technological growth." Preservationist groups, which were largely excluded from Connaughton's multiple-use planning process and which refused to compromise their concern for the future if environmental harmony were not restored, such as the Sierra Club, led the way. The traditional conservationism of men like Connaughton, who believed in the old Progressive's "gospel of efficiency" of Gifford Pinchot's day, which espoused a utilitarian strategy for exploiting resources and also believed that the Forest Service unilaterally knew what was best for the public good, no longer worked. The Forest Service and Region 5 were thereafter dragged into this environmental movement based on collaborative stewardship with an "expanding group of young, affluent, idealistic, and highly educated professionals who read widely, traveled, and appreciated wilderness" (Rice, Bullough and Orsi 1996: 600-602).

Loading logs on a truck using a "Shovel" loader. Eldorado National Forest

Chapter X: 1967-1978

Region 5 Conservation Contested

California: A Not So Golden State

The controversies and events that emerged in California in the early to mid-1960s – civil rights, confrontation on college campuses tied to sharp divisions over the Vietnam War, conditions in the state's agricultural fields, pollution and congestion associated with urbanization, ghetto riots, anxiety over and the rejection of the California lifestyle by some – unraveled the fabric of California's "golden" society. The state's economy, which had promised an improved quality of life, staggered erratically under rising inflation and taxes, due in part to the Vietnam War and the costs associated with President Lyndon B. Johnson's Great Society program. The consequences of the postindustrial pattern of metropolitan growth, coupled with California's rapid population growth also prompted the state to respond to pressure for clean air, unpolluted water and unspoiled natural areas. By 1966, reaction to these and other developments led to the rise of "cautious conservatism" and the election of actor Ronald Reagan as governor.

Governor Reagan's first term in office disappointed many fiscal conservatives. His "creative society" program, which sought to reduce California's growing deficit and tax burden by transferring responsibility for social programs to local communities and the private sector, failed to take into account that many programs requiring state expenditures were mandated by law or tied to federal policies on which the state depended. Governor Reagan also discovered that many special interests in California had a stake in perpetuating these programs, but with widespread popular support in his first term, Reagan was able to make substantial cuts in higher education, mental-health programs and welfare. Facing rising inflation during his second term, Governor Reagan continued his efforts to cut entitlement programs, such as reducing the welfare rolls through "work-fare" programs, producing significant savings.

On social issues, Governor Reagan's second term closely followed the policies of fellow Californian Richard M. Nixon, who was elected to the presidency in 1968. Both President Nixon and Governor Reagan made special efforts to identify themselves with the "silent majority" or "middle America," opposed to militants, radicals and youthful dissenters. Under Reagan, problems in race and community relations intensified, leading to the creation of militant revolutionary groups such as the Black Panther party in Oakland, which nourished and promoted the development of black pride and black culture, the Chicano protest movements for greater economic and cultural rights, and the Native American protests over the problem of alienated lands and the relocation policies of the Bureau of Indian Affairs, which led to the occupation Alcatraz Island by Native American militants from 1969 to 1971. Despite these upheavals, however, Governor Reagan appeared more interested in economic than social reform. Near the end of his second term, he turned to tax reform as the final solution to the growing California state budget deficits. In 1972, the Reagan-Moretti tax bill was passed, which raised sales taxes to 6 percent and increased corporation taxes to offset homeowners' exemptions. This policy created a treasury surplus for California. Then, on his own initiative, Governor Reagan introduced Proposition 1 in 1973. This was a constitutional amendment to prohibit legislatures from raising tax rates above a stipulated percentage of Californians' cumulative income. Though Proposition 1 was defeated, it left an indelible mark on California politics that led to a tax revolt and the passage of Proposition 13 in 1978 (Rice, Bullough, Orsi 1996: 548-553).

In the election of 1974, Democrat Jerry Brown, the son of former California Governor Edmund "Pat" Brown, replaced Ronald Reagan, who chose not to run for a third term. Governor Brown represented California's liberal-progressive tradition, and his appointment policies represented a major break with historical practice in California. By this time, "a majority of women held jobs outside their homes, but few held leadership positions in business and fewer still in politics or government." Governor Brown modified this situation by appointing women to state government departments usually led by men, and supported the creation of the Commission on the Status of Women to advise him and the legislature. By the end of his two terms in office, Governor Brown had appointed more than 1,500 women to state positions. Moreover, for the first time, members of minorities also found places in the higher echelons of state government, including the State Supreme Court and the University of California Board of Regents. Governor Brown also successfully contributed to social reform by his actions on behalf of farm laborers. In his first term, he help draft California's Agricultural Labor Relations Act, which gave farm workers the right to organize unions, thwarting the state's powerful agribusiness interests (Rice, Bullough, Orsi 1996: 554-557).

Governor Brown's Era of Limits Program was based in large part on British economist E. F. Schumacher's book, *Small Is Beautiful: Economics As If People Mattered* (1973). Schumacher wrote that "contraction of production and consumption, rather than expansion, provides the soundest basis for a humane society." Governor Brown applied his "era of limits" philosophy to slowing down and limiting the growth of California, which by 1970 had become not only the most urbanized state in the Union but the most populous as well. He applied this concept to many issues, including conservation of the environment. Toward this end, Governor Brown committed himself and his administration to preserving California's natural settings and its resources, and to ending their exploitation. The governor appointed conservationists to important environmental agencies and created the California Conservation Corps to provide a workforce for wilderness and urban improvement projects, thereby also alleviating unemployment among youth. He also supported the founding of the California Coastal Commission as well as proposed increases in the state's protected parklands. Many Californians concerned with the degeneration of the quality of life in the state – from air and water pollution and urban and suburban sprawl – applauded the governor and joined him in celebrating the first Earth Day in March 1970 (Rice, Bullough, Orsi 1996: 557).

Clean air was a major concern of Californians, and the outcry against smog grew intense in the 1960s. By this time, automobile traffic and other polluters belched more than 12,500 tons of petroleum-based contaminants a day into the air above Los Angeles, making it the smog capital of the state. Up until that time, state agencies had relied on voluntary compliance by local governments, companies and individuals as a means to keep pollution down. Eventually, some progress was made on muzzling stationary polluters, but the automobile continued to be the primary contributor to urban smog. In response to the unhealthy situation, in 1960, the California legislature passed the nation's first law establishing air-quality standards for vehicles sold in the state after 1966. Soon thereafter, because it was a national and not just a state problem, California pressed Congress to enact air-pollution legislation. Congress took action and passed the Clean Air Acts of 1967 and 1970, both based upon California's smog control experience (Rice, Bullough, Orsi 1996: 603, 605).

Water pollution control efforts in California followed a similar pattern. California's record of keeping its waters clean was spotty until well into the middle of the twentieth century. For decades, local governments, farmers, industries and individual households casually released a variety of chemicals and pollutants into nearby streams and onto the ground, including sewage, fertilizers, toxic pesticides and solvents. For example, untreated sewage had made San Francisco Bay waters and the ocean beyond the Golden Gate Bridge unhealthy for public use. Another example was in the harbor of Los Angeles, where hundreds of industries and some cities, as well as an untold number of ships, dumped pollutants into the ocean that killed practically all marine life for miles. Eventually, in the 1960s, California established the Water Resources Control Board, which established minimum water-quality standards, required regions to develop implementation plans and to set up enforcement agencies and allocated federal and state funds for local water purification (Rice, Bullough, Orsi 1996: 599-600, 603, 605).

Then there was the issue of urban and suburban sprawl caused by California's dramatic postwar growth. A clear example of this problem centered on the region surrounding San Francisco Bay. After World War II, competitive local governments and businesses, which were bent on making more room for industries, airports, warehouses and even garbage dumps, participated in a "land rush" to dike and fill the bay's wetlands. In 1846, San Francisco Bay was one of the largest estuaries in the world, stretching for fifty miles and covering 680 square miles of tidal flats and marshlands. By the 1960s, filling had cut the Bay to a mere 400 square miles of highly contaminated water, with only ten miles of shoreline open to the public. Outraged, local environmentally sensitive citizens founded the Save San Francisco Bay Association, which led to formation of the Bay Conservation and Development Commission (BCDC) whose mission was to develop a protection plan. Benefiting from overwhelming public support and emboldened by the public reaction to the recent oil spill at Santa Barbara, the state legislature passed a bill empowering the BCDC, making it the first regional government entity imposed upon an urban area by legislative fiat. Under the BCDC's guidance, San Francisco Bay's estuary recovered quickly, as the BCDC arrested the Bay's shrinkage and opened up hundreds of acres of formerly filled lands. The BCDC became a model program for managing dynamic environmental conditions by balancing conservation and development. Other regional planning and regulatory agencies patterned after the BCDC would follow to solve difficult environmental dilemmas (Rice, Bullough, Orsi 1996: 609-611).

Conservation Meets Environmentalism

On June 4, 1967, John W. "Jack" Deinema officially succeeded C. A. Connaughton as regional forester. Jack Deinema first entered the Forest Service as smokejumper in Idaho, later serving in ranger, staff and supervisor assignments on the Payette, Teton, and Challis National Forests in Region 4. Prior to service in the Washington office (WO), Regional Forester Deinema

John W. Deinema, Region 5's eighth regional forester (1967-1970)

was assistant regional forester in Region 4. Once in Washington, he was assigned to work with the Office of Economic Opportunity (OEO) on the Conservation Center program of the Job Corps (*The Log* 1967: August). Deinema was the fourth California regional forester who had no previous experience in the region, which reflected a preference for broad and regionally diverse experience over in depth knowledge of California. It also reverberated a bias against sectionalism that stemmed from S. B. Show's domination as regional forester for twenty years (USDA Forest Service 1968b).

Immediately upon coming to Region 5, Regional Forester Deinema visited most of the national forests. He then decided to call a supervisors' meeting in October to announce there would be few major policy changes, at least until he had more time to make a thorough analysis. He firmly communicated that his responsibilities were threefold: first, to give attention to major administrative problems; second, to give leadership and guidance on major Forest Service policy; and third, to coordinate all activities. To his Deputy Regional Forester Charles Yates, he delegated the tasks of leading and directing resource administration, including coordinating the requirements of multiple-use planning and correlating functional plans. This did not mean that Deinema was not interested in multiple-use management issues, for he was, but other issues occupied his short stay in Region 5 (1967-1970) (USDA Forest Service 1967a; *The Log* 1967: October). During those years, Deinema was extremely busy addressing administrative and conservation policy issues.

One of the first major administrative problems tackled by Regional Forester Deinema involved the Job Corps centers in Region 5. The Economic Opportunity Act of 1964 established Job Corps as one of several direct programs to help the less fortunate in President Johnson's "War on Poverty." The Job Corps was a national voluntary program to give poor youngsters a chance to obtain and keep jobs in which they could advance. Within a year after the program started, five conservation camps were operating on the Mendocino, Eldorado, Stanislaus, Angeles and Cleveland National Forests. Young men from these camps worked on a number of projects, such as "Operation Rehab" on the Shasta-Trinity National Forest, a flood damage repair program wherein the boys repaired damaged roads, trails and campgrounds. In 1967, under Regional Forester Deinema, Region 5 took over responsibility for personnel, education and corpsman programs previously handled by Office of Economic Opportunity employees. In 1968, the name Job Corps Centers was changed to Job Corps Civilian Conservation Centers. The title was reminiscent of the CCC program of the 1930s, which was held in high regard (*The Log* 1965: February; 1967: August; USDA Forest Service 1968c: 14; *California Log* 1969: June; Grace 2004: 33). Unfortunately, in January 1969, the four-year program ended because of federal budget cuts under the new Richard M. Nixon administration. In the end, the value of the work projects on the national forest land in Region 5 was more than $4 million. Over all, the Job Corps was appreciated within Region 5, and some youths eventually came to work for the Forest Service (*California Log* 1969: June; Grace 2004: 33; Grosch 2004: 41 Hill 2004: 3-4; Leisz 2004: 84-85).

Another major administrative problem confronting Regional Forester Deinema concerned equal employment opportunity (EEO). In 1969, Deinema appointed Region 5's first EEO advisory committee to ensure that minorities and women were not discriminated against in Forest Service hiring practices and to fill the need to develop and advance all employees, especially women and minorities. The advisory committee could be contacted by any employee of the Forest Service who felt he or she was being discriminated against because of race, color, religion or sex (*The California Log* 1969: November). Deinema thereafter gave his division chiefs and forest supervisors definite direction toward implementing a region-wide effort to conduct EEO training and involve personnel in developing unit action plans (*The Log* 1970: April), but Region 5 had a long way to go when it came to women. For instance, a *Log* article regarding women delivering supplies during fire work on the Stanislaus National Forest could not resist commenting on how "pleasing" they were to look at when they delivered the supplies (*The Log* 1968: September).

Public relations was a third major administrative problem faced by Regional Forester Deinema. During the years 1967-1970, the Forest Service continued to bask in the favorable light of the *Lassie* television show. One episode, which related the adventures of a group of blind children along the Whispering Pines Braille Trail on the San Bernardino National Forest, was talked about for an Emmy nomination, and *Lassie* books, such as *The Mystery of the Bristlecone Pine*, put across the Forest Service story nationwide

(*The California Log* 1968: January; 1969: December). To continue to reach younger people, Region 5 turned to conservation education, a provocative subject covered at a conference sponsored by the California Conservation Council in 1967. At this conference, conservation questions such as "What are the issues?", "Who cares about conservation?" and "What do we do about it?" were discussed. Soon thereafter, Regional Forester Deinema hired Jane Westenberger, a former California Conservation Council Director, to educate Region 5 staff on how to implement conservation education programs in school districts within their forest activity zones. Westenberger was one of the earliest women hired at this level of the organization. The objective of conservation education was to develop in the young people of California and Hawaii appreciation and understanding of their forests and enable them to plan and participate in the constructive actions needed to protect and perpetuate these and related resources. Regional Forester Deinema supported the program as

Students training in woodland skills by ranger, Klamath National Forest. 1961

"one of the most important responsibilities we have in the Forest Service." Later, Jane Westenberger reflected on her program. In trying to define conservation and its message to an interviewer, she said that "some people in the Forest Service believed that the Conservation Education program ought to somehow convince and persuade the public that what the Forest Service was doing was more correct more times than not and that somehow they ought to accept that and let us do our job." To other people, its job was to "help people understand resource management by the basic principles, not to be a specialist that they could go out and manage forests, but to make it easier for them to judge accurately how well the Forest Service was doing its job." The public was indeed concerned, and Regional Forester Deinema seemed to realize that Region 5 needed to start some kind of formal program dealing with the public. Conservation education was one such avenue, and was a program that eventually went national (*The Log* 1967: November; 1968: December; 1969: October; Westenberger 1991: 10-25; USDA Forest Service 1968a).

The last major administrative problem that consumed Regional Forester Deinema's time involved the inspection system. Prior to Deinema's taking office, Region 5 was busy trying to complete its fifth round of individual forest inspections, and in 1968, shortly after Deinema was made regional forester, Region 5 underwent its seventh and last GII. Researched and reported by Deputy Chief E. M. Bacon and General Inspector H. E. Howard, it was yet another variation of Forest Service inspections of this type. Programs in Hawaii were discussed but not included in the itinerary. That notwithstanding, Region 5's Division of State and Private Forestry continued to maintain close ties with the Hawaii division of forestry. For instance, under the Clarke-McNary Act of 1924, a short-term objectives and cooperative action program for Hawaii was undertaken by Region 5 in 1968. This program provided the Hawaii Division of Forestry assistance in developing a sound forestry program – including watershed protection and management, forest management and products utilization, range and wildlife, recreation and fire control – and building a strong professional organization to administer it (*The Log* 1968: September; 1969: June; USDA Forest Service 1968d).

The Bacon-Howard GII cited highlights in Region 5's performance as well as persistent problems. In their appraisal, Bacon and Howard commended Region 5 for the effective negotiations associated with Mineral King, a proposed winter sports area on the Sequoia National Forest, for its emphasis on multiple-use planning, for its effectiveness in working with segments of the California government, for its reduction in the number of human-caused fires and total acreage burned, for its fuel break program and for progress made in acquiring land for the Whiskeytown-Shasta-Trinity National Recreation Area under the recently passed Land and Water Fund Conservation Act. On the other hand, Bacon and Howard commented that the cost of protecting California's forest resources from fire remained the highest in the nation, that intensive silviculture was practiced on only a relatively small portion of the land devoted to the production of timber and that gaps existed in the coordination of land-use planning and zoning with other federal agencies, and at state, county and local levels. Bacon and Howard also recommended that the region find ways to bring the full range of "public" input into the decision-making process and suggested public meetings for special planning areas, district multiple-use plans and impact studies. They also suggested that Region 5 restructure the regional advisory committee to make it more closely reflective of the full range of public interests (USDA Forest Service 1968b).

In response to the latter criticism, Deinema, in Region 5's *Progress in 1968*, its annual report, invited greater public involvement in land management planning decisions and assured the public that Region 5 was seeking the "widest possible expression of public opinion as well as broad factual information upon which to make long-term decisions" (USDA Forest Service 1968c: 1). Deinema followed up this statement with action. Outside speakers, such as nationally known conservationist and writer Michael Frome, were brought to Region 5 in order to "tell it like it is." Frome did just that at a forest supervisors-division chiefs meeting in April 1970. At this meeting, Frome declared that the "day when the Forester alone knows what is best for the land and the people is over." In his speech, entitled "Commitments and the Public Confidence," the author continued by pointing out that the Forest Service must broaden its thinking and its actions, listen to the public, involve them in land management decisions and make use of the many talents available in the public. Following this meeting, Region 5 issued an "action plan" to improve public involvement in decision making by determining the public's needs and desires before decisions were made; by using all available techniques, including public meetings; by assuring better public understanding and public involvement by providing alternatives for public acceptance; and by inviting employee comments (*The California Log* 1970: April; June).

Meanwhile, multiple-use guides and impact studies, generated under former Regional Forester Connaughton, were completed without any discernible public input. They included the *Guide for Multiple Use Management of National Forest Land and Resources – Southern California Region* (1968) and *Multiple Use Management Plan for National Forest Lands – Lake Tahoe Basin* (1968). The *Southern California Region Guide* superseded and replaced the one issued in 1961, matching the format and approach of *A Guide for Multiple Use Management of National Forest Land and Resources: Northern California Subregion* (1964). The *Lake Tahoe Basin Multiple Use Management Plan* was also an update of a previous plan. By 1968, recreation and development strained the environmentally sensitive lake basin. The clear blue lake straddling the California-Nevada border in the High Sierra seemed about to succumb to erosion silt, sewage and chemical runoff, not to mention unsightly developments including hotels, vacation homes and shopping centers crowding the shores as well as smog as dense as San Francisco's from lines of creeping automobiles. "Many felt that development was happening too fast, too loose, and with little regard to the effects on the environment, or the community charms that brought them to Tahoe in the first place" (Rice, Bullough, Orsi 1996: 613: Norman, Lane, and McDowell 2004: 7). With jurisdiction over the area divided, an interstate approach was necessary, and fortunately, by this date in California, community and regional planning organizations to manage growth and give guidance to the pace and form of development had been established. After extensive negotiations, California, Nevada and the federal government agreed to establish the bi-state Tahoe Regional Planning Agency (TRPA). Made up of local and statewide representatives from both states, TRPA sought to manage growth and halt environmental decay with a land-use plan (Rice, Bullough, Orsi 1996: 613; Schmidt 2004: 36, 38-39). Since almost 50 percent of the land area in the Basin lay within the boundaries of the Eldorado, Tahoe and Toiyabe (Region 4) National Forests, Region 5 closely followed the agency, and Regional Forester Jack Deinema was appointed as the only federal person serving on TRPA at the time. Region 5 was ready to support and coordinate planning efforts with TRPA. In 1968, Region 5 also opened the Lake Tahoe Visitor Center, which featured picnic areas and a unique stream chamber that profiled the bank of a living stream and became an immediate visitor attraction (*The California Log* 1970: March, June; USDA Forest Service 1968c: 4; Peterson 2004: 70-72; Rice 2004: 11-13).

While visitors enjoyed the Lake Tahoe Visitor Center, the Sierra National Forest celebrated the seventy-fifth anniversary of its being set aside by President Benjamin Harrison. In 1968, it was much smaller than the original Sierra Forest Reserve. From its boundaries (and some added lands), the Sequoia and Kings Canyon national parks were created, along with Stanislaus, Inyo and Sequoia National Forests. In that same year, the San Bernardino and Cleveland National Forests also celebrated the anniversaries of their establishment as well as their historic accomplishments. For instance, the Cleveland National Forest noted among its many "firsts" that it recognized the potential for use of aircraft in fire protection (*The Log*: 1968: March, April; *The California Log* 1969: April, August). Other national forests in California

Naturalist with visitors on a nature walk, Sierra National Forest. 1963

at this time, such as the Angeles and Plumas National Forests, reviewed their early history as well. For instance, the Angeles National Forest bragged that the first U.S. ranger station constructed with government funds ($70.00) still stood on that forest (*The Log* 1967: November; 1968: October). Altogether, Region 5 became very aware of its past and began to gather historic photographs and other items for its files (*The California Log* 1969: June). In fact, at this time, many in the nation were becoming aware of the necessity to protect cultural resources such as historic and prehistoric sites. In 1966, Congress passed the National Historic Preservation Act (NHPA) to update the American Antiquities Act (1906), which initially authorized protection of antiquities and features of scientific or historical interest on land owned or controlled by the government. Four years prior to the passage of NHPA, the Forest Service was already exhibiting interest in its early history. At that time, the service recognized that a great deal of unrecorded early Forest Service history was fast disappearing with the deaths of the men who helped to make it, and began an oral history program with individuals who made significant contributions to early range management, land surveys and classification, timber surveys and early silviculture, fire control, administration and research (USDA Forest Service 1964). Not long after the passage of NHPA, the Forest Service demonstrated its commitment to historic preservation, and as will be seen, Region 5 was in the forefront of this recognition. For now, Regional Forester Deinema had his hands full dealing with present-day problems in resource management and the burgeoning environmental movement.

Environmental Management – The Nixon Years

In the late 1960s, increasing volumes of timber were sold and harvested in Region 5. In 1967, the largest 25 percent fund payment in Region 5's history was made. Close to $45 million was distributed as the state's share of the gross receipts received for the many uses of the national forests. Of this amount, California counties received $6.4 million (*The Log* 1967: September; 1968: September), derived mostly from timber sales and grazing use permits. California, not often thought to be a lumber state by the nation, was one of the top three timber producers, surpassed only by Washington and Oregon. The next year, Region 5 receipts set yet another new record by distributing $8.6 million to California counties. One reason for the financial boon was a dramatic increase in log exportation to Japan (*The Log* 1968: August). By 1969, Region 5 receipts set yet another new record for distribution to the counties – $15 million – again caused by lumber price increases associated with log exportation to Japan (*The Log* 1969: March; September).

To meet this growing demand, Regional Forester Deinema, along with Charles Connaughton, Regional Forester for Region 6, mutually agreed to explore ways to increase timber supplies from national forest lands in the Douglas-fir region of Oregon, Washington and California. After some study, they decided that they could increase supplies through intensive forest management rather than through shortening rotations (*The California Log* 1969: June). Following this announcement, Region 5's timber management policies gradually turned to clear-cutting over selective cutting. Region 5's timber management professionals recommended patch cutting, which was commonly known as clear-cutting, for these types of stands because Douglas-fir required maximum sunlight to regenerate. However, environmentalists began to wonder out loud about these increased sales and especially about clear-cutting, which they saw as "raping" the land (*Pacific/Southwest Log* 1985: July).

"Make love not lumber" protest.

In 1966, the California Region issued guidelines for mitigating the impacts of patch cutting by establishing categories of view-shed (e.g., near view, far view, near natural appearance) and providing detailed direction to preserve scenic quality; regional officials felt confident that the landscape could be properly managed while clear-cutting. Following the issuance of these guidelines, training sessions for forest officers were held on those forests that held significant stands of Douglas-fir, such as the Shasta-Trinity, Klamath, Six Rivers and Mendocino National Forests. Furthermore, at this time, there was increased logging of Douglas-fir, propelled by development of new products that used these species. Region 5 produced a number of glossy brochures to inform the public regarding this silvicultural practice, but little of it seemed to resonate with the general public (USDA Forest Service 1966a; 1966b; Smart 2004: 22; *California Log* 1972: February). On the other hand, the timber industry rose up against these guidelines. In 1968, they pressed Congress to pass a timber supply bill, recognizing the industry's dominant role in national forest management. Groups like the Western Wood Products Association (WWPA) opposed this "special and costly harvesting requirements to meet aesthetic objectives" and "abandoning timber management in some areas in order to promote preservation of scenery." But Chief Edward P. Cliff convinced Agriculture Secretary Orville Freeman of the value of landscape management, and the Forest Service thereafter produced a series of landscape management handbooks (*Pacific/Southwest Log* 1985: July). For the time being, the clear-cutting issue remained dormant in Region 5, but some in Region 5 were disappointed when the Forest Service adopted clear-cutting as a standard practice. They thought it was an error that cost the Forest Service a lot of public good will (Schmidt 2004: 52).

In addition to landscape management, Region 5 also utilized pesticides to protect timber from insect infestation and herbicides as an efficient and economical way of promoting the regeneration of trees, maintaining fuel breaks, and destroying noxious weeds. Prior to World War II, Forest Service usage of chemical pesticides and herbicides was limited. However, with the passage of the Forest Pest Control Act (1947), the Forest Service's confidence in scientific management, continued and increased. Rachel Carson's *Silent Spring* tempered the use of pesticides such as DDT elsewhere, but the Forest Service continued their use until DDT use was banned in 1972 by the Environmental Protection Agency (EPA). Thereafter, Region 5 turned to other pesticides not specifically banned, such as Malathion, Zectran, Sevin 4 oil and Orthene. On the other hand, aerial application of toxic herbicides continued. During the Vietnam War, the military used Agent Orange to defoliate the hardwood canopy in Vietnam and deny the enemy safe haven. In 1970, use of Agent Orange was discontinued in Vietnam, largely because it proved ineffective for military purposes. However, the Forest Service continued to use 2,4,5-trichlorophenol (2,4,5-T), one component of Agent Orange, and did so without sufficient study of the long-term effects

Dennis Blunt, forestry aid, applying herbicide to Bear Clover on Miami Creek Timber sale reforestation area, Sierra National Forest. 1966

and health risks of using this chemical agent. Region 5's usage of 2,4,5-T led to a "herbicide war" between the region and environmentalists. The debate subsided in 1984, when an Oregon judicial ruling required the federal government to study any potential risks under NEPA. Shortly thereafter, Zane G. Smith Jr., the regional forester in California, issued a moratorium on the use of 2,4,5-T and related herbicides, which was lifted in 1991 after an EIS process determined it could be safely sprayed on small areas, typically between 5,000 and 20,000 acres (Lewis 2005: Chapter 8, 2-6).

In areas of policy other than timber management, important decisions were also made during Regional Forester Deinema's administration. C. A. Connaughton's ten-year fire control plan for the national forests of California (1961) operated effectively. Watershed, as well as range and wildlife management appeared to have no immediate problems, although concern for saving the California condor, North America's largest bird, was mounting. In 1964, Region 5, in cooperation with the Audubon Society, Bureau of Sport Fisheries and Wildlife, and the California Department of Fish and Game, began a program to keep an official count of California condors at the Sespe Condor

Earth First protestors concerned about California condors demonstrate in front of the Los Padres National Forest headquarters, Goleta, California. 1986

Sanctuary on the Los Padres National Forest. In 1966, that number stood at fifty-one. Subsequent counts indicated that the great birds were barely holding their own. By 1969, bird watchers tallied only fifty-three. These relatively few condors were all that remained of a population that numbered into the hundreds in the mid-1800s, and the Forest Service and State wildlife agencies began to consider the condor in danger of becoming extinct. In response, Region 5 offered additional protection for the condors by declaring a moratorium on the issuance of additional gas and oil leases near the Sespe Sanctuary because studies indicated that blasting and other noises associated with gas and oil drilling were a major detriment to successful condor reproduction (*The Log* 1966: October; 1967: November; 1968: November, December;

1969: November; 1970: April). Meanwhile, public awareness of other wildlife species considered rare and endangered in California grew as well, including concern for the Piute cut-throat trout found in the White Mountains on the Inyo National Forest and the Kern River golden trout on the Sequoia National Forest (*The California Log* 1969: July; Schneegas 2003: 7).

Elsewhere, Region 5 took additional protective environmental measures. For instance, following the passage of the Wild and Scenic Rivers Act (1968), Region 5 took action to preserve free-flowing rivers that possessed outstandingly remarkable wild, scenic, recreational and similar values. Region 5 carried out a systematic inventory, and the Middle Fork of the Feather River became one of the first such rivers designated in California (Leisz 1990: 104). This occurred after a nine-year battle led by the Forest Service and the California Department of Fish and Game to prevent a local irrigation company from damning it (Cermak 2005). Region 5 also protected land within one-quarter mile of any wild river segment from mineral exploration and location. Furthermore, with the passage of the National Trails Act (1968), Region 5 integrated its trails, such as the Pacific Crest Trail, which passed through twenty-four national forests in three Forest Service regions, with a national system of recreation and scenic trails (*The Log* 1967: August; *California Log* 1970: November).

Wilderness designation was also very important to Region 5 during the Deinema years. Following the passage of the Wilderness Act of 1964, Region 5 conducted studies to reclassify its primitive areas into wilderness status. The San Rafael Wilderness (1968) within the Los Padres National Forest was the first new wilderness to be established under the Act, and later that same year the San Gabriel Wilderness became the second. The San Rafael Wilderness was located in the San Rafael and Sierra Madre mountains north of Santa Barbara, and along its highest ridges were forests of ponderosa, Jeffrey, sugar, Coulter, gray or foothill, and pinyon pine, with some big-cone Douglas-fir, white fir and incense cedar. The creation of the San Rafael Wilderness took a while to work through the regional office and the Washington office, but it almost doubled the 74,000 acre San Rafael Primitive Area, which also included the 1,200-acre sanctuary for the California condor. One reason why it took longer than expected to create was that a bitter and complex dispute arose among the Forest Service, conservationists and Congress over 2,200 acres of natural grass openings known as *portreros*, which also contained very important Chumash Indian pictographs. The 2,200 acres assumed a political importance far beyond their inherent value as wilderness, and the quarrel left a bitter taste in the mouths of everyone involved (*The Log* 1967: August; USDA Forest Service 1976a; Alfano 2004: 7: Pendergrass 1985: 183-187). On the other hand, there was no ill will regarding the newly established San Gabriel Wilderness. This wilderness was just thirty-five miles away from Los Angeles on the Angeles National Forest. Some of the most scenic country in Southern California was preserved in this chaparral-covered wilderness with rugged and challenging hiking and horse trails. Under the 1964 Wilderness Act, the Forest Service added some 1,300 acres to the former Devils Canyon-Bear Canyon Primitive area of 35,000 acres (*The Log* 1967: August, October; 1968: May/June; USDA Forest Service 1976a).

Horse packing in the Marble Mountain Wilderness Area. Klamath National Forest. 1960

Next, Region 5 submitted proposals for the Desolation and Ventana Wildernesses, which were approved. The Desolation Wilderness (1969) on the Eldorado National Forest consisted of approximately 64,000 acres, of which some 40,000 had previously been set aside as the Desolation Valley Primitive Area. Dominated by Pyramid Peak, it was one of the most northerly sections of the glaciated High Sierra-type scenic areas. The Ventana Wilderness (1969) included approximately 53,000 acres from the Ventana Primitive Area plus nearly 42,000 contiguous areas having suitable vegetative cover and remoteness. This hiker's paradise stretches along both sides of the Santa Lucia Range south of Monterey on the Los Padres National Forest (*The Log* 1967: September; *The California Log* 1969: June, August; USDA Forest Service 1976a). Finally, during Regional Forester Deinema's term, public hearings were held to consider the 97,000-acre Emigrant Basin Primitive Area on the Stanislaus National Forest for wilderness status (*The California Log* 1969: August;). As a result of the hearings, the broad expanses of glaciated granite, towering lava-capped peaks, numerous alpine lakes and meadows and deep granite-walled canyons were accepted into the National Wilderness

Preservation System in 1975, with a total acreage of about 104,000 acres (USDA Forest Service 1976a).

In 1966, Region 5 wilderness use measured 37 percent of the national total for all national forests. But wilderness use was only one part of an overall pattern of national forest recreation. By 1966, the California Region carried the largest percentage of recreation use in the nation, including 65 percent of the recreation residence, 57 percent of the organization camp use, 39 percent of the winter sport use and 25 percent of the camping use of the national total. And, in the late 1960s, encouraged by publications listing all Forest Service public campgrounds, recreational use continued to rise. New publications contained familiar warnings, such as "Be Careful with Fire," information on "Fishing and Hunting" and a section entitled "Aid to Campers," which admonished the forest traveler to register at ranger stations and other Forest Service stations designated on the map (*The Log* 1966: January, July, October; 1968: March). Recreation was "in," and the California Department of Parks and Recreation recognized the important role that California's national forests played in the state. However, one mixed blessing of this status was a growing trend for subdivision developers to locate nearer and nearer to the recreational opportunities of the national forests and to advertise this as an extra added attraction (*The Log* 1968: July). At this early date, no one quite realized that this exponential growth of urban interface would quickly translate to immense fire risk to the homes of people who wanted to have national forest views out their back door.

To pay for maintenance and construction of new recreational facilities, the Forest Service instituted the Golden Eagle Passport charge program in 1966-1967, and recreation started to pay its own way in other ways. User fees were collected for various recreation facilities at the entrances for campgrounds, picnic grounds and boat ramps. Fee increases were also applied to bring summer home permits more in line with their fair market value. Some forest users accepted the charge program without question because they felt that it would lead to better conditions (Feuchter 2004: 26, 37; Leisz 1990: 47), but others protested. In one case, a camper staged a "sit-in" in a Forest Service vehicle – a situation not covered in the *Forest Service Manual*. Others, such as counterculture "hippies," simply ignored all Forest Service rules and authority. By 1969, the "hippie problem" along the Monterey Coast of the Los Padres National Forest and adjacent areas seemed to accelerate with the coming of each summer and school vacation. Problems associated with the "flower children" included indecent conduct in public, narcotics use, public health problems, illegal occupancy, illegal use of fires, stream pollution, littering and general public nuisance (*The Log* 1966: July; 1967: October; 1968: July, August, October; 1969: June; Bachman 1990: 50).

In the final analysis, of all forest uses, Region 5's clientele was most sensitive to recreation. While the average citizen was somewhat concerned about how Region 5 managed timber, range, watersheds and wildlife habitat, those were peripheral compared with the central issue of recreational opportunities, where the Forest Service dealt directly with the consumer. Much needed support came from the private sector, which provided concessions, marinas, travel trailer courts, organization camps, service centers and winter sports areas, to name a few. But these "mom-and-pop" operations were growing increasingly insufficient to meet demand, and Region 5 turned to the large-investment corporation to meet these needs, especially when it came to winter sports areas (USDA Forest Service 1968e; Leisz 1990: 46). One example was the 41-million-dollar development of the Kirkwood Meadows Ski Area on the Eldorado National Forest, slated to open in the winter of 1972-1973 (*The California Log* 1969: June). Another was Walt Disney's Mineral King development. The Sierra Club would challenge both developments in 1969. While the club lost its suit to halt winter sports development of Kirkwood Meadows, it eventually stopped the Walt Disney Corporation's development of Mineral King.

On December 16, 1966, Walt Disney passed away. Heartfelt tributes poured out to Walt Disney, who had brought the gift of laughter and love of wildlife to America and the world. Only a few days before his death, he was engrossed in the planning for two major undertakings, Disney World in Florida and Mineral King in California. United States Senator Thomas Kuchel (R-California) proposed that the Forest Service build a public information center at Mineral King and name it "The Walt Disney Memorial Conservation Center" in tribute to this unique American genius (*Congressional Record* 1967: 12 January).

Despite the loss of its leader, in November 1967, Walt Disney Productions moved ahead with the planning stages of its Mineral King project, which called for the initial facilities to open in 1973 (Walt Disney Productions 1967), with visitation projected to reach 2.5 million annual visitor days by 1978 (USDA Forest Service 1967b). For the moment, the National Park Service (NPS) seemed to be on board with the project, as long as every effort was made to protect its values. Along this line, the NPS requested that alternative means of access, such as tramways, monorails and tunnels be fully considered, that an analysis of the results of road and other construction activities upon the soil drainage near Sequoia trees be addressed and that careful consideration be given to important scenic values. Pending completion of these studies, the NPS stated it would not issue a permit for the eleven-mile section of road within Sequoia National Park (U.S. Department of Interior: 1967). The Forest Service interpreted this statement as NPS approval for the project. Accordingly, in January 1968, a preliminary permit was granted for the project, after a possible obstacle to the development was eliminated when the Interior Department reportedly agreed to permit construction of a needed access road through Sequoia National Park to the Mineral King site in Sequoia National Forest.

Artist's conception for the Mineral King Development. 1967

In exchange for state support from California Governor Reagan for creation of Redwood National Park, President Lyndon B. Johnson offered a guarantee of highway access across Sequoia National Park to Mineral King. The California Division of Highways planned to make this road a model of the best practices in protecting the natural environment. Region 5's regional advisory committee also visited the site and gave the project their stamp of approval. Region 5 cooperated with the project by programming funds from the Land and Water Conservation Act to acquire certain parcels that were needed for proper development and preservation of the scenic beauty of the area, and by assigning a team of engineers and recreation specialists to help coordinate with Walt Disney Productions the planning and development of the self-contained alpine village, ski lifts and access road – now with a price tag of $60 million. Region 5 was convinced that this development by private capital would result in one of the finest recreation areas in the United States, so in January 1969, the Forest Service approved the Walt Disney Productions Master Development Plan. Highlights included a sub-level automobile reception center outside of the main Mineral King Valley, with parking for 3,600 cars. From this point, a cog-assisted railway would transport recreationists to an "American Alpine" motif village. By 1978, twenty-two ski lifts, reaching five bowls, were expected to be in operation. All seemed to go as planned until June 1969, when the Sierra Club, as the sole plaintiff, filed a suit in federal court challenging the authority of the Forest Service to issue a thirty-year term permit to Walt Disney and the authority of the NPS to issue a permit to the state for the access road over national park lands. The suit sought a permanent injunction restraining the Forest Service from issuing a term permit. The injunction was granted in July, temporarily halting the project (*The California Log* 1967: November; 1968: January, February, August, September; 1969: February, June, July; Pendergrass 1985: 164-172; Robinson 1975: 132).

The Sierra Club filed this suit to stop the Forest Service from building its showcase recreation complex for many reasons. The club's principal objection to the proposed development was the absence of ecological studies for the project. They felt that the Forest Service had put the cart before the horse by supporting the building of a large recreational resort without seriously considering the environmental costs through studies. Such environmental studies as were done came after Region 5 decided to develop the area. The Sierra Club wanted analysis that included factors other than just the number of users to be accommodated. In this regard, the club was motivated by a real desire to defend the public interest. On the other hand, the Sierra Club viewed Mineral King as a test case for its agenda, not merely important in and of itself but as a symbol of the continuing preservation-development conflict going back to the days of Hetch Hetchy and Gifford Pinchot. After four years of publicly voicing its objections to the proposed development – in the process missing an opportunity to negotiate with the Forest Service when Regional Forester Deinema was brought in to the California Region – the Sierra Club used legal issues as a means of achieving its larger objective, namely to force the agency to adopt practices consistent with their own policy of multiple-use management. The Sierra Club's legal stratagem also served other club goals. To win conservation fights, according to a spokesman for the club's Legal Defense Fund at the time, the club found lawsuits useful in that they bought the time necessary to rally support, and the courtroom provided a forum in which the facts could be obtained and aired in the public. Besides, a favorable decision often created a major obstacle for opponents by giving them the burden of having to obtain passage by Congress of a bill if they still wanted to prevail (Nienaber 1973: 106-123; Robinson 1975: 132).

The immediate reaction of Walt Disney Productions, which had won numerous awards for effectively communicating to the public the drama and beauty of nature and the need to conserve the nation's natural beauty, was to form an independent, blue-ribbon advisory council of conservationists to advise and consult with the company to make Mineral King a model for conservation-oriented recreational development for the future. The council members included NPS Associate Director Eivind T. Scoyen, former NPS Director Horace M. Albright, National Wildlife Executive Director Thomas L. Kimball and former President of the Sierra Club Bestor Robinson (1935-1959). Robinson, who until 1965 represented the majority of the Sierra Club, sat on the Disney advisory committee because he hoped to influence development in Mineral King. The Department of Justice, acting in behalf of the Forest Service, filed in United States District Court of Appeals a thirty-five-page brief stating that the Sierra Club had no standing to legally challenge such a project solely on the basis of a policy disagreement relating to management of federal lands. The government position prevailed in October 1970, but by that time, Regional Forester Jack Deinema had been promoted and transferred to Washington (*The California Log* 1969: December; 1970 June, July; Nienaber 1973: 110-115; Pendergrass 1985: 173).

The Mineral King problem would fall into the lap of Douglas "Doug" R. Leisz, who officially replaced J. W. Deinema on September 6, 1970. The *Lassie* era of the 1960s – when people sincerely trusted the Forest Service and supported its actions and policies – devolved thereafter into a period of controversy regarding just about anything the Service did. The days when many a poker-playing ranger's motto was to "work hard, play hard, drink hard" and when the Forest Service used to "do a lot of things and didn't tell anybody about anything," (Buck 2004: 18; Beardsley 2004: 20-24; Cermak 2004: 72) ended when environmental groups such as the Sierra Club began to scrutinize the agency's every action. To meet the situation, rangers oriented themselves from managing resources such as timber to managing people and their concerns at public meetings. They also resentfully looked over their shoulders to see if the Sierra Club was "on their tail" for some questionable action (Beardsley 2004: 28; Grace 2004: 35-36). For a time, it seemed that anything the Sierra Club advocated, the Forest Service was reflexively against, and vice versa (Radel 1990: 59).

Conservation Works with Environmentalism

At the time of his appointment, Doug Leisz was well known throughout Region 5 because he had worked there for seventeen years at various assignments. A native Californian, Regional Forester Leisz received his forestry degree from the University of California at Berkeley (1950). He began work within Region 5 as a nursery assistant on the Shasta National Forest. Subsequently, he held staff assistant and district ranger positions on the Mendocino, Six Rivers and Sequoia National Forests. On the latter forest, he was named staff officer for recreation, lands and timber. Following this, he served as forest supervisor of the Eldorado National Forest. In 1967, he transferred to Region 6 and soon served as Connaughton's assistant regional forester in charge of lands and minerals. Then in 1969, he left that post and became the director of manpower and youth conservation programs in Washington until his appointment as regional forester. Colleagues, who welcomed him back to Region 5, described him as having a lot of charisma, a person whom people liked, and one who spent time with individual forest supervisors and rangers. This native son of California, whose management style sought a team approach to problem solving and decision making, served Region 5 until June 1978, when Leisz was named deputy chief for administration for the Forest Service in Washington. Zane G. Smith Jr. replaced him as regional forester (*California Log* 1970: September; 1978: June; Cermak 2004: 77; Leisz 1990: 84; Leisz 2004: 128).

Regional Forester, Doug Leisz (right) with California Lieutenant Governor Ed Reinecke (left) signing first Wilderness Permit in Sacramento. 1971

Many critical issues confronted Regional Forester Leisz upon taking office as the ninth regional forester for the California Region, including, but not limited to, environmental planning and management mandated by legislation, such as the National Environmental Policy Act (NEPA) of 1969, and coordination with the California Environmental Quality Act (CEQA) of 1970. Conservation needed to work with environmentalism, for following the passage of NEPA, the Nixon administration called for the nation to begin to repair the damage we had done to our air, land and water and for a "New American Experience" that included more government responsiveness to citizens' needs. The American public demanded top-quality management of its natural resources, and the Forest Service's program for the 1970s needed to reflect this attitude. To meet this need, the Forest Service, under Chief Edward P. Cliff, issued its *Framework for the Future: Forest Service Objectives and Policy Guide*. This document recognized that the Forest Service's programs were out of balance in meeting the public's needs for the environmental 1970s and sought to correct this imbalance. As Chief Cliff stated. "We can no longer afford to emphasize programs that produce revenues at the expense of others. If we do, we will not be providing the proper service to the public…" In order to have a better-balanced, quality program, the Forest Service began to reduce timber sales, roading for sales and structural improvement items, and increasing wildlife, watershed, recreation and pollution control programs (*California Log* 1970: October). *Framework for the Future* was a nationwide program that led to a whole series of ecosystem training sessions around the country (Peterson 2004: 104). Soon thereafter, Woodsy Owl's "Give a Hoot. Don't Pollute" motto was introduced. Woodsy Owl became the Forest Service's nationwide symbol dedicated to combating pollution and improving the environment (Coutant 1990: 25).

The Forest Service also began to listen to the public and its own employees more closely. For instance, in 1972, Eldorado National Forest Supervisor Erwin Bosworth (father of future Chief Dale Bosworth) set up public "listening sessions" to gather public feelings and ideas concerning overuse and the future protection and management of unique areas before the Forest Service developed long-range management plans for sensitive areas (*California Log* 1972: January). In 1950, Regional Forester P. A. Thompson, in an article entitled "We Need An Informed Public," warned that Region 5 needed public support and would only get it through a public that was both involved and aware. Somehow during the Hendee, Connaughton and Deinema years, Region 5 had lost that important tenet of Region 5 policy (*California Region-Administrative Digest* 1950: June; *California Log* 1972: May). In addition, employees were allowed to once again voice their opinions openly. For instance, in 1974, employee comments were openly sought on the *Environmental Program for the Future* (EPFF). Chief John R. McGuire, who replaced Chief Cliff in 1972, stressed that as much information as possible must be received in reaching decisions for the future and stated, "I feel that comments from Forest Service personnel will provide a vital dimension to the public involvement process…I encourage all Forest Service employees to submit comments on EPFF directly to my office" (*California Log* 1974: October). A period of open dialogue was reached in 1975, when the *California Log* first printed a critical letter from Laurence Rockefeller of the National Resources Defense Council, in which he stated, "The Good Forester has nearly gone; he is being replaced by The Sales Agent. The Sales Agent listens to the timber industry, which has found it more profitable to cut down our forests than to replant its own." It then printed a letter from a district ranger on the Plumas National Forest, who responded, "It bothers me that intelligent and influential people felt a need to file lawsuits and write letters such as the one below. It bothers me that many of the people I talk with today express similar feelings about the Forest Service. Somewhere along the way, we as part of the bureaucracy that manages the Forest Service have failed to reach an understanding with part of our public" (*California Log* 1973: April).

In the meantime, Congress passed a series of legislative acts along these lines. Regional Forester Leisz, in his eight-year term in office, nimbly negotiated Region 5 through three major pieces of environmental legislation – the Threatened and Endangered Species Act (1973), the Forest and Rangeland Renewable Resources Planning Act (1974) and the National Forest Management Act (1976). Besides shifting Region 5 policy to respond to this legislation, other difficulties confronted the Leisz administration. Timber issues related to the public perception of Forest Service clear-cutting policy, roadless and wilderness area management and relations with the Sierra Club over Mineral King tested the new regional forester. Each of these policy concerns will be discussed later in greater detail.

By the time Regional Forester Leisz assumed office, Region 5 had become a major force in California and a major bureaucracy. In 1970, Region 5 administered close to 20 million acres of land within its unit boundaries, whose resources (timber, grazing, minerals, etc.) produced receipts of more than $60 million – of which 25 percent, or more than $13 million, was returned to the counties related to the collections. To manage this area – which produced water, forest products, wildlife and recreation opportunities for approximately one-fourth of the nation's most populous state – Region 5 engaged about 5,500 permanent and temporary employees, and had an operating budget of $80 million. These employees not only regulated resource use but also maintained a substantial array of infrastructure and administrative facilities, including 35,903 miles of dirt roads and 1,450 miles of paved roads, 1,106 road bridges, 14,304 miles of trails, 797 homes, 138 offices, 270 barracks and mess halls, 961 shops, warehouses and other storage buildings, 233 lookouts and seventy-eight heliports. Pertaining to recreational facility use, close to 45 million visitors used Region 5's 1,480 camp and picnic sites, 108 resorts, 212 organizational camps, 389 recreational residence tracts and thirty-two large, winter

Girl Scouts of America Encampment, Inyo National Forest. 1967

sports sites. Under wildlife, Region 5 managed an impressive 13,301 miles of fishing streams and 1,884 lakes, hosting more than 3.15 million angler visitor days. Visitor days related to big game hunting totaled 1.72 million and included 2,622 visitor days hunting for antelope, 11,809 for bear, 1,445 for bighorn sheep, 593,130 for deer and 1,140 for elk. Finally, Region 5 also protected more than 24 million acres in California from fire. This was an area larger than the entire state of Indiana and included seventeen wilderness areas totaling 1,561,472 acres of national forest land, and four primitive areas containing approximately 359,315 acres (USDA Forest Service 1970a).

While tending this great empire, Regional Forester Leisz dealt with many administrative policy issues. One growing issue was Nixon's New Federalism policy, which in 1970 shook up the organization of Region 5 with threats of reorganizing and restructuring the Forest Service, of relocating the regional office elsewhere and of consolidating various ranger districts and adjusting national forest boundaries. In 1971, Region 5 conducted a regional office study, which sought a location central to all national forests in California. Sites under consideration outside the San Francisco area included Sacramento, the central coast area in the vicinity of San Luis Obispo, Paso Robles-Morro Bay and the Santa Monica-Los Angeles International Airport area. Eventually plans to relocate the regional office were deferred. By 1974, Forest Service reorganization was called off completely as well (*California Log* 1970: November; USDA Forest Service 1971a). One offspring of this policy initiative was that in 1971 Region 5 replaced the general integrating inspections (GIIs) with a new approach proposed by Regional Forester Leisz, which de-emphasized the control aspects of inspection. Instead, emphasis was placed on resolution of critical issues that affected individual forests. In some cases, these issues were of regional or zone-wide influence. This team approach to problem solving was again explored in 1973 under the theme, "Framework for Change." As part of this initiative, Region 5 explored redefining ranger district, supervisor office and regional office roles in order to be more responsive to changing and increasing demands and desires (*California Log* 1971: May; October; 1973: July).

Another growing issue was the role of women in the Forest Service. In 1970, Congress passed the Equal Rights Amendment (ERA), which was then sent to the states for ratification. Perhaps inspired by the ERA fervor, in July 1971 Sue Alexandre, a new employee, fomented a revolt of clerical employees. In an open letter in the *California Log* to all Region 5 employees, Alexandre openly and forthrightly stated that the clerical employees, almost all of whom were women, were being treated as second-class citizens. Among some of her complaints were that clerical staff were expected, frequently without a "thank you," to make coffee, buy cigarettes, go to the store, organize birthday parties, clean up the office and occasionally type personal letters for supervisors; that clerical staff were expected to suffer the consequences of supervisors' mistakes, while not receiving any credit for their successes; that clerical employees were not allowed to attend staff meetings, conferences or trips in the field and that written communications about what was going on in the division or in the

Forest Service as a whole were not routed to clerical employees. She also felt that clerical employees felt generally powerless to change the situation or even to communicate their feelings about it (*California Log* 1971: July, August, October). Thereafter, letters to the editor from clerks and personnel officers poured into the information and education (I&E) division. These letters, with a few exceptions, supported Alexandre's position. Nevertheless, little seemed to come from this incident. In 1973, a women's program coordinator position was created within Region 5 to set up career ladders to provide for bridging the gap from clerical to technician and paraprofessional positions (*California Log* 1973: April). Initially, other than a few token positions – such as when one woman was placed in charge of the Plumas National Forest warehouse and then subsequently called a "warehousemaid" (*California Log* 1973: January) – women made little progress in breaking the glass ceiling, from the basement to only the first-floor level of much higher management levels, which kept them in clerical positions within Region 5 and out of professional and supervisor roles. Inexperienced "male" EEO counselors, whose responsibility was to argue for women in such cases, and who were eager to defend a "sweet, innocent young lady before them in a miniskirt, with legs crossed, tears running down her cheeks and a frog in her throat," were too often unwittingly sexist themselves (*California Log* 1972: February).

However, the pace of equality for women marginally quickened. Starting in the early 1970s, women were employed in Region 5's fire organization, but they had to cut their hair. The attitude, according to one Forest Service retiree, was that the Forest Service would not hire a man with long hair – such men were automatically labeled as hippies – and therefore they would not hire a woman with long hair (Buck 2004: 27-28). By 1974, Region 5 led the nation in appointing women to higher professional positions. By that date, Region 5 had employed eleven women foresters, close to half of the women foresters in the Forest Service nationwide, and women were feeling less discriminated in firefighting, which once was considered "for men only" (*California Log* 1976: February). Additionally, Region 5 had also begun to hire professional women for staff positions in San Francisco, such as Jane Westenberger, who ran the conservation education program (Westenberger 1991: 56-57).

Cultural resource management (CRM) was another administrative issue needing increasingly greater attention by the Forest Service and Region 5. The passage of the National Historic Preservation Act in 1966 altered the direction of historic preservation in the United States. Within a decade, additional laws, implementing regulations and other measures mandated that California's national forests be surveyed for the presence of cultural resources and that these resources be evaluated for their eligibility as National Register of Historic Places (NRHP). Public concern over environmental deterioration had prompted the passage of NEPA, which specifically recognized the federal government's responsibility for preserving important cultural resources. Section 102 required that an EIS must include both anticipated adverse effects and mitigation measures for all federally-assisted projects. In response to fears that significant properties might be lost before they could be placed on the register, President Nixon issued Executive Order 11593 on May 13, 1971, which asserted that the federal government would provide the leadership in preserving, restoring and maintaining the historic and cultural environment of the nation and ensured that federal agencies such as the Forest Service would establish policies and procedures to preserve and protect sites located on federally-owned or -controlled properties. Ultimately, federal agencies were obliged to inventory and nominate all eligible properties under their jurisdiction to the National Register (Hata 1992: 113-116; Boyd 1995: 4-6).

Hopper of old gold mill. Noble Canyon, Cleveland National Forest. 1966

Based on the issuance of Executive Order 11593 and the requirement for environmental review and compliance with Section 106 of NHPA, in 1969, Region 5 hired Donald S. Miller as Region 5's first regional archaeologist. His position was seen as a priority in spite of personnel ceiling problems. The Forest Service recognized that if disturbed sites were to be salvaged and undisturbed areas located, protected and interpreted under professional standards, prompt action was necessary. Because of the close ties between archaeological research and the interpretive work of the Forest Service, Miller's position was assigned to the division of I & E, but he was to work closely with the divisions of land and recreation (Miller 1998: 20-21; USDA Forest Service 1969). By 1970, Region 5 established a Forest Service history program, also attached to the Division of Information and Education but not considered as an urgent need (USDA Forest Service 1970c). In 1971, these two programs were combined, and a regional committee on archaeology-history was created to recommend both goals and programs for the next three to five years, and budgets for implementation. At this time, archaeology and history were moved into the Division of Lands and Minerals. The first goal was to rapidly develop a cultural resource survey program that was both responsive and professional. Part of this program entailed hiring full-time, professional archaeologists to be stationed in the supervisor's office of each forest. Other goals included establishing laboratories for processing artifacts recovered in the field, research facilities, and training, information and education programs (USDA Forest Service 1971b; 1971c).

Regional Archaeologist Miller followed through on these goals. By 1973, he established the first phase of a region-wide program to increase the protection and interpretation of thousands of prehistoric sites. Region 5's cultural resource program gained staff, assistants and contract archaeologists on several of the national forests within Region 5, as well as a laboratory and research facility. The objectives of the program were to implement and comply with federal laws and regulations, to preserve significant paleontological, archaeological and historic sites, and to use them for scientific, educational, recreational and other purposes. In the initial years, Region 5 CRM programs necessarily focused on reviewing applications and research designs for antiquity permits, surveying and recording sites for timber sales and other projects that had potential impacts on cultural resources, negotiating and overseeing cultural resource contracts and responding to "emergencies" dictated by various Forest Service projects such as timber access roads. Often it was a race to save the remains of sites; pages of history that ranged from prehistoric Indian encampments to turn-of-the-century logging camps (*California Log* 1971: March; 1972: June). When a significant site was discovered, the Forest Service mitigated it either by avoidance or excavation. To help with site identification, description and protection, the regional program developed paraprofessional training that enlisted other Forest Service personnel in CRM activities (*California Log* 1971: March; 1972: June; 1973: August). Interestingly, Regional Forester Leisz attended one of these early CRM training sessions as a trainee (Leisz 2005). Thereafter, the cultural resource management program was considered an integral part of the Forest Service's Environmental Program for the Future, or EPFF (USDA Forest Service 1971d).

As the decade of the 1970s rolled on, Region 5's cultural resource program grew in importance, especially after Congress passed new federal legislation affecting archaeological and historic resources on federal property. In 1974, Congress passed the Archaeological and Historical Conservation Act (AHCA), which amended earlier legislation and extended provisions for the protection of archaeological and historical data to cover any alteration of the terrain as a result of any federal construction project or activity (Boyd 1995: 6). To meet these new obligations, Region 5 recruited additional archaeologists. In 1974, Regional Forester Leisz felt it necessary that each forest have available the services of in-service professional archaeologists. By this date, the Modoc, Six Rivers, Plumas, Stanislaus and Angeles National Forests had full time, professional forest archaeologists. To meet the needs of the other twelve forests and the Lake Tahoe Basin, Leisz suggested hiring two zone (southern and northern) archaeologists to be stationed on the Mendocino and Sierra National Forests as needed. But by 1978, when Leisz moved on to the WO, there were fewer and fewer dollars to invest in professional archaeological services as well as administration and planning. Thereafter, more and more work was outsourced to private CRM firms, and Region 5 CRM personnel concentrated their efforts on the basic functions of management, including professional and technical advisory services to management and cooperators (USDA Forest Service 1974b; 1978a).

Then, in 1978, the American Indian Religious Freedom Act (AIRFA) was signed into law, which added a new dimension to Regional Archaeologist Miller's job. AIRFA set forth a policy to protect and preserve for Native Americans their inherent rights of freedom to believe, express and exercise the traditional religions (Boyd 1995: 6). Up until this time, relations between Native Americans and Region 5 were quite tenuous. In 1970, at the height of the seizure of Alcatraz Island in San Francisco Bay, a group of Pit River Native Americans tried to occupy Lassen National Park. In doing so, they were trying to claim about 3.4 million acres of land in Northern California that they believed still belonged to them and to protect certain portions of these lands for their religious significance. Thwarted in this effort, about fifty Native American men and women then occupied a site on the Lassen National Forest. When they were confronted, a small riot broke out, and people on both sides received minor injuries. Fourteen Native Americans and two non-Native Americans, one a local reporter, were arrested and prosecuted (*California Log* 1970: November).

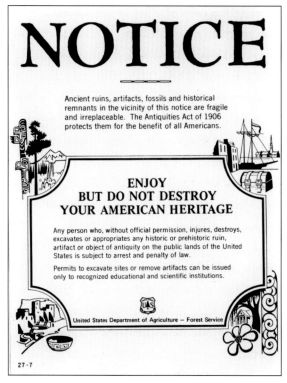

Forest Service notice educating the public on the protection and preservation of all cultural resources by law.

The "Sugarloaf incident," as it was called, set off a series of similar minor incidents on the Lassen National Forest over the next two years. In the summer of 1973, a group of Pit River tribal members began threatening to close down timber sales on the forest. They also once again tried to occupy lands that they considered sacred. Forest officers made repeated attempts to persuade them to leave and not interfere with any authorized logging operations. Refusing to comply, the leaders of the insurgent group were eventually arrested peacefully and released (USDA Forest Service 1974c). The Sugarloaf incident was indicative of a growing conflict between Native Americans and the Forest Service flowing from indigenous concerns for traditional cultural properties on federal lands. The passage of AIRFA now required that the Forest Service and other federal agencies consult directly with appropriate Native American groups whenever dealing with Native American sites found on forest lands. A new "legal" working relationship thus was forged between the Forest Service and Native Americans. This relationship was not always without differences of opinion, especially when it came the disposition of potentially significant human and artifactual remains claimed by the tribes (USDA Forest Service 1978b; U.S. Department of Interior 1980).

Up until this time, Forest Service law enforcement personnel were tasked with enforcing applicable laws and regulations for protecting natural and cultural resources and public improvements, along with occasional police work on special problems, such as investigating incendiary forest fires. But the Sugarloaf incident and other incidents added a new dimension, protection of people and their property. This new law enforcement task assumed many different dimensions. For instance, in the 1970s, Forest Service enforcement duties included curbing a campout by the Gay Liberation Front; regulating various gatherings of the Rainbow Family gatherings, a group that gravitated to the large, free, public areas offered by national forests and whose participants partook of nudity and outdoor love making; and preventing occasional violence and intimidation of other users by Hell's Angels during their motorcycle campouts (*California Log* 1970: July; Feuchter 2004: 47-49; Smith 1990: 50-53). Another new duty for the Forest Service began during the turbulent 1970s: patrolling national forests for marijuana plantations. Dealing with such situations became a new aspect of managing California's national forests, which led to special law enforcement people carrying firearms, which they had seldom done in the past (Alfano 2004: 15).

Besides the above internal administrative problems, Regional Forester Leisz's time was also diverted by the destiny of the Tahoe Lake Basin and cooperation with Lake Tahoe Region Authority (TRPA). In 1970, Regional Forester Deinema appointed a team of specialists, men skilled in ecology, hydrology, geology, forestry, and landscape and recreational planning, to cooperate with TRPA (*California Log* 1970: March). Now the fate of federal cooperation in the Tahoe Basin fell into the hands of Doug Leisz, when President Nixon appointed him to the TRPA board in 1972 (*California Log* 1972: November). In the meantime, the Forest Service purchased a 10,452-acre tract of eastern Lake Tahoe shoreline property using Land and Water Conservation Act funds, which became part of the Toiyabe National Forest. A year later, an important change in the Tahoe Basin came when the Forest Service completed an agreement to consolidate portions of the Toiyabe, Eldorado and Tahoe National Forest lands within the Tahoe Basin into one single area named the Lake Tahoe Basin Management Unit (LTBMU). Overnight, the LTBMU became the largest single land management entity within the basin. Over the next few years, as LTBMU worked with TRPA, as one source stated, "study after study, plan after plan was developed, debated, rejected, opposed, promoted or discarded. It was an era of crisis and confusion, with lawsuits lobbed in all directions." By the end of Leisz's tenure, the "flaws of the bi-state compact were obvious," which led the states of California and Nevada to revise it in their legislatures. In the meantime, the clarity of Lake Tahoe continued to degrade (Norman, Lane, McDowell 2004: 8, 10-11: Schmidt 2004: 39-47; Smith 2004: 34).

Overnight, national forest multiple-use planning efforts for Region 5 also appeared to be murky. Under the Multiple-Use Sustained-Yield Act of 1960, Region 5 had developed submarginal and national forest, as well as ranger-district-level, multiple-use plans. However, the passage of NEPA and the Forest Service shift to environmentalism contained in the *Framework for the Future* and portions of the *Environmental Program for the Future* upset and confused the comprehensive land-use planning process, making many of these plans seem now noncompliant. Multiple-use and impact surveys, as well as broad programs and major project proposals, were required to include "environmental statements" that were reviewed by the Council of Environmental Quality. Therefore, Region 5 suggested that "each Forest review those on-going projects that appear to have greatest potential for impact and are likely to be challenged" (USDA Forest Service 1970b). But NEPA also pitted Forest Service personnel against one another. For instance, as the Forest Service started writing environmental assessments, hydrologists didn't like the way timber was cut, and soil scientists didn't like the way roads were laid out (Kennedy 2004a: 29). Moreover, now the preparation of planning documents became an open public process. Up until this time, Forest Service planning had been geared toward the commodity approach, as analyzed by a cost/benefit system. The passage of NEPA in 1969 added a different dimension and a necessary and new ingredient to planning – public involvement. Like other federal land management entities, Region 5 plans were no longer internal documents but were prepared for public comment, which slowed the process considerably and frustrated many in the Forest Service, who were accustomed to acting with little or no formal public input.

A good example of the "new environmental planning approach" happened when Region 5 attempted to update its Northern California planning guide. Region 5 began its review of *A Guide for Multiple Use Management of National Forest Land and Resources: Northern California Subregion* in 1973. The 1966 *Guide* was designed to provide direction for Forest Service administrators in their multiple-use management of these national forests, and was prepared internally. The new *Guide* was due in 1975, and was part of the whole planning process. In previous multiple-use planning, the region had recognized broad physical and social differences, and unique land management situations by the establishment of multiple-use planning subregions. In 1973, there were five such subregions: Northern California (NORCAL), Southern California (SOCAL), Lake Tahoe Basin (LTBMU), Forest Service-administered units of the Whiskeytown-Shasta-Trinity National Recreation Area, and the Middle Fork Feather River Wild and Scenic River (USDA Forest Service 1973b). However, when the region began to update the NORCAL *Guide*, there immediately were questions about whether providing management direction would result in a need for an environmental impact statement, which was likely to delay the *Guide's* completion date. This question was soon answered in the negative, as long as the final document submitted to NEPA was in the form of a guide and no land-use decisions were made in it. At the same time, Region 5 also needed clarification concerning who outside the Forest Service would use the *Guide*, since this had a bearing on how much detail would go into some sections. This question was soon answered as well – the users would be the general public (USDA Forest Service 1973a; 1973c; 1974a).

To meet this situation, in early 1975, Region 5 worked out an environmental analysis report for planning area guides that documented the overall expected effects of the *Guide* on the human environment and alternative courses of action. Then Region 5 created a "working group" to help the Forest Service prepare the documents, especially the "assumptions" section. Wisely, the Forest Service selected representatives from different walks of life, including state and county officials, as well as representatives from the Western Timber Association, the National Wildlife Federation and the Sierra Club (USDA Forest Service 1975a).

After more than a year's deliberation, which included a review of the document by the forest supervisors of Northern California, in late 1976, Region 5 completed its environmental analysis for Northern California (NORCAL), which included a list of five alternatives. The five management alternatives were: I – management emphasizing natural, cultural and amenity values; II – management emphasizing maximum Northern California economic benefit; III – management emphasizing maximum national economic benefit; IV – intensive management emphasizing those resources most suitably provided by the national forest system lands; and V – extension of present management direction (USDA Forest Service 1976b). The analysis of the various management alternatives was responsive to the

Forest and Rangeland Renewable Resources Planning Act (RPA). Passed in 1974, RPA authorized long-range planning by the Forest Service to ensure the future supply of forest resources while maintaining a quality environment. RPA required that a renewable resources assessment and a Forest Service program be prepared every ten and five years, respectively, to plan and prepare for the future of our national resources. RPA also required inventories and assessments of wood, water, wildlife and fish, forage and outdoor recreation available on private and public lands.

By late 1976, the NORCAL *Guide* was sent out for comment to various federal, state and county public agencies, as well as private and nonprofit organizations and private citizens. In doing so, Region 5 supported the "new" concept of public involvement. Region 5 officials, however, learned that inviting comment in this way – essentially giving the public the opportunity to participate in the development of this program – was not necessarily a very pretty or effective process. Eighty-two groups and individuals commented on the NORCAL *Guide*. Their comments spanned the entire range of proposed alternatives, and there was no clear consensus regarding any one alternative. Some people commented on specific items that were either left out or needed to be included – in other words, comments directly related to whatever interest group they represented. However, other people had definite opinions on the various alternatives. In the final analysis, these people were divided between utilitarian conservationists who supported Alternative #1, which seemed to favor resource and economic development such as timber production and range improvements, and preservationists who supported Alternative #4, which seemed to favor wilderness and protection of watershed over resource development (USDA Forest Service 1976c: Various letters).

By June 1977, progress on the NORCAL *Guide* staggered to a halt. William H. Covey, the NORCAL *Guide* team leader, commented that it had not turned out as well as expected and that the regional office needed to give it more priority since it was past its 1975 due date (USDA Forest Service 1976d). Regional Forester Leisz encouraged its completion by the end of the year (USDA Forest Service 1976e), but Forest Service planning efforts were moving away from such types of documents. In 1976, Congress passed the National Forest Management Act (NFMA), which some thought would substitute for the NEPA process (Peterson 2004: 95). NFMA restated Forest Service commitment to responsible use of natural resources, set guidelines for timber management, required prompt reforestation of lands, assured public participation in the planning and management of the national forest system, formalized the RPA land management planning process and set a 1985 deadline for completion of national forest plans. The NORCAL *Guide* continued to languish, and by September 1977, no alternative had yet been selected. Regional Forester Leisz assured the public and special interest groups such as the Sierra Club that following the analysis of public comment, an additional alternative would be developed and that he expected a decision soon (USDA Forest Service 1977). However, the NORCAL *Guide* fell by the wayside after Leisz left for Washington in June 1978.

Meanwhile, NFMA became the new standard for the Forest Service. NFMA was the "advent of land management planning as a professional exercise involving one plan for national forests involving all possible disciplines and uses." However, as will be seen when the decade of the 1980s is discussed in the next chapter, "land management planning" according to one retiree, "just seemed to be a process that never ended. There was always another thing that needed to be considered or another change that needed to be made, and as a result, it just got on and on until everybody kind of lost faith in it, that it ever would get passed, or it ever would be completed" (Kennedy 2004a: 33).

Environmental Management Interruptous, 1970-1978

When President Nixon assumed office in 1969, the nation was beginning a major period of controversy over forest management. Many issues confronted the Forest Service all at once. First and foremost, people were very upset over the practice of clear-cutting, an issue that began in 1964, when the Forest Service began an even-aged management plan for the Monongahela National Forest in West Virginia. Concern shifted to Montana when the service began bulldozing terraces into steep, clear-cut hillsides on the Bitterroot National Forest; and even hit Alaska, where the Forest Service proposed a 8.75-*billion*-board-foot sale on the Tongass National Forest, for which clear-cutting would be the primary harvest method (Hirt 1994: 245-246).

In 1970, the clear-cutting controversy erupted in California, when the Columbia Broadcasting Station (CBS) aired a "special" piece on clear-cutting on the Six Rivers National Forest by Richard Threlkeld on CBS anchorman Walter Cronkite's hour-long evening news program. The Threlkeld piece claimed that "whole mountainsides" of virgin timber on the forest were being cut away, leaving nothing but rocks and dirt. The piece also claimed that though the Forest Service was making a conscientious effort to replant all clear-cut areas, by its own figures it was running behind almost five million acres over the entire forest system. To illustrate these claims, CBS televised aerial footage showing a patch cut of approximately eighty acres as a "representative" site on one of California's national forests. As it turned out, the Threlkeld piece was highly inaccurate and was simply sensational television journalism, as Regional Forester Leisz pointed out in a running battle with CBS as he attempted to get the network to recant the story. Three-quarters of the land shown in the piece was private land adjoining Forest Service land; both the private land and the twenty acres of Forest Service land had been restocked with 13,000 trees in January, 1969, which were too small to be visible at the distance from which Threlkeld chose to film; and finally, the 5-million-acre reforestation backlog was almost all acreage needing planting as the result of historic forest fires, insect and disease epidemics, lands denuded by strip mining, logging cuts as much as a century old and subsequently added to the national forest, or a combination of these factors. Regional Forester Leisz encouraged CBS President Richard Salant and Threlkeld to come back to Region 5 if they were truly interested in presenting the facts about national forest timber harvesting practices and spend some time on the ground with him. Over time, CBS hemmed and hawed over Leisz's invitation, and finally Threlkeld did return to the "scene of the crime." For three days, Leisz and Threlkeld, and his film crew, toured the Six Rivers and Shasta-Trinity national forests. In the end, according to Leisz, Threlkeld admitted that mistakes were made in the piece and that he had perhaps been duped by the Sierra Club. There was no public recantation by CBS, nor a "special" on the matter (*California Log* 1971: June, October; Leisz 2005). Despite this public relations setback, Region 5's timber management division hoped to move public opinion toward a more favorable view of specified clear cuts as a long-term, legitimate forestry practice, a practice that did not produce a pleasing appearance in the short run.

Meanwhile, the national debate over clear-cutting continued. In the summer of 1971, Senator Frank Church (D-Idaho) held oversight hearings on clear-cutting, which resulted in a number of studies by panels, committees, commissions and study groups – all of which contained some criticism of Forest Service timber harvesting practices. In November 1971, the hearings and studies on the subject led the Senate Subcommittee on Public Lands to issue a report entitled *Clearcutting on Federal Timberlands*. The Forest Service felt that the Church report was a balanced, objective review of the dispute over clear-cutting (*California Log* 1971: November; Hirt 1994: 247-251). In the face of these challenges, the Forest Service had produced its *Environmental Program for the Future*, which replaced the Kennedy administration's *A Development Program for the National Forests*. But EPFF only added fuel to the fire in the minds of environmentalists who opposed clear-cutting and believed that the Forest Service's logging program had overstepped its multiple-use principles and overreached biological capabilities. EPFF predicted that within ten years the national allowable cut would be increased by 70 percent through limiting wilderness and other withdrawals for watershed, recreation or wildlife in areas considered viable for commercial logging, by investments in reforestation and intensive silviculture, and by squeezing every last potential board foot out of the forests, whether there was a market for the timber or not. Furthermore, the Forest Service promised it would develop publicly acceptable timber harvesting and access practices that would be compatible with the preservation of scenic beauty, watershed management and other intangible environmental values (Hirt 1994: 251-253).

In response to this promise, in 1971, the California Region held an environmental conference in San Diego with the Sierra Club, the timber industry, private landscape architects and engineers in attendance in order to develop a visual resource management system. Two years later, the California Region began a visual resource management training program for employees. Additionally, Region 5 began experimenting with helicopter logging. This method offered the potential to harvest timber on fragile soils, to access scattered trees on steep slopes or on areas adjacent to streams, lakes and highways, and to log while maintaining or restoring aesthetically pleasing landscapes, which was a primary consideration. In May of 1971, the first log to be removed by "choppers" was taken from the Plumas National Forest (*California Log* 1971; May). A year later, the first helicopter logging in Southern California took place on the San Bernardino National Forest, when a huge Sikorsky 61 helicopter was used to remove salvage trees from the recent Bear Fire, which could not be logged by conventional methods (*California Log* 1972: February). By 1973, the Sikorsky 64E Skycrane commercial helicopter, the largest and most powerful crane helicopter in the world at that time, was put into action there. With the right conditions, it had the ability to fly all day and haul up to 400,000 board feet per day (*California Log* 1973: April). Helicopter logging allowed harvesting of areas where the topography prevented typical logging practices, but it was extremely expensive (Leisz 1990: 128-129).

In the meantime, national developments regarding the clear-cutting controversy resulted in legislation prescribing clear-cutting as a viable forest management practice. In May 1973, the Izaak Walton League successfully

Block cutting of Douglas-fir, Shasta-Trinity National Forest. 1963

sued the Department of Agriculture in district court over the clear-cut logging practices on the Monongahela National Forest as being contrary to the Organic Act of 1897, which stated that only "dead, physically mature, and large growth trees *individually marked for cutting* could be sold." The Forest Service appealed the district court's decision in *Izaak Walton League v. Butz*, but on August 21, 1975, the appeals court upheld the lower court (Hirt 1994: 260-263).

The Monongahela decision caused panic in Region 5. In a full-page story, the *California Log* described the impacts of the decision on Region 5 forests and on California forest products industries. First, the decision's interpretation would increase sale preparation costs to mark each tree to be cut, and the decision would deny commercial thinnings and intermediate cuts that were a significant proportion of Region 5's total harvest level. Second, the story pointed out that if the ruling applied to its current program, it would result in a 50 percent reduction of timber harvested, reducing sale volume to 950 million board feet. Finally, if California national forest production were reduced by 50 percent, California's rural communities would suffer sharply from loss of jobs in the primary manufacturing industries and there would be severe economic impacts on rural county governments in California. A reduction of national forest sales volume would find private forestlands becoming the primary source of timber. But since private forestry could not meet the demand, the United States would become increasingly dependent on imports, particularly Canadian (*California Log* 1976: April).

Because of the dire consequences of this decision on forestry and the timber industry, the Forest Service, along with the timber industry and their congressional allies, immediately drafted legislation to repeal or revise the Organic Act of 1897. On October 22, 1976, Congress passed the National Forest Management Act (NFMA), which Region 5 forest officers lobbied for throughout California. Prior to NFMA's passage, Regional Forester Leisz earnestly explained its provisions to audiences heavily representative of the forest industries to garner their support (*California Log* 1977: January). NFMA repealed major portions of the Organic Act of 1897, and adopted Senator Church's clear-cutting guidelines promulgated in the report *Clearcutting on Federal Timberlands*. NFMA also provided for forest planning and sets standards for clear-cutting that insured that "cut blocks, patches, or strips are shaped and blended to the extent practicable with the natural terrain…and [that they] be carried out in a manner consistent with the protection of soil, watershed, fish, wildlife, recreation, and esthetic resources and the regeneration of the timber resource." This provision was no less than silviculture prescription by legislation, but kept Forest Service discretion intact because they applied mainly to visual restrictions on clear-cutting (Hirt 1994: 260-263). More importantly, NFMA mandated intensive long-range planning for national forests. Congress no longer looked upon the Forest Service budget on a year-to-year basis but now appropriated funds for long-range comprehensive planning. NFMA also specifically required the Forest Service to preserve minimum viable populations of native wildlife and to protect multiple-use values and environmental quality. Furthermore, NFMA provided for opportunities for public involvement in that planning process (Williams 2000: 122-123; Hirt 1994: 263).

On the timber management reform side, NFMA adopted a restrictive definition of sustained yield called "non-declining even-flow," or NDEF, which stated simply that a "forest's output of timber must be capable of being sustained perpetually without declines." NDEF reaffirmed the Forest Service's commitment to sustained yield and was thought to result in immediately reduced harvests. However, loophole language in NFMA allowed the agency to "earn" higher harvest levels through additional investments in intensive management. In fact, according to one source, it allowed forest managers to increase current harvests levels entirely on the *expectation* of success for intensive management, instead of making the agency "earn" the increased harvests through documented successes. Thereafter, "unsustainable harvest levels and accelerated old-growth liquidation continued through the late 1970s and on into the 1980s, partly through the auspices of the earned harvest effect." As a result, the "first generation of forest plans developed under NFMA and released between 1982 and 1992 would uniformly adopt unjustifiably optimistic assumptions to support high timber harvest targets" (Hirt 1994: 263-265).

As a rigorous new forest planning effort began, requiring each national forest to initiate long-range planning, Region 5 undertook intensive management measures throughout California. NFMA authorized $200 million per year for eight years for reforestation processes, with the stipulation that all national forests capable of supporting growth be reforested. Region 5 estimated that 500,000 acres of marginal lands in the region could be reforested (*California Log* 1977: January). Tree nurseries in Region 5, which numbered only two (Placerville and Humboldt) due to the Mount Shasta Nursery being closed in 1970 because of frost problems, thereafter increased seedling production. There was also hope that California nurseries would be expanded so that Region 5's backlog of unstocked lands could be reforested by 1985 (*California Log* 1975: August). As a result of NFMA, a major reforestation effort took place on Region 5, and the timber management division was confident in its ability to regenerate stands that were cut and certainly to regenerate the brush fields that had been created (Leonard 2004: 16-17). However, in the period 1976-1977, California and much of the West experienced a major drought, affecting timber management and their management. The drought killed an estimated eight billon board feet of timber in the Sierra Nevada. In a region-wide effort with the

Brush clearing in preparation for tree planting, Eldorado National Forest

cooperation of the National Aeronautics and Space Administration (NASA), which provided aerial photographs, the region salvaged about two billion board feet in 1978-1979. Much of the timber was old growth, and much of the salvage was by helicopter. This effort was perhaps the largest salvage program in Region 5 on record, and it confirmed experimental data and wood experience that old growth was highly susceptible to drought (Cermak 2005). Timber stands are not particularly affected by months of dry weather, but when those months turn to years, even the hardiest trees slow their growth rate and become more susceptible to disease (*California Log* 1977: April).

... and fear of fire hung in the air.

Prior to the 1976-1977 droughts, Region 5 had one major fire season, which came in 1970, greeting Douglas Leisz, the new regional forester. Known as the "Big One," it burned literally the entire region. (Leisz 2004: 98). Tinder-dry fuels and Santa Ana winds blowing up to seventy miles per hour, combined to produce the greatest number of large-scale fires burning out of control at one time in California's history. During one week in late September to early October, more than 524,000 acres of valuable watershed, timber and recreation lands were destroyed by separate blazes in Central and Southern California. Of this acreage, 207,000 were national forest land destroyed on five national forests (Los Padres, San Bernardino, Angeles, Cleveland and Sequoia). During the height of the Laguna Fire on the Cleveland National Forest, which burned 185,000 acres, more than 2,000 men combated the blaze, including Forest Service and Native American crews from Regions 1, 3, 6 and 8. Thirteen persons lost their lives in fire-related incidents. Five of the thirteen were killed in a helicopter crash on the Fork Fire on the Angeles National Forest (*California Log* 1970: October).

The 1970 fire season was a watershed year for forest fire research in Region 5 as it pertained to Southern California. Region 5 learned that things just didn't work well when the Forest Service had that many fires coupled with an extreme shortage of resources, and it realized that the Forest Service had problems communicating with other agencies because each agency had its own radio frequency and terminology (Leisz 2004: 99). One spinoff of the "near disaster" of 1970 was the FIRESCOPE (Firefighting Resources of Southern California) program – one of the most ambitious studies undertaken by the Forest Fire Laboratory at Riverside, California. In 1975, Leisz authorized a regional policy change directed at promoting a Southern California fire management plan, which gave ample support to FIRESCOPE and related programs. One objective of the FIRESCOPE program was to provide fire managers at all levels with the information needed to make prompt decisions. Another objective was to coordinate the efforts of Southern California fire protection agencies and facilitate their working together on major wildland fires, rather than protecting their own areas. Essentially, the FIRESCOPE program developed the methods and the vocabulary for interagency cooperation, resulting in an incident command system (ICS), which managed a wide range of activities when multiple groups of firefighters needed to work together in complex fire situations. This required that the terminology for all agencies be consistent and, when fires occurred, that the resources closest to the fires, regardless of the agency's protective responsibility, be utilized. FIRESCOPE also led to computerized dispatch centers shared by the Forest Service and the California Department of Forestry and Fire Protection. No program in the Forest Service ever started smaller and got bigger than FIRESCOPE. ICS, which was the primary FIRESCOPE product, has today been adopted throughout the country where multiple agencies or multiple jurisdictions are involved in earthquakes and other emergency responses. "It revolutionized and brought a quantum jump forward," explained Leisz, "in both organizing for and using various technological advances to bear in tackling the fire suppression job" (*California Log* 1975: April; Wilson and Davis 1988: 11-12; Irwin 2004: 22-26; Leisz 1990: 81-84; Leisz 2004: 100-106; Millar 2004: 7; Stewart 2004: 10-11; Tyrell 2004: 19-21).

Another important fire management related program to come out of the 1970 fire season was SAFETY FIRST, which Regional Forester Leisz promoted following a bad fire on the Los Padres National Forest which burned three men (James 2004: 28). SAFETY FIRST, a cooperative program between the Forest Service and the California Division of Forestry (CDF), was meant to promote closer cooperation between the two agencies in the fire safety area and reduce fatalities. The Southern California fire plan funded the implementation of SAFETY FIRST (*California Log* 1973: November). Weary of experiencing deaths from fire suppression activities on a recurring basis, the program developers admitted that in a few cases the Forest Service was simply going to back off from fires where the risk to firefighters was unacceptable (Leisz 2004: 112-113). After implementation of this program – whose recommendations were based upon interviews with people at every level of the organization who had been connected with fire death – Region 5 did not have another fire fatality for almost a decade (Millar 2004: 9; Smart 2004: 17-18). However, some people felt that the SAFETY FIRST made Region 5's fire organization too safety conscious (Righetti 2004: 5).

Region 5's Southern California fire plan, which was eventually expanded to cover the entire region, also funded and implemented FOCUS (Fire Operation Characteristics Use Simulation), which was one of the Forest Service's first efforts to develop a computer-simulation model to evaluate alternative fire management organizations. It was designed to model all possible fire agency configurations and augment FIRESCOPE's incident command system or ICS. The Sequoia was one of the first national forests to test FOCUS in the field and helped that forest configure the best organization of resources: air tankers, helicopters,

Aerial tanker making a drop on the Shasta-Trinity National Forest. 1967

crews, engines and the like for fire management, saving the Forest Service much time and money in fire suppression efforts. (Wilson and Davis 1988: 12; Irwin 2004: 6-9).

Finally, after the numerous large Southern California fires in 1970, several congressmen asked why military aircraft were not used in assisting the Forest Service in fire suppression. The congressmen requested that the Air Force cooperate with the Forest Service to see if some form of assistance could be arranged in times of emergency. The result was the development of the Modular Airborne Fire Fighting System (MAFFS). This program became operational in June 1974, and Air Force crews were trained in the use of the modules for retardant dropping; by that date, five air bases in Region 5 were approved for air tanker operation (Ontario, Santa Barbara, Fresno, Stockton and Redding). MAFFS could be activated only when all regular Forest Service and CDF heavy air tankers were committed, and MAFFS aircraft could be used only with a Forest Service lead plane (*California Log* 1975: April).

By the time of the 1976-1977 drought, the California Region and the State of California were adequately prepared to fight the expected fires that came to California that year. Despite all precautions, firefighting agencies found themselves attacking more than 120 blazes from Humboldt County to San Bernardino. Forest Service air tankers, firefighters and helicopters were spread thin throughout Region 5, and crews were rushed in from all parts of the United States to help out. Under the MAFFS program, Air Force C-130 military aircraft, using slip-on units to convert them to air tankers, were called in as well to augment the air attack. They were immediately put to use on several fires. Major fires to hit Northern California were the Hog Fire on the Klamath National Forest (53,500 acres) and the Scarface Fire on the Modoc National Forest (90,000 acres). These fires were touched off by fierce lightning storms. By far the largest fire was the mammoth Marble Cone Fire near Monterey and Carmel Valley, which burned close to 175,000 acres on the northern part of the Los Padres National Forest. Although more than 344,000 acres burned in California, the miles of fire line and new fire plan strategies paid off (*California Log* 1977: September).

The 1976-1977 drought that brought the above conflagrations also hurt other Region 5 resources such as rangeland and threatened California's entire livestock industry. National forest range supported a major portion of each year's calf crop, and the well- being of those operations depended on forest ranges. During this trying time, the Forest Service was put under pressure to make use of forest range to the fullest extent possible because it was the last resource of natural forage in the state. However, cattle and sheep could not be the only consideration of the Forest Service, which was responsible for seeing that wildlife was able to use the scarce range resources as well during this severe drought. Fish and wildlife were hard pressed for survival in some areas. Conditions for survival were most critical for those species with limited territories that live in arid or semi-arid areas (*California Log* 1977 April).

With the passage of the Endangered Species Act in 1973, the Forest Service aimed to insure the survival of all native species of fish and wildlife, but its program was more reactive than proactive (Kennedy 2004a: 31-32). This act authorized survival programs for those species threatened with extinction such as bighorn sheep on the Inyo (Schneegas 2003: 5) or the California condor on the Los Padres National Forest. In 1971, the U.S. Postal Service issued a commemorative stamp featuring the California condor, with the theme, "Let Them Alone – Let Them Live" (*California Log* 1971: June). But despite the attention, the condors were not doing well: the 1971 census turned up only thirty-four in the mountains around the southern end of the San Joaquin Valley (*California Log* 1971: November), but the Forest Service optimistically estimated condor population holding at fifty to sixty birds. The status of the condor was becoming critical because of low recruitment of young birds into the breeding population, habitat problems and human disturbance (*California Log* 1972: June). Therefore, in 1973, the Forest Service, along with state and private agencies that had formed a condor advisory committee, launched a new recovery program. Biologists embarked on an experimental program of supplemental feeding of the condors to bring more young birds into the breeding population (*California Log* 1973: November; Leisz 1990: 62). One hundred pounds of carrion were fed to the condors in this supplemental feeding program. Additionally, by 1975, in a national recovery plan, Region 5 assisted the endangered birds in several ways. It acquired additional pieces of private land within or adjacent to the Sisquoc and Sespe Condor Sanctuary on the Los Padres National Forest, continued a moratorium on mining use and on all oil drilling within certain areas of the condor reserve and tried to enforce a 3,000-foot air corridor over the Sespe Sanctuary (*California Log* 1975: May, October). Finally, to aid in the survival of the California condor and other endangered species located in the California Region, Region 5 employed a resource forester and biologist in 1976 to give technical advice to forests trying to protect habitats of various endangered plants and animals. As endangered species coordinator, the first job of this type set up in the nation within the Forest Service, this person helped determine critical habitats and coordinated needs of endangered species with other resource programs, primarily timber management (*California Log* 1976 June). Eventually, the Forest Service researcher and others concluded that the condor was heading to extinction and that to save the bird, captive breeding would be necessary (Leisz 1990: 63).

One thing that occurred out of the efforts to save the condor was that Region 5 became more and more sensitive to inventorying wildlife populations and working with the California Department of Fish and Game. Thanks to the Threatened and Endangered Species Act, Region 5 began to look into various forms of wildlife, ranging from large animals like bighorn sheep to the full range of bird life, irrespective of whether the species had any value for hunting or fishing. A decade later, these studies would lead to the northern spotted owl controversy of the 1990s, which impacted forest management in the states of Washington and Oregon and parts of northwestern California such as the Shasta-Trinity National Forest (Leisz 1990: 64-67; Tyrell 2004: 25), and to the California spotted owl controversy, which heavily impacted the Sierra Nevada National Forest areas (Stewart 2004: 17).

Meanwhile, during the drought, Region 5 took action to perpetuate endangered species by reviewing water permits that threatened or conflicted with threatened, rare or endangered species and modifying them to ensure species survival. Wherever necessary, Region 5 also helped the wildlife situation by developing and retaining water for them. This was not an easy task given the general dependence of Californians on national forest watersheds. With declining flows from California's watershed to the highly technical system of dams and canals used to irrigate farmland and provide water to the people throughout the state, Region 5 received ever-increasing applications for water resource developments and for special-use permits for water use, some of which were denied because of the adverse impact they would have had on threatened species (*California Log* 1977: April).

Recreation on California's national forests was also directly affected by the 1976-1977 drought. First, continued drought conditions limited ski opportunities throughout the Sierra Nevada because of a lack of snowpack. This, coupled with decreased incomes over the previous two drought years, caused some ski areas to defer major expansion. Developed sites were adversely affected as well. Water supplies were reduced and, in some instances, interrupted. Some forests examined the feasibility of hauling water to campgrounds if the drought continued. Reduced stream flow releases from impoundments also adversely affected numerous recreationists, such as river running activities. The situation was so serious that the Forest Service considered closing the popular Taylor Creek stream profile chamber at Lake Tahoe to maintain the water level at Fallen Leaf Lake (*California Log* 1977: April).

The drought affected recreation within the California Region at an inopportune time because in 1976, Region 5 was in the midst of its roadless area review and evaluation, otherwise known by the acronym RARE (later it was referred to as RARE I, after a court decision in 1972 "overturned" the RARE EIS, calling it inadequate). A key feature of the Wilderness Act of 1964 provided that the Forest Service inventory its undeveloped (or roadless) areas for possible consideration as wilderness. Each identified area needed to contain 5,000 acres or more, to contain no roads except trails and to be generally undeveloped. The overall objective of this review was to identify those areas that justified further study for possible management as wilderness areas. Such identification would allow prompt recognition of wilderness values and would permit protective management of these values without hampering management of other areas unqualified for consideration. The first step – inventory – was completed in April 1972. Thereafter, Region 5 held public meetings in Pasadena, Fresno, Oakland, Sacramento and Redding to receive comments from interested individuals and group representatives (*California Log* 1972: April; Coutant 1990: 25). Public response was overwhelming. Nearly 900 oral and written statements were received at the public meetings. According to the Forest Service analysis, more than half (54 percent) stated there was enough wilderness. Nearly one-third (31 percent) recommended that multiple use was the best alternative. Others (20 percent) wanted more wilderness, and still others (10 percent) wanted more time to study the situation. Comments also ranged from those who wished to close off wilderness from any visitors to those who wanted roads built into wilderness to afford everyone the chance to see wilderness scenery. Several environmental groups wanted a moratorium on logging and road construction until complete multidiscipline studies were made. The results were sent off to the WO for national input (*California Log* 1972: June).

Region 5 took all comments under consideration and in February 1973 produced a list of sixteen roadless areas covering 0.75 million acres in twelve national forests in California. These roadless areas were included in the 235 new study areas chosen by Chief Forester McGuire for further consideration for possible inclusion into the National Wilderness Preservation System. Of the sixteen roadless areas listed, the following were located in Northern California: Klamath, Shasta-Trinity and Six Rivers National Forests – Salmon-Trinity Alps Primitive Area additions (201,643 acres); Klamath National Forest – Johnson (4,400 acres), Snoozer (20,000 acres), Shackleford (4,440 acres), Etna (10,600 acres) and Portuguese (31,878 acres); and Shasta-Trinity National Forest – Mount Shasta (24,740 acres). Proposed new wilderness study areas for national forests in Central California included Eldorado and Stanislaus National Forests – Mokelumne addition (9,818 acres); Sierra National Forest – North Fork San Joaquin area (39,980 acres); Sequoia and Sierra National Forests – High Sierra Primitive Area Addition (24,365 acres); and Inyo National Forest – White Mountains and Upper Kern areas (130,625 acres). In Southern California, the list included the Los Padres National Forest – Madulce area (32,000 acres) and Angeles National Forest – Sheep Mountain (31,680 acres) and Cucamonga (3,500 acres) areas. Thereafter, the chief filed a draft environmental impact statement with the Council on Environmental Quality (CEQ), and invited public review of the new study area selection process and the adequacy of the tentative list. The public was invited to participate in this process as Region 5 studied the various land management alternatives for these areas before the final recommendations were presented to Congress (*California Log* 1973: February).

By this date, the Forest Service was quite familiar with the difficulties of filing an EIS. One issue of the *California Log* joked about the process with the following item, entitled "Good News…Bad News." The piece read, "God appeared before Moses, the story went, and said that first he had some good news: God would lead the chosen people out of bondage, lay low their enemies, part the waters of the sea, and deliver them to the promised land. And the bad news? 'You,' God told Moses, 'will have to write the environmental impact statement'" (*California Log* 1974: October).

Needless to say, many wilderness proponents expressed disappointment over RARE I and the Forest Service evaluations in the EIS process, especially regarding how the Forest Service determined how much wilderness was needed now and in the future, what criteria the service used for wilderness designation and how the agency balanced diverse and often diametrically opposed public views on wilderness. Wilderness had become a very confusing and controversial issue. The general public was confused over what met Congressional intent for wilderness, and there was a certain amount of confusion even among resource managers. By 1977, disappointment mounted over the results of RARE I, so much so that the Forest Service conducted a second-generation evaluation and inventory known as RARE II. This review was undertaken to speed up location decisions. RARE II was a more extensive review and allowed more time for various parties to participate. The RARE II process triggered a tremendous amount of input from the public, more than any previously experienced by any federal agency. The office of information in San Francisco had a major role in setting up public meetings to gather the input and then analyze it and present it to the regional forester and division staff so that they could decide which areas to recommend to Congress for wilderness designation. The RARE II inventory was completed and published in November 1978. (Coutant 1990: 29; Kennedy 2004b: 11-13; Leisz 1990: 72-73; Westenberger 1991: 122-123).

Instead of using the restrictive definition of "roadless areas" from RARE I, the RARE II guidelines allowed slight traces of human impact, such as limited fencing, fire towers, unimproved roads and the like, provided that the marks of man's activities did not preclude an undeveloped ambiance. Furthermore, RARE II's evaluation procedure met exacting requirements of both "biocentric" and "anthropocentric" views of wilderness. In other words, RARE II emphasized wilderness use and natural conditions. To meet this goal, the Forest Service developed a wilderness attributes rating system to assess how areas met the criteria of the 1964 Wilderness Act. The primary attributes in the rating system included natural integrity, apparent naturalness, outstanding opportunities for solitude and opportunities for primitive recreation (*California Log* 1978: March, May).

When the Wilderness Act was passed in 1964, the Forest Service was directed by Congress to study existing primitive areas and to advise Congress as to which of the areas should be reclassified as wilderness areas. Of these primitive areas, eight were in Region 5. By 1971, four areas, San Rafael, San Gabriel, Ventana and Desolation, had already been accepted into the wilderness system. The four remaining California primitive areas yet to be considered were the High Sierra on the Sierra and Sequoia National Forests; the Salmon-Trinity Alps on the Shasta-Trinity, Klamath and Six Rivers National Forests, the Agua Tibia on the Cleveland National Forest and the Emigrant Basin on the Stanislaus (*California Log*: 1971: March). Each had its own unique history.

The Stanislaus Emigrant Basin Primitive Area, covering 105,000 acres, and the 12,000-acre Agua Tibia Primitive Area on the Cleveland National

Forest were reclassified without much controversy. In February 1972, President Nixon sent the two new Region 5 wilderness proposals to Congress for addition to the National Wilderness Preservation System. After they were approved, the California Region had a total of nineteen wilderness areas, leaving two remaining primitive areas, Salmon-Trinity Alps and the High Sierra (*California Log*: 1972: February).

Reclassification of these remaining primitive areas took more time, consideration and difficult debate. For the Salmon-Trinity Alps Primitive Area, the Forest Service held a series of ten public meetings on reclassification. A total of 1,300 people participated in the 1971 meetings, with 169 people expressing their opinion regarding the size and boundaries of the proposed wilderness. The reclassification study encompassed three national forests, the Shasta-Trinity, Klamath and Six Rivers, and was designed to gather the views of people living near the proposed wilderness. The Forest Service heard from residents, local government representatives, the forest products industry and the mining industry. Next, the Forest Service held public hearings to get the opinions of people and groups living in places further away from the Alps study area (*California Log*: 1971: March). However, it was not until December 1974 that the Salmon-Trinity Alps Primitive Area was reclassified as wilderness under President Gerald R. Ford. At that time, the proposed Salmon-Trinity Alps Wilderness included more than 290,000 acres. Mountain ridges and deep glacier-cut canyons comprised this vast California coastal range wilderness between the Trinity River and the Salmon River. Trout fishing was excellent in the many streams. Bear were plentiful, and occasionally a wildcat could be seen in this latest addition to the National Wilderness Preservation System (USDA Forest Service 1976a). The reclassification of the High Sierra Primitive Area into wilderness was not as difficult because much of it had been reclassified prior to the 1970s. In 1964, the greater portion of the High Sierra Primitive Area was designated as the John Muir Wilderness Area. However, 13,000 acres of land that was not included in the Muir Wilderness Area retained the name High Sierra Primitive Area. Additionally, in 1965, about 1,300 acres of this High Sierra Primitive Area became part of Kings Canyon National Park. In March 1971, public meetings were held to study the possible reclassification of this area within the Sierra and Sequoia National Forests (*California Log* 1971: March). However, this extremely rough mountainous area, possibly the most wild in California, was not reclassified until 1974. At that time, more than 30,000 acres were recommended as the

Monarch Wilderness to Congress, named so after the vicinity's Monarch Divide, a rough terrain with few travel routes (USDA Forest Service 1976a).

With the increased wilderness acreage came a wilderness permit system. In March 1971, Region 5 set up an ad hoc committee, composed of individuals from different interest groups and the Forest Service, to review a proposed wilderness permit system. The permit, a combination wilderness entry and campfire permit, was obtained by a visitor prior to entry into a wilderness or primitive area. Ostensibly, the objective of the permit was not to restrict use but to obtain better public understanding of wilderness use and more reliable visitor data. On June 25, 1971, Region 5's wilderness permit system went into effect, requiring individual hikers traveling into any of the seventeen wilderness or four primitive areas to obtain a permit before entry. After careful review, the ad hoc committees recommended continuing the wilderness permit system the next year. In the meantime, the wilderness permit program gained broad public acceptance, and the system was deemed a successful educational effort by Region 5, which continued it permanently (*California Log* 1971: March, July, November; 1972: March; Leisz 1990: 48-51).

However, the growing population of California demanded outdoor recreation to a degree and of a kind not satisfied by opening up a few backpacking trails. Region 5 had a greater problem of satisfying the huge and growing demand for outdoor recreational activity on California's national forests. In the intervening time, while wilderness permits and reclassification of primitive areas into wilderness occurred, Walt Disney Productions awaited a decision by the United States Supreme Court on the fate of their controversial ski project at Mineral King. As will be remembered, in October 1970, the government position prevailed in the Court of Appeals, but thereafter, the Sierra Club on November 5, 1970, sought a review of the case before the Supreme Court. On February 22, 1971, the Supreme Court granted the Sierra Club a hearing on the Mineral King development for its fall term. On November 17, 1971, the Sierra Club and the government argued the case before the Supreme Court, after which the seven justices reserved decision (*California Log* 1970: November; 1971: March, November). With a Supreme Court decision pending on Mineral King, Regional Forester Leisz, Associate Regional Forester Slim Davis and Sequoia National Forest Supervisor Pete Wyckoff met with Walt Disney Productions to discuss the state of affairs. Even with continual delays, officials representing the Disney organization were steadfast for development. If necessary, they expressed a willingness to further modify the previously approved development plan so as to incorporate additional environmental compatibility (*California Log* 1972: April).

On April 19, 1972, the Supreme Court, by a vote of 4 to 3, rejected the Sierra Club's suit aimed at blocking construction of Disney's all-season resort at Mineral King. However, the decision was only a temporary setback for the club, because though the high court held that the conservation group lacked standing to sue, they added that the Sierra Club could return to a lower court and attempt to amend the suit. The Sierra Club did just that and in July won the standing argument in the Mineral King case before a district court judge. The club also broadened its claim under NEPA at this time. The Forest Service had not filed an EIS because the Mineral King development process was underway prior to NEPA's passage. However, many believed that the Sierra Club's action was an "exercise in frustration" because the Court of Appeals had already ruled in favor of the Forest Service on the allegations made by the Club (*California Log* 1972: May, August; Robinson 133; Pendergrass 1985: 179).

With the Supreme Court decision in hand, the Forest Service and Walt Disney Productions picked up the development program where it had been shelved three years earlier. By this time, many of the main components of the project had been revised. At a major press conference held in Visalia on May 3, 1972, and attended by fifty television, radio and newspaper reporters, Walt Disney Productions explained how it intended to eliminate miles of improved road by extending the electrically-powered, cog-assisted railroad westward across Sequoia National Park to a gateway point outside of its boundaries. This railroad would be publicly owned, and Disney would operate it on a nonprofit basis. This new plan would eliminate the need for a parking structure, provide greater control over the number of visitors allowed into the area and cut down on the number of recreation facilities that would have to be developed (*California Log* 1972: May). These major revisions were developed over the previous two years under the guidance of Disney's authoritative conservation advisory committee, which was made up of nationally recognized conservationists. Walt Disney Productions felt that they had responded in good faith to the government's requirements and to conservationists (Walt Disney Productions 1972). Regional Forester Leisz at this time reaffirmed the need for quality ski opportunities like Mineral King, and was delighted to support the new concept of a non-polluting, environmentally compatible transportation system that he hoped would assure access to the ordinary family of skiers at a reasonable price (USDA Forest Service 1972).

Leisz felt that the master plan for Mineral King was still sound. Even though the Forest Service was not required to do so, Region 5 prepared and filed an environmental impact statement covering the entire project, a procedure that would provide for full public involvement. Region 5 staff and Leisz felt that because this was a fairly major undertaking, it should undergo rigorous environmental analysis (Leisz 2004: 108). For the next few years, Walt Disney Productions and the Forest Service worked on the draft EIS for Mineral King. According to Regional Forester Leisz, many people in the Forest Service were impatient for the development to go ahead and criticized the action. Preservation groups were pleased at Region 5's action, but they also wished that the Forest Service would simply drop the proposal, an improbable outcome considering the time and money spent by all parties by this date (Leisz 1990: 55).

During the ninety-day review period, the Forest Service received more than 2,000 comments, many of which pointed out a number of problems in the draft EIS, including insufficient information to assess environmental impacts, inadequacy of mitigation measures, alternatives not clearly identified and evaluated, inconsistency with the purposes of the Sequoia National Game Refuge and restrictive recreational user costs. During 1975, additional studies were undertaken to provide information to address these concerns, and on February 26, 1976, the *Mineral King Final Environmental Statement* was filed with the Council on Environmental Quality in Washington, D.C. In releasing it, Regional Forester Leisz stated that the "public comments were extremely beneficial in pointing out areas of concern. We believe that the revised proposal effectively blends the needs of the American public with our concerns to protect the environment.... The year-round recreational complex will be unique to California. It will provide a rare opportunity to introduce a largely urban population to a high-country setting" (*California Log* 1976: March; Leisz 1990: 56).

Even with these changes, the Sierra Club continued to oppose Mineral King and made it their battle cry. By this time, the Sierra Club was under the leadership of J. Michael "Mike" McCloskey, and it's members were no longer willing to compromise (Pendergrass 1985: 180). Off the record, McCloskey reportedly stated that the Sierra Club was in financial trouble and they were desperately seeking to find ways to make money, to get more members and to attract donations to the club. Fighting Mineral King was a popular cause, one that continued to identify the Sierra Club as the protector of the environment and that would get them additional funding (Leisz 2004: 110). The Sierra Club and the Wilderness Society eventually earned the label extreme environmentalists among some Region 5 personnel (Schmidt 2004: 51-52).

Notwithstanding, in early 1978, the Mineral King issue, a perpetual point of contention between environmentalists and heavy recreationists, took an interesting turn of events. Following the Mineral King EIS, the original Disney plan was cut nearly in half by arbitrators attempting to cool the battle between the two groups. Disney representatives thereafter declared the scaled-down version as not economically feasible. Thereafter, California Representative John Krebs of Fresno introduced legislation to transfer the 15,600-acre area to Sequoia National Park (*California Log* 1978: March). As early as 1975, the National Park Service reportedly became opposed to the Mineral King project, and some saw the "conscientious" Park Service as the savior in this prolonged contest between the Forest Service and the Sierra Club (Robinson 1975: 134-135). Ultimately, the Sierra Club triumphed. In 1979, Congress passed legislation transferring the area to Sequoia National Park – an action that closed the twelve-year effort of the Forest Service to establish a year-round recreational complex. (Pendergrass 1985: 180-181).

California's National Forests: Today and Tomorrow

Regional Forester Leisz left California's national forests in the hands of Regional Forester Zane Smith Jr. at a critical juncture in the state's history, when the State of California itself was finishing an "era of limitations" under Governor Jerry Brown. By 1978, Californians in general had gained a sense of conservation, which led to a heightened sense of the importance of California's national forests. Many also realized that the national forests would be more valuable in future years, substantially so, but only if the Forest Service strove for a better balance of uses on the forests, made the necessary investments to assure continued productivity and maintained a sense of responsibility for the future. Some Californians believed that the Forest Service was up to the task, but others feared that the Service needed guidance from the general public on how to manage one-fifth of the state's resources. These Californians argued that with rising demand for forest resources, more leisure time, accelerated urbanization, escalating conflicts among forest users and a growing concern for preservation of the "natural" forest, the Forest Service needed help. With Californians' desire for more wildlife, more wood products, a primeval forest, ski resorts, hiking trails, quiet and solitude, land for grazing, off-road vehicle trails and minerals, Region 5 faced the difficult task of balancing these demands in its multiple-use planning.

In an intriguing move, California's Resources Agency pulled together a citizens committee from diverse backgrounds to review Forest Service policy and programs in Region 5 in order to provide leadership with constructive criticism. Committee members included academics, timber industry leaders, environmentalists and a few politicians such as Tom Bradley, the first African-American mayor of the City of Los Angeles. To gather information, the citizens committee held hearings in Sacramento and Los Angeles, where witnesses testified regarding their concerns. These witnesses ranged from state government officials (e.g. California's Department of Forestry, Fish and Game, and California Department of Water Resources) to timber and mining industry organizations (e.g., Western Timber Association, Society of American Foresters and California Mining Association) to recreation groups (e.g., California Association of Four-Wheel Drive Club, California High Sierra Packers and Sierra Ski Areas Association) to environmental groups (e.g., Sierra Club, California Wilderness Coalition, Friends of the River and the California Wildlife Federation) (California Resources Agency 1979: 1-4).

The committee's report, *Today and Tomorrow: Report of Citizens Committee on U.S. Forest Service Management Practices in California*, concluded that the basic statutes that governed the administration of the national forests were sound, with the exception of the 1872 Mining Law and amendments, and should be retained. They were confident that the Forest and Rangeland Renewable Management Act of 1974 and the National Forest Management Act of 1976 had placed Region 5's general multiple-use management programs into a clearer budgeting and reporting framework and had established more specific requirements for Forest Service planning. Otherwise, *Today and Tomorrow* offered a number of observations, one of the most important being that the Forest Service in California was emphasizing timber harvesting to the detriment of other resources. This problem was caused by four factors. Forest Service budgets enacted by Congress emphasized timber and mining programs in order to produce cash flow into the General Treasury and into the affected counties' 25 percent funds. This left little funding for outdoor recreation, watershed and particularly wildlife and fish management in California. National quotas for timber production were set by the Washington office and Congress and were not regionally based. Pressures "to get out the cut" prevented Region 5 from giving proper attention to sale planning, geologic constraints or good multiple-use management principles. Finally, Region 5 professional staff consisted chiefly of foresters. In 1979, fully

59 percent of the professional occupations in Region 5 were related to forestry. Because of management emphasis on timber production, other resources were being neglected (California Resources Agency 1979: 3, 16-24).

Other criticisms outlined in *Today and Tomorrow* were more specific. The citizens committee felt that reforestation efforts needed to be increased by the Forest Service; that wildlife and fish habitat were being adversely affected by timber harvesting, fire suppression and eradication of hardwoods and brushfields; that the Forest Service was not identifying potential erosion, landslide and slope stability, and sedimentation problems to prevent them from occurring or mitigate for unavoidable impacts; and that the Forest Service should train and equip personnel for effective law enforcement on national forests or reach agreements with and fund local enforcement agencies to provide protection. There were management suggestions as well. *Today and Tomorrow* recommended that the regional forester assure that management objectives were met on all national forests of California, that the Forest Service work toward consensus building and mediation or negotiation to resolve differences of opinion regarding major management decisions and that the Forest Service and the State of California should sign a memorandum of understanding (MOU) for Forest Service cooperation with state environmental programs (California Resources Agency 1979: ii-iii).

In the end, *Today and Tomorrow* made one unassailable conclusion about the management of the national forests in California: that many decisions that needed making were potentially controversial, that some of those decisions were not being made and that when made they were not being implemented. The report warned that controversies would continue because now there were contending views as to the proper management and "wise" use of the national forests. Interests such as the timber industry, downhill skiers, wilderness advocates, fishermen, naturalists and outdoor recreationists all had a legitimate claim under the law as long as their requests were not detrimental to the basic values of the forests. At the same time, these interests had the ability to frustrate management decisions that were unsound procedurally or completely unacceptable to an interest group. Future conflicts would be avoided only by adequate Region 5 procedures for involving the public and by agency response to this involvement. The next Regional Forester, Zane Smith, Jr. had his work cut out for him in the problematic decade of the 1980s as environmentalism evolved into the ecological movement.

Chapter XI: **1978-1987**

Recommitment and Roots of Ecosystem Management

*C*alifornia: Eden or Wasteland?

By the late 1970s, many Californians had come to believe that California was a lost Eden with a troubled future. At the end of President Jimmy Carter's administration in 1981, California, as well as America, was in domestic trouble. Persistent unemployment and the pernicious problem of inflation plagued his administration. American workers were losing their jobs to foreign competition as more and more manufacturers moved their operations to cheap-labor countries where there was less restrictive environmental and labor laws. Agricultural exports held up fairly well and did not harm California's agribusinesses, but one of the nation's most pressing economic problems, the energy crisis, was most urgent in California. Because of the Arab oil embargo of 1973-1974, constrained oil sources resulted in a fuel shortage in the United States. To address the problem, President Carter presented to Congress a comprehensive energy program hinged on energy conservation and development of alternative energy sources. Carter's program encouraged the use of solar energy for heating and included a tax on large, fuel-inefficient cars and an increase in the regulated oil and gas prices. California, which relied on petroleum products to fill more than half of the state's energy needs and had for many years acted as if these resources were infinite, considered the situation and began to spend money on developing alternative sources of power, such as geothermal, solar and nuclear energy. Many residents and businesses also encouraged oil companies to reopen abandoned fields and to explore for additional sources, especially beneath California's coastal waters. Environmentalists resisted further dependence on oil as well as nuclear energy after the near nuclear disaster at Three Mile Island. Meanwhile, rising costs forced Californians to slow their demands on energy supplies and to demand more energy-efficient cars and homes (Rice, Bullough, Orsi 1996: 563-566).

In addition to an energy crisis, there was also a water crisis, which also led many Californians to reconsider whether or not they were squandering away California's lifeblood – water. Issues related to watershed problems and the severe drought of 1975-1977 best illustrated this worry. These concerns gave impetus to a project known as the Peripheral Canal. Initially, this project was designed to allow more water to be shipped south from Northern California to agribusinesses in the San Joaquin Valley in order to meet such emergencies. As the Peripheral Canal project developed, project supporters proffered familiar economic arguments: the project would increase irrigation opportunities for growers, would foster continued expansion of suburban growth in Southern

California and would compensate for future losses of Colorado River water. However, in doing so, the Peripheral Canal, studies showed, would also have dire environmental consequences, the most serious of which was that it would leave insufficient water to flush the Sacramento and San Joaquin delta region. Without this water, according to environmentalists, there would be serious saline intrusions of the delta from the Pacific Ocean, making parts of it a wasteland. That notwithstanding, in 1979, Governor Jerry Brown signed off on the Peripheral Canal bill, stating that California could either have problems without water, or have water with problems. Despite Governor Brown's support, however, Californians felt that they would rather have problems without water, as he discovered when the legislative act was forced to a referendum in 1982. Californians voted down Proposition 9, the Peripheral Canal referendum, the first rejection of a big water project at the polls in California's history. Traditional booster arguments, such as prosperity depended on continued economic and population growth and that Southern California was running out of water, failed. The delta water crisis illustrated the classic confrontation over resource utilization and the environment (Rice, Bullough, Orsi 1996: 592-597).

Another example of the struggle between preserving Eden and losing it involved the crusade to save Mono Lake, a unique water body in the volcanic basin east of Yosemite. During World War II, the City of Los Angeles had acquired the water rights to the streams feeding Mono Lake and began diverting them for municipal needs. By the early 1980s, the lake's level had dropped fifty feet, severely damaging the lake's ecosystems. Trout fisheries in the streams supplying the lake were decimated. Lake water became saltier, and brine shrimp died off. Offshore rookery islands were exposed to predators that devoured hatches of birds. Finally, the basin's harsh winds blew across exposed lakebeds, creating harmful alkali dust storms. These conditions led the Audubon Society and the Friends of the Earth in 1979 to challenge Los Angeles' water rights. In a landmark decision, the California State Supreme Court ruled against the city. The court agreed with the environmentalists' contention that the common-law principle of public trust did not allow water rights to be modified or abolished if water diversion caused major environmental harm. On the basis of this public trust doctrine, a 1989 injunction prohibited the city from further diversion of water until the lake had significantly recovered (Rice, Bullough, Orsi 1996: 617-619).

The above examples typify a growing crisis in California over the environment and society's needs. In the meantime, by the end of the 1970s,

California appeared to be falling apart, reflecting the unsettled nature of the nation's affairs. Three presidents had led the country through the agonies of an unpopular war, a continuing energy crisis, a decline in United States prestige abroad, a decline in the dollar, a weakening of the country's cities and unprecedented inflation. Inflation and rising operating costs, increased foreign competition, limited watershed and other problems, such as the end of the Vietnam War and the Cold War that had hurt the aerospace and computer industries, slowed down California's momentum for the first time. By the beginning of the 1980s, California showed the strains of overdevelopment. After decades of prosperity and expansion following World War II, California's natural resources – water, air, farmland, scenery, open space – approached exhaustion. As one source put it, "Like the overused older states of the east, much of the state was congested, littered, polluted, and disfigured. Californians were destroying the unique natural heritage that had caused so much of their success. Smog, traffic jams, urban sprawl, overcrowded parks, and polluted water were only the most obvious manifestations of profound environmental transformations" (Rice, Bullough, Orsi 1996: 575-576).

During the problematic decade of the 1980s, Californian society began to fragment as well. The Golden State had lost its luster to those in the nation seeking their Eden. As immigration rates declined precipitously, the state's economy, which historically thrived on population growth, faltered. New immigration patterns with Hispanics and Asians emerged that made California increasingly a region populated by an aggregate of minorities. Their growing assertiveness in politics, business and society clashed with the interests of those who tried to maintain control of the state. The new emigrants also vied with one another for economic opportunity, cultural dominance and living space. Furthermore, in the wake of a tax revolt and the passage of Proposition 13 in 1978 – a ballot initiative that resulted in a cap on property taxes in California – state and local governments were increasingly strapped financially as crime rose, public education declined and public roads, parks and libraries deteriorated. At the same time that minority and ethnic groups and partisan groups tried to effectively deal with these problems, in the 1980s California women took their place in the vanguard of the struggle to win equality with men in the workplace (Rice, Bullough, Orsi 1996: 576).

Faced with inflation, high interest rates and other economic problems, and confronted with social disruptions, Americans elected Ronald Reagan to the presidency, attracted by his promises to greatly reduce federal income taxes and cut spending on social programs, while at the same time undertaking a massive military buildup. Given the high projected national deficits, the Reagan administration also pushed hard to cut the costs of doing business. He promised to harness government and free businesses from burdensome regulations. In his administration's view, basic problems in federal agencies included both excessive planning and disproportionately high administrative costs that were rooted in a tendency to proliferate staff at higher levels. The Reagan administration's solution was cutbacks and austerity in programs and streamlining federal staff and operating procedures through greater use of business methods. Whereas the Carter administration was sympathetic to environmentalism, the Reagan administration was not very friendly. When Reagan was California governor, he rarely appeared more than apathetic toward environmental protection, and under the Reagan presidency, the staffs and budgets of federal environmental agencies were cut and the agencies were discouraged from vigorous enforcement of regulations (Rice, Bullough, Orsi 1996: 621-626).

In California, inflation, recession and falling government revenues caused by Proposition 13 made the public less receptive to costly environmental regulations, which many felt added to consumer and government costs. When conflicts in the Middle East in the early 1980s caused a rise in the price of crude oil, the federal government, in order to increase energy supplies, weakened environmental standards, reversing many prior regulations. Meanwhile, throughout the 1980s some environmental groups in California experienced declines in memberships and contributions. They had sought to preserve the environment as a whole and criticized unrestrained economic and technological growth at the expense of the environment, but their apocalyptic predictions for the future if environmental harmony was not restored and their refusal to compromise tarnished their legacy in the eyes of those who were pro-business and others who questioned this radical approach to environmentalism. Many saw environmental battles over plans to build a Dow Chemical Company factory complex in the delta and the Standard Oil of Ohio (SOHIO) terminal for Alaskan oil at Long Beach harbor as increasingly anti-business. Pro-development, conservative political leaders who came to power on Reagan's coattails were convinced that environmentalism was a threat to the state's economy and supported George Deukmejian for governor. Deukmejian won election in 1982, and within a year, he sought to revive the Peripheral Canal project, abolish the Coastal Commission and weaken the California Environmental Quality Act (CEQA). With the support of

President Reagan, Deukmejian also tried to open up environmentally sensitive offshore waters to oil drilling. In 1986 he was reelected on the promise of holding down new government spending, improving the climate for business and fighting crime (Rice, Bullough, Orsi 1996: 621-624).

As a result of Defense Department policies, California prospered economically in the 1980s. President Reagan's huge increases in federal spending for military projects dramatically stimulated the economy, and California continued to dominate the nation's defense and aerospace industries. By the mid-1980s, California companies had garnered almost 20 percent of Defense Department expenditures and one-third of the Department of Energy and National Aeronautics Space Administration (NASA) budgets. In Southern California, Los Angeles became the largest manufacturing center in the United States and the foremost complex of aerospace industries in the world. At the same time, in Northern California, and closely associated with California institutions of higher education at Stanford and Berkeley, a personal computer and software revolution was well underway in the famed Silicon Valley in the so-called clean computer and related telecommunications fields. New firms such as Apple Computer, Inc., joined giants and well-established companies such as Intel, Fairchild, National Semiconductor and IBM in making Santa Clara County the center of the technological world. But even this high-tech revolution was a mixed blessing for California's environment when it was discovered that acids and other toxins used in the high-tech manufacturing process were seeping into the water tables and wells surrounding manufacturing plants (Rice, Bullough, Orsi 1996: 626-629).

As the California public became dismayed over these incidents and others, and wanted action to protect their environment, they turned not to national environmental organizations but to local groups, for by the end of the 1980s, a general ecological movement had arisen in California that exhibited a concern for a common natural world – a movement that spanned class, regions, gender, ethnicity and political lines in the state. Since that time, Californians have faced many vital ecological questions that even today they struggle to resolve. One source posed the following questions, which summarizes the direction of this ecological movement: "How can nature preservation, the public interest, and private property rights be reconciled? With soaring demand and diminishing supply, who should allocate resources, establish environmental policy, and mediate among contending parties? Should the public assert more control to assure that resources serve the best interests of the citizens? If so, what should be the limits of public power, and should local, regional, state, or federal officers wield it? Which regions have the superior right to resources, especially water – those of origin, or those with the greatest need? What constitutes 'beneficial use' of resources – economic development or preservation of natural conditions and species? And should beneficial use be measured in the short term, or the long, term?" (Rice, Bullough, Orsi 1996: 624-625). Many of these questions would also be pertinent to laying the groundwork for ecosystem management in California's national forests in the 1990s.

A Problematic Decade Ahead

In July 1978, friends and well-wishers congratulated Regional Forester Douglas Leisz as he left for Washington to become the new deputy chief of administration. In October, Zane G. Smith Jr., a confident professional from the Washington office (WO), was named head of Region 5. In the WO, he had served as director of recreation management since 1972 and had recently spearheaded the second phase of the roadless area review and evaluation, otherwise known as RARE II. Prior to his WO experience, "Zane," as he liked to simply sign his correspondence, had served as deputy supervisor and then supervisor of the Sierra National Forest from 1968 to 1970. In returning to Region 5, he

Zane G. Smith, Jr., District 5's tenth regional forester (1978-1987)

became the tenth regional forester serving the California Region. However, shortly after his arrival, Region 5 officially changed its name to the Pacific Southwest Region (R-5) to better reflect closer cooperation with the Hawaiian Islands and the Territory of the Pacific Islands. Accordingly, the *California Log* thereafter became the *Pacific/Southwest Log* (1979) and then the *Pacific Southwest News Log* (1987) (*California Log* 1978: July, September, October).

Zane Smith Jr. served the Pacific Southwest Region from 1978 to 1987, the last regional forester to serve more than four years in the position. In his opening remarks to the region, he stated that the Forest Service had a "particularly sensitive leadership role." He went on to say, "All of you are aware of the public's concern for the environment and the social values associated with Forest Service programs and services. In many instances, we have been challenged. Often these challenges strike at the very essence of our policies, and sometimes at our professionalism. It is important that you and I view this climate constructively – as something that will help us strengthen our performance and thereby strengthen the public's confidence in us…. In our work, we need to emphasize service to the public above all" (*California Log* 1978: October).

Smith's tenure as regional forester was not an easy one. Fortunately, at approximately the same time Smith was appointed regional forester, Max Peterson succeeded John R. McGuire as chief of the Forest Service. Peterson began his career with the Forest Service on the Plumas National Forest (1949-1953) under Regional Foresters Perry Thompson and Clare Hendee and then went on to work on the Cleveland National Forest (1953-1955) and on the San Bernardino National Forest (1955-1959) under Charles Connaughton. Before becoming the deputy regional forester and regional forester in the Southern Region, Peterson became regional engineer in Region 5 (1966-1971). All in all, Chief Peterson spent the better part of fifteen years in California. He was well aware of the challenges and the great variety of conditions in California before Regional Forester Smith – everything from the heavily used Southern California watershed forests to the more remote Northern California forests (*California Log* 1979: July).

On the administrative level, Regional Forester Smith faced significant

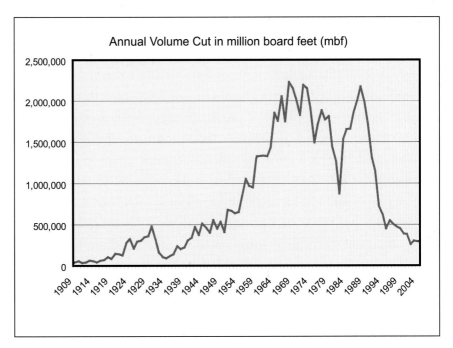

Timber Cut in District/Region 5, 1908-2004

difficulties, such as budget issues associated with the Reagan era. A major concern of this era was the reduction of agency funding. Starting in 1979, under the Carter administration, Congress began cutting the Forest Service's budget. Timber management received only 93 percent of its budget request for 1980. The range program got only 51 percent of the money it said it needed, and recreation program money was shaved as well (*Pacific Southwest Log* 1979: May). At first, part-time employment was seen as a solution to staffing needs, and management increasingly used such appointments to fill and cover certain essential tasks. These people possessed the very skills that were most in demand and included persons with certain family responsibilities, students, some disabled persons, senior employees seeking a gradual transition into retirement and others unable to work full time (*Pacific Southwest Log* 1980: May).

However, part-time employment was only a stop-gap measure, as inflation, higher interest rates and other problems plagued the economy. The ongoing growth that Forest Service programs had experienced during most of the 1970s came to screeching halt. In June 1981 Associate Chief Doug Leisz came to the Pacific Southwest Region to discuss budget issues, cuts in funding and the maintenance of program priorities. Leisz indicated at a Pacific Southwest regional meeting that the Forest Service under the Reagan administration would be the focus of greater expectations for minerals, timber, recreation and other benefits that the national forests could provide. The challenge, according to Leisz, would be to deliver the goods and services while accepting cuts in funding and limits on personnel ceilings (*Pacific Southwest Log* 1981: June). John B. Crowell Jr., President Reagan's new assistant secretary of natural resources and the environment, followed up on Leisz's comments in October of that year. Speaking before a meeting of regional office employees, Crowell stated that the Forest Service needed to find ways to increase production of goods and services from national forests, and he believed that outputs could be increased for all resources within the principles of multiple use. He also said that the most important Forest Service activity over the next four years would be to complete the national forest plans that would identify and analyze where and how outputs could be increased. "Unrestrained computer runs" were to be used to "estimate the maximum potential outputs for each resource." In Crowell's estimation, minerals from the national forests were to be considered a major potential source of benefits to the economy over the next decade; range could be improved to produce more forage for cattle and sheep and grazing fees adjusted to make them commensurate with the costs incurred in range management; and accelerated harvest of the "old growth" timber inventory on national forests could add significantly to the lumber supply and bring down material costs in housing construction (*Pacific Southwest Log* 1981: October). In the end, Crowell and the Reagan administration stressed commodity values (e.g., timber and minerals) and industry over amenity values (e.g., wilderness) (Frome 1984: 8-9; Radel 1991: 67-69). Crowell, who had been legal counsel for the Louisiana Pacific Corporation, a lumber giant, considered the "biggest problem in the Forest Service to be too much old growth in the Pacific Northwest." This perspective was quite different from that of the previous Carter administration and his Assistant Secretary of Agriculture Rupert Cutler, a biologist who had a strong environmental background and a "preservationist" outlook on things (Harn 1990: 65-67; Leisz 1990: 74). One retiree, a past Inyo National Forest forest supervisor, strongly believed that in terms of conservation, the Reagan administration was the poorest since Warren G. Harding, when they had the Teapot Dome scandals in public lands (Radel 1991: 68-69).

By 1983, many believed that hard economic times were not temporary, and ironically, despite the Crowell's call for more timber, the recession of the early 1980s dried up demand for wood products, and the Reagan administration soon found itself in the awkward position of allowing timber companies to renege on their existing contracts. In January of that year, the Pacific Southwest Region received its final budget allocation, which was $6 million less than expected. Regional Forester Smith warned that the region could expect further reductions in fiscal year 1984 in response to the overall economic recovery program, and that lower levels of program funding would probably prevail for some time thereafter. The cuts were unexpected since traditionally Congress adds to the president's budget before it is approved. Smith nevertheless affirmed his commitment to a "No RIF Policy" for the region. He felt the region could accomplish needed re-staffing without RIFs, or reductions in force. In the end, a RIF policy cost more, since salaries tended to remain high because senior employees with higher grades and step levels were retained, often forcing out younger employees, thereby disrupting the normal progress from recruitment and training through experience on the job and career advancement. To implement the no reduction in force policy, staff were required to take a second look at how they organized, classified and assigned work. Organizational changes occurred; no longer could the Pacific Southwest Region afford specialists with a narrow range of activities. Now employees were needed who could do a little bit of everything and were flexible, while still maintaining their expertise in their specialty. The Pacific Southwest Region also learned to work smarter. Providing staff with clear, concise and easily understood directions when making assignments and insistence on completed work eliminated wasted staff and management time. A third area of opportunity was to look at new ways to work. Alternatives included job sharing, extended details, working from long distance, quality circles, ad hoc work groups, term appointments, term promotions, full use of computer technology and voluntary reduced work time programs. Smith hoped that this adaptability would allow the region to achieve cutbacks in personnel through normal attrition (*Pacific Southwest Log* 1983: April).

Timber and engineering received the greatest reductions. Because of budget cuts, Pacific Southwest Region timber sales were reduced from two billion board feet and harvests from 1.6 billion to just 800 million board feet. Other programs were affected as well. For instance, fire protection on the Mendocino National Forest was cut by 20 percent, and the San Bernardino National Forest's recreation budget was cut 35 percent – at a time when that national forest had more recreationists than any other in the nation. At the same time, the number of permanent employees in the Pacific Southwest Region decreased from 7,098 in 1980 to 6,639 in 1983 – a 6.7 percent drop in workforce numbers. Further staff reductions would come when an expected 250 employees region wide would retire by the end of 1983, and another 645 by 1987 (*Pacific Southwest Log* 1983: April).

While juggling budget cuts and the maintenance of program priorities, Regional Forester Smith was confronted with an issue that had as great an impact on the Pacific Southwest Region as any – charges of sex discrimination. In June 1981, when former Regional Forester Leisz visited the region, he pointed out that the Forest Service remained fully committed to upward mobility, equal opportunity and affirmative action (*Pacific Southwest Log* 1981: June). Leisz's last comment was a reference to equal employment opportunity (EEO) issues and a transformation in the role of women, minorities (African-Americans, Asians, Hispanics and Native Americans) and the disabled in the Forest Service. Leisz's comment was an indirect reference to what many referred today as the "consent decree." In the retrospective opinion of Ron Stewart, a former regional forester, this issue would drive everything in the region, and not just hiring decisions (Stewart 2004: 16-17).

In the early 1970s, many women and minorities in the Forest Service began to complain about discrimination in job evaluations, claiming that the Forest Service was a white man's organization and that women and minorities were not being given equal opportunity for job advancement (Smith 1990: 18). This problem was exacerbated by the fact that for a long period of time, women and minorities – particularly women – were discouraged from going into forestry. Women tried to apply to forestry schools across the nation, but they were often flatly turned down (Stewart 2004: 15-16) or they were accepted into some schools but not allowed into the field once they were hired.

These complaints came to a head in June 1972, when Gene Bernardi, a Forest Service employee at the Pacific Southwest (PSW) Forest and Range Experiment Station in Berkeley, filed a class action suit against the Forest Service alleging sex discrimination in both the hiring and the promotional practices of the Forest Service. Bernardi alleged that she was denied promotion because of her sex and that discrimination on the basis of sex was a common practice in the Forest Service. The lawsuit *Gene Bernardi et al. v. Earl Butz* was filed in federal court under the Civil Rights Act of 1964. However, the case did not go to trial. Instead, in June 1981, the parties negotiated and approved a settlement: the Bernardi Consent Decree (No. C-73-11110 SC). While denying that any of its practices were discriminatory, the Forest Service pledged under this consent decree to strive to eliminate under-representation of women within each GS job series and each grade level (*Pacific Southwest Log* 1984: March).

Under the consent decree between the Forest Service and the women employees in the Pacific Southwest Region, a five-year program to increase representation of women in the workforce was begun. Under Phase I, a needs assessment was prepared, analyzing the causes, extent and means of eliminating under-representation of women in each job series and grade level. Phase II of the consent decree called for the Forest Service to develop interim goals and timetables to aid in meeting the long-range goal of the decree. To accomplish this end, a special task force composed of a cross section of regional office and field employees was formed, along with a steering committee to provide technical guidance to the task force. Phase III called for an action plan to implement the final, approved recommendations. Although Chief Max Peterson was responsible for the overall implementation of the decree, Regional Forester Zane Smith and Pacific Southwest Range and Experiment

Station Director Robert Z. Callaham were responsible for the day-to-day implementation of the decree (*Pacific Southwest Log* 1982: July).

Over the course of the next few years, Regional Forester Smith committed the Pacific Southwest Region to the consent decree and the goals of civil rights EEO programs in general. He endeavored to meet the spirit as well as the letter of the Forest Service's commitment to women and minorities. However, during a period of contraction, it would prove difficult to accomplish (*Pacific Southwest Log* 1983: April), and the region for the most part ignored it, despite Smith's efforts. Top management thought they could just tell the organization to follow the consent decree and it would be done, without the need for follow-up to ensure that the decree was being followed. As it turned out, the organization did not share in all of these values and did not move in the desired direction (Leonard 2004: 31-32; Smart 2004: 45-46).

There were, however, some breakthroughs. In the summer of 1983, Diane Pryce became the first woman smokejumper in the region. Although barely qualifying in height, weight and experience categories, through lots of stamina and hard work, Pryce qualified for the smokejumper group at the Northern California Service Center in Redding, which had been established in 1957 (*Pacific Southwest Log* 1983: October). At the same time, however, the region also lost experienced women like Dorothy Wothe, who in 1983 stepped down from her post as Smith Peak Lookout for the Stanislaus National Forest after more than thirty years of service (*Pacific Southwest Log* 1983: January).

In November 1983, the Pacific Southwest Region completed its action or implementation plan to reach its long-term objective of making the percentage of women in the region's workforce comparable to the percentage in the civilian workforce at large. The implementation plan, which was submitted to Chief Peterson, focused mainly on institutional barriers to hiring, training and advancement of women in the region. These barriers involved personnel procedures, classification of positions, training opportunity and, more generally, certain traditional organizational and personal attitudes that were not formally established but were nonetheless influential. To achieve the general objectives, the implementation plan called for a removal of these barriers – not setting quotas. Smith optimistically hoped to overcome the "good old boy" network by appealing to fairness in hiring and promoting all employees, not just women and minorities. "No program will be perfect," said Smith. "But as the intent of the decree is understood and methods are given a chance to work, I believe it will deserve our support because it is equitable and advantageous to all employees." Essentially, the regional office sought to change employee behavior by appealing to a sense of fairness instead of establishing fixed quotas, and then hoped that things would change naturally (*Pacific Southwest Log* 1984: March).

Between 1981 and 1984, there seemed to be little visible progress toward fulfilling the promise of the consent decree, although there was growing resentment and hostility among many male employees who felt discriminated against by the process. Some white males complained that they were not even going to bother applying for jobs because they knew only women are going to be selected. Deputy Regional Forester Warren Davies tried to console them with a by-the-book approach, stating, "There have been times when a woman was selected, and it's legitimate. The Supreme Court has said that, all things being equal, a selection based on correction of under representation of past practices is fully legal" (*Pacific Southwest Log* 1984: March).

In January 1984, the implementation plan was widely publicized and discussed in the region. Two programs were established to speed things along. The first program was called the focused placement program (FPP) and opened up positions at the GS-11 through GS-13 level for the next year to substantially qualified individuals who lacked the experience and exposure that the Forest Service was typically looking for when filling these vacancies. The region hoped to bring women with high potential into future management positions, and 30 percent of these vacancies were set aside for FPP. The second program was called the accelerated development program (ADP), which essentially was designed to train individuals over a one- or two-year period to fill future known vacancies. People under ADP would be automatically promoted to the first available vacancy that fit their needs as well as the Forest Service's (*Pacific Southwest Log* 1984: March). As one retiree related, "When things were left too late in the process, you ended up with, instead of creating training programs that enabled women to move up in the fire organization, you had to make a training program that was exclusively for women. Instead of giving women a reasonable share of the promotions, you had to give women all the promotions in order to address the goals. It became very difficult. It became very difficult for white males" (Leonard 2004: 32). The *laissez faire* approach to EEO, however, was not working, and soon the region moved to a quota system, but shied away from actually calling it that.

By 1984, there was great anxiety over delays in the process, which put the Forest Service at least a year or so behind the court-ordered goal of a 43 percent, across-the-board (grades and series) hiring of women by the year 1986. When an outside monitor was asked how the Forest Service was doing at this point, she answered truthfully that it "took them an awfully long time to get things in place," but she also felt that the Forest Service had responded in good faith and that the service had a history of doing what it sets out to do (*Pacific Southwest Log* 1984: March).

Criticism in the field was harsher, and morale dropped for both men and women. One male ranger commented that a lot of women in the Forest Service simply did not have the background or education to become engineers or foresters – essentially "you can't make a GS-3 clerk into a Forest Ranger," and another male ranger expressed fears that the organization would deteriorate because the best candidate for a management position might be overlooked in order to move a less experienced woman upward (*Pacific Southwest Log* 1984: March). These types of statements show the level of frustration and fear within the region by some rangers, but they also did not reflect reality – how many GS-3 clerks were being promoted to rangers. Furthermore, they made the underlying assumption that clerks were always women, and district rangers always men (Conners 2005). Still others were afraid that overzealous supervisors would put undue pressure on men to leave the Forest Service in order to move women into those jobs so they could meet their expected quotas. On the other hand, though many women felt that the idea was a good one and took advantage of programs such as FPP and ADP, they were also disillusioned by the slowness of the process.

Furthermore, qualified women were apprehensive that promotions based upon accomplishments would thereafter be perceived as undeserved in some way (*Pacific Southwest Log* 1984: March) – a stigma that even former Chief Peterson believed existed (Peterson 2004: 85-86). To alleviate some of these misperceptions between the sexes, the regional office initiated a "Changing Roles of Men and Women" program, which put it at the forefront in programs relating to organizational and personal development in the Forest Service. This program, and others, evolved from the self-actualization movement of the 1960s and 1970s and heightened the awareness of how traditional definitions of men's and women's roles influence decisions Forest Service employees make and their expectations of others on the job; it examined the impacts that change was having on these roles in their work lives; and it identified strategies to realize the full potential of all human resources available on each Forest. Many considered these sessions as just another trendy, touchy-feely program, while others thought that they were relevant to their lives and attitudes about other people (*Pacific Southwest Log* 1984: June/July). One related side issue that came out of the "Changing Roles for Men and Women" sessions was the subject of sexual harassment, which the regional office took a firm stand on. In an editorial written for the *Pacific Southwest Log*, Regional Forester Smith covered the topic. In this piece, he emphatically stated that there was no "major" or "minor" forms of sexual harassment – all forms were "major," were a violation of the law and constituted a threat to the harmonious professional working environment, and would not be tolerated. Thereafter, the pages of the *Log* were filled with educational information on the legal background of problem, profiles of harassment, proper standards of behavior, manager responsibilities and what an individual should do if harassed (*Pacific Southwest Log* 1985: September).

Despite the above actions, and with little consideration for a 16 percent reduction in workforce and a 25 percent reduction in budget under the Reagan administration, in June 1985, the attorney for the class brought a complaint of non-compliance against the Forest Service, charging that the Forest Service had not completed actions to which it had committed in the regional Consent Decree Implementation Plan. Faced with the threat of non-compliance, Regional Forester Smith was determined to catch up with unmet goals. Toward this end, he held weekly briefings on consent decree activities, and he even made and sent out a videotape to all the national forests that explained what progress had been made and how the region needed to recommit itself to reaching full representation of women in the workforce and eliminate grade gaps (*Pacific Southwest Log* 1985: September, October). However, by the end of 1985, the Pacific Southwest Region fell short of its goal of 43 percent across-the-board hiring of women by the year 1986. By year's end, the number of women in the total permanent workforce had only reached 28.9 percent. Women were getting a larger percentage of promotion opportunities than they had gotten in the past. They also were getting a larger percentage of opportunities than their proportion in the workforce. To illustrate progress, the *Pacific Southwest Log* carried a new section called "Personnel Notes," which listed individual promotions and reassignments, with one issue congratulating four woman and six men as new district rangers. However, when broken down by category, women made up 40.7 percent of the administrative positions but only 21.3 percent of the technical positions and a mere 12.8 percent of the ranks of the professionals (*Pacific Southwest Log* 1985: November, December).

While the Pacific Southwest Region awaited a court decision on the complaint of non-compliance, Regional Forester Smith pushed on with attempting to fulfill the Region's obligations. In 1986, the process used by selection officials to assess job candidates' qualifications for supervisory positions was improved by adding selection criteria that addressed the candidate's human resource management skills. "Women," according to one source, "have traditionally played the role of 'humanizers,' the one that cares about people, who are attuned to other's needs, who listen, who accommodate and work out compromises." Studies of successful managers valued these traits and, indirectly, the people who possessed them. To assist current and future supervisors in understanding and rating human resource management skills, the RO developed a *Rating Guide for Evaluation for Supervisory Positions*. This action opened more supervisory jobs to women and allowed the region to better utilize its workforce (*Pacific Southwest Log* 1986: March, April/May).

Despite some gains in closing representation and grade gaps for women in Region 5, in May 1987, the U.S. district judge in San Francisco signed an order extending the consent decree for an additional three-year period. The court determined that the region had failed to make full use of programs that were key to ending under-representation, such as accelerated development, focused placement and bridge positions, that the region had failed to respond in a timely and quality fashion to the monitor's request for information and that the region failed to accurately manage the consent decree funds. The extension order required both the RO and the PSW station to complete all of the outstanding obligations from the original consent decree. This work fell to the next regional forester, Paul F. Barker, to complete (*Pacific Southwest Log* 1987: May). However, photographs of the RO officers and supervisors on Pacific Southwest management team taken in 1987 indicate just how far the region still had to go. In the RO, there were only two women on the eighteen-member team of regional, deputy and assistant foresters and officers – Jane Westenberger (information officer) and Joan Brechbill (consent decree officer). Furthermore, only one of California's eighteen Forest Supervisors was a woman, Geri Larson. Forest Supervisor Larson started out working for the Forest Service in 1962 as a research forester at the PSW station in Berkeley, and thereafter served as a public information specialist and an environmental analysis specialist in the regional office. In 1978, Larson was promoted by Regional Forester Leisz to deputy forest supervisor on the Tahoe National Forest, and then in 1985, she was selected as the new forest supervisor for the Tahoe – the

Geri Larson, first female forest supervisor in Forest Service history, discussing project with Regional Forester Doug Leisz. 1978

first female forest supervisor in Forest Service history (*Pacific Southwest Log* 1985: February; Leisz 2004: 123-124; Williams 2000: 128; Frome 1987: 67).

Ultimately, the consent decree succeeded and failed. It forcibly brought to the attention of the Forest Service that it did not have a balance of either gender or race in its makeup (Smart 2004: 47). Eventually, it did accelerate the advancement of women in a variety of positions, and at the same time an increased number of minorities and disabled were hired for a variety of Forest Service positions. But some Forest Service employees, namely men, resented the consent decree, and found it demoralizing (Smith 1990: 21; Westenberger 1991: 83-87). In one person's opinion, "Women were put into jobs that they had never been trained for, that they didn't have a chance of being able to do successfully, and they would be frustrated because they couldn't do the job, and they would prove to the hardcore men that had lost the job that women can't do that job." Many chose early retirement because "they saw a number of people who had been working for ten years on a job who couldn't be promoted because they were the wrong sex" (Kennedy 2004a: 35-36).

Enforcement of the consent decree was a time-consuming task for the Pacific Southwest Region. Complying with the Forest and Rangeland Renewable Resources Planning Act (RPA) of 1974 and the National Forest Management Act (NFMA) of 1976 also took up a great deal of time and

Firefighters in and around a fire camp in the 1980s.

energy in the hectic 1980s. RPA and NFMA provided national direction for Forest Service management of renewable natural resources.

The Resources Planning Act required the Forest Service to periodically assess the condition and productive capability of all private and public forest and range lands in the United States. Under RPA, the agency was required to prepare a program that defined the share of national needs for goods, service, and amenities the Forest Service should provide through its management of national forests, state and private cooperative programs, and research. The RPA program also defined specific levels of outputs and investments for each of the nine regions of the Forest Service. To be sure outputs and investments reflected current needs and capabilities, every five years the RPA program was reviewed, and every ten years an RPA assessment was conducted. The initial RPA assessment and program were completed in 1975, and the RPA program updated in 1980. In developing the national RPA program, the Forest Service conducted nationwide public involvement to identify major public issues, concerns and opportunities to be addressed and established targeted outputs. The mission of the Pacific Southwest Regional planning was thereafter to assign these RPA national program targets to the individual national forests and to provide leadership in sustaining and improving the flow of goods and services from forest lands in California, Hawaii, Guam and the Pacific Trust Territories (USDA Forest Service 1980b: 1-2, 5, 7).

NFMA, which amended RPA, on the other hand, required that individual plans be prepared for the management of the land and resources of each national forest, including determination of timber harvest levels. In response to NFMA, the Forest Service issued planning regulations on September 17, 1979, that was mostly prepared by a committee of scientists. The regulations required that each national forest undergo a long-range planning process with mandated public involvement. The challenge was getting everyone on the same planning page.

In 1980, several individual forests began working on their forest plans. The RO had issued standards and guidelines for forest planning (USDA Forest Service 1981b), but within a year, they were amended four times, leading to little conformity in the approach, format or terminology in these early forest plans. A Washington office program review of land and resource management planning in the Pacific Southwest Region recommended that the region work with the WO land management planning staff to establish minimum planning process criteria and standards to improve uniformity and integration of resources and to assure that a satisfactory range of alternatives was made available for the public and line officers (USDA Forest Service circa 1980).

To remedy the situation, in June 1981, the RO issued *Land Management Planning (LMP) Direction*, which replaced the previous RO guideline. This document delineated all of the components the RO expected to find in each forest plan. Besides addressing questions regarding each resource, each forest plan was expected to have four key elements: a section summarizing the analysis of the management situation (essentially a picture of the national forest as it exists); a section depicting forest-wide multiple-use goals and objectives (a description of the direction of policies, standards and requirements to be applied to the total Forest); a section providing multiple-use management prescriptions for each management area (on-the-ground management direction to be applied to resource use, development, maintenance and protection of specific areas of land); and a monitoring program to provide periodic determinations of the effects of the management practices. However, several of the forests – the Sierra, Six Rivers, Klamath, and Shasta-Trinity National Forests – were excused from total compliance with *Land Management Planning (LMP) Direction* because they had progressed to a stage at which strict adherence to the new standards would seriously delay their issuance of draft environmental impact statements (DEIS) (USDA Forest Service 1981c).

Another problem associated with some of these early individual forest plans was unintended bias. Conclusive statements regarding results of the forest planning process were made before the completed analysis, giving such diverse interest groups as the Western Timber Association (WTA) and Sierra Club the perception that current planning was no more than an exercise to support predetermined decisions. When Regional Forester Smith got wind of this problem, he clearly chastised the forest supervisors, stating, "Our planning must be an open process, which encourages input and support from a wide range of interests. We must not prejudice this process with premature judgments" (USDA Forest Service 1980a).

In the meantime, with the change of national direction from the Carter to the Reagan administration, the chief of the Forest Service directed that draft forest plans be delayed until a draft regional plan had been prepared. It was rationalized that since the forest plans were the responsibility of the regional forester, there needed to be consistency of content and approach between all of these plans (USDA Forest Service 1981a). The purpose of the regional plan was to provide direction, standards and guidelines for national forest, state and private, and research programs. The regional plan also served as a link between national direction and forest planning by allocating programs to the forest level, by resolving regional public issues and management concerns and by identifying needed research. The forest plans, on the other hand, would provide information needed to adjust regional and national direction.

In June 1981, in compliance with RPA, the Pacific Southwest Region issued its *Regional Land and Resource Management Plan* (draft). This document analyzed the management situation in the region as it pertained to each resource (timber, range, water, minerals, wildlife and recreation) (USDA Forest Service 1981d: 5). At the same time, in compliance with NFMA and NEPA, the Pacific Southwest Region distributed its *Draft Environmental Impact Statement (DEIS) for the Pacific Southwest Region*. The purpose of the DEIS was to obtain comment on the proposed standards, guidelines and planning goals of the management of Forest Service activities in the region. First, there were nine NFMA-required standard guideline items that needed consideration, which centered on how the silvicultural systems should be applied in the Pacific Southwest Region. For the DEIS, a regional interdisciplinary planning team also developed a preliminary list of regional issues and initiated public involvement. As a result of screening the 2,200 responses received during the formal ninety-day public comment period, a total of seventy-nine potential issues were identified. This number was eventually reduced to just twelve selected regional issues and concerns: sensitive species habitat, water yield, water quality, chaparral management, mineral development, prescribed fire use, structural fire use, land use for fire protection, seasonal closures for fire protection, compatible land management, dams and water diversions, and research natural areas. Other issues were deferred until forest plans were completed (USDA Forest Service 1981e: 1). In the end, this regional planning process, which emphasized the individual parts of the forest, would eventually give way to a new conceptual framework – ecosystem management that distinguished the ecosystem itself as the context for management rather than just these individual parts (USDA Forest Service 1995: xi). By around 1983, all regional plans that were reconstituted from regional guides essentially disappeared.

Recommitment and Roots of Ecosystem Management, 1978-1987

Timber management in the late 1970s and throughout the 1980s was the focus of a major public debate in California. Although landscape management and clear-cutting issues were resolved thanks in part the Pacific Southwest

Region integrating visual quality management in its forest planning after 1980 (*Pacific Southwest Log* 1985: July), the primary issue was timber harvest volumes. The *Regional Land and Resource Management Plan*'s stated goal was to increase total timber supply from the national forests by intensively managing those lands where timber production was cost-effective (USDA Forest Service 1981d: 66). High interest rates in the early 1980s reduced housing starts and demand for timber such that the amount harvested from California's national forests declined to a low of 876 million board feet in 1982 (*Pacific Southwest Log* 1984: February). Budget cuts in timber sales administration and engineering road construction lowered timber sale targets up until 1985 (*Pacific Southwest Log* 1985: July), but starting in the mid-1980s, Congress actively imposed timber harvest targets on the Forest Service. Many in the Forest Service, including Chief Max Peterson, recognized that these timber targets were unrealistic. In fact, Randal O'Toole, a forest economist and critic of the Forest Service, exposed problems in planning within the Pacific Southwest Region. In 1986, his analysis of the timber yield tables used for seven of California's national forest plans found that the 1977 timber yield model had overestimated growth rates for existing California forests by approximately one-third. Furthermore, the forest plans had underestimated the time it would take for second-growth forests to reach desirable harvest age. "Such errors resulted," one historian wrote, "in higher estimates of future timber volume, which made the currently excessive harvest levels seem sustainable." In the end, the California Region reviewed its timber yield tables, acknowledged errors in assumptions and held up releasing forest plans pending revisions (Hirt 1994: 272-273).

Meanwhile, the Reagan administration continued to promise the political establishment that the high harvests levels could be achieved under multiple-use, sustained-yield principles if the agency received the needed fiscal support through appropriations. However, renewed timber sales and harvesting increasingly conflicted with recreation, fisheries, wildlife, soils and water resources, and vice versa. These conflicts led to increased activism by environmentalists, appeals and lawsuits, especially at the project level, to force substantial cutbacks. Key areas of conflict that magnetized the controversies involved roadless areas and wildlife habitat.

As noted earlier, prior to coming to the region, Zane Smith Jr. had worked extensively on the roadless area review and evaluation, otherwise known as RARE II – nicknamed so by an environmentalist who suspected that the

Caterpillar tractor and operator. Soup Creek Timber Sale, Modoc National Forest. 1963

study was a "slightly modified version of the original." The Forest Service implemented RARE II to ameliorate pressure from both environmentalists and commodity users. In California, the Forest Service recommended for wilderness sixty-nine roadless areas representing almost 900,000 acres, 176 non-wilderness areas covering almost 2.5 million acres and 118 areas for further planning. This recommendation was heavily influenced by non-wilderness sentiment among small Northern California communities that were suffering from depressed economies tied to a stagnating timber industry. Timber spokesman in this part of California argued that they would lose much needed timber, housing costs would rise and employment opportunities would decline if too much land was set aside for roadless areas. More importantly, they were joined in their opposition to roadless areas by some segments of the outdoor recreation community, specifically those who enjoyed using motorbikes, jeeps and snow machines. Even the elderly and disabled communities opposed more roadless areas because they wanted motorized access to these scenic areas (Pendergrass 1985: 194-195, 200). At this time, the Pacific Southwest Region was employing the disabled in greater numbers, and experimenting with barrier-free designs for campgrounds and recreation areas. Nonetheless, many disabled persons advocated an open access policy even on roadless areas (*Pacific Southwest Log* 1979: May; 1983: February).

Of course, California environmentalists were not pleased with these limited results, which they saw as a rush to exploit wilderness areas by the Reagan administration. However, this time legal action was not taken up by the Sierra Club but by the State of California. In its suit filed in U.S. District Court, the State of California charged that the Forest Service's EIS process was faulty because it solicited comments from only the Northern California counties and had "failed to ask the 97.5 percent of the population in California urban areas what they thought about the plans for logging, mining, and recreation." Regional Forester Smith conceded this point but thought that the service's plans still reflected the interests of California's citizens. The district court disagreed. It held that RARE II had not included sufficient information to qualify its judgments and that the non-wilderness areas, as well as areas planned for further study, could not be developed before site-specific impact statements were prepared (Pendergrass 1985: 198-199). Personally, Regional Forester Smith was disappointed in this legal action. Trying to mend fences on this issue, in a reply to one letter Smith stated "RARE II was not a complete failure as you might suggest, although I admit there were weaknesses…Perhaps the greatest deficiency was forcing RARE II to accomplish too much. I would have been more comfortable if RARE II had simply decided the more obvious and universally accepted determinations…. On balance, however, I think RARE II was a success. For the first time it put wilderness in a national perspective. For the first time somebody thought about what the wilderness system ought to look like when it was complete. For the first time somebody took the time to establish criteria and characteristics of a complete National Wilderness Preservation system. RARE II developed more information and brought the question of land use, particularly wilderness, to the attention of more people in this country than any other effort" (USDA Forest Service 1982). In the end, RARE II did resolve much of the land allocation problem in California. The court challenge, as well as forest data inventories for forest plans in the 1980s, caused the restudy of many areas for roadless designation. The old 1978 RARE II database was eventually revised into a new 1983 roadless area database, which proved more satisfactory to all parties (USDA Forest Service 1983). Even so, at this point, finding a agreement between all parties would have taken a miracle. The Forest Service considered federal legislation to adopt all of the RARE II areas that had been proposed, but the issue became too controversial in many states. RARE II languished and finally ended up being accomplished on a state-by-state basis. In 1984, California passed a Wilderness Act that incorporated the Forest Service's RARE II recommendations and added a few additional areas proposed by environmentalists. Senators from California, such as Pete Wilson and Alan Cranston, hoped that the areas added by the California Wilderness Act would end the wilderness issue for the state, but the battle for additional wilderness continued (Leisz 1990: 75-76).

In the interim, several new wilderness areas in California were added to the National Wilderness Preservation System. They were the Golden Trout and the Santa Lucia Wildernesses, both created in the last days of the Carter administration. At 306,000 acres, the Golden Trout Wilderness was the larger of the two new wildernesses and was located in the southern Sierra Nevada on the Sequoia and Inyo National Forests. The wilderness was named after California's state fish, the golden trout. The golden trout and a subspecies, the little Kern golden trout, were found in the Little Kern and South Fork of the Kern River that flowed through this area. The Santa Lucia Wilderness was located in the Los Padres National Forest just east of San Luis Obispo. Lopez Canyon was the heart of this new 21,250-acre area (*California Log* 1978: May). Neither the Golden Trout nor the Santa Lucia Wildernesses stirred much controversy, but the proposed designation of the San Joaquin Wilderness in 1984 was a different matter. This proposed 110,000-acre area of rugged river and timber country northeast of Fresno, California, was blocked from wilderness designation because of two factors. At issue were approximately 13,000 acres of harvestable timber (old-growth red and white fir) near Pincushion Peak, and the plans for at least a dozen hydroelectric projects, including four diversionary reservoirs, for the rivers and streams in the proposed wilderness (Pendergrass 19885: 200).

Though timber interests were able to stop the proposed San Joaquin Wilderness, another important issue that would place constraints on timber management was just on the horizon. This one involved endangered wildlife and wildlife habitat preservation. More than 600 of the 800 species of mammals, birds, reptiles, amphibians and fish native to California live on the 20 million acres of national forest land, making the Forest Service the single largest wildlife habitat manager in the state. By the mid-1980s, estimated national forest game populations included antelope (3,456), bear (10,313), elk (612) and deer (385,227), and habitat management programs were in place for all of these animals. Additionally, a wildlife sanctuary was established for the California bighorn sheep on the Inyo National Forest, and the Sespe and Sisquoc Sanctuaries for the California condor. The Forest Service, in cooperation with the U.S. Fish and Wildlife Service and the State of California, also worked to restore habitat for anadromous fish – those that swim upstream from the sea to fresh water to spawn. Cooperation measures included clearing blocked stream channels, installing fish ladders to improve upstream migration and restoring eroded areas in watersheds to prevent sedimentation of stream gravels needed for spawning (*Pacific Southwest Log* 1984: February).

In the early 1970s, Congress passed several key pieces of legislation to protect threatened and endangered species. For instance, in 1971, Congress approved the Wild Horses and Burros Protection Act to protect and manage wild, free-roaming horses and burros as components of the public lands. Forest supervisors were directed to count the population of horses and burros on their national forests, and that count became the "official" managed population. Each year, wild horse roundups took place, such as on the Modoc National Forest, where in 1978 more than 1,000 wild horses – the largest such population anywhere in the United States – roamed on the Devil's Garden Plateau.

Helicopters were used to help roundup wild horses and burros.

The program was an outstanding example of cooperation between the Forest Service and interested private organizations and individuals committed to the survival and well-being of that incomparable animal, the wild horse (*California Log* 1978: January).

But by far the most important piece of legislation that affected the California Region was the Threatened and Endangered Species Act of 1973. At first, the impact of this legislation in California was not evident. To meet the management objectives of the act, the Pacific Southwest Region cooperated with many federal and state agencies to survey threatened species. For instance, in 1978, the region participated in a nationwide census of the American bald eagle. The most important bald eagle winter habitat in the state was the Klamath Basin area, which straddles the California-Oregon border. Here almost half of the wintering bald eagles in California were found. The Shasta-Trinity National Forest was important as a wintering spot, where 10 percent of the state's winter population of eagles nested (*Pacific Southwest Log* 1981: May). Another important bird species in the California Region was the osprey, or fish hawk – a bird of prey found along seacoasts, lakes and rivers. In 1971, prior to the Endangered Species Act, the Forest Service placed the osprey on its special watch list. At that time, the Lassen National Forest set aside 1,200 acres of land on the west shore of Eagle Lake as an osprey management area. An estimated 10 percent of California's total nesting population used Eagle Lake. Over time, the Eagle Lake ospreys were significantly helped by Forest Service actions. For instance, when a serious problem arose when most of the snags used as nesting sites were found to be deteriorating, the Forest Service stepped in and replaced them with artificial nesting poles, which were quickly accepted by the osprey. These "high-rises" for the osprey not only increased the overall nesting population to 23 percent by 1983 but also increased public awareness of the need to help these majestic birds in their struggle for survival (*Pacific Southwest Log* 1983: February; 1985: July; Schneegas 2004: 15). The unarmored three-spine stickleback, a small scaleless fish found in the upper reaches of the Santa Clara river system, was also fully protected. This fish was once abundant in the Los Angeles, San Gabriel and Santa Ana River systems of the Los Angeles Basin, but by the 1980s, these rivers had been reduced to concrete-lined drains, leaving the only surviving population in sections of the Santa Clara River and a few of its small tributaries. Because of this precarious situation, the Forest Service and the Fish and Game Commission of the State of California worked cooperatively to save this fish from extinction and protect the ecological stability of this river system (*Pacific Southwest Log* 1984: June/July).

By 1984, there were dozens of federally listed endangered, threatened and sensitive species. Endangered species at that time included the San Joaquin kit fox, brown pelican, California condor, bald eagle, peregrine falcon, blunt-nosed leopard lizard, Owens River pupfish and the unarmored three-spine

Chick Martin of Lake Tahoe (left) and Doug Leisz, Forest Supervisor, identifying wildflowers. Eldorado National Forest. 1965

stickleback. Threatened species included the southern sea otter, Lahontan cutthroat trout, Paiute cutthroat trout and the Little Kern golden trout. Finally, the Forest Service list of sensitive species included the tule elk, Nelson bighorn sheep, Mt. Pinos chipmunk, Mt. Pinos blue grouse, prairie falcon, osprey, goshawk, redband trout, summer steelhead and the Karok Indian snail (*Pacific Southwest Log* 1984: February).

Another sensitive species was the spotted owl. In 1981, the Pacific Southwest Region began to study this secretive woodland raptor. At that time, studies had shown that spotted owls were abundant in many areas of California, such as on the Eldorado National Forest (Harn 1990: 53). However, Regional Forester Smith had declared them a sensitive species because forest uses and management activities on the national forests could reduce their habitat to below levels needed to maintain a viable population. In California there were two subspecies of the spotted owl, each having a different home range. The northern spotted owl inhabited the North Coast, Cascade and Sierra Nevada mountain ranges. Their preferred habitat, older coniferous forests, was being severely disturbed by heavy timber harvesting of old-growth stands. On the other hand, the southern spotted owl nested in the central and southern parts of the state and the inner Great Basin region. Like their cousins, their preferred habitat included old-growth conifers, but they were also known to inhabit broadleaf trees, mostly along rivers and streams (*Pacific Southwest Log* 1981: September).

At first, studies indicated that part of the problem in the decline of the spotted owl was the absence of snags (dead or diseased trees) for nesting. In the past, the attitude toward snags was that the "only good snag was a long snag" – in other words, one that was on the ground. Since they were often hosts of insects and disease, snags were removed during timber harvests as a fire and safety precaution when the Forest Service went from a protection phase to a production management mode (Harn 1990: 82-83). Private forests did the same. However, in 1977, the Forest Service developed a national snag policy to provide habitat to cavity-nesting bird populations dependent on them, including swallows, wrens, woodpeckers and spotted owls. Meanwhile, the 1981 *Draft Environmental Impact Statement (DEIS) for the Pacific Southwest Region* raised the status of spotted owl habitat to a major issue. At that time, Forest Service direction for commercial timber lands called for maintaining at least 5 percent of each forest type in older mature stands of timber, exclusive of wilderness areas, as the minimum requirement for maintaining plant and animal diversity. Timber industry leaders such as the Western Timber Association supported alternatives that provided an economically optimum level of timber production while maintaining viable populations of owls, but some wildlife advocates feared that maintaining only a minimum viable population levels in commercial timbers could, in the long run, reduce the number of spotted owls by as much as 80 percent (*Pacific Southwest Log* 1981: September).

The Pacific Southwest Region thereafter experimented with man-made cavity nests for spotted owls, as they had done with the osprey, but with limited success (*Pacific Southwest Log* 1986: April/May). In the interim, the spotted owl controversy intensified, especially in Oregon and Washington. In 1987, to give resource managers in the Pacific Southwest Region information to help determine the spotted owls' habitat needs, a number of owls were trapped and equipped with small radio transmitters on the Sierra, Tahoe, Klamath and Six Rivers National Forests so that the region could better understand the owls' daily movements and estimate the size of their ranges. The spotted owl study would take five years to complete (*Pacific Southwest Log* 1987: June). What they found was that the spotted owl was dependent on old growth for its survival. Some in the region realized immediately that this could be a big controversy down the line and started trying to inform people, mostly within the agency, that this had the potential to become a major problem. Many in Forest Service inner circles, however, did not want to hear this, including Regional Forester Smith. One retiree remembered telling the boss that "probably the spotted owl was going to look like Tyrannosaurus rex before it was all over with," and was disappointed that people like Smith could not seem to understand, or did not want to understand, that wildlife habitat issues were going to become important. Upon reflection, according to this retiree, the issue was not the spotted owl; the real issue was old growth, and how much old growth needed to be saved. If it wasn't the spotted owl, it would have been some other species (Harn 1990: 54-57, 73).

Meanwhile, a court injunction forced the Forest Service to reduce logging in its remaining patches of old-growth forest along the Pacific Coast. The spotted owl was seen as an "indicator species" of old growth, and to environmentalists, its plight represented that of the many species associated with that increasingly rare habitat (Hirt 1994: 277). Consideration of wildlife habitat soon thereafter became an essential part of the Pacific Southwest Region decision-making process. It was not formally put into the directive system and manuals, although there was some thought it should have been at this time. One Forest Service employee, a biologist and ecologist named Bill Laudenslayer Jr., did crystallize this concern in a coauthored work entitled *Guide to Wildlife Habitats of California* (1988) (Harn 1990: 43-44). Ultimately, the spotted owl issue impacted timber harvests in the Pacific Southwest Region, reducing them from the 1.5 billion board feet in the 1980s to just 500 million or fewer by the early 1990s (Stewart 2004: 17).

Rounding up wild horses, protecting endangered, threatened, and sensitive species such as the unarmored three-spine stickleback, and tracking spotted owls were part of a new and growing enforcement aspect of California's regional mission that came out of the environmental 1970s. Law enforcement became vitally important to the Forest Service as well and ranged from policing mineral exploration to arresting people for arson, theft and unlawful occupancy.

In 1978, the U.S. Bureau of Mines calculated that each American citizen required 40,000 pounds of new mineral materials (including petroleum and other energy sources) annually (*California Log* 1978: May). California ranked third among the states in the value of annual mineral production and owed its prosperity not to just gold but unromantic minerals such as tungsten for light bulb filaments and diatomite for filtering wine. Minerals discovered and commercially produced on California's national forests included nickel, cobalt, chromium, copper and gold, and sources of energy such as oil, natural gas, uranium and geothermal. A number of laws and regulations ensured that mining development on California's national forests was compatible with the multiple use of other resources. For instance, the Wilderness Act of 1964 precluded further exploration and mining on designated wilderness areas after 1983, the Wild and Scenic River Act precluded any mining within one-quarter mile of designated rivers and the Threatened and Endangered Species Act of 1973 required that mineral exploration and development not jeopardize the existence of fish and wildlife in danger of extinction. Following the passage of the latter act, the Forest Service published *Title 36, Code of Federal Regulations, Section 252,* which dealt with the use of the surface of national forest lands by anyone operating under the Mining Law of 1872. This regulation required minimal adverse environmental impacts on national forest resources, mandated that an environmental assessment and determination be made whether a NEPA environmental impact statement (EIS) was required or not, and held the operator responsible for complying with all environmental laws (*Pacific Southwest Log* 1981: May).

During the 1980s, new emphasis was placed on developing the nation's mineral capabilities. Faced with economic problems, the Reagan administration encouraged the Forest Service to generate more output from its forests, especially mineral production, which Assistant Secretary for Natural Resources Crowell believed was a major potential source of benefits to the economy over the next decade (*Pacific Southwest Log* 1981: October). Crowell's viewpoint was supported by a 1979 Office of Technology Assessment study, which reported, "Mineral production is the best use of any tract of land and thus makes mineral activity the preferred use on any federal land that is open to such activity." During the 1980s, the Forest Service cooperated with the mineral industry. To acquire permits, mining companies had to submit an approved reclamation plan if they conducted surface mining operations involving the removal of overburden in the amount of 1,000 cubic yards or more at any one location of one acre or less. The Forest Service monitored companies to see that their operations were consistent with standards set by federal, state and local governments and to assure that mining operations resolved environmental problems arising from the development of large projects (King 1994: 17-20).

One perpetual problem associated with mining was the law enforcement issue and occupation of Forest Service lands based on mining claims, such as communes of counterculture people from San Francisco living along the Salmon River on the Klamath National Forest (Smart 2004: 30-31). This issue came to a head in the 1970s on the Shasta-Trinity National Forest in the small community of Denny on the Big Bar Ranger District, and persisted into the 1980s. In the distant past, this community developed on private land intermixed with federal land and was composed mostly of gold miners living on their claims. Mining activity continued up until the Depression years. Since the 1930s, mining had been essentially nonexistent. A few occupants remained; others sold claims and buildings to newcomers who seemingly did not understand the 1872 mining law and its requirements, or simply disregarded them. A few buildings and claims were abandoned and later reclaimed. Other people just constructed new buildings without any permit. By the 1970s, the Forest Service considered most of people living in Denny as trespassers because the residents could show no authority for their occupancy. In 1971, a shooting and beating incident took place during a mineral examination in this isolated area where law enforcement was hours away. The Forest Service made subsequent attempts to establish law and order, but they were met with armed resistance and with overt threats of violence. In the late 1970s, the Forest Service finally turned to the courts. More than fifty Denny residents were served with "Complaints in Ejectment." It was hoped that this action in U.S. District Court would result in the determination of whether the occupants were trespassers or not (*California Log* 1976: June).

The courts finally resolved the Denny issue in the Forest Service's favor, but a new and more dangerous breed of unauthorized occupancy emerged on Shasta-Trinity National Forest land and elsewhere in the early 1980s – marijuana growing. Throughout the national forests, but particularly on forests in Northern California, growing marijuana became a lucrative business after 1978. The three-forest area of the Klamath and Shasta-Trinity National Forests was being irreverently referred to the "golden triangle" because of the heavy growth of marijuana there. In the remote backcountry, drug dealers could grow their illegal crops with less risk of detection by law enforcement officials. Raids on these "plantations" resulted in the recovery and destruction of thousands of plants worth millions of dollars. One raid in 1981 on the Shasta-Trinity National Forest recovered marijuana worth $1.5 million. Authorities also found a cache of automatic weapons, dynamite, handguns and rifles – an ominous trend indicating a willingness by growers to protect crops from intruders whether they be Forest Service officials or the public who visited the national forests (*Pacific Southwest Log* 1981: July/August; Rice 2004: 27-28). In 1981 alone, the Pacific Southwest Region received 800 reports of confrontations between growers and visitors to the national forests. Forest Service employees had even been warned to "mind their own business," or they and their homes and families would be attacked. Up until 1984, in response to the situation on the Shasta-Trinity National Forest and elsewhere, the Forest Service did little more than report any detected growing locations to the law enforcement agencies, spray some herbicides on plantations and yank plants out of the ground, and try to educate the public to report such cultivation as well (*Pacific Southwest Log* 1982: July; Rice 2004: 27-28).

By 1984, the Denny area in western Trinity County had become the most lawless rural area in the state. It had become a place where shootings, arson and physical violence were becoming commonplace against those who legitimately used this 115,000-acre part of the Shasta-Trinity National Forest. The Denny unlawful occupation, aggravated by the illegal growing of marijuana, became intolerable. Armed Forest Service personnel joined with the Trinity County Sheriff's Department in a three-year program to clean up the area and to prevent the establishment of marijuana plantations (*Pacific Southwest Log* 1984: September). Another objective of bringing "gun toters" into the area was to protect the public and to regain forest resources from these criminal elements (Smart 2004: 51-52).

Public safety and law enforcement on the Shasta-Trinity National Forest was only the tip of the iceberg of violations of regulations and laws related to national forests in California. Violations had risen from 800 in 1965 to more than 40,000 by 1981. Of that number, approximately 10,000 led to written citations for infractions such as rowdy conduct at campgrounds, which could result in arrest. In most cases, the Forest Service relied upon the tact and insight of its employees to explain regulations to forest visitors without causing undue resentment, and to know when a warning was sufficient. By the 1980s, the region had five special agents and two claims examiners. Additionally, nine national forests not only had special agents but also about fifty employees called Level IV law enforcement officers, trained to handle specific types of law enforcement situations as adjunct duties. Unlike the National Park Service, which had its own park police force, the Forest Service combined enforcement duties with other duties (*Pacific Southwest Log* 1981: July/August).

A major crime committed on the national forests each year was arson. Of the 1,200 wildland fires caused by people each year, arsonists started approximately 200. On these fires, only 3 percent of the arsonists were ever identified, tried and convicted, due mainly to the difficulty in obtaining evidence and witnesses. In the case where a wildland fire was started through carelessness, the person responsible was found to be liable for fire suppression and damage costs. Timber theft was another major crime in terms of potential dollars lost to the federal treasury. Theft involved everything from illegal marking of trees for harvest to direct theft (*Pacific Southwest Log* 1981: July/August).

The Forest Service also prosecuted robbers of the past. For decades, a major portion of the record of more than 20,000 years of human history of California's national forests had been unwittingly destroyed and severely damaged by arrowhead, bottle and other relic collectors. To curtail this alarming destruction on national forests and other federal lands, Congress passed the Archaeological Resources Protection Act (ARPA) of 1979. The major provisions of this legislation made it a felony punishable by a maximum penalty of $20,000 and/or one year in prison to sell, purchase, transport, exchange, damage, remove or excavate archaeological resources without a valid permit. Thereafter, the Forest Service beefed up its law enforcement efforts against "pot hunters" and tried to discourage vandalism that eroded many prehistoric and historic sites containing unique national heritage information (*Pacific Southwest Log* 1980: November). The public aided the Forest Service enforcement efforts by reporting illegal digs, such as an incident in which hikers witnessed three men digging in Native American burial sites and sifting the earth for artifacts. Thanks to their help, the men were convicted and fined under ARPA (*Pacific Southwest Log* 1982: June).

In the 1980s, cultural resource management had become a major and successful activity within the California Region. Issues of the *Pacific Southwest Log* regularly covered cultural resource management, including archaeological and historic research and achievements within the region. For instance, one issue discussed the role African-Americans played in the Civilian Conservation Corps (CCC), such as those in the Piedra Blanca CCC motor pool. These young men were the undisputed champions of all the California CCC camps, driving more than 220,000 miles over narrow, unpaved, sometimes precipitous road and trails on the Los Padres National Forest (*Pacific Southwest Log* 1982: June). Another issue described the nomination to the National Register of Historic Places (NRHP) of the La Moine Lumber and Trading Company railroad logging system on the Shasta-Trinity National Forest. The site included sixty-seven archaeological sites along the railroad grade, including mill sites, tanks and towers, dumps, boilers, camp sites, single structures, trestles and flumes (*Pacific Southwest Log* 1982: February). There were also articles on Native American cultures in California. They explained the great linguistic diversity among native Californians and that, when speaking of the Chumash or the Hoopa, one was identifying a culture or a way of life that had existed for thousands of years (*Pacific Southwest Log* 1982: June). Pieces were also written about significant archaeological discoveries such as one of the highest known Native American villages sites in North America on the Inyo National Forest at 11,800 feet (*Pacific Southwest Log* 1983: January).

In the 1980s, relations with various Native American groups were variable. For instance, no doubt the Maidu of Northern California were pleased when the Forest Service saved Soda Rock, an important "power spot" in Maidu mythology. According to Maidu traditional beliefs, Soda Rock was of cultural importance to them, as with waterfalls, stream crossings, cliffs and large rocks that were abodes of malevolent beings who took the form of rattlesnakes, water imps, biting ants and water witches. The Forest Service managed this as a cultural landmark, located within the Plumas National Forest, after years of negotiations between the Forest Service and the Soda Rock quarry operators (*Pacific Southwest Log* 1987: April).

On the other hand, the Yurok and other native groups strongly opposed Forest Service plans to build the Gasquet-Orleans Road through this stunningly beautiful place on the Six Rivers National Forest. In the late 1960s, the Forest Service began the forty-mile long "G-O" or "GO Road." But in the mid-1970s, when the road began to penetrate Native American holy grounds, they, along with environmentalists, fought a seven-year legal and bureaucratic battle to stop the completion of the road. The GO Road was designed to open up competition to a wider variety of logging mills by allowing timber to be hauled from the Klamath River Basin up over the top of the divide and down onto the coastal side, into Eureka or Crescent City (Kennedy 2004b: 8). The Forest Service thought the road would revitalize that area's flagging lumber industry while at the same time open up a spectacular recreational area. The problem was that the Yurok, Karuk, Tolowa and Hoopa tribes had used the area for centuries. Tribal leaders warned that if this sacred high country was desecrated by the GO Road, the Great Creator would punish the perpetrators. Meanwhile, the Sierra Club, the Audubon Society and the California State Attorney General's Office banded together to request the chief of the Forest Service to keep the road from passing through this sacred land around Chimney Rock. They made a last stand in federal court under provisions of the American Indian Religious Freedom Act. They won in the 9th Circuit Court, but lost in the Supreme Court. This and other court decisions propelled the creation of a national and Region 5 Tribal Relations Program.

Within the Forest Service, relations between the cultural resource management (CRM) staff and other resource managers were not always harmonious either. During the 1970s, the cultural resources program was largely considered an "unfunded mandate." Nevertheless, Regional Archaeologist Don Miller managed to develop a program with several archaeologists working in zones throughout the region, with each national forest having a forest archaeologist, who also often provided historian services as well. In these early years and into the 1980s, they not only surveyed, recorded and protected archaeological and historic sites on a project level, but some also managed to prepare cultural resource overviews, prime examples being James A. McDonald's *Cultural Resources Overview of the*

Klamath National Forest, 1826-1941 (1979), James Johnston and Elizabeth Budy's *Cultural Resource Management Overview of the Lassen National Forest, 1848-1945* (1982) and E. R. Blakley and Karen Barnette's *Historical Overview of the Los Padres National Forest, 1542-1984* (1985). Some national forests contracted out the work to private CRM firms, which produced works such as WESTEC Services' *Cultural Resources Overview of the San Bernardino National Forest* (1982) or W. Turrentine Jackson, Rand Herbert, and Stephen Wee's *History of the Tahoe National Forest, 1840-1940* (n.d.).

In the 1980s, with money and personnel tight, the CRM program in the Forest Service slowed down, even though it still had to meet its legal responsibilities regarding cultural resources. At this time, the initial objectives of the program were twofold: first, to complete an inventory of cultural resources on California's national forest system lands by 1985 to provide a database for land management planning and second, to complete an inventory of all cultural resources on national forest land by 1990. Until these inventories were completed, CRM staff urged other resource managers to exercise caution to

Historic mining cabins near Tioga pass, Inyo National Forest

ensure cultural resources were not damaged, destroyed or transferred, and to involve cultural resource professionals in the decision-making process. Often cultural resources were at the center of confrontational duels between the CRM people and timber managers, whose personnel did not quite understand that obeying federal laws regarding cultural resource management, such as ARPA and AIRFA, was not discretionary (Rock 1981: 3; Boyd 1994: 16-17).

Finally, in the early 1980s, the Pacific Southwest Region came to realize that it was not properly implementing and fully integrating history into its CRM program. A critical analysis of the practice of history in the region indicated that materials of historical resource value (including correspondence, memoranda, reports, photographs, maps and related material) were not being protected and that historical resources and research were not being used by Forest Service staff in their analysis and decision making. In 1982, while there were many CRM staff trained in archaeology and anthropology, there were only two "historians" on the CRM staff. The consequence was that the report stated that knowledge and use of historical resources by CRM personnel was "niggardly and required historical work is unsupervised, usually uninspired, and relegated to tertiary importance behind archaeology and anthropology." On some forests, if there was any history program, it was delegated to volunteers with the "older American" program, and involved conducting oral history programs and collecting historical photographs in scrapbooks. The Pacific Southwest Region did not take its history program seriously until after 1982, when the Washington office mandated that each region develop a history program, to include the writing of an administrative history of each forest, and one of the region as well (Public History Services 1982: Preface, 5-7). In 1984, during the tenure of the Regional Forester Zane Smith, the Forest Service contracted for a book-length regional history (*Pacific Southwest Log* 1984: February), which was not published, and because of budget limitations, a full history program was never fully institutionalized in the Pacific Southwest Region at the time.

Roots of Ecosystem Management

In July 1987, Regional Forester Zane Smith was appointed special assistant to Chief Dale Robertson for recreation planning, and Paul F. Barker was made the new regional forester. Barker was immediately greeted by yet another severe summer fire season – the worse fire fuels situation in California since 1977 (*Pacific Southwest Log* 1987: July, September). It seemed as though the Pacific Southwest Region was set to continue along a well-rutted historical path of considering land use as a series of single-use allocations to address specific problems or to address issues of the most vocal constituencies. But this freight train to yet another battle between utilitarian conservation versus preservation was derailed by the introduction of the concept of ecosystem management to the Forest Service by Chief Dale Robertson. His action precipitated a change in the way Forest Service approached decision making (Williams 2000: 142). In response to this drive for ecosystem management from the WO, in April 1995, the Pacific Southwest Region produced *Sustaining Ecosystems: A Conceptual Framework*, which provided an analysis process to be used as a step in implementing ecosystem management within the region. Based on the interrelationship of land, water, plants, animals and people, and the cross-pollination of ideas from a number of disciplines, *Sustaining Ecosystems* sought to strive for balance among these elements in a framework that integrated physical, biological and cultural/social dimensions (USDA Forest Service 1995: passim).

Ecosystem management was not a radical departure from the past, but more like a merging of traditions. In looking back over its history, the Pacific Southwest Region had been working toward this holistic goal from the very beginning but had never quite understood it. In many ways, Californians had been trying to heal what they had destroyed in the past and utilize resources to their sustainable limits, while at the same time preventing the rest of the environment from becoming a wasteland for future generations. Essentially, ecosystem management was the harmonization of the Forest Service's utilitarian conservation ethic and its tenets of taking care of the land and using resources wisely, with the preservationists' or environmentalists' concern for addressing the cultural/social needs of people, including aesthetic beliefs and lifestyles of all cultural and social groups (USDA Forest Service 1995: passim).

By the late 1980s, the Forest Service had this perspective and enough management tools to journey down this new path of ecosystem management. The environmental legislation of the 1960s and 1970s forced the Forest Service to slow its cutting and assess the land, which gave it time to gain a new perspective. The broader perspective was needed in order to understand when an ecosystem was healthy or under stress because the conservation of diversity was in disharmony. The strength of legislation like the Endangered Species Act, National Environmental Policy Act, National Forest Management Act and others was that they recognized that public lands could play an important role in the conservation of biodiversity for the nation. On the other hand, legislation such as the American Indian Religious Freedom Act and the National Historic Preservation Act acknowledged that diversity was important for cultural and social as well as biophysical dimensions. When the biological and cultural/social dimensions were taken together with understanding the constraints of the physical dimension (topography, geology, climate, nutrients and hydrology), a framework for ecosystem management was born. With it, Forest Service management shifted to focusing on issues and concerns relevant to larger spatial and temporal scales (e.g., species viability). It also offered hope for the future resolution of key ecologic questions such as what is sustainable, what do we want, what do we have, and how do we get there? (USDA Forest Service 1995: passim).

Epilogue

After Regional Forester Zane Smith's departure in 1987, Region 5 underwent many changes and was affected by many events—too many to mention in this epilogue, and too close to the present to be properly analyzed historically. In fact, in the post-1987 period, six different regional foresters were appointed as Pacific Southwest Regional Forester: Paul F. Barker (1987-1990), Ronald E. Stewart (1991-1995), G. Lynn Sprague (1995-1998), Bradley E. Powell (1998-2001), Jack A. Blackwell (2001-2005), and Bernard Weingardt (2005 to the present). Nonetheless, the author surveyed a small sample of current and retired Forest Service personnel, including regional foresters and Forest Service scholars, to provide a list of at least three actions that influenced and shaped the region. The following epilogue is the author's interpretation of those comments, which has been enhanced by examination of recent scholarly sources (Ruth 2005; Lewis 2005; Stewart 2004), and its purpose is to give the next historian some direction when the sequel to this history is written.

First, and perhaps foremost, Region 5 history in the post-1987 period continued to be strongly influenced by the "Consent Decree" (No. C-73-11110 SC), a proposed settlement agreement for a 1981 class-action lawsuit against the Forest Service filed in 1972 (*Gene Bernardi, etc. et al. Vs.et al. v. Earl Butz*). The suit alleged sex discrimination in both hiring and promotional practices of the agency in California, and on April 3, 1991, a final settlement was approved between the Court and the affected parties. Under the terms of the Consent Decree, Region 5 and the Pacific Southwest Station were required to establish certain hiring goals and affirmative action requirements. In the late 1980s and early 1990s, the Bernardi Consent Decree was the single largest internal issue facing Region 5, and was often a powerfully divisive issue among employees. In the short run, some believed that the Bernardi Consent Decree had negative effects on the agency. They felt that the decree put the "accelerated" women in a terrible position, forcing them to succeed or be judged as failures. Some did succeed and the Forest Service greatly benefited. Others did not, and both they and the Forest Service "lost.". Another problem was that because of the settlement, Forest Service hiring and promotion practices dramatically shifted away from the concept of meritocracy. Despite these problems, however, it was generally believed that thanks to the decree and the court-ordered monitoring, Region 5 moved toward a healthier, more equitable workplace with greater diversity of gender. Women increasingly found leadership positions, including positions in fire and law enforcement, that had been denied to them in the past.

These achievements may have ended a twenty20-year lawsuit regarding the employment of women in Region 5, but the Bernardi Consent Decree spawned subsequent lawsuits and agreements, and resulted in a series of actions by minorities and other under-represented groups. For example, Hispanic employees filed a class-action suit that eventually reached a resolution agreement in 1992. This agreement required Region 5 to eliminate any barriers to hiring, promoting, and retaining of Hispanics in the Region 5 workforce. A second agreement, effective February 2003, required Region 5 to undertake and actively continue specific measures designed toward reaching these goals. The litigious atmosphere created by the Bernardi Decree also spilled over into other areas. For instance, female employees filed the class-action suit, *Donnelly vs. Veneman*. The Donnelly class- action suit was directed at sexual harassment problems in Region 5 and sought to resolve them in a timely and effective manner by eliminating sexual harassment and hostile environments. The settlement implemented a zero tolerance policy against sexual harassment and ensured that persons committing or contributing to sexual harassment were held accountable for their actions. The Donnelly suit was finally settled in February 2002. To guarantee successful implementation of the provisions of the Donnelly action, monitoring of Forest Service compliance was extended to February 2006.

In the long run, as painful as it was, the Consent Decree, and the other legal actions drove the biggest increase in workforce diversity in the Region and changed it for the better.

According to a consensus of opinion, the second major event that characterized and influenced the history of Region 5 in the post 1987 period was the Sierra Nevada Framework decision, the largest planning effort ever undertaken by the region. Signed in January 2004 by Regional Forester Jack Blackwell, the Sierra Nevada Forest Plan Amendment culminated a fourteen-year effort at planning and research aimed at preserving old- growth forests and wildlife in the Sierra. The decision affected 11.5 million acres on eleven different national forests in the 430-mile-long Sierra Nevada mountain range, from the northeast border of Oregon to the Sequoia National Forest in the south.

Many controversies and concerns led to the Sierra Nevada Framework decision. They began in the late 1980s, when conservation groups raised concerns over timber sales in old-growth forests, which contributed to destruction of critical habitat for the California spotted owl. At the same time, Region 5 had completed land and resource management plans (LMPs) for the region that laid out the future course for resource use within each national forest in the Sierra Nevada. These LMPs, despite the controversy over clear-cutting and related practices, were designed to increase timber production and utilization on individual national forests with little regard for regional factors or characteristics. These two events clashed, leading to public controversy. Thereafter, the research of Forest Service and academic scientists, along with environmental and industry politics, dictated the course of events.

First, sophisticated activist groups, such as the Sierra Club, the Wilderness Society, the Natural Resources Defense Council, the Audubon Society, and the National Wildlife Federation, followed Region 5's land management planning process and critically pointed out the adverse ramifications of each LMP on the respective environments. They were not alone. Private businesses, such as the timber, grazing, and recreation industries, also paid close attention to the Sierra Nevada planning process and offered their criticisms as well. Conflict and disagreements over land management planning between the two groups resulted in rancorous discussions and debates. Environmentalists wanted most areas left untouched by modern society, or even restored to pre-settlement, Eden-like conditions. On the other hand, private industry looked to the forests for commodity production - whether it was for timber, grazing, mining, or recreation. Caught in the middle of the argument between these diverse interests and a public awakened to environmental issues, Region 5 conducted a number of scientific studies regarding the conservation of the California spotted owl. They eventually led to a broader purpose—the protection of the Sierra Nevada ecosystem.

As a response to growing scientific and public concern about the status of the California spotted owl, in 1991, Region 5 formed an assessment and planning team to develop a successful conservation strategy for the endangered bird. The steering committee for this team was a multi-agency planning group that included the Forest Service, National Park Service (NPS), Bureau of Land Management (BLM), Fish and Wildlife Service (FWS) and resources agencies of the State of California, along with representatives from county governments, environmental groups, forest products industries, and several other organizations. At the same time, ecosystem management had been introduced by the Forest Service. This "new" Forest Service land ethic, influenced by the writings of Aldo Leopold, entailed gathering better knowledge of the landscapes, resources, and ecological dynamics, leading to increased support for research on these elements of national forests. The steering committee created two teams, a "technical team" to provide expertise in avian biology and ecology, and a "policy implementation team" to provide policy and economic analysis. By the end of the process, these teams presented their analysis and recommendations, known as the CASPO (California spotted owl) study. It concluded that current management direction as proposed by the LMPs was detrimental to the long-term well being of the species. Ultimately, they found that habitat protection for the owl could not be achieved while simultaneously allowing clear-cutting or otherwise permitting the removal of large, old- growth trees in these forests.

The Forest Service responded by implementing the 1992 CASPO guidelines in order to provide interim protection for the California spotted owl. This resulted in a dramatic decline in timber harvests in the national forests of the Sierra Nevada, which the CASPO and independent sources predicted. Meanwhile, the Forest Service, under direction from Congress, gathered additional scientific information under a new and broader scientific assessment of the health and sustainability of the Sierra Nevada. In 1993, Congress authorized funds for a scientific study by an independent panel of scientists to study and analyze the entire Sierra Nevada ecosystem. The main objective of this study was to include an assessment of all lands and resources, both public and private, and to present to Congress a plan with a spectrum of alternatives for future consideration. This study built on existing scientific research and conducted assessments of the region's ecosystems and natural resources, and examined in depth their relationship to society. It became known as the Sierra Nevada Ecosystem Project (SNEP), which published its findings in three volumes in 1996, with an addendum in 1997.

But regardless of the promise of ecosystem management and Forest Service attempts to develop a "scientifically credible" conservation strategy, public dissatisfaction and political and social activism over national forest management in Region 5 continued. In the end, it reshaped resource planning and management in the Sierra Nevada entirely. After completion of CASPO and SNEP, and subsequent studies, the new emphasis of Forest Service planning in the Sierra Nevada centered not just on the protection of the spotted owl, but on the conservation of biological diversity and resource sustainability for the entire area. In January 1998, Region 5 launched the "Sierra Nevada Framework" to further revise its management policy in the region toward these ends. By year's end, the Forest Service had identified five problem areas that needed immediate attention: old forest ecosystems; aquatic, riparian and meadow ecosystems; fire and fuels management; noxious weeds; and, lower hardwood ecosystems on the west side of the Sierra. Three years later, the 2001 Framework decision addressed these concerns. Key aspects of the decision were a commitment to restore and protect several million acres of old- growth forest habitat, with core area protections for the California spotted owl and the goshawk, protection of trees greater than twenty inches in diameter on Forest Service lands within the eleven national forests, protection of critical aquatic refuges through a stream buffer system, and a fuel-reduction program that focused on small-diameter trees, brush, and surface fuels. Another objective of the Framework was to amend forest plans in conformance with the National Forest Management Act (NFMA) and the National Environmental Policy Act (NEPA).

The Sierra Nevada Framework was a major shift of larger Forest Service policy and society toward "ecosystem management" as an expression of the multiple-use management policies of the past. Region 5 planning efforts for national forests in the Sierra Nevada mountain range, which in the early 1990s focused only on the California spotted owl, evolved into a plan that addressed all components of the ecosystem of this area in a balanced and scientific manner. The Sierra Nevada Framework committed Region 5 (Pacific Southwest), Region 4 (Intermountain), and the Pacific Southwest Research Station to integrate new science into the management of the national forests of the Sierra Nevada. It also committed the Forest Service to long-term cooperation and coordination with Native American tribes, local governments and communities, and people generally concerned about the health of the Sierra.

The third and final activity that influenced Region 5 history in the post-1987 period was a combination of Forest Service "analysis paralysis" and a rapidly shifting national administration policy that thrust Region 5, and other regions, into a whirlpool of indecision. Like most of the Forest Service regions in the late 1980s and 1990s, Region 5 got bogged down writing "defensible" plans and related environmental documents, and too often, projects could not get done because of the commitment of resources to the planning process. Related to this was the rise of "professional" appellants—groups and individuals who filed administrative appeals, and sometimes threatened lawsuits, for just about any proposed Region 5 project anywhere in the region. By the end of the decade, some groups were appealing projects—particularly timber sales—as a matter of principle, or as a part of their broader agenda, rather than in specific opposition to an individual plan. These tactics caused Region 5 to take ten to fifteen years to complete LMPs for some individual national forests. This situation forced Region 5 to produce legally defensible environmental documents, which required a great diversion of resources away from the "doing" to the "planning," and significantly delaying the completion of some projects. On the other hand, this fear of appeals forced the Forest Service to build stronger scientific or economic justifications for its decisions and to be more sensitive to public concerns for proposed projects and led to important modifications of procedural requirements. Coupled with this "analysis paralysis" was an increasing interest and involvement in land management by individual administrations. This polarized the public even more over issues and sometimes led to decisions not by Forest Service officials in California, but by officials in the US Department of Agriculture. News media would refer to these as "Clinton Administration" or "Bush Administration" decisions.

What was already a complex planning process involving "heavy doses of scientific uncertainty and clashing public preferences and values" had become even more difficult in the region because of "analysis paralysis" and its links to the question of "active management" and treatment of fire fuels. During the last week of October and the first week of November 2003, thirteen wildfires swept through Southern California, creating a disaster on an unbelievable scale. Close to 12,000 firefighters fought these blazes, which burned 750,000 acres, resulted in $120 million in suppression costs, caused billions of dollars in damage, destroyed 4,000 homes, and resulted in the loss of 22 human lives. By far, it was the most costly and destructive fire in the region's history. According to *Southern California Firestorm 2003: A Report for the Wildland Fire Lessons Learned Center*, the catastrophic fire resulted from drought that had extended over several years. Other contributing factors to the unprecedented firestorm were widespread tree mortality resulting from insect infestation, high fuel loads resulting from years of full suppression of wildfires in both forests and chaparral and the inability of the Forest Service to use thinning and burning to reduce fuel loads because of environmental concerns, and the continuing expansion of the wildland interface in places like Southern California (Mission-Centered Solutions, Inc. and Guidance Group, Inc. 2003: 1).

The Southern California urban conflagration generated intense political interest in national forest management, which was not lost on Congress. Although the issue of "overstocked" forests was not a major factor in these fires, Congress moved quickly to debate and pass the Healthy Forest Restoration Act of 2003, which had languished in the legislative body for many years. This bipartisan legislation sought to expedite fuel reduction projects on national forest and other federal lands and was seen by some as a major victory in the natural resource area for the George W. Bush administration.

Passage of the Healthy Forest Restoration Act broke through what former Regional Forester Ronald E. Stewart (1991-1995) differentiated as the "Age of Wicked Problems." In a paper presented at the Pacific Southwest Centennial Forum on California, Hawaii and the Pacific Islands (Sacramento, California November 5-6, 2004), Stewart identified the analysis paralysis components that have defined the 1995 to present time period. In the "Age of Wicked Problems," Stewart's seven characteristics of a wicked problem are:

1) The definition of the problem or issue is in the "'eye of the beholder.'" How each individual person chooses to explain the problem determines the scope of the search for resolution. This leads, more or less, to another four characteristics.

2) Because each individual defines the problem in their [sic] own terms, there is no single correct formulation for a wicked problem, only more or less useful ones.

3) This also suggests that solutions are generally good or bad, rather than true or false. The validity of any solution cannot be tested in an objective way.

4) Each wicked problem concerns an assemblage of resources combined with effective demands in ways that are unique in time and space.

5) Consequently, any solution developed is a one-shot operation, with little or no chance to learn by direct trial and error

6) We also cannot know when all possible solutions have been explored, because there is no stopping rule.

7) And, finally, each wicked problem is extremely important and each solution significant. The decision maker cannot be wrong, even occasionally, and so must choose solutions only after agonizing appeal.

How, and whether or not, Region 5 and the Forest Service worked its way out of the analysis paralysis described by Stewart's "Age of Wicked Problems" is left to the next historian.

Anthony Godfrey, Ph.D.
U.S. West Research, Inc.
Salt Lake City, Utah

Appendix A: Timeline

1540	Melchior Diaz crosses the Colorado near Yuma on his trek across Arizona to become the first European in Alta California.
1542	Joao Rodriguez Cabrillo sails into San Diego harbor.
1602	Captain Vizcaino discovers and names Monterey Bay.
1760	Visitador-general Jose de Galvez, disturbed to learn that the Russians were pushing down the Pacific Coast from Alaska, decides its time that the Spanish organize and settle California as bulwark against further penetration. Baja California Governor Gaspar de Portola leads men northward and discovers the Santa Cruz redwoods.
1769	July 16: Father Serra formally founds first mission in California at San Diego.
1776	Juan Bautista de Anza leads settlers to California from northern Mexico overland to Monterey and founds Presidio San Francisco.
1780	The total population of Alta California stands at approximately 600 settlers.
1781	Families settle on the Porciuncula (Los Angeles) River, and the town of Los Angeles is founded.
1785	A land ordinance established a survey system consisting of townships six miles square, divided into thirty-six square sections of 640 acres each. Five sections were reserved: four for minerals and one for elementary public education. The remaining thirty-one sections were sold at auction, at a minimum of $1 per acre.
1800	The total population of Alta California stands at approximately 1,200 settlers.
1812	The General Land Office (GLO) is established as part of the Treasury Department to sell, grant or otherwise convey public lands, and prepare and maintain all records. District land offices are established to conduct auctions and sale of lands in territories and states. Minimum price: $1.25 per acre. Russians build Fort Ross north of San Francisco Bay.
1813	Spanish government decree orders Alta California reforestation measures.
1825	California becomes a territory of the Mexican Republic.
1826	Jedediah S. Smith and trappers arrive at Mission San Gabriel, the first Americans to travel overland to California from the Southwest.
1833	The Indian mission system ends abruptly when secularization became the law of the land. This event signals the first land rush to California and shift in population. In the next decade, more than 300 ranchos were granted to Mexican citizens and carved out largely out of mission-held land. Eventually the ranchos numbered more than 800.
1834	California's first commercial sawmill is built at Molino in Sonoma County.
1835	The total population of Los Angeles stands at approximately 1,500 people.
1841	The first known formal request to cut timber on the public domain in California. The request to Governor Bautista Alvarado is for ponderosa pine in the future San Gabriel Reserve.
1846	June 14: Republic of California proclaimed.
1847	James Wilson Marshall builds Sutter's Mill at Coloma gold discovery site. While building sawmill on the American River, Marshall discovers gold in 1848, which starts the Gold Rush to California. California Gold

	Rush draws thousands of gold seekers from around the world.
1848	Mexico cedes 338 million acres to U.S., including California, Nevada, Utah, and Arizona, and parts of New Mexico, Colorado and Wyoming.
1849	The Department of the Interior (DOI) is established. GLO is transferred to new department.
1850	September 9: California is admitted into the Union. Total population of California stands at approximately 92,500 people.
	The state's first forest fire control laws are passed by legislature.
1851	The surveying district of California is established.
1853	The first district land offices are established in Los Angeles and Benicia, California, and the survey of public lands in California begins.
1861	June 28: Central Pacific Railroad Co. of California is organized.
1862	May 15: U.S. Department of Agriculture (USDA) is established.
	The Homestead Act provides unrestricted settlement of public lands requiring only residence, cultivation and some improvements on a tract of 160 acres. After living on and farming the acres for six months, a settler could buy the land at $1.25 per acre or, after five years, for only $15 total. Under this act, more than a million pioneers settled the West during the next seventy years.
	The Transcontinental Railroad Act grants to Union Pacific and Central Pacific Railroad companies rights of way and title to five alternate sections to a depth of ten miles on each side of the line plus additional lands for stations, shops and other property, amounting to more than 20 million acres.
	The Morrill Land Grant Act authorizes grants of public lands to establish and support state vocational colleges teaching agriculture and mechanical arts. Distinguished list of colleges and universities is established, including the University of California.
1864	Congress grants Yosemite Valley and the Mariposa Grove of Big Trees to the State of California for "public use, resort, and recreation." The two tracts are to be held "inalienable for all time," the first areas in the country to be preserved specifically for future generations.
1865-1871	During this period, at least six acts grant ten alternate, old, numbered sections of land per mile on each of the roads to various railroad companies in California.
1866	The General Mining Act of 1866 authorizes the exploration and occupation of mineral lands in the public domain, both surveyed and unsurveyed. Essentially opens all public lands to mineral exploration and patent.
	Yosemite Valley becomes the first state park in the nation.
1869	May 10: First Transcontinental railroad is completed at Union/Central Pacific junction, Promontory Point, Utah.
1872	The first national park, Yellowstone National Park, is established from public domain lands.
	The General Mining Law of 1872, intended to settle western lands, declares that mineral exploration and development would have priority over all uses of the land and allowed free entry into public domain land to explore, develop and produce locatable minerals.

1873	Dr. Franklin B. Hough becomes the first federal forest officer. Four years later, the Commissioner's Office became the Division of Forestry in the Department of Agriculture. Hough was succeeded by Dr. Egleston, Dr. Bernard Fernow and later by Gifford Pinchot.
1875	September 10: American Forestry Association is formed.
1877	The Desert Lands Act authorizes sale of 640-acre tracts of arid public domain lands at $1.25 per acre upon proof of reclamation of lands by irrigation. Area is reduced to 320 acres in 1891.
1878	The Timber and Stone Act provides that anyone willing to pay $2.50 an acre and swear not to amass any parcel greater than 160 acres could acquire timbered public land that was unfit for agriculture. This law applied directly to California, as well as to Oregon, Nevada and the Territory of Washington.
	The Free Timber Act allows homesteaders to cut lumber on public land that heretofore had been used exclusively for mining.
1879	The U.S. Geological Survey (USGS) is established.
	Dr. Hough, commenting in his report on timberland devastation, states,: "…it is clearly evident that the absence of any provision tending to protection of future growths or to the prevention of waste will, in future time, be regarded with unavailing regret."
1881	The Division of Forestry is established in the USDA, with Dr. Hough still in charge.
1885	The California State Board of Forestry is created, which conducted the forestry business of the state for the next eight years until it was abolished in 1893.
1889	The Department of Agriculture is raised to cabinet status.
1890	In an attempt to save the "big trees," the General Grant and Sequoia national parks are set aside in California.
	The total population of California stands at approximately 1.2 million people, of whom 250,000 are rural residents.
1891	March 3: The Forest Reserve, or Creative Act (16 U.S.C. 471) is passed, giving the president power to establish forest reserves from forest and range lands in the public domain. This important piece of legislation slipped through Congress without question and without debate. It became the basis for the nation's national forest system and was the seed from which the national forests grew.
1892	Conservationist John Muir founds the Sierra Club.
	December 20: San Gabriel Timberland Reserve is created (later part of Angeles National Forest), the second reserve in the nation embracing more than 550,000 acres.
1893	President Harrison leaves office after creating fifteen reserves totaling 13 million acres.
	February 14: The 4-million-acre Sierra Forest Reserve is created, which encompassed the southern half of the Sierra Nevada.
	February 25: The Trabuco Canyon Forest Reserve (later part of Cleveland National Forest) and the San Bernardino Forest Reserve are created, totaling about 800,000 acres.
1894	Cary Act. Provides grants for reclamation of arid public lands of 1 million acres to states if settlers would occupy and irrigate 160 acres, with at least twenty acres under cultivation.

1897	February 22: President Cleveland creates thirteen forest reserves. National forest reserves are designated under DOI.
	February 22: San Jacinto Forest Reserve (later part of Cleveland National Forest) and the Stanislaus Forest Reserve are created.
	June 4: Congress passes the Organic Administration Act, which specifies the purposes for which forest reserves can be established, their administration and protection. The act allowed hiring employees to administer the forests and opened the reserves for use. Response to rapid and wasteful deforestation of western lands. Essentially, it opened up the forest reserves to use, and it cleared the road to sound administration, including the practice of forestry. Management was carried out by the DOI, GLO and USGS.
1898	March 2: The Pine Mountain Forest Reserve (later part of Los Padres National Forest), and the Zaca Lake Forest Reserve (later part of Los Padres National Forest) are created.
	July 1: Gifford Pinchot succeeds Bernard Fernow as chief of the Division of Forestry and is given the title of forester. When Pinchot became chief of the Division of Forestry (changed to the Forest Service in 1905), he requested that his title be changed from chief to "forester," as there were many chiefs in Washington, but only one forester. The forester title remained in effect until 1935, when the title "chief" was readopted.
1899	February 28: The Mineral Springs Act is passed, allowing the leasing of sites on forest reserves for health or pleasure.
	April 13: The Tahoe Forest Reserve is created.
	October 2: The Santa Inez Forest Reserve (later part of the Los Padres National Forest) is created.
1900	Greening of Imperial Valley through irrigation begins.
	The total population of California stands at approximately 1.5 million people, of whom 300,000 are rural residents.
1901	A new Division of Forestry is created in the GLO under the DOI.
1902	The Minnesota Forest Reserve is established. It was the first forest created by Congress and not by presidential proclamation.
	Reclamation Act. Establishes a system of water development projects for irrigation of arid lands and a evolving fund based on sale of public lands in Arizona, California, Nevada and other western states. Authorizes homesteading of up to 160 acres if lands are reclaimed through irrigation and the cost of water paid by each homesteader.
	The General Land Office begins major timber sale work on the reserves in California. The contracts contained twenty-five stipulations. A year later, timber sale contracts contained fire protection clauses and required brush piling for slash disposal.
1903	The first federal wildlife refuge in Florida is created.
	A joint state-federal survey report on forests and water situation in California is approved by Governor George C. Pardee and Gifford Pinchot and results in the California legislature creating the office of state forester and a new Board of Forestry.

	December 22: The Santa Barbara Forest Reserve is created (a combination of Pine Mountain, Zaca Lake and the Santa Inez forest reserves—later part of Los Padres National Forest).
1904	November 29: The Warner Mountains Forest Reserve (later part of the Modoc National Forest) and Modoc Forest Reserve (later part of the Modoc National Forest) are created.
1905	The first Forest Service manual, The Use Book, is published, which codifies laws, regulations and standards for administration and defines the purpose of the forest reserves: "Forest Reserves are for the purpose of preserving a perpetual supply of timber for home industries, preventing destruction of the forest cover which regulates the flow of streams, and protecting local residents from unfair competition in the use of forest and range…the prime object of the forest reserves is use."
	The California State Board of Forestry is reconstituted. The Forest Protection Act is passed, and E. T. Allen is appointed as first California state forester.
	The total acreage of forest reserves in California is approximately 14.4 million acres.
	February 1: Forest reserves are transferred from DOI to USDA, giving it full administrative control and responsibility over 63 million acres of public forest.
	March 3: An act is passed renaming the Bureau of Forestry as the Forest Service effective July 1, with Gifford Pinchot as chief forester.
	March 27: The Plumas Forest Reserve is created.
	April 26: The Trinity Forest Reserve is created.
	May 6: The Klamath Forest Reserve is created.
	June 2: The Lassen Peak Forest Reserve is created.
	July 24: The Diamond Mountain Forest Reserve is created.
	October 3: The Shasta Forest Reserve is created.
	October 3: Lake Tahoe Forest Reserve is changed to Tahoe Forest Reserve.
	November 11: The Yuba Forest Reserve is created.
1906	Pinchot organizes the forest reserves into three inspection districts.
	Coal lands are withdrawn from entry and patent under the General Mining Law of 1872.
	First step in revenue sharing is taken when Congress provides that 10 percent of money derived from the sale of timber or use fees go to states in which the income originated, for schools and roads. In 1908, the amount was raised to 25 percent.
	Yosemite Valley is returned by the state of California to the federal government for protection and administration and is added to Yosemite National Park.
	June 8: The American Antiquities Act is passed, authorizing protection of antiquities and features of scientific or historical interest on land owned or controlled by the government.
	June 11: The Forest Homestead Act is passed, allowing agricultural lands within forest reserves to be available for homesteading purposes.
	June 25: The San Luis Obispo and Monterey forest reserves are created.

	July 18: The Pinnacles Forest Reserve is created.
	September 17: The Yuba Forest Reserve is incorporated into the Tahoe Forest Reserve.
1907	The three inspection districts are reorganized into six districts. District 5's headquarters are in San Francisco.
	February 6: The Stony Creek Forest Reserve is created.
	March 4: All forest reserves are renamed national forests.
	May 25: The Inyo National Forest is created.
	October 26: The San Benito National Forest is created.
1908	December: California District 5 is created, with headquarters in San Francisco. Frederick E. Olmstead becomes first California District 5 forester. The formation of the District 5 office was the beginning of the development of forestry policies and procedures geared to California conditions. The primary responsibilities of the District 5 office were to provide regional direction to forest management programs on the national forests.
	The Office of Silviculture (1908-1919) of California District 5 establishes timber marking rules that provided for 200-foot scenic corridors along roads, lakes and river fronts. Marking was to be light and aimed at improving the appearance of the forest.
	July 1: The Angeles National Forest is established from San Bernardino National Forest and parts of the Santa Barbara and San Gabriel national forests.
	July 1: The San Luis Obispo National Forest is renamed the San Luis National Forest.
	July 1: Mono National Forest is established from parts of Inyo, Sierra, Stanislaus and Tahoe national forests.
	July 1: The Sequoia National Forest is established from the southern portion of the Sierra National Forest.
	July 1: Lassen Peak National Forest changed to Lassen National Forest.
	July 1: The California National Forest is established from parts of the Trinity and Stony Creek national forests.
	July 1: The Cleveland National Forest is established by consolidating Trabuco Canyon and San Jacinto national forests.
	July 1: Warner National Forest is incorporated into the Modoc National Forest.
	July 1: Diamond Mountain National Forest is incorporated into the Plumas and Lassen national forests.
	July 1: Pinnacles and San Benito national forests are incorporated into the Monterey National Forest.
1909	Homesteading acreage is reduced to 320 acres in dry farming lands where irrigation is not possible.
	The redwood lumber industry begins a reforestation program; the California Forest Protective Association is organized.
	February 18: The Calaveras Bigtree National Forest is created.
1910	Gifford Pinchot supports development of the Hetch Hetchy Reservoir near Yosemite National Park as a prime example of "full utilization" and is opposed by John Muir. This conflict raised the issue of the role of scenic quality and recreation within the spectrum of multiple use. The Sierra Club

withdrew support for the Forest Service management of national parks and pressed for the creation of a National Park Service. To offset this move, Chief Forester Henry Graves asked district foresters to identify areas in national forests that would be better suited to parks. Those areas would remain under Forest Service management. The Forest Service already managed the national parks and national monuments within the national forest system (NFS).

January 7: Gifford Pinchot is fired by President Taft and replaced by Henry S. Graves (1910-1920).

July 1: The San Luis National Forest is incorporated into the Santa Barbara National Forest.

July 1: The Kern National Forest is established from part of the Sequoia National Forest.

July 28: The Eldorado National Forest is established from part of the Tahoe and Stanislaus national forests.

1911   March 1: The Weeks Act authorizes the USDA Forest Service cooperative efforts with states to provide forest fire protection and purchase lands needed to regulate the flow of navigable streams or for production of timber. It directs that 25 percent of national forests' receipts be returned to the states to fund public schools and public roads in counties where the forests are located and leads to numerous additions to and eliminations of national forest lands. Forest boundaries are moved to ridgelines.

June: Frederick E. Olmsted, first California District 5 Forester (1908-1911), resigns.

July: Coert Du Bois is appointed second California District 5 forester.

1912   August: The Appropriations Act provides that 10 percent of all forest receipts from fiscal year 1912 be used for roads and trails within the national forests in the states from which the receipts came.

1914   August: War breaks out in Europe.

A forestry school is established at University of California.

1915   Based on strong support for recreational development, a Forest Service branch of recreation is established.

March 15: The Term Lease Law allows permits for stores, hotels, summer homes and other structures on national forests, not to exceed eighty acres and thirty years.

July 1: The Kern National Forest is transferred back to the Sequoia National Forest.

1916   The National Park Service (NPS) is created. Despite Chief Forester Graves' efforts to show that the Forest Service could manage both national forests and national parks, he could not overcome the concerns raised by the Hetch Hetchy controversy. Mistrust of "multiple use" led to withdrawing parks from Forest Service management and creation of the NPS.

The first Forest Service campground is constructed.

The Stock Raising Homestead Act Authorizes the sale of 640 acres of public domain land suitable only for grazing livestock, provided some range improvements are installed.

July: The Road Act aids states with construction of rural roads and enlarges the scope of the forest road program through 10 percent fund expenditures, in cooperation with local communities.

1917   The Forest Service hires a consulting landscape architect to prepare a national study of recreation uses on the national forests.

April: America declares war against Germany and the Central Powers.

1918	November 11: The armistice with Germany is signed, ending World War I.
1919	The Forest Service hires its first full-time landscape architect, Arthur Carhart, who develops the idea of wilderness and primitive areas of very limited development that exclude roads and summer homes.

February: The Appropriations Act provides funding for direct construction, without local cooperation, of such roads and trails as necessary for national importance.

August 18: The Monterey National Forest is incorporated into the Santa Barbara National Forest.

November: Coert Dubois, second California District 5 Forester (1911-1919), resigns.

November: Paul G. Redington is appointed third California District 5 Forester (1919-1926).

1920	The Federal Water Power Act creates the Federal Power Commission and provides for the improvement of navigation, the development of waterpower and the use of public lands in relations to it.

The Mineral Leasing Act removes phosphate, sodium, potassium, native asphalt, sulfur and fuel minerals from location under the Mining Law of 1872. Such deposits are subject to exploration and disposal only through a prospecting permit and leasing system. The act specifies royalty rates, lease size and lease term for each kind of leasable mineral.

Aldo Leopold, in an article for the Journal of Forestry, suggests establishing a wilderness of at least 500,000 acres in each of the eleven western states.

The Forest Service and National Park Service agree to review national forests for sites best suited to transfer to national parks. The Forest Service also agrees to protect entrance areas to parks situated within national forests and to take special care with respect to logging and grazing near parks.

The District 5 Office of Silviculture is changed to Branch of Forest Management (1920-1935).

March: Forester Henry S. Graves (1910-1920) resigns. William B. Greeley (1920-1928) is named chief forester.

1921	The Federal Highway Act provides that the United States aid the states in construction of rural post roads, and for other purposes.

Arthur Carhart pushes for more commitment to recreation in the Forest Service, and resigns the following year in frustration from what he viewed as lack of support for recreation and scenic values by the Forest Service. Thereafter, the Forest Service operated without a full-time landscape architect until 1933.

1922	March 20: The General Land Exchange Act allows the exchange of tracts of federal land in national forests for private land within forest boundaries when of equal value (not necessarily of equal acreage).
1924	The Forest Service head of engineering endorses the use of scenic strips to screen harvested areas.

June 3: The first administrative wilderness area (574,000 acres) is established on the Gila National Forest in New Mexico.

June 7: The Clarke-McNary Act provides for the protection of forest lands, for the reforestation of denuded areas, for the extension of national forests and for other purposes, in order to promote the continuous production of timber on lands chiefly suitable there for. Also expanded the 1911 Weeks Act's authority for federal-state cooperation in fire protection and forestry efforts and allows purchases of forestlands in watersheds, not just the headwaters of navigable streams.

1925	The Eddy Tree Breeding Station, now known as Institute of Forest Genetics, Placerville, is established.

September 30: The San Bernardino National Forest is re-established from parts of the Angeles and Cleveland national forests.

1926 Forest Chief William B. Greeley orders an inventory of all undeveloped national forest lands larger than 230,400 acres (10 townships) in order to withhold these areas from unnecessary road building and forms of special use of a commercial character that would impair their wilderness character.

Stuart Bevier Show is appointed fourth California District 5 forester (1926-1946).

July 1: The California Forest Experiment Station is established in Berkeley, California, with Edward I. Kotok as director.

1928 Robert Y. Stuart is named chief forester (1928-1933).

The Woodruff-McNary Act provides money for additional land purchases.

The McSweeney-McNary Act establishes a ten-year forestry research program and survey of forestry resources and establishes regional experiment stations.

1929 May 1: Forest Service "districts" are renamed "regions" to avoid confusion with ranger districts.

1930 The Knutson-Vanderburg Act requires purchasers of national forest timber to deposit money to cover the cost of replanting harvested areas and removing undesirable trees from the areas.

The Forest Service issues Regulation L-20, which defines management priorities for primitive areas.

1931 Forest Chief Robert Stuart calls on forest supervisors to rate timber, watershed, grazing and recreation resources on their forests, in order of present and future importance.

The California Forest Experiment Station changes its name to the California Forest and Range Experiment Station.

1932 July 12: The California National Forest is renamed the Mendocino National Forest.

1933 A National Plan for American Forestry. The Forest Service sends the Copeland Report to the Senate (in March), calling for a comprehensive management plan for the national forests, including plans for trails, recreation facilities, administrative facilities and lookouts.

Chief Forester Stuart dies in office, and Ferdinand A. Silcox (1933-1939) becomes his successor.

April 5: The Office of Emergency Conservation Work (ECW) is established. It is later called Civilian Conservation Corps (CCC).

April 17: The first ECW or CCC camp is established on the George Washington National Forest near Luray, Virginia.

May 12: The Federal Emergency Relief Administration (FERA or ERA) is established, which includes a works division that later became the WPA.

1934 June 28: The Taylor Grazing Act establishes grazing districts on 80 million acres of public domain for use of the livestock industry, and grazing permits are issued within each district. Intended to conserve grazing lands from overgrazing. Consists of unreserved and unappropriated lands in 10 western states and Alaska. Essentially ends unregulated grazing on national forests.

The California Region hires David Muir as a landscape engineer in its Division of Engineering to work on roadside treatments.

1935 The Works Progress Administration is created from the works division of FERA.

President Roosevelt withdraws public lands from public entry. End of the Homestead Act of 1862.

The Wilderness Society is formed by Bob Marshall, Aldo Leopold and Arthur Carhart.

The Region 5 Branch of Forest Management is changed to the Division of Timber Management (1935-1974).

April 8: The Emergency Relief Appropriations (ERA) Act is passed, permitting funding and operation of ECW or CCC camps.

1936   The Omnibus, or Upstream, Flood Control Act (1936) recognizes that flood control is a national rather than a local problem.

December 3: The Santa Barbara National Forest is renamed the Los Padres National Forest.

1937   The Bankhead-Jones Farm Tenant Act authorizes federal purchase of privately- owned farmlands no longer capable of producing sufficient income. Owners and family were relocated and the submarginal lands retired from agricultural use. Eventually about 20 million acres were purchased, with 20,000 acres in California.

The California Region hires six forest landscape architects to work with CCC program.

June 28: Emergency Conservation Work is officially renamed Civilian Conservation Corps.

1938   The Small Tract Act authorizes sale or lease to U.S. citizens of tracts not exceeding five acres of public domain lands for use as home cabin, camp, recreation or business sites. Mineral rights were reserved. The act did not apply to national forests.

1939   Chief Forester Silcox dies in office, and Earle H. Clapp (1939-1943) is made acting chief.

September: Outbreak of World War II in Europe.

1940   The U.S. Fish and Wildlife Service is created.

1941   The War Production Board undertakes to coordinate the procurement programs of the armed forces and to allocate materials between military and civilian needs.

December 7-11: The United States declares war against Japan, Germany and other Axis Powers.

1942   Under Executive Order 9066, Japanese-Americans are removed to relocation centers.

June 30: The CCC is eliminated.

1943   Lyle F. Watts (1943-1952) is named chief, replacing acting chief Clapp.

1944   The Sustained Yield Forest Management Act authorizes the Forest Service to enter into long-term, noncompetitive contracts with local lumber mills in timber-dependent communities to assure a continuous supply of wood products.

1945   Smokey Bear is officially introduced as the symbol of fire prevention.

The California Forest Practice Act divides California into four forest districts: redwood, north Sierra pine, south Sierra pine, and coastal range pine and fir. The central purpose of the act is to conserve and maintain the productivity of the timberlands in the interests of the economic welfare of the state and continuance of the forest industry.

May 7:   Germany formally surrenders.

	July 1: The Mono National Forest is incorporated into the Inyo and Toiyabe national forests.
	September 2: Japan formally surrenders.
1946	Stuart Bevier Show retires. Perry A. Thompson is appointed fifth California District 5 forester (1946-1950)
	The GLO and the Division of Grazing, both in the DOI, are combined to form the Bureau of Land Management (BLM).
	Reorganization Plan #3 transfers mineral leasing authority for minerals in acquired lands (including national forests and grasslands) from the Department of the Agriculture to the Department of the Interior.
	January 1: The Central Sierra Snow Laboratory, Soda Springs, California is established.
1947	The Mineral Leasing Act for Acquired Land states that fertilizer and fossil fuel minerals on acquired lands are subject to permit and lease by the secretary of the interior, with consent from the surface managing agency, to ensure that land is utilized for the purposes for which it was acquired.
	The Materials Act authorizes disposal of such materials as sand, grave, stone and common clay from public lands through a sales system.
	The Forest Pest Control Act directs the secretary of agriculture to take measures to prevent, retard, control, suppress or eradicate incipient, threatening, potential or emergency outbreaks of forest insect pests and tree diseases.
	June 3: Six Rivers National Forest established from parts of Klamath, Siskiyou, and Trinity National Forests.
1950	Perry A. Thompson retires. Clare W. Hendee is appointed sixth California District 5 forester (1950-1955).
	April 24: The Granger-Thye Act, upholds Forest Service authority to regulate and collect grazing fees.
1952	Chief Lyle F. Watts resigns, and Richard E. McArdle (1952-1962) is named chief.
1953	The Sierra Club opposes Forest Service plans to harvest a Jeffrey pine stand at Deadman Creek on the Inyo National Forest.
1954	The Multiple Mineral Development Act reserves the use of the surface area of mining claims to the public land agency for the purpose of managing non-minerals surface resources so long as such management does not materially interfere with legitimate mining operations.
1955	Charles Arthur Connaughton is appointed seventh California District 5 forester (1955-1967).
	Golden Anniversary of the Forest Service.
	There is further opposition to planned harvests on the Kern Plateau in the Inyo and Sequoia national forests.
	A timber resource review of Forest Service reveals timber growth exceeds removal in the nation's forests for the first time.
	July 23: The Multiple Use Mining Act establishes a program to examine all national forest lands for mining claims and to resolve occupancies on invalid mining claims.
1957	A wilderness bill is introduced by Senator Hubert H. Humphrey.
	The Forest Service follows the lead of the National Park Service's Mission 66 program by launching "Operation Outdoors," a five-year expansion and renovation plan for recreation facilities. In conjunction

with this, many landscape architects were hired over the next five years.

The California Department of Water Resources sires the California Water Plan.

1958   The Townsite Act allows conveying 640 acres per application to towns near national forests where no other national forest use is overriding.

1959   The California Water Plan (CWP) envisions a network of dozens of reservoirs, pumping stations and electrical generating plants, linked by thousands of miles of aqueducts and pipelines. The CWP's central objective was to impound the runoff of the Sacramento River tributaries to transport water through a north-south artery to irrigate the dry western and southern San Joaquin Valley, and then to pump it over the Tehachapi Mountains into Southern California.

The California Forest and Range Experiment Station is renamed the Pacific Southwest Forest and Range Experiment Station, which is later shortened to Pacific Southwest Research Station.

1960   The Multiple-Use and Sustained-Yield Act authorizes and directs the secretary of the agriculture to develop and administer the renewable resources on the national forests (outdoor recreation, range, watershed, timber, wildlife and fish) for multiple use and sustained yield of the several products and services obtained there from.

Recreation is explicitly recognized as one of the multiple uses for which national forests should be managed.

The Mining Act is based on an inventory and analysis of the surface resources of national forests in which the Forest Service expanded multiple-land classification and management efforts.

1962   Edward P. Cliff (1962-1972) is named chief.

1964   September 3: The Wilderness Act establishes a National Wilderness Preservation System to protect and preserve the primeval character of selected areas. Environmental concerns gain increasing national attention.

Mining is precluded on designated wilderness lands after 1983.

1966   The Landscape Management and Clear-cutting Issue: The California Region issues guidelines entitled "Handling the Impacts of Patch Cutting." Categories of view-shed are being developed—near view, far view, near natural appearance.

1967   Charles Arthur Connaughton retires. John W. Deinema is appointed eighth California District 5 forester (1967-1970)

The Landscape Management and Clear-cutting Issue. The California Region issues detailed direction to preserve scenic quality, stating, " Forest resources are to be managed to provide protection of scenic values."

September 11: The Forest Fire Laboratory, Riverside, California, is established.

October 3: The Institute of Pacific Islands Forestry, Honolulu, is established.

1968   The Wild and Scenic Rivers Act preserves free-flowing rivers that possess outstandingly remarkable wild, scenic, recreational and similar values.

Land within a quarter mile of the bank of any wild river segment is withdrawn from mineral exploration and location.

The National Trails Act institutes a national system of recreation and scenic trails.

The timber industry presses Congress for a Timber Supply Bill that recognizes the dominant role of timber in national forest management.

Studies begin on a Walt Disney-proposed winter sports development at Mineral King on the Sequoia National Forest. This study was among the first to use computerized analysis for visual resources.

1969 Civil rights and minorities are a major concern, culminating in 1969 when Native Americans take over Alcatraz Island in protest.

1970 Douglas R. Leisz appointed ninth California District 5 forester (1970-1978)

The Geothermal Steam Act authorizes the leasing of geothermal resources along with associated byproducts from public lands through noncompetitive and competitive leasing systems.

California becomes the most urban state in the nation as well as most populous.

January 1: The National Environmental Policy Act is signed into law, requiring evaluation of the environmental impact of federally funded projects and programs and generally requiring an environmental assessment and/or an environmental impact statement be submitted to the federal government before a project can begin.

February: Forest Service issues Framework for the Future: Forest Service Objectives and Policy Guides.

May: Senator Jennings Randolph of West Virginia calls for the Council on Environmental Quality (CEQ) to evaluate clear-cutting on the Monongahela National Forest.

1971 The Wild Horses and Burros Protection Act protects and manages wild, free-roaming horses and burros as components of the public lands.

The slogan "Give a Hoot! Don't Pollute" is created, and Woodsy Owl is introduced as a symbol of anti-pollution.

January: Senator W. J. McGee of Wyoming introduces the Moratorium on Clear-cutting Bill.

May: The California Region holds an environmental conference in San Diego with Sierra Club, timber industry, private landscape architects and engineers to develop a visual resource management system.

1972 John R. McGuire (1972-1979) is named chief.

The Coastline Initiative authorizes a commission to regulate all coast development, which is considered a major conservation development in California history.

A class-action lawsuit is filed against the Forest Service (Gene Bernardi et al v.. Earl Butz). The suit alleges sex discrimination in both hiring and promotional practices of the agency in California.

March: A Senate subcommittee led by Senator Church issues its report "Clear-cutting on Federal Timberlands."

June: The Forest Service agrees to the Senate subcommittee's report.

1973 In Izaak Walton v. Butz, the U.S. District Court rules that clear-cut logging on the Monongahela National Forest is contrary to the Organic Act of 1897, which stated that only "dead, physically mature, and large growth trees" individually marked for cutting could be sold.

The Threatened and Endangered Species Act is aimed at ensuring the survival of all native species of fish and wildlife, and authorizes survival programs for those species threatened with extinction.

Mineral exploration and development is not to jeopardize the existence of fish and wildlife in danger of extinction.

April: The California Region begins visual resource management training for employees.

1974 The Forest and Rangeland Renewable Resources Planning Act (RPA) authorizes long-range planning by the Forest Service to ensure future supply of forest resources and maintain a quality environment. The RPA requires that a renewable resources assessment and a Forest Service program be prepared every ten and five years, respectively, to plan and prepare for our national resource future. The RPA requires inventories and assessments of the wood, water, wildlife and fish, forage, and outdoor recreation available on private and public lands.

The RPA provides criteria for clear-cutting similar to the Church guidelines, which in turn had borrowed heavily from the CEQ proposals of 1971.

The Forest Service publishes Title 36 Code of Federal Regulations, Section 252., which is concerned with the use of the surface of national forest system lands by anyone operating under the Mining Law of 1872 and minimizes adverse environmental impacts on national forest system resources. Each plan requires an environmental assessment and that a determination be made whether an environmental impact statement is required. The plan may be modified to mitigate environmental impacts, and once it is approved, the operator is expected to comply with all the environmental laws that apply.

1975 The State of California passes the Surface Mining Reclamation Act (SMARA), whose function is to locate, classify and designate minerals so that they can be protected and considered in planning and to provide for reclamation plans.

The Region 5 Division of Timber Management changed to the Timber Management Staff.

1976 The National Forest Management Act (NFMA) restates the Forest Service's commitment to responsible use of natural resources, sets guidelines for timber management, requires prompt reforestation of lands, assures public participation in the planning and management of the national forest system, formalizes the RPA land managment planning process and sets a 1985 deadline for completion of national forest plans. It repeals the Homestead Act of 1862.

NFMA provides for forest planning and sets standards for clear-cutting that ensure "cut blocks, patches, or strips are shaped and blended to the extent practicable with the natural terrain… and be carried out in a manner consistent with the protection of soil, watershed, fish, wildlife, recreation, and esthetic resources and the regeneration of the timber resource."

May 21: The Redwood Sciences Laboratory, Arcata, California, is established.

1977 Work begins on writing regulations for NFMA. Region 5 develops visual quality index (VQI) to respond to visual quality concerns in the act.

1978 Zane G. Smith Jr. is appointed tenth California District 5 forester (1978-1987).

1979 R. Max Peterson (1979-1987) is named chief.

February: The Forest Service and BLM sign a memorandum of agreement (MOA) with the State of California Resources Agency. Under this MOA, submission and approval of a reclamation plan leading toward a permit is required to conduct surface mining operations involving removal of overburden in the amount of 1,000 cubic yards or more at any one location of one acre or less.

	September 4: The Forestry Sciences Laboratory, Fresno, California, is established.
1980	Region 5 integrates a visual quality management system into FORPLAN for use in forest planning and establishes direction on the existing visual condition (EVC) inventory and requirements on the use of the use of the VQI.
1981	A proposed settlement for the Gene Bernardi et al. v. Earl Butz class-action lawsuit is filed against the Forest Service and is issued as a consent decree (No. C-73-11110 SC). On April 3, 1991, a final settlement was approved between the court and the affected parties. Under the terms of the consent decree, Region 5 and the Pacific Southwest Station were required to establish certain hiring goals and affirmative action requirements.
1983	The Small Tracts Act provides for sale or exchange of NFS land less than $150,000 in value and amounting to no more than 10 acres to settle boundary and other minor land line problems.
1987	F. Dale Robertson (1987-1993) is named chief.
	Paul F. Baker is appointed eleventh California District 5 forester (1987-1990).
	August 31: The Silviculture Laboratory, Redding, California, is established.

Appendix B:

USDA Forest Service
Pacific Southwest Region
Administratively known as Region 5

Forest Service Leadership in California from 1908–2005

District 5 Foresters — Years Served

Frederick E. Olmsted	1908-1911
Coert DuBois	1911-1919
Paul G. Redington	1919-1926

Title changed to Regional Forester in 1929

Stuart Bevier Show	1926-1946
Perry A. Thompson	1946-1950
Clare W. Hendee	1951-1955
Charles A. Connaughton	1955-1967
John W. Deinema	1967-1970
Douglas R. Leisz	1970-1978
Zane G. Smith, Jr.	1978-1987
Paul F. Barker	1987-1990
Ronald E. Stewart	1991-1995
G. Lynn Sprague	1995-1998
Bradley E. Powell	1998-2001
Jack A. Blackwell	2001-2005
Bernard Weingardt	2005-present

Appendix C:

District/Regional Office Locations

California District 5	First National Bank Building Post and Montgomery Streets San Francisco, California	December 1908
	Adams-Grant Building 114 Sansome Street San Francisco, California	April 1914
	Ferry Building Market Street and The Embarcadero San Francisco, California	July 1920
California Region 5	Wells Fargo Building Second and Mission Streets San Francisco, California	December 1933
	Wentworth Smith Building (Temporary Location) 45 Second Street San Francisco, California	April 1935
	Phelan Building 760 Market Street San Francisco, California	June 1935
	U.S. Appraisers Building (Appraisers Stores and Immigration Station) 630 Sansome Street San Francisco, California	August 1944
Pacific Southwest Region	U.S. Appraisers Building (Appraisers Stores and Immigration Station) 630 Sansome Street San Francisco, California	1959
Pacific Southwest Region	Mare Island 1323 Club Drive Vallejo, California	1999 - Present

References Cited

Chapter I: Prehistory to 1889

Bigelow, R.L.P.
    1926   "Tahoe National Forest." Pacific Southwest Archives, Pacific Southwest Region, USDA Forest Service, Vallejo, California.

Blackburn, Thomas and Kat Anderson
    1993   *Before the Wilderness: Environmental Management by Native Californians.* Menlo Park, California: Ballena Press.

Brown, William S.
    1945a   "History of the Los Padres National Forest, Compiled by William S. Brown, Senior Information Specialist." San Francisco, California. Pacific Southwest Archives, Pacific Southwest Region, USDA Forest Service, Vallejo, California.

    1945b   "History of the Modoc National Forest, United States Forest Service, California Region." San Francisco, California. Pacific Southwest Archives, Pacific Southwest Region, USDA Forest Service, Vallejo, California.

Burlingame, Merrill G.
    1977   "Lansford Warren Hastings." In *The Reader's Encyclopedia of the American West.* Edited by Howard R. Lamar. New York: Thomas Y. Crowell Company.

Caughey, John W.
    1973   "The Californian and His Environment." In *Essays and Assays: California History Reapprised.* Edited by George H. Knoles. San Francisco: California Historical Society.

Cermak, Robert W.
    n.d.   ""Fire in the Forest: A Narrative History of Forest Fire Control in the California National Forests, 1898-1956."" Public Affairs & Communication Office, Pacific Southwest Region, USDA Forest Service, Pacific Southwest Region, Vallejo, California.

Clar, C. Raymond
    1959   *California Government and Forestry: From Spanish Days until the Creation of the Department of Natural Resources in 1927.* Sacramento: Division of Forestry, Department of Natural Resources, State of California.

Cook, Cheryl A.
    1991   "List of Important Dates in Conservation History." Pacific Southwest Archives, Pacific Southwest Region, USDA Forest Service, Vallejo, California.

Conners, Pamela Ann
    1989   "Patterns and Policy of Water and Hydroelectric Development on the Stanislaus National Forest, 1850-1920." Master's thesis. Public History Department, University of California, Santa Barbara.

1992	"Influence of the Forest Service on Water Development Patterns in the West." In *Origins of the National Forests: A Centennial Symposium*. Edited by Harold K. Steen. Durham, North Carolina: Forest History Society.

Dassman, Raymond F.
1999	"Environmental Changes Before and After the Gold Rush." In *A Golden State: Mining and Economic Development in Gold Rush California*. Edited by James J. Rawls and Richard J. Orsi. Berkeley: University of California Press.

Dupree, A. Hunter
1957	*Science in the Federal Government: A History of Policies and Activities to 1940*. Cambridge, Massachusetts: Belknap Press of Harvard University Press.

Friedhoff, William H. (Ole)
1944	"Mining Activities in National Forests of California." Pacific Southwest Archives, Pacific Southwest Region, USDA Forest Service, Vallejo, California.

Gates, Gerald R.
n.d.	"Cultural Resource Overview: Modoc National Forest." Alturas, California: Modoc National Forest, Pacific Southwest Region, USDA Forest Service.

Gibson, Arrell Morgan
1980	*The American Indian: Prehistory to the Present*. Lexington, Massachusetts: D.C. Heath and Company.

Godfrey, Anthony
1994	"Historic Resource Study: Pony Express National Historic Trail." Denver, Colorado: National Park Service, U.S. Department of the Interior.
2003	"No Longer in the Shadows of the Tetons: Administrative Facilities of the Caribou-Targhee National Forest, 1891-1960." FS Report No. TG-CB-03-722. Salt Lake City, Utah: U.S. West Research, Inc.

Gutierrez, Ramon A. and Richard J. Orsi, eds.
1998	*Contested Eden: California Before the Gold Rush*. Berkeley: University of California Press.

Heizer, Robert F.
1978	*California Indians*. Volume 9 of *Handbook of North American Indians*. General editor William C. Sturtevant. Washington D.C.: Smithsonian Institution.

High, James
1951	"Some Southern California Opinion Concerning Conservation of Forests, 1890-1905." *Historical Society of Southern California Quarterly* Vol. XXXIII (December): 291-312.

Jackson, W. Turrentine, Rand Herbert, and Stephen Wee
n.d.	"History of the Tahoe National Forest, 1840-1940: A Cultural Resources Overview History." David, California: Jackson Research Projects. Pacific Southwest Archives, Pacific Southwest Region, USDA Forest Service, Vallejo, California.

Johnston, James and Elizabeth Budy
    1982   "Cultural Resource Management Overview : Lassen National Forest." Volume One: Cultural History. Pacific Southwest Archives, Pacific Southwest Region, USDA Forest Service, Vallejo, California.

Lewis, Henry T.
    1973   *Patterns of Indian Burning in California: Ecology and Ethnohistory.* Ramona, California: Ballena Press.

Limerick, Patricia Nelson
    1987   *The Legacy of Conquest: The Unbroken Past of the American West.* New York: W.W. Norton & Company.

Lockmann, Ronald F.
    1981   *Guarding the Forests of Southern California: Evolving Attitudes Toward Conservation of Watershed, Woodlands, and Wilderness.* Glendale, California: The Arthur H. Clark Company.

McDonald, James A.
    1979   "Cultural Resource Overview: Klamath National Forest, California." Klamath National Forest, USDA Forest Service. Pacific Southwest Archives, Pacific Southwest Region, USDA Forest Service, Vallejo, California.

Palmer, Kevin
    1992   "Vulcan's Footprints on the Forest." In *Origins of the National Forests: A Centennial Symposium.* Edited by Harold K. Steen. Durham, North Carolina: Forest History Society.

Preston, William
    1998   "Serpent in the Garden: Environmental Change in Colonial California." In *Contested Eden: California Before the Gold Rush.* Edited by Ramon A. Gutierrez and Richard J. Orsi. Berkeley: University of California Press.

Pyne, Stephen J.
    1982   *Fire in America: A Cultural History of Wildland and Rural Fire.* Princeton, New Jersey: Princeton University Press.

Rawls, James J. and Richard J. Orsi, eds.
    1999   *A Golden State: Mining and Economic Development in Gold Rush California.* Berkeley: University of California Press.

Rice, Richard B., William A. Bullough and Richard J. Orsi
    1996   *The Elusive Eden: A New History of California.* 2nd Edition. New York: McGraw Hill Companies, Inc.

Robinson, Douglas
    1943   "Inyo National Forest." Pacific Southwest Archives, Pacific Southwest Region, USDA Forest Service, Vallejo, California.

Rolle, Andrew
    1973   "Brutalizing the California Scene." In *Essays and Assays: California History*

Reapprised. Edited by George H. Knoles. San Francisco: California Historical Society.

Rose, Judy   Personal communication. 17 May 2004.

Rowell, Galen
- 2002   *California the Beautiful.* New York: Welcome Enterprises, Inc.

Savage, Christine E.
- 1991   "Six Rivers National Forest: A Contextual Cultural Resources Chronology of Events On or Near Forest Lands." Pacific Southwest Archives, Pacific Southwest Region, USDA Forest Service, Vallejo, California.

Shipek, Florence Connolly
- 1987   *Pushed into the Rocks: Southern California Indian Land Tenure, 1769-1986.* Lincoln, Nebraska: University of Nebraska Press.

Spence, Mary Lee
- 1977   "John Charles Fremont." In *The Reader's Encyclopedia of the American West.* Edited by Howard R. Lamar. New York: Thomas Y. Crowell Company.

Stewart, Omer C.
- 2002   *Forgotten Fires: Native Americans and the Transient Wilderness.* Edited by Henry T. Lewis and M. Kat Anderson. Norman, Oklahoma: University of Oklahoma Press.

Strong, Douglas H.
- 1973   *These Happy Grounds: A History of the Lassen Region.* Red Bluff, California: Walker Lithograph, Inc.

Supernowicz, Dana Edward
- 1983   "Historical Overview of the Eldorado National Forest." Master's thesis. History Department, University of California, Irvine.

USDA, Forest Service
- 1952   *Highlights in the History of Forest Conservation.* Washington, D.C.: Superintendent of Documents, Government Printing Office.

Watkins, T.H.
- 1973   *California: An Illustrated History.* New York: Weathervane Books.

Williams, Gerald W.
- 2000   *The USDA Forest Service—The First Century.* FS-650. Washington, D.C.: USDA Forest Service.

Chapter II: 1890-1904

Anonymous
    n.d.   "Sierra National Forest: The Story of Its Beginning as a Forest Reserve; Its Growth, Management, Difficulties, and Relation to People, ..." Pacific Southwest Archives, Pacific Southwest Region, USDA Forest Service, Vallejo, California.

Ayres, R. W.
    1911   "History of the Stanislaus." Box 1: Historical Manuscripts, 1902-1913, Record Group 95, National Archive Record Center, San Bruno, California.

    1938   "Origin of Name and Date of Creation of the National Forests of the California Region." Box 3: Selected Alphabetical Subject Files, 1909-1974, Record Group 95, National Archive Record Center, San Bruno, California.

Bean, Walton
    1968   *California: An Interpretive History.* New York: McGraw-Hill Book Company.

Brown, William S., compiler
    1945a   "History of Angeles National Forests." San Francisco, California: California Region, United States Forest Service.

    1945b   "History of the Los Padres National Forest." San Francisco, California: California Region, United States Forest Service.

    1945c   "History of the Modoc National Forest." San Francisco, California: California Region, United States Forest Service.

Browning, Peter, compiler and editor
    1988   *John Muir in His Own Words: A Book of Quotations.* LaFayette, California: Great West Books.

Buck, John M.
    1974   "History of the Division of Timber Management." San Francisco, California: Region 5, U.S. Forest Service.

Cermak, Robert W.
    n.d.   "Fire in the Forest: A Narrative History of Forest Fire Control in the California National Forests, 1898-1956." Public Affairs & Communication Office, Pacific Southwest Region, USDA Forest Service, Pacific Southwest Region, Vallejo, California.

    1994   "Range of Light-Range of Darkness: The Sierra Nevada, 1841-1905." Revised October 1994. Public Affairs & Communication Office, Pacific Southwest Region, USDA Forest Service, Pacific Southwest Region, Vallejo, California.

Clar, C. Raymond
    1959   *California Government and Forestry: From Spanish Days until the Creation of the Department of Natural Resources in 1927.* Sacramento: Division of Forestry, Department of Natural Resources, State of California.

Cleland, Robert Glass
    1962   *From Wilderness to Empire: A History of California.* New York: Alfred A. Knopf.

Clepper, Henry, editor
    1971   *Leaders of American Conservation.* New York: The Ronald Press Company.

Dupree, A. Hunter
    1957   *Science in the Federal Government: A History of Policies and Activities to 1940.* Cambridge, Massachusetts: Belknap Press of Harvard University Press.

Gates, Gerald R.
    n.d.   "Cultural Resource Overview Modoc National Forest." Alturas, California: Modoc National Forest, Pacific Southwest Region, United States Department of Agriculture.

Hays, Samuel P. Hays
    1969   *Conservation and the Gospel of Efficiency: The Progressive Conservation Movement, 1890-1920.* New York: Atheneum.

High, James
    1951   "Some Southern California Opinion Concerning Conservation of Forests, 1890-1905." *Historical Society of Southern California Quarterly* Vol. XXXIII (December): 291-312.

Jackson, W. Turrentine, Rand Herbert and Stephen Wee
    n.d.   "History of the Tahoe National Forest, 1840-1940: A Cultural Resources Overview History." Davis, California: Jackson Research Projects. Pacific Southwest Archives, Pacific Southwest Region, USDA Forest Service, Vallejo, California.

Lockmann, Ronald F.
    1981   *Guarding the Forests of Southern California: Evolving Attitudes Toward Conservation of Watershed, Woodlands, and Wilderness.* Glendale, California: The Arthur H. Clark Company.

Lux, Linda
    2004   Personal communication, 7 October 2004.

Muhn, James
    1992   "Early Administration of the Forest Reserve Act: Interior Department and General Land Office Policies, 1891-1897." In *Origins of the National Forests: A Centennial Symposium.* Edited by Harold K. Steen. Durham, North Carolina: Forest History Society.

Pisani, Donald J.
    1977   "Lost Parkland: Lumbering and Park Proposals in the Tahoe-Truckee Basin." *Journal of Forest History* Volume 21 (No. 1): 4-17.

Rice, Richard B., William A. Bullough and Richard J. Orsi
    1996   *The Elusive Eden: A New History of California.* 2nd Edition. New York: McGraw Hill Companies, Inc.

Robbins, William G.
    1985   *American Forestry: A History of National, State & Private Cooperation.* Lincoln, Nebraska: University of Nebraska Press.

Robinson, Glen O.
    1975   *The Forest Service: A Study in Public Land Management.* Baltimore, Maryland: The John Hopkins University Press.

Robinson, John W.
    1991   *The San Gabriels: The Mountain Country from Soledad Canyon to Lytle Creek.* Arcadia, California: Big Santa Anita Historical Society.

Robinson, John W. and Bruce D. Risher
    1993   *The San Jacintos: The Mountain Country from Banning to Borrego Valley.* California: Big Santa Anita Historical Society.

Rolle, Andrew F.
    1963   *California: A History.* New York: Thomas Y. Crowell Company.

Rose, Gene
    1993   *Sierra Centennial: One Hundred Years of Pioneering on the Sierra National Forest.* Clovis, California: Sierra National Forest.

Smith, Darrell Hevenor
    1930   *The Forest Service: Its History, Activities and Organization.* Washington, D.C.: The Brookings Institution.

Strong, Douglas H.
    1967   "The Sierra Forest Reserve: The Movement to Preserve the San Joaquin Valley Watershed." *California Historical Society Quarterly* Vol. XLVI (March 1967): 3-18.

    1981   "Preservation Efforts at Lake Tahoe, 1880 to 1980." *Journal of Forest History* Volume 25 (No. 2): 78-97.

Watkins, T. H.
    1973   *California: An Illustrated History.* New York: Weathervane Books.

Who Was Who
    1962   *Who Was Who in America: A Companion Volume to Who's Who in America.* Volume 1: Biographies of the Non-Living with Dates of Deaths Appended. 5th Printing. Chicago: A. N. Marquis Company.

Williams, Gerald W.
    2000   *The USDA Forest Service—The First Century.* FS-650. Washington, D.C.: USDA Forest Service.

Chapter III: 1905-1911

Abbey, Robert Harvey
    1968   "Early Day Experiences in the U.S. Forest Service." Part I (1905-1920). Public Affairs & Communication Office, Pacific Southwest Region, USDA Forest Service, Pacific Southwest Region, Vallejo, California.

Anonymous
    n.d.   "Sierra National Forest: The Story of It's Beginning as a Forest Reserve." Pacific Southwest Archives, Pacific Southwest Region, USDA Forest Service, Vallejo, California.

Ayres, R. W.
    1938   "Origin of Name and Date of Creation of the National Forests of the California Region." Box 3: Selected Alphabetical Subject Files, 1909-1974, Record Group 95, National Archive Record Center, San Bruno, California.

    1941   "History of the Forest Service in California." Box 4: Selected Alphabetical Subject File, 1909-1974, Record Group 95, National Archive Record Center, San Bruno, California.

    1958   "History of Timber Management in the California National Forests, 1850 to 1937." San Francisco: Forest Service, U.S. Department of Agriculture. Pacific Southwest Archives, Pacific Southwest Region, USDA Forest Service, Vallejo, California.

Bachman, Earl E.
    1967   *Recreation Facilities…A Personal History of Their Development in the National Forests of California.* Berkeley, California: Pacific Southwest Forest and Range Experiment Station, California Region, Forest Service, U.S. Department of Agriculture.

Boyd, Anne
    1995   "A Brief History of the U.S. Forest Service Cultural Resources Management Program: A Look at the Laws and Two Forest Perspectives." Pacific Southwest Archives, Pacific Southwest Region, USDA Forest Service, Vallejo, California.

Brown, William S., cCompiler
    1945a   "History of Angeles National Forests." San Francisco, California: California Region, United States Forest Service. Public Affairs & Communication Office, Pacific Southwest Region, USDA Forest Service, Pacific Southwest Region, Vallejo, California.

    1945b   "History of the Los Padres National Forest, Compiled by Wm. S. Brown, Senior Information Specialist." San Francisco, California. Pacific Southwest Archives, Pacific Southwest Region, USDA Forest Service, Vallejo, California.

Buck, John M.
    1974   "History of the Division of Timber Management." San Francisco, California: Region 5, U.S. Forest Service.

California District News Letter
    1924   "Letters of Appreciation." Volume V, No. 26.

    1925   "Frederick Erskine Olmsted, 1872-1925." Volume VI, No. 8.

Cermak, Robert W.
    n.d.   "Fire in the Forest: A Narrative History of Forest Fire Control in the California National Forests, 1898-1956." Public Affairs & Communication Office, Pacific Southwest Region, USDA Forest Service, Pacific Southwest Region, Vallejo, California.

Clepper, Henry, editor
    1971   *Leaders of American Conservation.* New York: The Ronald Press Company.

Cook, Cheryl A.
    1991   "List of Important Dates in Conservation History." Pacific Southwest Archives, Pacific Southwest Region, USDA Forest Service, Vallejo, California.

Conners, Pamela A.
    1989   "Patterns and Policy of Water and Hydroelectric Development on the Stanislaus National Forest, 1850 to 1920." M.A. thesis. University of California, Santa Barbara.

    1992   "Influence of the Forest Service on Water Development Patterns in the West." In *Origins of the National Forests: A Centennial Symposium.* Edited by Harold K. Steen. Durham, North Carolina: Forest History Society.

    2004   Personal communication, 8 November 2004.

Dana, Samuel Trask and Sally K. Fairfax
    1980   *Forest and Range Policy: Its Development in the United States.* Second edition. New York: McGraw-Hill Book Company.

Davies, Gilbert W. and Florice M. Frank
    1992   *Stories of the Klamath National Forest, The First 50 Years: 1905-1955.* Hat Creek, California: HiStory ink Books.

Dempsey, Stanley
    1992   "Cautious Support: Relations Between the Mining Industry and the Forest Service, 1891-1991." In *Origins of the National Forests: A Centennial Symposium.* Edited by Harold K. Steen. Durham, North Carolina: Forest History Society.

Fishlake-Fillmore *News* (Utah)
    n.d   Document in collection of U.S. West Research, Inc., Salt Lake City, Utah.

Fox, Charles E. and Clyde M. Walker
    n.d.   "Grazing on Range—Watersheds in California." (Sacramento: Conservation Education Section, California Department of Natural Resources). Pacific Southwest Archives, Pacific Southwest Region, USDA Forest Service, Vallejo, California.

Friedhoff, William H. (Ole)
    1944   "Mining Activities in National Forests of California." Pacific Southwest Archives, Pacific Southwest Region, USDA Forest Service, Vallejo, California.

Hata, Nadine Ishitani
    1992   *The Historic Preservation Movement in California, 1940-1976.* SacramenntoSacramento, California: California Department of Parks and Recreation.

Hays, Samuel P. Hays
    1969   *Conservation and the Gospel of Efficiency: The Progressive Conservation Movement, 1890-1920.* New York: Antheneum.

Headley, Donn E.
    1992   "The Cooperation Imperative: Relationships Between Early Forest Administration and the Southern California Metropolis, 1892-1908." In *Origins of the National Forests: A Centennial Symposium.* Edited by Harold K. Steen. Durham, North Carolina: Forest History Society.

Hofstadter, Richard
    1955   *The Age of Reform: From Bryan to F.D.R.* New York: Knopf.

Hundley, Norris, Jr.
    2001   *The Great Thirst, Californians and Water: A History.* Revised Edition. Berkeley, California: University of California Press.

Jackson, W. Turrentine, Rand Herbert, and Stephen Wee
    n.d.   "History of the Tahoe National Forest, 1840-1940: A Cultural Resources Overview History." David, California: Jackson Research Projects. Pacific Southwest Archives, Pacific Southwest Region, USDA Forest Service, Vallejo, California.

Jarvi, Sim E.
    1961   "Outdoor Recreation, Conservation and Land use —Angeles National Forest, U.S. Forest Service." Alphabetical Subject Files, Box 5, Angeles National Forest, Record Group 95, National Archive Record Center, Laguna Niguel, California.

Kahrl, William L.
    1983   "Water for California Cities: *Origins of the Major Systems.*" *Pacific Historian* Volume 27 (No. 1): 17-23.

Kroeber, Theodora
    1976   *Ishi in Two Words: A Biography of the Last Wild Indian in North America.* Berkeley, California: University of California Press.

Lux, Linda, et. al.
    2003   "Recreation Residence Tracts in the National Forests of California from 1906 to 1959." Public Affairs & Communication Office, Pacific Southwest Region, USDA Forest Service, Pacific Southwest Region, Vallejo, California.

Martin, Ilse B., David T. Hodder, and Clark Whitaker
    1981   "Overview of the Cultural Historic Resources of Euro-American and Other Immigrant Groups in the Shasta-Trinity National Forest." Prepared for USDA, Forest Service, Shasta-Trinity National Forest, Redding, California Contract No. 53-9A28-9-2974. Playa del Ray, California: Geoscientific Systems and Consulting.

Morris, Edmund
    1979   *The Rise of Theodore Roosevelt.* New York: Ballantine Books.

Mowry, George Edwin
    1951   *The California Progressives.* Berkeley, California: University of California Press.
    1958   *The Era of Theodore Roosevelt, 1900-1912.* New York: Harper & Row Publishers.

Munns, Edward N.
    1913   "Reforestation in Southern California." Box 1: Historical Manuscripts, 1902-1913, Record Group 95, National Archive Record Center, San Bruno, California.

Palmer, Kevin
    1992   "Vulcan's Footprints on the Forest." In *Origins of the National Forests: A Centennial Symposium.* Edited by Harold K. Steen. Durham, North Carolina: Forest History Society.

Pendergrass, Leo F.
    1985   "The Forest Service in California: A Survey of Important Developments and People from 1905 to the Present." PenSec, Inc., Edmonds, Washington. Pacific Southwest Archives, Pacific Southwest Region, USDA Forest Service, Vallejo, California.

Price, Frank
    1946   "Early Day History of the Mendocino National Forest, California." Pacific Southwest Archives, Pacific Southwest Region, USDA Forest Service, Vallejo, California.

Rice, Richard B., William A. Bullough and Richard J. Orsi
    1996   *The Elusive Eden: A New History of California.* 2nd Edition. New York: McGraw Hill Companies, Inc.

Robbins, William G.
    1985   *American Forestry: A History of National, State & Private Cooperation.* Lincoln, Nebraska: University of Nebraska Press.

Robinson, Douglas
    1933   "Inyo National Forest." Pacific Southwest Archives, Pacific Southwest Region, USDA Forest Service, Vallejo, California.

Robinson, Glen O.
    1975   *The Forest Service: A Study in Public Land Management.* Baltimore, Maryland: The John Hopkins University Press.

Rose, Gene
    1993   *Sierra Centennial: One Hundred Years of Pioneering on the Sierra National Forest.* Clovis, California: Sierra National Forest.

Show, Stuart Bevier
    n.d.   "The Development of Forest Service Organization, Personnel, and Administration in California." Unpublished manuscript.

    1963   "Oral History S. Bevier Show: National Forests in California." Interview Conducted by Amelia Roberts Fry. Transcribed 1965. Regional Cultural History Project. Berkeley, California: University of California.

Smith, Darrell Hevenor
    1930   *The Forest Service: Its History, Activities and Organization.* Washington, D.C.: The Brookings Institution.

Sterling, E. A.
    1906   "The Striking Features of the Forest and Water Situation in California." Paper Presented at Society of American Foresters. Box 2: Historical Manuscripts, 1902-1913, Record Group 95, National Archive Record Center, San Bruno, California.

Strong, Douglas H.
    1973   *These Happy Grounds: A History of the Lassen Region.* Red Bluff, California: Walker Lithograph, Inc.

    1981   "Preservation Efforts at Lake Tahoe, 1880 to 1980." *Journal of Forest History* Volume 25 (No. 2): 78-97.

Supernowicz, Dana Edward
    1983   "Historical Overview of the Eldorado National Forest." Master's thesis. History Department, University of California, Irvine.

Tweed, William C.
    1980   *Recreation Site Planning and Improvement in National Forests, 1891-1942.* Forest Service Publication FS-354. Washington D.C.: Government Printing Office.

USDA Forest Service
    1910   "Proceedings of the Supervisor's Meeting, District 5. San Francisco, California, December 13-19. Box 1: Historical Manuscripts, 1902-1913, Record Group 95, National Archive Record Center, San Bruno, California.

Waugh, Frank A.
  1918   *Recreation Uses on the National Forests.* Washington D.C.: Government Printing Office.

Weekly Bulletin: Forest Service, District 5
  1916   "Happenings." Volume 1, No. 1.

Williams, Gerald W.
  2000   *The USDA Forest Service—The First Century.* FS-650. Washington, D.C.: USDA Forest Service.

Williams, Gerald D., cCompiler
  1993   "Biographical Information about the USDA Forest Service Chiefs (Foresters), 1905-Present." Public Affairs & Communication Office, Pacific Southwest Region, USDA Forest Service, Pacific Southwest Region, Vallejo, California.

Chapter IV: 1911-1918

Ayres, R. W.
  1941   "History of the Forest Service in California." Box 4: Selected Alphabetical Subject File, 1909-1974, Record Group 95, National Archive Record Center, San Bruno, California.

  1942   "Development of Fire Protection in the California Region." Box 4, Alphabetical Subject Files, 1909-1974, Record Group 95, National Archive Record Center, San Bruno, California.

  1958   "History of Timber Management in the California National Forests, 1850 to 1937." San Francisco: Forest Service, U.S. Department of Agriculture. Pacific Southwest Archives, Pacific Southwest Region, USDA Forest Service, Vallejo, California.

Benedict, M. A.
  1930   "Twenty-One Years of Fire Protection in the National Forests of California." *Journal of Forestry* Volume XXVIII: 707-710.

Burnett, Bruce B.
  1933   "Forest Highways in California with Connecting State and County Highways." Box 5: Alpha-Numeric Series-E, Roads and Trails, Record Group 95, National Archive Record Center, San Bruno, California.

California District News Letter
  1919   "From Messrs. Lackey, Brown, Gracey and Harris." Volume 2, No. 46.

Cermak, Robert W.
  n.d.   "Fire in the Forest: A Narrative History of Forest Fire Control in the California National Forests, 1898-1956." Public Affairs & Communication Office, Pacific Southwest Region, USDA Forest Service, Pacific Southwest Region, Vallejo, California.

  2004 Personal communication, 25 September 2004.

Clar, C. Raymond
  1959 *California Government and Forestry: From Spanish Days until the Creation of the Department of Natural Resources in 1927.* Sacramento: Division of Forestry, Department of Natural Resources, State of California.

Cleland, Robert Glass
  1962 *From Wilderness to Empire: A History of California.* New York: Alfred A. Knopf.

Cochran, Thomas C.
  1977 *200 Years of American Business.* New York: Dell Publishing Company.

Conners, Pamela Ann
  2004 Personal communication, 8 November 2004.

Dana, Samuel Trask
  1956 *Forest and Range Policy: Its Development in the United States.* New York: McGraw-Hill Book Company, Inc.

Dana, Samuel Trask and Sally K. Fairfax
  1980 *Forest and Range Policy: Its Development in the United States.* Second edition. New York: McGraw-Hill Book Company.

Fox, Charles E. and Clyde M. Walker
  n.d. "Grazing on Range—Watersheds in California." (Sacramento: Conservation Education Section, California Department of Natural Resources). Pacific Southwest Archives, Pacific Southwest Region, USDA Forest Service, Vallejo, California.

Godfrey, Anthony
  2003 "Historic Preservation Plan: Placer and Hard Rock Mining Resources in Montana." 3 Volumes. Prepared for the Bureau of Land Management, Montana State Office. Purchase Order ESP000053. Salt Lake City, Utah: U.S. West Research, Inc.

Hays, Samuel P.
  1957 *The Response to Industrialism, 1885-1914.* Chicago, Illinois: University of Chicago Press.

Hoffman, Abraham
  1968 "Angeles Crest: The Creation of a Forest Highway System in the San Gabriel Mountains." *Southern California Quarterly.* Volume L, No. 3: 309-345.

Link, Arthur S.
  1963 *Woodrow Wilson and the Progressive Era, 1910-1917.* New York: Harper & Row, Publishers.

Lux, Linda, et al.
  2003 "Recreation Residence Tracts in the National Forests of California from 1906 to 1959." Public Affairs & Communication Office, Pacific Southwest Region, USDA Forest Service, Pacific Southwest Region, Vallejo, California.

Palmer, Kevin
 1992 "Vulcan's Footprints on the Forest." In *Origins of the National Forests: A Centennial Symposium*. Edited by Harold K. Steen. Durham, North Carolina: Forest History Society.

Pendergrass, Lee F.
 1985 "The Forest Service in California: A Survey of Important Developments and People from 1905 to the Present." PenSec, Inc., Edmonds, Washington. Pacific Southwest Archives, Pacific Southwest Region, USDA Forest Service, Vallejo, California.

 1990 "Dispelling Myths: Women's Contributions to the Forest Service in California." *Forest and Conservation History* Volume 34, No. 1.

Rice, Richard B., William A. Bullough and Richard J. Orsi
 1996 *The Elusive Eden: A New History of California*. 2nd Edition. New York: McGraw Hill Companies, Inc.

Rose, Judy Personal communication. 17 May 2004.

Rose, Gene
 1993 *Sierra Centennial: One Hundred Years of Pioneering on the Sierra National Forest*. Clovis, California: Sierra National Forest.

Rowley, William D.
 1985 *U.S. Forest Service Grazing and Rangelands*. College Station, Texas: Texas A&M University Press.

Show, Stuart Bevier
 1963 "Oral History S. Bevier Show: National Forests in California." Interview Conducted by Amelia Roberts Fry. Transcribed 1965. Regional Cultural History Project. Berkeley, California: University of California.

Steen, Harold K.
 1976 *The U.S. Forest Service: A History*. Seattle: University of Washington Press.

Supernowicz, Dana Edward
 1983 "Historical Overview of the Eldorado National Forest." Master's thesis. History Department, University of California, Irvine.

Supervisor's News Letter
 1918 Volume 1, No. 1-10.

Tweed, William C.
 1980 *Recreation Site Planning and Improvement in National Forests, 1891-1942*. Forest Service Publication FS-354. Washington D.C.: Government Printing Office.

USDA Forest Service
 1910 "Proceedings of the Supervisor's Meeting, District 5. San Francisco, California, December 13-19. Box 1: Historical Manuscripts, 1902-1913, Record Group 95, National Archive Record Center, San Bruno, California.

	1912	"Proceedings of the Supervisor's Meeting, District 5. San Francisco, California. Pacific Southwest Archives, Pacific Southwest Region, USDA Forest Service, Vallejo, California.
	1916	"Grazing Management Reports—Annual Working Plans." Box 1: Alphabetical and Numeric Subject Files, 1908-1946. Record Group 95, National Archive Record Center, San Bruno, California.
	1917	"Annual Grazing Studies and Working Plans—Circular by District Forester Du Bois." Box 1: Alphabetical and Numeric Subject Files, 1908-1946. Record Group 95, National Archive Record Center, San Bruno, California.

Watkins, T.H.
    1973   *California: An Illustrated History.* New York: Weathervane Books.

Waugh, Frank A.
    1918   *Recreation Uses on the National Forests.* Washington D.C.: Government Printing Office.

Weekly Bulletin: Forest Service, District 5
    1916   Volume 1, No. 1.
    1917   Volume 1, No. 15

Who Was Who
    1962   *Who Was Who in America: A Companion Volume to Who's Who in America.* Volume 1: Biographies of the Non-Living with Dates of Deaths Appended. 5th Printing. Chicago: A. N. Marquis Company.

Wiebe, Robert H.
    1967   *The Search for Order, 1877-1920.* New York: Hill and Wang.

Williams, Gerald W.
    2000   *The USDA Forest Service—The First Century.* FS-650. Washington, D.C.: USDA Forest Service.

Williams, Gerald W., compiler
    1993   "Biographical Information about the USDA Forest Service Chiefs (Foresters), 1905-Present." Public Affairs & Communication Office, Pacific Southwest Region, USDA Forest Service, Pacific Southwest Region, Vallejo, California.

Chapter V: 1919-1932

American Tree Association
    1924   *Forestry Almanac.* Washington, D.C.: The American Tree Association.

Ayres, R. W.
    1941   "History of the Forest Service in California." Box 4: Selected Alphabetical Subject File, 1909-1974, Record Group 95, National Archive Record Center, San Bruno, California.

1942   "Development of Fire Protection in the California Region." Box 4, Alphabetical Subject Files, 1909-1974, Record Group 95, National Archive Record Center, San Bruno, California.

1958   "History of Timber Management in the California National Forests, 1850 to 1937." San Francisco: Forest Service, U.S. Department of Agriculture. Pacific Southwest Archives, Pacific Southwest Region, USDA Forest Service, Vallejo, California.

Ayres, R. W. and Wallace I. Hutchinson
1931   *Forest Rangers' Catechism: Questions and Answers on the National Forests of the California Region.* Washington D.C.: Government Printing Office.

Bean, Walton
1968   *California: An Interpretive History.* New York: McGraw-Hill Book Company.

Benedict, M.A.
1930   "Twenty-One Years of Fire Protection in the National Forests of California." *Journal of Forestry* Volume XXVIII: 707-710.

Browning, Peter, compiler and editor
1988   *John Muir in His Own Words: A Book of Quotations.* LaFayette, California: Great West Books.

Burnett, Bruce B.
1933   "Forest Highways in California with Connecting State and County Highways." Box 5: Alpha-Numeric Series-E, Roads and Trails, Record Group 95, National Archive Record Center, San Bruno, California.

California District News Letter
1920-30   Various Numbers

California Ranger Region Five
1930-32   Various Numbers

Caughey, John Walton
1940   *California.* New York: Prentice-Hall, Inc.

Cermak, Robert W.
n.d.   "Fire in the Forest: A Narrative History of Forest Fire Control in the California National Forests, 1898-1956." Public Affairs & Communication Office, Pacific Southwest Region, USDA Forest Service, Pacific Southwest Region, Vallejo, California.

1991   "Pioneering Aerial Forest Fire Control: The Army Air Patrol in California, 1919-1921." *California History: The Magazine of the California Historical Society.* Volume LXX, No. 3 (Fall): 290-305.

Clar, C. Raymond
    1959   *California Government and Forestry: From Spanish Days until the Creation of the Department of Natural Resources in 1927.* Sacramento: Division of Forestry, Department of Natural Resources, State of California.

Cleland, Robert Glass
    1962   *From Wilderness to Empire: A History of California.* New York: Alfred A. Knopf.

Cook, Alistair
    1973   *Alistair Cooke's America.* New York: Alfred A. Knopf.

Dana, Samuel Trask and Sally K. Fairfax
    1980   *Forest and Range Policy: Its Development in the United States.* Second edition. New York: McGraw-Hill Book Company.

Dempsey, Stanley
    1992   "Cautious Support: Relations Between the Mining Industry and the Forest Service, 1891-1991." In *Origins of the National Forests: A Centennial Symposium.* Edited by Harold K. Steen. Durham, North Carolina: Forest History Society.

Fox, Charles E. and Clyde M. Walker
    n.d.   "Grazing on Range—Watersheds in California." (Sacramento: Conservation Education Section, California Department of Natural Resources). Pacific Southwest Archives, Pacific Southwest Region, USDA Forest Service, Vallejo, California.

Friedhoff, William H. (Ole)
    1944   "Mining Activities in National Forests of California." Pacific Southwest Archives, Pacific Southwest Region, USDA Forest Service, Vallejo, California.

Godfrey, Anthony
    2003   "Historic Preservation Plan: Placer and Hard Rock Mining Resources in Montana." 3 Volumes. Prepared for the Bureau of Land Management, Montana State Office. Purchase Order ESP000053. Salt Lake City, Utah: U.S. West Research, Inc.

Hoffman, Abraham
    1968   "Angeles Crest: The Creation of a Forest Highway System in the San Gabriel Mountains." *Southern California Quarterly* Volume L (No.3): 309-345.

Kirkendall, Richard S.
    1974   *The United States 1929-1945: Years of Crisis and Change.* New York: McGraw-Hill Book Company.

Kotok, Edward I.
    1928   "Forests and Water." California Forest Experiment Station. Pacific Southwest Archives, Pacific Southwest Region, USDA Forest Service, Vallejo, California.

Kneipp, L. F.
    1930   "Recreational Use of the National Forests." *Journal of Forestry* Volume XXVIII: 618-625.

Kraebel, Charles J.
    1927   "First Progress Report on the Work in Southern California." California Forest Experiment Station. Pacific Southwest Archives, Pacific Southwest Region, USDA Forest Service, Vallejo, California.

Leuchtenburg, William E.
    1958   *The Perils of Prosperity, 1914-32.* Chicago: University of Chicago Press.

Lowdermilk
    1928   "Role of Chaparral and Brush Forests in Water and Soil Conservation." Pacific Southwest Archives, Pacific Southwest Region, USDA Forest Service, Vallejo, California.

Morison, Samuel Eliot
    1965   *The Oxford History of the American People.* New York: Oxford University Press.

Munns, Edward N.
    1919   "The Control of Flood Water in Southern California." *Journal of Forestry* Volume XVII, No. 4 (April 1919): 423-429.

News Letter [District 5]
    1919   Volume 1, Nos. 17-60, various.
    1920   Volume 1, Nos. 61-86, various

Pendergrass, Lee F.
    1985   "The Forest Service in California: A Survey of Important Developments and People from 1905 to the Present." PenSec, Inc., Edmonds, Washington. Pacific Southwest Archives, Pacific Southwest Region, USDA Forest Service, Vallejo, California.

Reyes, Jacinto Damien and John Edwin Hoggs
    1930   "Thirty Years Fighting Fires in Our Forests: Recollections of the Life of Forest Ranger…" *Tourist Topics* 52-56, 70-71.

Rice, Richard B., William A. Bullough and Richard J. Orsi
    1996   *The Elusive Eden: A New History of California.* 2nd Edition. New York: McGraw Hill Companies, Inc.

Rolle, Andrew F.
    1963   *California: A History.* New York: Thomas Y. Crowell Company.

Rowley, William D.
    1985   *U.S. Forest Service Grazing and Rangelands: A History.* College Station, Texas: Texas A & M University Press.

Shepherd, Jack
    1975   *The Forest Killers: The Destruction of the American Wilderness.* New York: Weybright and Talley.

Show, Stuart Bevier
    1963   "Oral History S. Bevier Show: National Forests in California." Interview Conducted by Amelia Roberts Fry. Transcribed 1965. Regional Cultural History Project. Berkeley, California: University of California.

Smith, Darrell Hevenor
    1930   *The Forest Service: Its History, Activities and Organization.* Washington, D.C.: The Brookings Institution.

Swain, David C.
    1963   *Federal Conservation Policy, 1921-1933.* Berkeley, California: University of California Press.

Supervisor's News Letter (Confidential)
    1919   Volume 1, Nos. 11-16, various.

USDA Forest Service
    1919-29   "Grazing Management Reports—Annual Working Plans." Box 1: Alphabetical and Numeric Subject Files, 1908-1946. Record Group 95, National Archive Record Center, San Bruno, California.

    1927   *Annual Investigative Report for 1926 and Investigative Program for 1927: District 5.* Pacific Southwest Archives, Pacific Southwest Region, USDA Forest Service, Vallejo, California.

    1930a   *Annual Investigative Report for 1929 and Investigative Program for 1930: District 5.* Pacific Southwest Archives, Pacific Southwest Region, USDA Forest Service, Vallejo, California.

    1930b   "Accomplishment Report: California National Forest Region." Washington, D.C.: Government Printing Office.

    1930c   *Forestry Handbook for California.* San Francisco: California Region, Forest Service, United States Department of Agriculture.

    1931   *Federal Activities in the National Forests of the California Region.* Pacific Southwest Archives, Pacific Southwest Region, USDA Forest Service, Vallejo, California.

    1948   *Tree Breeding at the Institute of Forest Genetics.* Miscellaneous Publication No. 659. Washington D.C.: Government Printing Office.

    1976   June 24 Memorandum: Region 5 Wilderness History: Charles R. Joy, "History of Wilderness in the National Forest System California (Region 5)." Pacific Southwest Archives, Pacific Southwest Region, USDA Forest Service, Vallejo, California.

Who Was Who
    1962   *Who Was Who in America: A Companion Volume to Who's Who in America.* Volume 3: Biographies of the Non-Living with Dates of Deaths Appended. 5th Printing. Chicago: A.N. Marquis Company.

Williams, Gerald W.
    2000   *The USDA Forest Service—The First Century.* FS-650. Washington, D.C.: USDA Forest Service.

    2005   Personal communication, 11 March 2005.

Williams, Gerald W., compiler
    1993   "Biographical Information about the USDA Forest Service Chiefs (Foresters), 1905-Present." Public Affairs & Communication Office, Pacific Southwest Region, USDA Forest Service, Pacific Southwest Region, Vallejo, California.

Wilson, Carl C. and James B. Davis
    1988   *Forest Fire Laboratory at Riverside and Fire Research in California: Past, Present, and Future.* Berkeley, California: Pacific Southwest Forest and Range Experiment Station.

Woodbury, T. D.
    1930   "Development of Silvicultural Practices in the California National Forests." *Journal of Forestry* Volume XXVIII: 693-700.

Chapter VI: 1933-1941

Angeles National Forest Grapevine
    1990

Ayres, R. W.
    1941   "History of the Forest Service in California." Box 4: Selected Alphabetical Subject File, 1909-1974, Record Group 95, National Archive Record Center, San Bruno, California.

    1942   "Development of Fire Protection in the California Region." Box 4, Alphabetical Subject Files, 1909-1974, Record Group 95, National Archive Record Center, San Bruno, California.

    1958   "History of Timber Management in the California National Forests, 1850 to 1937." San Francisco: Forest Service, U.S. Department of Agriculture. Pacific Southwest Archives, Pacific Southwest Region, USDA Forest Service, Vallejo, California.

Burnett, Bruce B.
    1933   "Forest Highways in California with Connecting State and County Highways." Box 5: Alpha-Numeric Series-E, Roads and Trails, Record Group 95, National Archive Record Center, San Bruno, California.

Bean, Walton
    1968   *California: An Interpretive History.* New York: McGraw-Hill Book Company.

California Ranger Region Five
    1933-41   Various Numbers

Clar, C. Raymond
    1969   *California Government and Forestry-II during the Young and Rolph Administrations.* Sacramento, California: Division of Forestry, Department of Conservation, State of California.

Cole, Olen Jr.
    1999   *The African-American Experience in the Civilian Conservation Corps.* Gainesville, Florida: University Press of Florida.

Coutant, Gerald J.
    1990   *A Chronology of the Recreation History of the National Forests and the U.S.D.A. Forest Service: 1940-1990.* Pacific Southwest Archives, Pacific Southwest Region, USDA Forest Service, Vallejo, California.

Fox, Charles E.
    1948   *Know Your National Forests: A Story of Conservation Through Wise Use.* Box 8: Selected Alphabetical Subject File, 1909-1974, Record Group 95, National Archive Record Center, San Bruno, California.

Friedhoff, William H. (Ole)
    1944   "Mining Activities in National Forests of California." Pacific Southwest Archives, Pacific Southwest Region, USDA Forest Service, Vallejo, California.

Godfrey, Anthony
    2003   "Historic Preservation Plan: Placer and Hard Rock Mining Resources in Montana." 3 Volumes. Prepared for the Bureau of Land Management, Montana State Office. Purchase Order ESP000053. Salt Lake City, Utah: U.S. West Research, Inc.

Kirkendall, Richard S.
    1974   *The United States 1929-1945: Years of Crisis and Change.* New York: McGraw-Hill Book Company.

Leisz, Douglas
    2005   Personal communication, 1 February 2005.

Leuchtenburg, William E.
    1963   *Franklin D. Roosevelt and the New Deal.* New York: Harper & Row Publishers.

Lux, Linda
    2005   Personal communication, 9 March 2005.

Lux, Linda, et al.
    2003   "Recreation Residence Tracts in the National Forests of California from 1906 to 1959." Public Affairs & Communication Office, Pacific Southwest Region, USDA Forest Service, Pacific Southwest Region, Vallejo, California.

Pendergrass, Lee F.
    1985   "The Forest Service in California: A Survey of Important Developments and People from 1905 to the Present." PenSec, Inc., Edmonds, Washington. Pacific Southwest Archives, Pacific Southwest Region, USDA Forest Service, Vallejo, California.

Rice, Richard B., William A. Bullough and Richard J. Orsi
    1996   *The Elusive Eden: A New History of California.* 2nd Edition. New York: McGraw Hill Companies, Inc.

Robinson, Cyril S.
    1940   "Notes on the California Condor Collected on Los Padres National Forest, California." Box 2: Selected Alphabetical Subject Files, 1906-1959. Record Group 95, National Archive Record Center, San Bruno, California.

Rolle, Andrew F.
    1963   *California: A History.* New York: Thomas Y. Crowell Company.

Rowley, William D.
    1985   *U.S. Forest Service Grazing and Rangelands: A History.* College Station, Texas: Texas A & M University Press.

Show, Stuart Bevier
    n.d.   "The Development of Forest Service Organization, Personnel, and Administration in California." Unpublished manuscript.

    1939   "Aircraft and Fire Control in National Forests." Box 3: Selected Alphabetical Subject File, 1909-1974, Record Group 95, National Archive Record Center, San Bruno, California.

    1940a   "Our View is From the Air: The Airplane Plays an Important Role in Uncle Sam's Forest Fire Fighting Army in California." *California Conservationist* Volume V (No. 11): 1-4.

    1940b   "Forestry and the People's Welfare." Box 3: Selected Alphabetical Subject File, 1909-1974, Record Group 95, National Archive Record Center, San Bruno, California.

    1963   "Oral History S. Bevier Show: National Forests in California." Interview Conducted by Amelia Roberts Fry. Transcribed 1965. Regional Cultural History Project. Berkeley, California: University of California.

Steen, Harold K.
    2004   *The U.S. Forest Service: A History* (Centennial Edition). Durham, North Carolina: Forest History Society; and Seattle, Washington: University of Washington Press.

Swift, Lloyd W.
    1936   *A Study of the Distribution of Grazing Privileges, Region 5.* Box 10: Selected Alphabetical Subject File, 1909-1974, Record Group 95, National Archive Record Center, San Bruno, California.

Swift, Lloyd W. and A. Fausett
    1938   *A Report on the Grazing Administrative Studies in the California Region.* Box 11: Selected Alphabetical Subject File, 1909-1974, Record Group 95, National Archive Record Center, San Bruno, California.

Tweed, William C.
    1980   *Recreation Site Planning and Improvement in National Forests, 1891-1942.* Forest Service Publication FS-354. Washington D.C.: Government Printing Office.

U.S. Senate
    1933   *A National Plan for American Forestry.* 2 Vols. Washington, D.C.: Government Printing Office.

USDA California Region 5
    1936   *Forests and Farms: A Pictorial Presentation of the Social and Economic Services of the National Forest to Agriculture.* Pasadena, California: California Region, Forest Service, United States Department of Agriculture.
    1940   "Put It Out." Box 3: Selected Alphabetical Subject File, 1909-1974, Record Group 95, National Archive Record Center, San Bruno, California.

USDA Forest Service
    1932   *Copeland Resolution Report: Senate Resolution 175.* Pacific Southwest Archives, Pacific Southwest Region, USDA Forest Service, Vallejo, California.
    1932-41   "Grazing Management Reports—Annual Working Plans." Box 1: Alphabetical and Numeric Subject Files, 1908-1946. Record Group 95, National Archive Record Center, San Bruno, California.
    1935   "Management-Condor, 1935-1953." Box 2: Selected Alphabetical Subject Files, 1906-1959. Record Group 95, National Archive Record Center, San Bruno, California.
    1937   *Region 5: Plan of Work—1937.* Pacific Southwest Archives, Pacific Southwest Region, USDA Forest Service, Vallejo, California.
    1940a   "High Sierra Ski School for Forest Rangers." Box 3: Selected Alphabetical Subject Files, 1909-1974. Record Group 95, National Archive Record Center, San Bruno, California.
    1940b   Earle H. Clapp to Regional Foresters and Directors, 30 March 1940. Box 3: Selected Alphabetical Subject File, 1909-1974, Record Group 95, National Archive Record Center, San Bruno, California.
    1940c   "Summary of Discussion on Forest Program Educational Campaign." Box 3: Selected Alphabetical Subject File, 1909-1974, Record Group 95, National Archive Record Center, San Bruno, California.
    1940d   *The Forest Problems of California.* Compiled by California Forest and Range Experiment Station. Box 3: Selected Alphabetical Subject File, 1909-1974, Record Group 95, National Archive Record Center, San Bruno, California.
    1940e   *The Forest Program and Its Application to California Problems and Conditions.* Prepared by California Region, Forest Service. Box 3: Selected Alphabetical Subject File, 1909-1974, Record Group 95, National Archive Record Center, San Bruno, California.

1940f	S. B. Show to Forest Supervisors, 5 December 1940. Box 3: Selected Alphabetical Subject File, 1909-1974, Record Group 95, National Archive Record Center, San Bruno, California.
1976	June 24 Memorandum: Region 5 Wilderness History: Charles R. Joy, "History of Wilderness in the National Forest System California (Region 5)." Pacific Southwest Archives, Pacific Southwest Region, USDA Forest Service, Vallejo, California.

Watkins, T. H.
    1973   *California: An Illustrated History.* New York: Weathervane Books.

Who Was Who
    1962   *Who Was Who in America: A Companion Volume to Who's Who in America.* Volume 1: Biographies of the Non-Living with Dates of Deaths Appended. 5th Printing. Chicago: A.N. Marquis Company.

Williams, Gerald W.
    2000   *The USDA Forest Service—The First Century.* FS-650. Washington, D.C.: USDA Forest Service.

Wilson, Carl C. and James B. Davis
    1988   *Forest Fire Laboratory at Riverside and Fire Research in California: Past, Present, and Future.* Berkeley, California: Pacific Southwest Forest and Range Experiment Station.

Wolf, Robert E.
    1990   *The Concept of Multiple Use: The Evolution of the Idea within the Forest Service and the Enactment of the Multiple-Use Sustained-Yield Act of 1960.* Prepared for the Office of Technology Assessment, Congress of the United States. Pacific Southwest Library, Pacific Southwest Region, USDA Forest Service, Vallejo, California.

Chapter VII: 1941-1945

Brown, William S.
    1943   "Sky Watchers of the Hinterlands: California Region, U.S. Forest Service." Pacific Southwest Archives, Pacific Southwest Region, USDA Forest Service, Vallejo, California.

California Region – Administrative Digest
    1942-45   Various

Cermak, Robert W.
    n.d.   "Fire in the Forest: A Narrative History of Forest Fire Control in the California National Forests, 1898-1956." Public Affairs & Communication Office, Pacific Southwest Region, USDA Forest Service, Pacific Southwest Region, Vallejo, California.

Granger, Christopher M.
    1965   "Forest Management in the United States Forest Service, 1907-1952." Edited by Amelia R. Fry. Berkeley, California: Regional Oral History Office, Bancroft Library.

Hirt, Paul W.
    1994   *A Conspiracy of Optimism: Management of the National Forests Since World War Two.* Lincoln, Nebraska: University of Nebraska Press.

Nash, Gerald D.
    1973   *The American West in the Twentieth Century: A Short History of an Urban Oasis.* Albuquerque, New Mexico: University of New Mexico Press.

Pendergrass, Lee F.
    1985   "The Forest Service in California: A Survey of Important Developments and People from 1905 to the Present." PenSec, Inc., Edmonds, Washington. Pacific Southwest Archives, Pacific Southwest Region, USDA Forest Service, Vallejo, California.

Price, Frank
    1948   "Record of Civilian Public Service Camp Program in California National Forest Region, 1942-1946." Pacific Southwest Archives, Pacific Southwest Region, USDA Forest Service, Vallejo, California.

Rice, Richard B., William A. Bullough and Richard J. Orsi
    1996   *The Elusive Eden: A New History of California.* 2nd Edition. New York: McGraw Hill Companies, Inc.

Show, Stuart Bevier
    1942a   "The National Forests in Wartime." Special Articles, 1942. Box 4: Selected Alphabetical Subject Files, 1909-1974. Record Group 95, National Archive Record Center, San Bruno, California.

    1942b   "Forest Service War Activities." Special Articles, 1942. Box 4: Selected Alphabetical Subject Files, 1909-1974. Record Group 95, National Archive Record Center, San Bruno, California.

    1942c   "Forest Protection in the War Zone." Special Articles, 1942. Box 4: Selected Alphabetical Subject Files, 1909-1974. Record Group 95, National Archive Record Center, San Bruno, California.

    1942d   "We Must All Prevent Forest Fires." Special Articles, 1942. Box 4: Selected Alphabetical Subject Files, 1909-1974. Record Group 95, National Archive Record Center, San Bruno, California.

    1942e   "Your Forest Playgrounds in Wartime." Special Articles, 1942. Box 4: Selected Alphabetical Subject Files, 1909-1974. Record Group 95, National Archive Record Center, San Bruno, California.

	1942f	"National Forests Remain Open." Special Articles, 1942. Box 4: Selected Alphabetical Subject Files, 1909-1974. Record Group 95, National Archive Record Center, San Bruno, California.
	1943a	"Wartime Forest Fire Prevention." Pacific Southwest Archives, Pacific Southwest Region, USDA Forest Service, Vallejo, California.
	1943b	"The National Forests of California: Their Social and Economic Resources and the Need for Protecting Them from Fire." Pacific Southwest Archives, Pacific Southwest Region, USDA Forest Service, Vallejo, California.
	1944	"Lumber, Stepchild of the Golden State." Special Articles, 1942. Box 5: Selected Alphabetical Subject Files, 1909-1974. Record Group 95, National Archive Record Center, San Bruno, California.

USDA Forest Service
- 1942 — National Defense, 1942. Box 4: Selected Alphabetical Subject Files, 1909-1974. Record Group 95, National Archive Record Center, San Bruno, California.
- 1945 — "A Report on Range Allotment Analysis, Region 5." Range Allotment Analysis, 1945. Box 9: Selected Alphabetical Subject Files, 1909-1974. Record Group 95, National Archive Record Center, San Bruno, California.
- 1946a — *Rubber from Guayule: Emergency Rubber Project.* June. Pacific Southwest Archives, Pacific Southwest Region, USDA Forest Service, Vallejo, California.
- 1946b — *Final Report: The Emergency Rubber Project, A Report on Our War-time Guayule Rubber Program.* December. Pacific Southwest Archives, Pacific Southwest Region, USDA Forest Service, Vallejo, California.
- 1946c — "Forestry in Wartime: Report of the Chief of the Forest Service, 1942." Pacific Southwest Archives, Pacific Southwest Region, USDA Forest Service, Vallejo, California.
- 1956 — "History of War Activities, 1942-1945." Pacific Southwest Archives, Pacific Southwest Region, USDA Forest Service, Vallejo, California.
- 1990 — *The History of Engineering in the Forest Service (A Compilation of History and Memoirs, 1905-1989).* Washington, D.C.: Forest Service Engineering Staff.

Watkins, T. H.
- 1973 — *California: An Illustrated History.* New York: Weathervane Books.

Williams, Gerald W.
- n.d. — "The Story of Smokey Bear." Attached to personal communication, 11 March 2005.
- 1998 — "Aircraft Warning Service in World War II." Attached to personal communication, 11 March 2005.
- 2000 — *The USDA Forest Service—The First Century.* FS-650. Washington, D.C.: USDA Forest Service.

2004	"Balloon Bombs of World War II." Attached to personal communication, 11 March 2005.	

Chapter VIII: 1946-1954

Administrative Digest-California Region
    1947-50   Various

Anonymous
    1947a   "San Gorgonio." *California Magazine of the Pacific*. March. Box 7: Selected Alphabetical Subject File, 1909-1974, Record Group 95, National Archive Record Center, San Bruno, California.

    1947b   "Skiing on San Gorgonio?" *Nature Magazine*. February. Box 7: Selected Alphabetical Subject File, 1909-1974, Record Group 95, National Archive Record Center, San Bruno, California.

Asplund, Rupert
    2004   Interview of Rupert Asplund by Bob Cermak. Region Five History Project. Public Affairs & Communication Office, Pacific Southwest Region, USDA Forest Service, Pacific Southwest Region, Vallejo, California.

Bachman, Earl Eugene
    1967   *Recreation Facilities…A Personal History of Their Development in the National Forests of California*. Berkeley, California: Pacific Southwest Forest and Range Experiment Station, California Region, Forest Service, U.S. Department of Agriculture.

    1990   Oral History Interview with Earl Eugene Bachman by Jerry Rouillard. USDA Forest Service Oral History Project 1991 Centennial, Forest Reserves Act. Capital Campus Public History Program and Oral History Program, California State University, Sacramento, California. Public Affairs & Communication Office, Pacific Southwest Region, USDA Forest Service, Pacific Southwest Region, Vallejo, California.

Bean, Leslie S.
    1951   "L.S. Bean Study of Projects in Region 5." Box 42: Selected Historical Files (Numeric Subject Files), 95-96-049, Record Group 95, National Archive Record Center, San Bruno, California.

Beardsley, Don
    2004   Interview with Don Beardsley. The Greatest Good: A History of The Forest Service Project Files. Public Affairs & Communication Office, Pacific Southwest Region, USDA Forest Service, Pacific Southwest Region, Vallejo, California.

Blanchard, George S.
    1949   *An Army Study on Program Control in the U.S. Forest Service, Department of Agriculture*. Washington, D.C.: Department of the Army. Public Affairs &

Communication Office, Pacific Southwest Region, USDA Forest Service, Pacific Southwest Region, Vallejo, California.

Branch, W. C. and W. F. Murray
    1950   "General Integrated Inspection-Hawaii." Box 12: Selected Alphabetical Subject File, 1909-1974, Record Group 95, National Archive Record Center, San Bruno, California.

Branch, W. C.
    1956   "Hawaii Inspection." Box 42: Selected Historical Files (Numeric Subject Files), 95-96-049, Record Group 95, National Archive Record Center, San Bruno, California.

California Region – Administrative Digest
    1946-47   Various

Cermak, Robert W.
    n.d.   "Fire in the Forest: A Narrative History of Forest Fire Control in the California National Forests, 1898-1956." Public Affairs & Communication Office, Pacific Southwest Region, USDA Forest Service, Pacific Southwest Region, Vallejo, California.

    2005   Personal communication, 14 February 2005.

Chapline, W. R.
    1951   "Range Management History and Philosophy." *Journal of Forestry* Vol. 49: 634-638.

Clark, Ella E.
    1951   "Smokejumpers—The Quickest Way to the Fire." *Pacific Discovery* (July-August): 4-14.

Clawson, Marion
    1950   *The Western Range Livestock Industry.* New York: McGraw-Hill Book Company, Inc.

Cliff, Edward P.
    1954   "Memorandum-Report, Inspection R-5." Box 0063-A-0123, Record Group 95, Federal Record Center, San Bruno, California.

Conners, Pamela A.
    1998   *A History of the Six Rivers National Forest…Commemorating The First Fifty Years.* Eureka, California: Six Rivers National Forest, Pacific Southwest Region, USDA Forest Service.

Coutant, Gerald J.
    1990   *A Chronology of the Recreation History of the National Forests and the U.S.D.A Forest Service: 1940 to 1990.* (Draft Copy for Review). Pacific Southwest Archives, Pacific Southwest Region, USDA Forest Service, Vallejo, California.

Craft, Edward C.
- 1954 "Memorandum-Report, Inspection R-5." Box 0063-A-0123, Record Group 95, Federal Record Center, San Bruno, California.

Feuchter, Roy
- 2004 Interview of Roy Feuchter by Steve Kirby. Region Five History Project. Public Affairs & Communication Office, Pacific Southwest Region, USDA Forest Service, Pacific Southwest Region, Vallejo, California.

Fox, Charles E.
- 1948a *Know Your National Forests: A Story of Conservation Through Wise Use.* Box 8: Selected Alphabetical Subject File, 1909-1974, Record Group 95, National Archive Record Center, San Bruno, California.
- 1948b *Where Rivers Are Born: The Story of California's Watersheds.* Box 8: Selected Alphabetical Subject File, 1909-1974, Record Group 95, National Archive Record Center, San Bruno, California.

Fox, Gordon D.
- 1952 "Inspection of Operation Phases of Administration-Region 5." Box 42: Numeric-Subject Files, Selected Historical Files, 1906-1959, Record Group 95, National Archive Record Center, San Bruno, California.

Hendee, Clare
- 1955a Letter: Clare Hendee to Forest Supervisors & Division Chiefs, 12 January. Box 53: R.O. Circulars, Numeric-Subject Files, Selected Historical Files, 1906-1959, Record Group 95, National Archive Record Center, San Bruno, California.
- 1955b Letter: Clare Hendee to Supervisors, 18 February. Box 53: R.O. Circulars, Subject Files, Selected Historical Files, 1906-1959, Record Group 95, National Archive Record Center, San Bruno, California.
- 1955c Letter: Clare Hendee to Division Chiefs and Supervisors, 17 August with attachment "Remarks of Howard Hopkins, W.O. G.I. Inspector, Meeting of Field-Going Personnel, Monday July 11, 1955." Box 53: R.O. Circulars, Subject Files, Selected Historical Files, 1906-1959, Record Group 95, National Archive Record Center, San Bruno, California.

Kirchner, Walt
- 2004 Interview of Walt Kirchner by Max Younkin. Region Five History Project. Public Affairs & Communication Office, Pacific Southwest Region, USDA Forest Service, Pacific Southwest Region, Vallejo, California.

Kraebel, C. J. and J. D. Sinclair
- 1947 "Some Aspects of Watershed Management in Southern California." Box 8: Selected Alphabetical Subject File, 1909-1974, Record Group 95, National Archive Record Center, San Bruno, California.

Leisz, Douglas
 2004 Interview of Douglas Leisz by Bob Smart. Region Five History Project. Public Affairs & Communication Office, Pacific Southwest Region, USDA Forest Service, Pacific Southwest Region, Vallejo, California.

 2005 Personal communication, 1 February 2005.

Littlefield, T. R.
 1947 "Watershed Management in the National Forests of California." Box 8: Selected Alphabetical Subject File, 1909-1974, Record Group 95, National Archive Record Center, San Bruno, California.

Loveridge, E. W. and W. L. Dutton
 1946 "Report to the Chief on General Integrating Inspection Including Follow-up of Previous G.I. Inspections: California Region." Pacific Southwest Archives, Pacific Southwest Region, USDA Forest Service, Vallejo, California.

Loveridge, E. W. and M. M. Nelson
 1951 "Region Five G.I.I., 1951" Box 11: Selected Historical Files, 1909-1974, Record Group 95, National Archive Record Center, San Bruno, California.

McKinsey & Company, Management Consultants
 1955 "Evaluation of Forest Service Timber Sales Activities: Department of Agriculture, Forest Service" Contract No. W-227-FS-55. Public Affairs & Communication Office, Pacific Southwest Region, USDA Forest Service, Pacific Southwest Region, Vallejo, California.

National Audubon Society
 1950 "Statement of John H. Baker, President of the National Audubon Society at the Public Hearing in Los Angeles on August 21, 1950 with Regard to the Proposed Withdrawal of Certain Lands….This for the Better Protection of the California Condor." Box 2: Selected Alphabetical Subject File, 1909-1974, Record Group 95, National Archive Record Center, San Bruno, California.

Pacific Southwest Log
 1985

Parkinson, Dana
 1954 "I & E Inspection: Region 5 and California Station." Pacific Southwest Archives, Pacific Southwest Region, USDA Forest Service, Vallejo, California.

Peterson, Max
 2004 Interview of Max Peterson by Bryan Payne. Region Five History Project. Public Affairs & Communication Office, Pacific Southwest Region, USDA Forest Service, Pacific Southwest Region, Vallejo, California.

Price, J. H.
 1931 "Inspection Report-Hawaii." Box 12: Selected Alphabetical Subject File, 1909-1974, Record Group 95, National Archive Record Center, San Bruno, California.

Nash, Gerald D.
　1973　　*The American West in the Twentieth Century: A Short History of an Urban Oasis.* Albuquerque, New Mexico: University of New Mexico Press.

Radel, Joseph T.
　1991　　Oral History Interview with Joseph T. Radel by Jerry Rouillard. USDA Forest Service Oral History Project 1991 Centennial, Forest Reserves Act. Capital Campus Public History Program and Oral History Program, California State University, Sacramento, California. Public Affairs & Communication Office, Pacific Southwest Region, USDA Forest Service, Pacific Southwest Region, Vallejo, California.

Redlands Daily Facts
　1947a　　"San Gorgonio…Wilderness or Ski Resort?" 3-6 February. Box 7: Selected Alphabetical Subject File, 1909-1974, Record Group 95, National Archive Record Center, San Bruno, California.

　1947b　　Editorial: "San Gorgonio Remains Primitive." 18 June. Box 7: Selected Alphabetical Subject File, 1909-1974, Record Group 95, National Archive Record Center, San Bruno, California.

Rice, Richard B., William A. Bullough and Richard J. Orsi
　1996　　*The Elusive Eden: A New History of California.* 2nd Edition. New York: McGraw Hill Companies, Inc.

Rice, Bob
　2004　　Interview of Bob Rice by Janet Buzzini. Region Five History Project. Public Affairs & Communication Office, Pacific Southwest Region, USDA Forest Service, Pacific Southwest Region, Vallejo, California.

Robinson, Glen O.
　1975　　*The Forest Service.* Baltimore, Maryland: John Hopkins University Press for the Resources for the Future, Inc.

Rolle, Andrew F.
　1963　　*California: A History.* New York: Thomas Y. Crowell Company.

Roth, Dennis M.
　1984　　"The National Forests and the Campaign for Wilderness Legislation." *Journal of Forest History* (July 1984): 112-125.

　1988　　*The Wilderness Movement and the National Forests.* College Station, Texas: Intaglio Press.

Rowley, William D.
　1985　　*U.S. Forest Service Grazing and Rangelands: A History.* College Station, Texas: Texas A&M University Press.

Saturday Evening Post
　1951　　Milton Silverman, "The Fabulous Condors' Last Stand." 7 April: 36, 148-150.

Show, S. B.
    1944   "Lumber, Stepchild of the Golden State." Box 5: Selected Alphabetical Subject File, 1909-1974, Record Group 95, National Archive Record Center, San Bruno, California.

Sieker, John H.
    1945   "White Gold." Box 5: Selected Alphabetical Subject File, 1909-1974, Record Group 95, National Archive Record Center, San Bruno, California.

Sierra Club Bulletin
    1946   "Wilderness Today, Resort Tomorrow?" No. 4, August. Box 7: Selected Alphabetical Subject File, 1909-1974, Record Group 95, National Archive Record Center, San Bruno, California.

Steen, Harold K.
    2004   *The U.S. Forest Service: A History* (Centennial Edition). Durham, North Carolina: Forest History Society; and Seattle, Washington: University of Washington Press.

Thompson, Perry A.
    1947   "Sierra Synthesis: Multiple-Resource Management of the National Forests of California." Box 7: Selected Alphabetical Subject File, 1909-1974, Record Group 95, National Archive Record Center, San Bruno, California.

U.S. House of Representatives
    1949   "Congressional Hearing, Modoc National Forest." Box 42: Selected Historical Files (Numeric Subject Files), 95-96-049, Record Group 95, National Archive Record Center, San Bruno, California.

USDA Forest Service
    1945a   "Copter Use for Firefighting." Box 5: Selected Alphabetical Subject File, 1909-1974, Record Group 95, National Archive Record Center, San Bruno, California.

    1945b   "A Report on Range Allotment Analysis, Region 5." Range Allotment Analysis, 1945. Box 9: Selected Alphabetical Subject Files, 1909-1974. Record Group 95, National Archive Record Center, San Bruno, California.

    1946a   "Use of the Helicopter on the Castaic Fire, Angeles National Forest, September 9-10, 1946." Box 7: Selected Alphabetical Subject File, 1909-1974, Record Group 95, National Archive Record Center, San Bruno, California.

    1946b   Letter: Lyle F. Watts to Regional Forester, San Francisco, California, 20 November. Box 7: Selected Alphabetical Subject File, 1909-1974, Record Group 95, National Archive Record Center, San Bruno, California.

    1946c   "Forest Service Announces Hearing on San Gorgonio Proposal." 11 December. Box 7: Selected Alphabetical Subject File, 1909-1974, Record Group 95, National Archive Record Center, San Bruno, California.

1946d   "Helicopters, 1946." Box 5: Selected Alphabetical Subject File, 1909-1974, Record Group 95, National Archive Record Center, San Bruno, California.

1947a   "Forest Land Problems in California." Box 73-A-808, Record Group 95, Federal Record Center, San Bruno, California.

1947b   "Special Articles—Helicopters and Fires." Box 7: Selected Alphabetical Subject File, 1909-1974, Record Group 95, National Archive Record Center, San Bruno, California.

1947c   "Forestry in the Golden State." Box 7: Selected Alphabetical Subject File, 1909-1974, Record Group 95, National Archive Record Center, San Bruno, California.

1947d   "Opening Statement by P. A. Thompson, Regional Forester at Public Hearing on San Gorgonio Primitive Area." 19 February. Box 7: Selected Alphabetical Subject File, 1909-1974, Record Group 95, National Archive Record Center, San Bruno, California.

1947e   "Watts Rules on Proposal for Changing San Gorgonio Wilderness Area Bounds." 10 June. Box 7: Selected Alphabetical Subject File, 1909-1974, Record Group 95, National Archive Record Center, San Bruno, California.

1947f   "Management Plan for Sespe Wildlife Preserve." Box 2: Selected Alphabetical Subject File, 1909-1974, Record Group 95, National Archive Record Center, San Bruno, California.

1950   "Public Campgrounds on the National Forests." Pacific Southwest Archives, Pacific Southwest Region, USDA Forest Service, Vallejo, California.

1951a   "Supervisor's Meeting, Region 5: Recommendations and Program of Action." Box 11: Selected Alphabetical Subject File, 1909-1974, Record Group 95, National Archive Record Center, San Bruno, California.

1951b   Information News Release. "Los Padres National Forest." 23 January. Box 2: Selected Alphabetical Subject File, 1909-1974, Record Group 95, National Archive Record Center, San Bruno, California.

1951c   Suggested Information News Release. "Oil and Gas Applications in the Santa Ynez Watershed." Box 11: Selected Historical Files, 1909-1974, Record Group 95, National Archive Record Center, San Bruno, California.

1952	"Forest Service Area Planning Guide: Region 5 (Revised)." Box 1, Records of the USFS—Los Padres National Forest, Alpha Subject Files, 1915-1963, Record Group 95, National Archive Record Center, Laguna Niguel, California.
1953a	"A Few Highlights on California National Forests." Pacific Southwest Archives, Pacific Southwest Region, USDA Forest Service, Vallejo, California.
1953b	"Northwestern California's Timber: A Report by the U.S. Forest Service, California Region." Box 53: R.O. Circulars, Numeric-Subject Files, Selected Historical Files, 1906-1959, Record Group 95, National Archive Record Center, San Bruno, California.
1954	*Facts About the Resources and Management of the National Forests of the California – Region Five* . Pacific Southwest Archives, Pacific Southwest Region, USDA Forest Service, Vallejo, California.
1955	Press Release: "First Classified Scenic Area Established by Forest Service." 18 August 1955. Pacific Southwest Archives, Pacific Southwest Region, USDA Forest Service, Vallejo, California.
1961	*Multiple Use Management Guides for the National Forests of Southern California*. Revised. Box 6, Records of the USFS—San Bernardino National Forest, Numeric Subject Files, Record Group 95, National Archive Record Center, Laguna Niguel, California.
1976	June 24 Memorandum: Region Five Wilderness History: Charles R. Joy, "History of Wilderness in the National Forest System California (R-5)." Pacific Southwest Archives, Pacific Southwest Region, USDA Forest Service, Vallejo, California.

Watkins, T. H.
1973	*California: An Illustrated History*. New York: Weathervane Books.

Watts, Lyle F.
1951	"Statement of Lyle F. Watts, Chief of U.S. Forest Service at Press Interview in San Francisco, July 23, 1951." Box 11: Selected Alphabetical Subject File, 1909-1974, Record Group 95, National Archive Record Center, San Bruno, California.

Wilderness Society
1947	"Why We Cherish San Gorgonio *Primitive*." *The Living Wilderness* Volume 12 (No. 20): 1-7.

Wilson, Carl C.
1991	Oral History Interview with Carl C. Wilson by Susan Douglass. USDA Forest Service Oral History Project 1991 Centennial, Forest Reserves Act. Capital Campus Public History Program and Oral History Program, California State University, Sacramento, California. Public Affairs & Communication Office, Pacific Southwest Region, USDA Forest Service, Pacific Southwest Region, Vallejo, California.

Wilson, Carl C. and James B. Davis
- 1988 *Forest Fire Laboratory at Riverside and Fire Research in California: Past, Present, and Future.* Berkeley, California: Pacific Southwest Forest and Range Experiment Station.

Chapter IX: 1955-1967

Bachman, Earl Eugene
- 1967 *Recreation Facilities…A Personal History of Their Development in the National Forests of California.* Berkeley, California: Pacific Southwest Forest and Range Experiment Station, California Region, Forest Service, U.S. Department of Agriculture.
- 1990 Oral History Interview with Earl Eugene Bachman by Jerry Rouillard. USDA Forest Service Oral History Project 1991 Centennial, Forest Reserves Act. Capital Campus Public History Program and Oral History Program, California State University, Sacramento, California. Public Affairs & Communication Office, Pacific Southwest Region, USDA Forest Service, Pacific Southwest Region, Vallejo, California.

Bean, Walton
- 1968 *California: An Interpretive History.* New York: McGraw-Hill Book Company.

Beardsley, Don
- 2004 Interview with Don Beardsley. The Greatest Good: A History of The Forest Service Project Files. Public Affairs & Communication Office, Pacific Southwest Region, USDA Forest Service, Pacific Southwest Region, Vallejo, California.

California Forest Industries
- 1959 *California Forest Facts.* Published in cooperation with American Forest Products Industries, Inc. Pacific Southwest Archives, Pacific Southwest Region, USDA Forest Service, Vallejo, California.

California Log
- 1955-63 Various

Church, Joseph B. "Joe" and Ginny Church
- 2004 Interview with Joseph B. "Joe" and Ginny Church by Bob Cermak. Region Five History Project. Public Affairs & Communication Office, Pacific Southwest Region, USDA Forest Service, Pacific Southwest Region, Vallejo, California.

Connaughton, Charles A.
- 1976 "Forty-three Years in the Field with the U.S. Forest Service: An interview with Charles A. Connaughton conducted by Elwood R. Maunder." Santa Cruz, California: Forest History Society.

Cleland, Robert Glass and Glenn S. Dumke
- 1962 *From Wilderness to Empire: A History of California.* New York: Alfred A. Knopf.

Clepper, Henry, editor
    1971   *Leaders of American Conservation.* New York: The Ronald Press Company.

Conners, Pamela Ann
    2005   Personal communication, 14 March 2005.

Dresser, Bill
    2004   Interview with Bill Dresser by Del Pengilly. Region Five History Project. Public Affairs & Communication Office, Pacific Southwest Region, USDA Forest Service, Pacific Southwest Region, Vallejo, California.

Feuchter, Roy
    2004   Interview with Roy Feuchter by Steve Kirby. Region Five History Project. Public Affairs & Communication Office, Pacific Southwest Region, USDA Forest Service, Pacific Southwest Region, Vallejo, California.

Grantham, Dewey
    1976   *The United States Since 1945: The Ordeal of Power.* New York: McGraw-Hill Book Company.

Gray, Bob
    2004   Interview with Bob Gray by Susana Luzier. Region Five History Project. Public Affairs & Communication Office, Pacific Southwest Region, USDA Forest Service, Pacific Southwest Region, Vallejo, California.

Grosch, Ed
    2004   Interview with Ed Grosch by Bob Smart. Region Five History Project. Public Affairs & Communication Office, Pacific Southwest Region, USDA Forest Service, Pacific Southwest Region, Vallejo, California.

Hirt, Paul W.
    1994   *A Conspiracy of Optimism: Management of the National Forests Since World War Two.* Lincoln, Nebraska: University of Nebraska Press.

James, Jim
    2004   Interview with Jim James by Gerald Gause. Region Five History Project. Public Affairs & Communication Office, Pacific Southwest Region, USDA Forest Service, Pacific Southwest Region, Vallejo, California.

LaLande, Jeff
    2003   "The 'Forest Ranger' in Popular Fiction." *Forest History Today* (Spring/Fall 2003): 2-29.

Leisz, Douglas
    2004   Interview with Douglas Leisz by Bob Smart. Region Five History Project. Public Affairs & Communication Office, Pacific Southwest Region, USDA Forest Service, Pacific Southwest Region, Vallejo, California.

    2005   Personal communication, 1 February 2005.

Lessel, Ralph
    2004   Interview with Ralph Lessel by Del Pengilly. Region Five History Project. Public Affairs & Communication Office, Pacific Southwest Region, USDA Forest Service, Pacific Southwest Region, Vallejo, California.

Lux, Linda
    2005   Personal communication, 9 March 2005.

Macebo, Lorraine
    2004   Interview with Lorraine Macebo by David Schreiner. Region Five History Project. Public Affairs & Communication Office, Pacific Southwest Region, USDA Forest Service, Pacific Southwest Region, Vallejo, California.

Millar, Richard
    2004   Interview with Richard Millar by John Grosvenor. Region Five History Project. Public Affairs & Communication Office, Pacific Southwest Region, USDA Forest Service, Pacific Southwest Region, Vallejo, California.

Nienaber, Jeanne O.
    1973   *The Politics and Policy of Environmental Decision-Making: A Case of the U.S. Forest Service.* Berkeley, California: Department of Political Science, University of California.

Norman, Rex, Don Lane, and Robert McDowell
    2004   *Perspectives on the Tahoe Story.* USDA Forest Service, Lake Tahoe Basin Management Unit. Public Affairs & Communication Office, Pacific Southwest Region, USDA Forest Service, Pacific Southwest Region, Vallejo, California.

Pacific/Southwest Log
    1982

Pendergrass, Lee F.
    1985   "The Forest Service in California: A Survey of Important Developments and People from 1905 to the Present." PenSec, Inc., Edmonds, Washington. Pacific Southwest Archives, Pacific Southwest Region, USDA Forest Service, Vallejo, California.

Peterson, Max
    2004   Interview with Max Peterson by Bryan Payne. Region Five History Project. Public Affairs & Communication Office, Pacific Southwest Region, USDA Forest Service, Pacific Southwest Region, Vallejo, California.

Radel, Joseph T.
    1991   Oral History Interview with Joseph T. Radel by Jerry Rouillard. USDA Forest Service Oral History Project 1991 Centennial, Forest Reserves Act. Capital Campus Public History Program and Oral History Program, California State University, Sacramento, California. Public Affairs & Communication Office, Pacific Southwest Region, USDA Forest Service, Pacific Southwest Region, Vallejo, California.

Rice, Bob
    2004   Interview with Bob Rice by Janet Buzzini. Region Five History Project. Public Affairs & Communication Office, Pacific Southwest Region, USDA Forest Service, Pacific Southwest Region, Vallejo, California.

Rice, Richard B., William A. Bullough and Richard J. Orsi
    1996   *The Elusive Eden: A New History of California.* 2nd Edition. New York: McGraw Hill Companies, Inc.

Smith, Glenn Spencer
    1990   Oral History Interview with Glenn Spencer Smith by Lyn Protteau. USDA Forest Service Oral History Project 1991 Centennial, Forest Reserves Act. Capital Campus Public History Program and Oral History Program, California State University, Sacramento, California. Public Affairs & Communication Office, Pacific Southwest Region, USDA Forest Service, Pacific Southwest Region, Vallejo, California.

Steen, Harold K.
    2004   *The U.S. Forest Service: A History* (Centennial Edition). Durham, North Carolina: Forest History Society; and Seattle, Washington: University of Washington Press.

Stewart, Ron
    2004   Interview with Ron Stewart by Fred Kaiser. Region Five History Project. Public Affairs & Communication Office, Pacific Southwest Region, USDA Forest Service, Pacific Southwest Region, Vallejo, California.

The Log
    1964-67   Various

USDA Forest Service
    1955a   "Management Direction for the National Forests of Sequoia Eastside Subregion." File Basic Data—Area Planning, Box 2, Los Padres National Forest Alphabetical-Subject Files, 1915-1963, Record Group 95, National Archive Record Center, Laguna Niguel, California.

    1955b   Letter: Clare Hendee to Regional Foresters, 14 February 1955. Annual Plan of Work. Box 0063-A-123, Regional Office, Record Group 95, National Archive Record Center, San Bruno, California.

    1955c   Annual Plan of Work for Region 5. Box 0063-A-123, Regional Office, Record Group 95, National Archive Record Center, San Bruno, California.

    1956a   Letter: Charles Connaughton to Supervisors, 21 February 1956. Box 53: R.O. Circulars, Numeric-Subject Files, Selected Historical Files, 1906-1959, Record Group 95, National Archive Record Center, San Bruno, California.

    1956b   Letter: Charles Connaughton to Supervisors, Depots, and Work Centers, 17 March 1956. Box 0063-A-123, Regional Office, Record Group 95, National Archive Record Center, San Bruno, California.

1956c	*Facts about the California Region*. Pacific Southwest Archives, Pacific Southwest Region, USDA Forest Service, Vallejo, California.
1957a	Edward C. Crafts and Russell B. McKennan, "General Integrating Inspection Report: California Region and California Forest and Range Experiment Station. Box 3/20, 95-94-1, Record Group 95, National Archive Record Center, San Bruno, California.
1957b	Annual Plan of Work for Region 5. Box 0063-A-123, Regional Office, Record Group 95, National Archive Record Center, San Bruno, California.
1957c	*Regional Forester's 1957 Report*. Box 12: Selected Alphabetical Subject File, 1909-1974, Record Group 95, National Archive Record Center, San Bruno, California.
1958a	Letter: Richard E McArdle to Regional Forester, Region 5, 22 May 1958. Box 3/20, 95-94-1, Record Group 95, National Archive Record Center, San Bruno, California.
1958b	*Regional Forester's 1958 Report*. Pacific Southwest Archives, Pacific Southwest Region, USDA Forest Service, Vallejo, California.
1959a	*Regional Forester's 1959 Report*. Pacific Southwest Archives, Pacific Southwest Region, USDA Forest Service, Vallejo, California.
1959b	"How Can We Resolve Conflicts in Forest Land Use?" by Charles A. Connaughton. Box 95-68A1270, Record Group 95, Federal Record Center, San Bruno, California.
1960a	*Regional Forester's 1960 Report*. Pacific Southwest Archives, Pacific Southwest Region, USDA Forest Service, Vallejo, California.
1960b	*Multiple Use Management Plan for Kern Plateau*. Revised. Box 8, 95-91-0003, Record Group 95, Federal Record Center, San Bruno, California.
1960c	*Multiple Use Management for the Northwest Subregion*. Box 1, 95-66-016, Record Group 95, Federal Record Center, San Bruno, California.
1961a	*Regional Forester's 1961 Report*. Pacific Southwest Archives, Pacific Southwest Region, USDA Forest Service, Vallejo, California.
1961b	*Multiple Use Management Guide for the National Forests of Southern California*. Revised 1961. Box 6, San Bernardino National Forest Alphabetical-Subject Files, Record Group 95, National Archive Record Center, Laguna Niguel, California.
1961c	*The National Forests of the California Region (R-5) What, Where & Why*. Pacific Southwest Archives, Pacific Southwest Region, USDA Forest Service, Vallejo, California.
1962	*Regional Forester's 1962 Report*. Pacific Southwest Archives, Pacific Southwest Region, USDA Forest Service, Vallejo, California.
1963	*Regional Forester's 1963 Report*. Pacific Southwest Archives, Pacific Southwest Region, USDA Forest Service, Vallejo, California.

1964	Letter: A.W. Greeley to Regional Forester, All Regions with attached chart "National Forest Multiple Use and Sustained Yield Management," 5 May 1965. Box 7, 95-91-0003, Record Group 95, Federal Record Center, San Bruno, California.
1964	*A Guide for Multiple Use Management of National Forest Land and Resources: Northern California Subregion.* Box 7, 95-91-0003, Record Group 95, Federal Record Center, San Bruno, California.
1966	George Vitas, General *Functional Inspection (I&E) Report: California Region and More Briefly the Pacific Southwest Forest And Range Experiment Station.* USDA Forest Service, Washington, D.C. Pacific Southwest Archives, Pacific Southwest Region, USDA Forest Service, Vallejo, California.
1976	June 24 Memorandum: R-5 Wilderness History: Charles R. Joy, "History of Wilderness in the National Forest System California (R-5)." Pacific Southwest Archives, Pacific Southwest Region, USDA Forest Service, Vallejo, California.

Watkins, T. H.
    1973   *California: An Illustrated History.* New York: Weathervane Books.

Williams, Gerald W.
    2000   *The USDA Forest Service—The First Century.* Washington, D.C.: USDA Forest Service.

    2001   "Wilderness Act and the Roadless Area Reviews." Attached to personal communication, 11 March 2005.

Wilson, Carl C.
    1991   Oral History Interview with Carl C. Wilson by Susan Douglass. USDA Forest Service Oral History Project 1991 Centennial, Forest Reserves Act. Capital Campus Public History Program and Oral History Program, California State University, Sacramento, California. Public Affairs & Communication Office, Pacific Southwest Region, USDA Forest Service, Pacific Southwest Region, Vallejo, California.

Chapter X: 1967-1978

Alfano, Sam
    2004   Interview with Sam Alfano by Larry Hornberger. Region Five History Project. Public Affairs & Communication Office, Pacific Southwest Region, USDA Forest Service, Pacific Southwest Region, Vallejo, California.

Bachman, Earl Eugene
    1990   Oral History Interview with Earl Eugene Bachman by Jerry Rouillard. USDA Forest Service Oral History Project 1991 Centennial, Forest Reserves Act. Capital Campus Public History Program and Oral History Program, California State University, Sacramento, California. Public Affairs & Communication Office, Pacific Southwest Region, USDA Forest Service, Pacific Southwest Region, Vallejo, California.

Beardsley, Don
    2004   Interview with Don Beardsley. The Greatest Good: A History of The Forest Service Project Files. Public Affairs & Communication Office, Pacific Southwest Region, USDA Forest Service, Pacific Southwest Region, Vallejo, California.

Boyd, Anne
    1995   "A Brief History of the U.S. Forest Service Cultural Resources Management Program: A Look at the Laws and Two Forest Perspectives." Pacific Southwest Archives, Pacific Southwest Region, USDA Forest Service, Vallejo, California.

Buck, Hurston and Joyce
    2004   Interview with Hurston and Joyce Buck by Larry Hornberger. Region Five History Project. Public Affairs & Communication Office, Pacific Southwest Region, USDA Forest Service, Pacific Southwest Region, Vallejo, California.

California Log
    1969-78   Various

California Resources Agency
    1979   *Today and Tomorrow: Report of Citizens Committee on U.S. Forest Service Management Practices in California.* Sacramento, California: The Resources Agency.

Cermak, Robert
    2004   Interview with Robert "Bob" Cermak by Larry Gause. Region Five History Project. Public Affairs & Communication Office, Pacific Southwest Region, USDA Forest Service, Pacific Southwest Region, Vallejo, California.
    2005   Personal communication, 14 February 2005.

Congressional Record
    1967   "The Walt Disney Memorial Conservation Center—Tribute to a Unique American Genius." 12 January 1967: S196-S203.

Coutant, Gerald J.
    1990   *A Chronology of the Recreation History of the National Forests and the U.S.D.A Forest Service: 1940 to 1990.* (Draft Copy for Review). Pacific Southwest Archives, Pacific Southwest Region, USDA Forest Service, Vallejo, California.

Feuchter, Roy
    2004   Interview with Roy Feuchter by Steve Kirby. Region Five History Project. Public Affairs & Communication Office, Pacific Southwest Region, USDA Forest Service, Pacific Southwest Region, Vallejo, California.

Grace, Harry
    2004   Interview with Harry Grace by Gene Murphy. Region Five History Project. Public Affairs & Communication Office, Pacific Southwest Region, USDA Forest Service, Pacific Southwest Region, Vallejo, California.

Grosch, Ed
    2004   Interview with Ed Grosch by Bob Smart. Region Five History Project.

Public Affairs & Communication Office, Pacific Southwest Region, USDA Forest Service, Pacific Southwest Region, Vallejo, California.

Hata, Nadine Ishitani
    1992   *The Historic Preservation Movement in California, 1940-1976.* Sacramento, California: California Department of Parks and Recreation.

Hill, Paul
    2004   Interview with Paul Hill by David Schreiner. Region Five History Project. Public Affairs & Communication Office, Pacific Southwest Region, USDA Forest Service, Pacific Southwest Region, Vallejo, California.

Hirt, Paul W.
    1994   *A Conspiracy of Optimism: Management of the National Forests Since World War Two.* Lincoln, Nebraska: University of Nebraska Press.

Irwin, Robert L.
    2004   Interview with Robert L. "Bob" Irwin by Glenn Gottschall. Region Five History Project. Public Affairs & Communication Office, Pacific Southwest Region, USDA Forest Service, Pacific Southwest Region, Vallejo, California.

James, Jim
    2004   Interview with Jim James by Gerald Gause. Region Five History Project. Public Affairs & Communication Office, Pacific Southwest Region, USDA Forest Service, Pacific Southwest Region, Vallejo, California.

Kennedy, John and Mary
    2004a   Interview with John and Mary Kennedy by John Fiske. Region Five History Project. Public Affairs & Communication Office, Pacific Southwest Region, USDA Forest Service, Pacific Southwest Region, Vallejo, California.

Kennedy, Jon
    2004b   Interview with Jon Kennedy by John Grosvenor. Region Five History Project. Public Affairs & Communication Office, Pacific Southwest Region, USDA Forest Service, Pacific Southwest Region, Vallejo, California.

Leisz, Douglas R.
    1990   Oral History Interview with Douglas R. "Doug" Leisz by Judy Allen. USDA Forest Service Oral History Project 1991 Centennial, Forest Reserves Act. Capital Campus Public History Program and Oral History Program, California State University, Sacramento, California. Public Affairs & Communication Office, Pacific Southwest Region, USDA Forest Service, Pacific Southwest Region, Vallejo, California.

    2004   Interview with Douglas Leisz by Bob Smart. Region Five History Project. Public Affairs & Communication Office, Pacific Southwest Region, USDA Forest Service, Pacific Southwest Region, Vallejo, California.

    2005   Personal communication, 1 February 2005.

Leonard, George
    2004   Interview with George Leonard by Bob Van Aken. Region Five History Project. Public Affairs & Communication Office, Pacific Southwest Region, USDA Forest Service, Pacific Southwest Region, Vallejo, California.

Lewis, James G.
    2005   *The Greatest Good: A USFS Centennial History* (draft). Attached to personal communication, 28 February 2005.

Millar, Richard
    2004   Interview with Richard Millar by John Grosvenor. Region Five History Project. Public Affairs & Communication Office, Pacific Southwest Region, USDA Forest Service, Pacific Southwest Region, Vallejo, California.

Miller, Donald S.
    1998   Oral History Interview with Donald S. Miller by Jacqueline S. Reiner. Pacific Southwest Archives, Pacific Southwest Region, USDA Forest Service, Vallejo, California.

Nienaber, Jeanne O.
    1973   *The Politics and Policy of Environmental Decision-Making: A Case of the U.S. Forest Service*. Berkeley, California: Department of Political Science, University of California.

Norman, Rex, Don Lane, and Robert McDowell
    2004   *Perspectives on the Tahoe Story*. USDA Forest Service, Lake Tahoe Basin Management Unit. Public Affairs & Communication Office, Pacific Southwest Region, USDA Forest Service, Pacific Southwest Region, Vallejo, California.

Pacific/Southwest Log
    1985

Pendergrass, Lee F.
    1985   "The Forest Service in California: A Survey of Important Developments and People from 1905 to the Present." PenSec, Inc., Edmonds, Washington. Pacific Southwest Archives, Pacific Southwest Region, USDA Forest Service, Vallejo, California.

Peterson, Max
    2004   Interview with Max Peterson by Bryan Payne. Region Five History Project. Public Affairs & Communication Office, Pacific Southwest Region, USDA Forest Service, Pacific Southwest Region, Vallejo, California.

Rice, Bob
    2004   Interview with Bob Rice by Janet Buzzini. Region Five History Project. Public Affairs & Communication Office, Pacific Southwest Region, USDA Forest Service, Pacific Southwest Region, Vallejo, California.

Rice, Richard B., William A. Bullough and Richard J. Orsi
    1996   *The Elusive Eden: A New History of California.* 2nd Edition. New York: McGraw Hill Companies, Inc.

Righetti, Bob
    2004   Interview with Bob Righetti by David Schreiner. Region Five History Project. Public Affairs & Communication Office, Pacific Southwest Region, USDA Forest Service, Pacific Southwest Region, Vallejo, California.

Robinson, Glen O.
    1975   *The Forest Service.* Baltimore, Maryland: John Hopkins University Press for Resources for the Future, Inc.

Schmidt, Andrew R.
    2004   Interview with Andrew R. Schmidt by Nordstrom "Nord" Whited. Region Five History Project. Public Affairs & Communication Office, Pacific Southwest Region, USDA Forest Service, Pacific Southwest Region, Vallejo, California.

Schneegas, Edward R.
    2003   Interview with Edward R. Schneegas by Bob Cermak. Region Five History Project. Public Affairs & Communication Office, Pacific Southwest Region, USDA Forest Service, Pacific Southwest Region, Vallejo, California.

Smart, Bob
    2004   Interview with Bob Smart by Doug Leisz. Region Five History Project. Public Affairs & Communication Office, Pacific Southwest Region, USDA Forest Service, Pacific Southwest Region, Vallejo, California.

Smith, Glenn Spencer
    1990   Oral History Interview with Glenn Spencer Smith by Lyn Protteau. USDA Forest Service Oral History Project 1991 Centennial, Forest Reserves Act. Capital Campus Public History Program and Oral History Program, California State University, Sacramento, California. Public Affairs & Communication Office, Pacific Southwest Region, USDA Forest Service, Pacific Southwest Region, Vallejo, California.

Steen, Harold K.
    2004   *The U.S. Forest Service: A History* (Centennial Edition). Durham, North Carolina: Forest History Society; and Seattle, Washington: University of Washington Press.

Stewart, Ron
    2004   Interview with Ron Stewart by Fred Kaiser. Region Five History Project. Public Affairs & Communication Office, Pacific Southwest Region, USDA Forest Service, Pacific Southwest Region, Vallejo, California.

The Log
    1967-68   Various

U.S. Department of Agriculture, Forest Service

1964   Letter: Gordon D. Fox to Regional Foresters, 15 June 1964. Box 2, 95-68a-1270, National Archive Record Center, San Bruno, California.

1966a   Letter: Charles Connaughton to Forest Supervisors, Shasta-Trinity, Klamath, Six Rivers, and Mendocino, 3 March 1966 with attachment "Handling the Impacts of Patch Cutting." Box 7, 95-91-003, National Archive Record Center, San Bruno, California.

1966b   "Handling the Impacts of Patch Cutting." Revised 8 July 1966, Box 7, 95-91-003, National Archive Record Center, San Bruno, California.

1967a   "Minutes of the Supervisory Meeting, October 5 & 6, 1967." Pacific Southwest Archives, Pacific Southwest Region, USDA Forest Service, Vallejo, California.

1967b   Letter: Peter J. Wyckoff to Regional Forester with attachment "Mineral King VIS Plan." 5 July 1967. Box 9, 95-96-049, Record Group 95, National Archive Record Center, San Bruno, California.

1968a   "Conservation Education" Region Five Manual Supplement No. 1. March 1969. General Integrated Inspection of Forest Service Activities in California (Bacon and Howard). Box 4: 95-94-1, Record Group 95, National Archive Record Center, San Bruno, California.

1968b   E. M. Bacon and H. E. Howard, *Report of the General Integrated Inspection of Forest Service Activities in California*. Box 4: 95-94-1, Record Group 95, National Archive Record Center, San Bruno, California.

1968c   *Progress in 1968: Forest Service, California Region*. Pacific Southwest Archives, Pacific Southwest Region, USDA Forest Service, Vallejo, California.

1968d   *U.S. Forest Service Program in Hawaii: Short Term Objectives and a A Cooperative Action Program*. Box 42: 95-96-049, Record Group 95, National Archive Record Center, San Bruno, California.

1968e   Bill Dresser, "National Forest Recreation." Presented at Forest Service Supervisor-Assistant Regional Forester's Meeting, Redding, California. 25 March 1969. Box 4: 95-94-1, Record Group 95, National Archive Record Center, San Bruno, California.

1969   Letter: Grant A. Morse to Joe Fynn and W. S. Davis with Attachment "Justification Statement Archeologist." 24 April 1969. Pacific Southwest Archives, Pacific Southwest Region, USDA Forest Service, Vallejo, California.

1970a   *Important Facts about the National Forests in California*. Pacific Southwest Archives, Pacific Southwest Region, USDA Forest Service, Vallejo, California.

1970b   Letter: B. K. Cooperrider to Forest Supervisor, Tahoe National Forest. 23 July 1970. Box 3, Angeles National Forest, Numeric-Subject Files, Record Group 95, National Archive Record Center, Laguna Niguel, California.

1970c   Letter: Edward P. Cliff to Regional Foresters, Directors, Area Directors, and WO Staff. 24 August 1970. Pacific Southwest Archives, Pacific Southwest Region, USDA Forest Service, Vallejo, California.

1971a   Letter: William T. Dresser to Regional Forester 29 June 1971. Box 1, Angeles National Forest, Numeric-Subject Files, Record Group 95, National Archive Record Center, Laguna Niguel, California.

1971b   "Report on the First Meeting of the Regional Archaeology-History Committee." Pacific Southwest Archives, Pacific Southwest Region, USDA Forest Service, Vallejo, California.

1971c   "Report on Perspectives of a Sufficient Antiquities Management Program, Region 5, United States Forest Service. Pacific Southwest Archives, Pacific Southwest Region, USDA Forest Service, Vallejo, California.

1971d   "An Antiquities Management Statement as it relates to The Forest Service Environmental Program for the Future." Pacific Southwest Archives, Pacific Southwest Region, USDA Forest Service, Vallejo, California.

1972   "Statement of Douglas R. Leisz, Regional Forester." Visalia, California. 3 May 1972. Box 1, Sequoia National Forest, Numeric-Subject Files, Record Group 95, National Archive Record Center, Laguna Niguel, California.

1973a   Letter: Douglas R. Leisz to Forest Supervisors, Division Chiefs, Station Director, 5 December 1973. Box 4, 95-91-0003, Record Group 95, National Archive Record Center, San Bruno, California.

1973b   "Land Use Planning Overview." Box 3, Angeles National Forest, Numeric-Subject Files, Record Group 95, National Archive Record Center, Laguna Niguel, California.

1973c   Letter: Larry Wade to Forest Supervisor & Staff. 17 May 1973. Box 3, Angeles National Forest, Numeric-Subject Files, Record Group 95, National Archive Record Center, Laguna Niguel, California.

1974a   Letter: Tahoe National Forest to Forest Supervisor, Shasta Trinity National Forest, 14 February 1974. Box 4, 95-91-0003, Record Group 95, National Archive Record Center, San Bruno, California.

1974b   Letter: Douglas R. Leisz to Forest Supervisors. 19 September 1974. Box 1, Los Padres National Forest, Numeric-Subject Files, Record Group 95, National Archive Record Center, Laguna Niguel, California.

1974c   "Sugarloaf Incident—Lassen National Forest." 8 January 1974. Public Affairs & Communication Office, Pacific Southwest Region, USDA Forest Service, Pacific Southwest Region, Vallejo, California.

1975a   "Mailing List for Working Group for Northern California Area Guide." 17 April 1975. Box 4, 95-91-0003, Record Group 95, National Archive Record Center, San Bruno, California.

1976a   June 24 Memorandum: Region Five Wilderness History: Charles R. Joy, "History of Wilderness in the National Forest System California (Region Five)." Pacific Southwest Archives, Pacific Southwest Region, USDA Forest Service, Vallejo, California.

1976b   Letter: Resources Agency of California to Douglas R. Leisz. 26 November 1976. Box 8, 95-91-003, Record Group 95, National Archive Record Center, San Bruno, California.

1976c   Various Letters Re: NORCAL *Guide*. Box 8, 95-91-003, Record Group 95, National Archive Record Center, San Bruno, California.

1976d   Letter: William H. Covey to Regional Forester, R-5. 17 June 1977. Box 8, 95-91-003, Record Group 95, National Archive Record Center, San Bruno, California.

1976e   Letter: Douglas R. Leisz to Forest Supervisors, LTBMU Administrator, Staff Directors. 15 July 1977. Box 8, 95-91-003, Record Group 95, National Archive Record Center, San Bruno, California.

1977   Letter: Douglas R. Leisz to A. R. Gutowsky, Mother Lode Chapter, Sierra Club. 27 September 1977. Box 8, 95-91-003, Record Group 95, National Archive Record Center, San Bruno, California.

1978a   Letter: Douglas R. Leisz to Forest Supervisors, LTBMU. 5 June 1978. Box 3, Cleveland National Forest, Numeric-Subject Files, Record Group 95, National Archive Record Center, Laguna Niguel, California.

1978b   Various Letters: Special Interest Areas. Box 3, Cleveland National Forest, Numeric-Subject Files, Record Group 95, National Archive Record Center, Laguna Niguel, California.

U.S. Department of Interior, National Park Service

1967   "The Mineral King Area as Related to Sequoia National Park." January 1967. Box 9, 95-96-049, Record Group 95, National Archive Record Center, San Bruno, California.

1980   *Sequoia/Kings Canyon: Research and Consultations for Purposes of Implementing the American Indian Religious Freedom Act and Ascertaining Such Residual Rights of Indians as Might Exist*. National Park Service. Pacific Southwest Archives, Pacific Southwest Region, USDA Forest Service, Vallejo, California.

Walt Disney Productions

1967   "Walt Disney's Hope for Mineral King." Brochure. Box 9, 95-96-049, Record Group 95, National Archive Record Center, San Bruno, California.

1972   "Remarks by E. Cardon Walker, President, Walt Disney Productions." 3 May 1972. Box 1, Sequoia National Forest, Numeric-Subject Files, Record Group 95, National Archive Record Center, Laguna Niguel, California.

Westenberger, W. Jane.
    1991   Oral History Interview with W. Jane Westenberger by Susan Douglass. USDA Forest Service Oral History Project 1991 Centennial, Forest Reserves Act. Capital Campus Public History Program and Oral History Program, California State University, Sacramento, California. Public Affairs & Communication Office, Pacific Southwest Region, USDA Forest Service, Pacific Southwest Region, Vallejo, California.

Williams, Gerald W.
    2000   *The USDA Forest Service—The First Century.* Washington, D.C.: USDA Forest Service.

Wilson, Carl C. and James B. Davis
    1988   *Forest Fire Laboratory at Riverside and Fire Research in California: Past, Present, and Future.* Berkeley, California: Pacific Southwest Forest and Range Experiment Station.

Chapter XI: 1978-1987

Blakley, E. R. and Karen Barnette
    1985   *Historical Overview of the Los Padres National Forest, 1542-1984.* Pacific Southwest Archives, Pacific Southwest Region, USDA Forest Service, Vallejo, California.

Boyd, Anne
    1995   "A Brief History of the U.S. Forest Service Cultural Resources Management Program: A Look at the Laws and Two Forest Perspectives." Pacific Southwest Archives, Pacific Southwest Region, USDA Forest Service, Vallejo, California.

California Log
    1978   Various

Conners, Pamela Ann
    2005   Personal communication, 15 March 2005.

Frome, Michael
    1984   *The Forest Service.* 2nd Edition, Revised and Updated. Boulder, Colorado: Westview Press.

Harn, Joseph Harry
    1990   Oral History Interview with Joseph "Joe" Harry Harn by Becky Carruthers. USDA Forest Service Oral History Project 1991 Centennial, Forest Reserves Act. Capital Campus Public History Program and Oral History Program, California State University, Sacramento, California. Public Affairs & Communication Office, Pacific Southwest Region, USDA Forest Service, Pacific Southwest Region, Vallejo, California.

Hirt, Paul W.
    1994   *A Conspiracy of Optimism: Management of the National Forests Since World War Two.* Lincoln, Nebraska: University of Nebraska Press.

Jackson, W. Turrentine, Rand Herbert, and Stephen Wee
    n.d.   "History of the Tahoe National Forest, 1840-1940: A Cultural Resources Overview History." David, California: Jackson Research Projects. Pacific Southwest Archives, Pacific Southwest Region, USDA Forest Service, Vallejo, California.

Johnston, James and Elizabeth Budy
    1982   "Cultural Resource Management Overview: Lassen National Forest." Volume One: Cultural History. Pacific Southwest Archives, Pacific Southwest Region, USDA Forest Service, Vallejo, California.

Kennedy, John and Mary
    2004a   Interview with John and Mary Kennedy by John Fiske. Region Five History Project. Public Affairs & Communication Office, Pacific Southwest Region, USDA Forest Service, Pacific Southwest Region, Vallejo, California.

King, Tom
    1994   "The National Forests: 100 Years of Managing Minerals." *Minerals Today* (August 1994): 16-21.

Leisz, Douglas R.
    1990   Oral History Interview with Douglas R. Leisz by Judy Allen. USDA Forest Service Oral History Project 1991 Centennial, Forest Reserves Act. Capital Campus Public History Program and Oral History Program, California State University, Sacramento, California. Public Affairs & Communication Office, Pacific Southwest Region, USDA Forest Service, Pacific Southwest Region, Vallejo, California.
    2004   Interview with Douglas Leisz by Bob Smart. Region Five History Project. Public Affairs & Communication Office, Pacific Southwest Region, USDA Forest Service, Pacific Southwest Region, Vallejo, California.

Leonard, George
    2004   Interview with George Leonard by Bob Van Aken. Region Five History Project. Public Affairs & Communication Office, Pacific Southwest Region, USDA Forest Service, Pacific Southwest Region, Vallejo, California.

McDonald, James A.
    1979   "Cultural Resource Overview: Klamath National Forest, California." Klamath National Forest, USDA Forest Service. Pacific Southwest Archives, Pacific Southwest Region, USDA Forest Service, Vallejo, California.

Pacific Southwest Log
    1980-87   Various

Pendergrass, Lee F.
    1985   "The Forest Service in California: A Survey of Important Developments and People from 1905 to the Present." PenSec, Inc., Edmonds, Washington. Pacific Southwest Archives, Pacific Southwest Region, USDA Forest Service, Vallejo, California.

Peterson, Max
    2004   Interview with Max Peterson by Bryan Payne. Region Five History Project. Public Affairs & Communication Office, Pacific Southwest Region, USDA Forest Service, Pacific Southwest Region, Vallejo, California.

Radel, Joseph T.
    1991   Oral History Interview with Joseph T. Radel by Jerry Rouillard. USDA Forest Service Oral History Project 1991 Centennial, Forest Reserves Act. Capital Campus Public History Program and Oral History Program, California State University, Sacramento, California. Public Affairs & Communication Office, Pacific Southwest Region, USDA Forest Service, Pacific Southwest Region, Vallejo, California.

Rice, Bob
    2004   Interview with Bob Rice by Janet Buzzini. Region Five History Project. Public Affairs & Communication Office, Pacific Southwest Region, USDA Forest Service, Pacific Southwest Region, Vallejo, California.

Rice, Richard B., William A. Bullough and Richard J. Orsi
    1996   *The Elusive Eden: A New History of California*. 2nd Edition. New York: McGraw Hill Companies, Inc.

Rock, James T.
    1981   "Cultural Resources Klamath National Forest Situation Statement." April 1981. Public Affairs & Communication Office, Pacific Southwest Region, USDA Forest Service, Pacific Southwest Region, Vallejo, California.

San Francisco *Chronicle*
    1982   "Indians Fighting to Block Road Into Holy Lands;" and "Indian Doctor Warns of Great Creator's Wrath." 12 July 1982.

Schneegas, Edward R.
    2003   Interview with Edward R. Schneegas by Bob Cermak. Region Five History Project. Public Affairs & Communication Office, Pacific Southwest Region, USDA Forest Service, Pacific Southwest Region, Vallejo, California.

Smart, Bob
    2004   Interview with Bob Smart by Douglas R. Leisz. Region Five History Project. Public Affairs & Communication Office, Pacific Southwest Region, USDA Forest Service, Pacific Southwest Region, Vallejo, California.

Smith, Glenn Spencer
    1990   Oral History Interview with Glenn Spencer Smith by Lyn Protteau. USDA

Forest Service Oral History Project 1991 Centennial, Forest Reserves Act. Capital Campus Public History Program and Oral History Program, California State University, Sacramento, California. Public Affairs & Communication Office, Pacific Southwest Region, USDA Forest Service, Pacific Southwest Region, Vallejo, California.

Stewart, Ron
   2004   Interview with Ron Stewart by Fred Kaiser. Region Five History Project. Public Affairs & Communication Office, Pacific Southwest Region, USDA Forest Service, Pacific Southwest Region, Vallejo, California.

US Department of Agriculture, Forest Service
   circa 1980   Al Lampi, John Leasure, Bob Van Aken, Bill Russell and Jon Kennedy, *Program Review of Land and Resource Management Planning in Region 5*. Washington D.C. Box 6, 95-91-0003, National Archive Record Center, San Bruno, California.

   1980a   Letter: Zane G. Smith Jr. to Forest Supervisors. 12 November 1980. Box 6, 95-91-0003, National Archive Record Center, San Bruno, California.

   1980b   *The Pacific Southwest Region Today*. Working Draft. 24 September 1980. Box 6, 95-91-0003, National Archive Record Center, San Bruno, California.

   1981a   Letter: Zane G. Smith Jr. to Forest Supervisors, Staff Directors. 21 January 1981. Box 6, 95-91-0003, National Archive Record Center, San Bruno, California.

   1981b   Letter: Zane G. Smith Jr. to Forest Supervisors, and Staff Directors with Attachment *Land Management Planning Direction*. 26 June 1981. Box 6, 95-91-0003, National Archive Record Center, San Bruno, California.

   1981c   Letter: Zane G. Smith Jr. to Forest Supervisors, Administrator, LTBMU, Staff Directors. 15 July 1980. Box 6, 95-91-0003, National Archive Record Center, San Bruno, California.

   1981d   *Regional Land and Resource Management Plan* (Draft). June 1981. Box 6, 95-91-0003, National Archive Record Center, San Bruno, California.

   1981e   *Pacific Southwest Region: Draft Environmental Impact Statement*. June 1981. Box 6, 95-91-0003, National Archive Record Center, San Bruno, California.

   1982   Letter: Zane G. Smith Jr. to Dr. Richard P. Gayle. 4 February 1982. Box 3, 95-91-0003, National Archive Record Center, San Bruno, California.

   1983   Letter: Zane G. Smith Jr. to Forest Supervisors, Resource Staff Directors, O.I. 19 August 1983. Box 3, 95-91-0003, National Archive Record Center, San Bruno, California.

   1995   *Sustaining Ecosystems: A Conceptual Framework*. April. Public Affairs & Communication Office, Pacific Southwest Region, USDA Forest Service, Pacific Southwest Region, Vallejo, California.

WESTEC Services
    1982   *Cultural Resources Overview of the San Bernardino National Forest.* Pacific Southwest Archives, Pacific Southwest Region, USDA Forest Service, Vallejo, California.

Westenberger, W. Jane.
    1991   Oral History Interview with W. Jane Westenberger by Susan Douglass. USDA Forest Service Oral History Project 1991 Centennial, Forest Reserves Act. Capital Campus Public History Program and Oral History Program, California State University, Sacramento, California. Public Affairs & Communication Office, Pacific Southwest Region, USDA Forest Service, Pacific Southwest Region, Vallejo, California.

Williams, Gerald W.
    2000   *The USDA Forest Service—The First Century.* FS-650. Washington, D.C.: USDA Forest Service.

Epilogue

Lewis, James G.
    2005   *The Greatest Good: A USFS Centennial History* (draft). Attached to personal communication, 28 February 2005.

Mission-Centered Solutions, Inc. and Guidance Group, Inc.
    2003   *Southern California Firestorm 2003: Report for the Wildland Fire Lessons Learned Center.* Prepared for the National Advanced Resource Technology Center. Parker, Colorado.

Ruth, Lawrence
    2005   "Conservation on the Cusp: The Reformation of National Forest Service Policy in the Sierra Nevada" (draft). Attached to personal communication, 1 March 2005.

Stewart, Ronald E.
    2004b   "PSW: Wood, Water, Fire, Recreation, and People: The Contributions of Research." Paper presented at Pacific Southwest Centennial Forum on California, Hawaii and the Pacific Islands, Sacramento, California. November 5-6, 2004.

Index

Index of People

A
Abbey, Robert Harvey, 82-84
Agassiz, Alexander, 44
Albright, Horace, 450
Alexandre, Sue, 455-456
Allen, Benjamin F., 37-39, 49, 52, 56, 110
Anza, Juan Bautista de, 5
Arguello, Luis, 71
Arnold, Henry H. "Hap," 186-187, 241-242
Arr, Ray, 122 *photo*
Audubon, John W., 58
Austin, Mary, 31
Ayres, R.W., 74, 98, 221, 223, 256, 321

B
Bacon, E.M., 436-437
Baldwin, Roger, 134
Ballinger, Richard, 96, 282
Barker, Paul F., 502, 523, 525
Barnette, Karen, 522
Barrett, Louis A., 83, 91, 215-218, 221, 232, 244, 252-253, 256
Bean, Leslie S., 339-340
Benedict, Maurice A., 215, 322
Bernardi, Gene, 497
Bierstadt, Albert, 22, 106
Bigelow, Richard L.P., 55, 69, 78, 78 *photo*, 80, 159, 200, 221, 256, 331
Blackwell, Jack A., 525
Blakely, E.R., 522
Bonner, F.E., 210
Boothe, Roy, 322
Bosworth, Dale, 452
Bosworth, Erwin, 452
Bowers, W.W., 38
Bradley, Tom, 485
Brandeberry, J.K., 263-264
Brandis, Dietrich, 75
Brandt, Robert, 423-424
Brewer, William Henry, 44
Brower, David R., 363, 366, 401
Brown, Edmund G. "Pat," 424, 430
Brown, Jerry, 430-431, 484, 488
Brownlow, Ranger, 98
Budy, Elizabeth, 522
Burcham, L.T., 358
Burnett, Bruce B., 246, 254, 258

C
Callaham, Robert Z., 498

Capper, Arthur, 181
Carhart, Arthur H., 200
Carnall, N.C., 106
Carson, Rachel, 379, 412
Carter, James "Jimmy," 487, 490, 506
Carter, Thomas H., 38-39, 42
Cecil, George H.,
Cermak, Robert "Bob," 58
Chaplin, Charlie, 165, 175
Charlton, Rushton H., 186
Chavez, Cesar, 378
Church, Frank, 466, 469
Clapp, Earle H., 289, 291-293, 325 *photo*
Clarke, John W., 183
Cleland, Robert Glass, 117, 377
Cleveland, Grover, 44-46
Cliff, Edward P., 341, 413, 441, 452-453
Colbert, Claudette, 165
Colby, Will, 215-216
Commoner, Barry, 412
Connaughton, Charles Arthur
    administration and leadership, 381-384, 493
    biography and career, 380-381
    conservation/preservation disputes, 426-428
    fire control plan, 443
    Kern Plateau multiple use plan, 387-389
    Mineral King winter sports development controversy, 369, 387, 420-425
    multiple use management, 382, 384, 387, 400, 414, 420, 425, 428
    Operation Multiple Use, 402
    photographs, 380 *photo*, 383 *photo*
    prescribed burning, 383
    public involvement and land management planning, 437, 452
    recreation, 420
    relationship with timber industry, 382, 387
    timber management policy, 382, 408, 425, 440
    timber sales, 385
    transfers to Region 6 (Pacific Northwest), 426, 451
    wilderness and, 413-414
    winter sports, 383
Conner, Pam, 103
Cooke, Alistair, 166
Coolidge, Calvin, 182, 194, 199
Copeland, Royal S., 237
Covey, William H., 464
Crafts, Edward C., 384
Craig, Malin, 241, 243
Cranston, Alan, 510
Crawford, Joan, 165
Cronemiller, Fred P., 272, 272 *photo*, 320, 330, 394

Cronkite, Walter, 465
Crowell, John B. Jr., 494-495, 517
Cutler, Rupert, 495

D

Daggett, Hallie Morse, 124, 124 *photo*
Dasmann, Raymond F, 380
Dasmann, William, 394
Davies, Warren, 499
Davies, William Heath, 31
Davis, Wilfred "Slim," 360, 481
Deinema, John W. "Jack,"
  administration and leadership, 432-435, 443, 446
  biography and career, 432-433
  equal employment opportunity (EEO) issues including Consent Decree, 434
  Mineral King winter sports development controversy, 447-450
  photograph, 433 *photo*
  public involvement and land management planning, 452
  Tahoe Regional Planning Agency (TRPA), 438, 461
  timber management policy, 440
  transfers to Washington office, 450
Deering, Robert L., 171, 240, 242, 247
DeMille, Cecil B., 179
Deukemejian, George, 490-491
Diaz, Melchior, 3-4
Diaz, Porfirio, 116
Disney, Walt, 377, 422, 424-425, 447
Dobson, J.W., 92
DuBois, Coert,
  administration and leadership, 89, 118-119, 144, 171, 249, 330
  Army Air Service patrols, 186-187
  biography and career, 74, 118-119, 170
  California Inspection District 5, 85
  economic fire protection policy, 140, 189-190
  fire control and protection plans and policy, 97-99, 136-138, 146, 264
  grazing fee system, 141-142
  ideology and philosophy, 201
  management or work plans, 185
  multiple use management, 130
  opinion of General Land Office, 75
  photograph of, 118 *photo*
  professionalization of staff, 172
  range reconnaissance, 144-145
  relationship with Frederick Olmsted, 118
  relationship with Gifford Pinchot, 74, 118
  salaries and working conditions, 168-170
  scientific management and Taylorism, 119-120, 160
  Systematic Fire Protection in the California Forests, 136, 138-139, 201
  timber reconnaissance surveys, 132-133

utilitarian conservation, 160
World War I, 153, 159, 169, 189
Dunham, Frederick E., 255
Dunning, Duncan, 183-185, 205
Dutton, W.L., 334
Dyer, Ernest I., 287

E

Eaton, Fred, 35
Edwards, John, 122 *photo*
Egleston, Nathaniel, 24
Eisenhower, Dwight D., 306, 373, 376, 394, 399-400
Elliott, Al, 285
Elliott, J.E., 250 *photo*
Elliott, Joseph Clinton, 221
Elliott, M.B., 142
Engen, Alf, 286
Engle, Clair, 357

F

Fairbanks, 175
Fall, Albert B., 282
Farquhar, Francis, 216
Fausett, A., 273
Fechner, Robert, 239
Fermi, Enrico, 299
Fernow, Bernhard, 24
Findlay, George, 91
Ford, Gerald R., 480
Fox, Charles E., 33,
Fox, Gordon D., 338, 406
Freeman, Orville, 414, 424, 441
Freemont, John C., 8, 22
Friedan, Betty, 406
Friedhoff, W.H., 280
Friedlander, Isaac, 27
Frome, Michael, 437
Frome, R.L., 89

G

Gable, Clark, 165
Gallagher, W.M., 132
Galvez, Jose de, 4
Gearhart, Bertrand "Bud," 284-285
George, Henry, 30
Gibbs, George, 252, 281-282
Gibbs, Oliver Wolcott, 44
Ginsberg, Allen, 377
Gish, Lillian, 165
Goddard, Paulette, 324, 325 *photo*

Godwin, David P., 137, 140
Gowen, George M., 264
Graves, Henry S., 118, 126, 181, 182, 201
    biography and career, 96, 170
    private forestry, importance of, 110
    recreation, 146
    relations with National Park Service, 149-150
    relationship with Gifford Pinchot, 97, 170
    salaries and working conditions, 169-170
    timber marking, 131
    timber work plans, 130
    25th anniversary of Forest Service, 220
    wildlife management, 151
    World War I, 153
Greeley, William B., 80, 200
    administration and leadership, 220, 249
    biography and career, 172
    grazing fees, 195
    1924 fire season, 193, 195
    1926 fire season, 204
    recreation management, 212
    relationship with National Park Service (NPS), 214-215
    resigns, 203
    wilderness area system, 215
Griswold, Norman W., 106

H

Hague, Arnold, 44
Hall, Harvey Monroe, 46
Hall, L. Glenn, 252, 281
Hammitt, R.F., 132, 201
Harding, Warren G., 181-182, 199, 495
Harriman, E.H., 110
Harrington, Michael, 404
Harrison, Benjamin, 36-38, 40-41, 438
Hastings, Lansford W., 8
Hatton, John H. 85, 89, 99-100, 142
Hayward, Susan, 324
Headley, Roy, 137, 140-141, 169, 189, 201
Hearst, William Randolph, 117
Heceta, Bruno, 71
Heller, Alfred, 380
Hendee, Clare W.
    administration and leadership, 337-340, 344, 347, 384, 493
    biography and career, 330, 337
    multiple use management, 373
    photograph, 337 *photo*
    public involvement and land management planning, 452
    range management, 358

    recreation management, 362-363
    timber sales, 385
    transfers to Washington office, 380
Herckinson, Gladys, 122 *photo*
Hermann, Binger, 76, 79
Hess, L.W., 267
Hodge, William Churchill, Jr., 56, 190
Hodge, William G., 85-86, 107, 137, 148
Homans, G.M., 89, 91
Hoover, Herbert C., 208, 208 *photo*, 220, 228, 233
Hoover, J. Edgar, 161
Hopkins, Harry, 235
Hopping, Ralph, 134-135
Hosmer, R.S., 56
Hough, Franklin B., 24
Houston, David Franklin, 170, 179, 186
Howard, H.E., 436-437
Howie, Louis, 240
Huber, Walter, 216
Hughes, Bennett O., 316
Humphrey, Hubert H., 400
Hutchinson, Wallace I.,
    1924 fire season, 193
    Kings Canyon National Park controversy, 283
    photograph of, 171 *photo*, 302 *photo*
    public relations, 171, 174-175, 206-207, 245, 254
    publications, 175, 206, 223
    retires, 338
    service record, 221
    World War II, 302

I

Ickes, Harold L., 281-282, 285
Ingoldsby, Mollie, 124

J

Jackson, Helen Hunt, 9, 30-31, 108
Jackson, W. Turrentine, 522
Jardine, James T., 207, 217
Johnson, Hiram W., 63, 115, 282-283
Johnson, Lyndon Baines, 412-413, 448
Johnson, Robert Underwood, 33, 39
Johnston, James, 522
Jolson, Al, 165
Jordan, Chester E., 242

K

Kaiser, Henry J., 298
Katz, Rosie, 122 *photo*
Keith, William, 22-23

Kelley, Evan, 240
Kelley, Harriet, 124
Kennedy, John F. 401-403, 411-412
Kerensky, Alexander, 158
Kerouac, Jack, 377-378
Kimball, Thomas L., 450
King, Harold C., 268
Kinney, Abbott, 21-22, 34-37, 39, 49, 94, 324
Knox, John, 91
Kotok, Edward I.,
    California Forest and Range Experiment Station, 204
    fire research studies, 192
    1924 fire season, 193
    photographs of, 171 *photo*, 192 *photo*
    publications, 205, 211, 247, 269, 291
    reforestation problem, 136
    relationship with S.B. Show, 201, 204
    transfers to Washington office, 294
Kraebel, Charles J., 210-211
Krans, Alma, 169
Krebs, John, 484
Kroeber, Theodore, 72
Kuchel, Thomas, 447

L

LaChappelle, Ed, 423
Lane, E.A., 89
Larson, Geri, 502-503, 503 *photo*
Laudenslayer, Bill Jr., 516
Lawrence, E.O., 299
LeConte, Joseph, 58
Leiberg, John B., 61
Leigh, Janet, 423
Leisz, Douglas "Doug,"
    administration and leadership, 451, 453, 455, 503
    biography and career, 451
    budget issues, cuts in funding, and maintenance of programs, 494
    CBS clear-cutting Threlkeld television segment, 465-466
    consent decree, 496
    cultural resource management, 459
    environmental planning, 451
    incident command system (ICS), 472
    Kern Plateau multiple use plan, 389
    Mineral King winter sports development controversy, 450, 453, 481-483
    multiple use management plans and guides, Northern California, 464
    photographs, 451 *photo*, 503 *photo*
    Tahoe Regional Planning Agency (TRPA), 461
    transfers to District 6, 451
    transfers to Washington office, 492

Lenin, Vladimir, 158
Leopold, Aldo, 200, 287, 527
Lincoln, Abraham, 19
Lippincott, J.B., 35
Loveridge, Earl, 334, 336
Lowerdermilk, W.C., 211
Lucas, "Barefoot" Tom, 55
Lukens, Theodore P., 34-36, 40 *photo by*, 77, 94
Lull, George B., 85-86
Lummis, Charles Fletcher, 31
Lupkin, T.P., 43

M

MacMurray, Fred, 324
Mainwaring, Ranger, 69
Mantz, Paul, 187
Margolin, Louis, 132
Marsh, George Perkins, 23
Marshall, James Wilson, 9
Marshall, Robert, 285
Martin, Glenn L., 117
Mason, Ira J., 347, 349
Mather, Stephen T., 150, 213-215
McArdle, Richard E.
    administration and leadership, 380-382, 384-385
    Multiple Use Sustained-Yield Act (MUSYA), 400
    wilderness and, 413-414
McCloskey, J. "Mike," 483
McDonald, James A., 521
McDuffie, Duncan, 216
McGuire, John R., 452, 478, 493
McKennan, Russell B., 384, 406
McKinley, William, 46-50, 63, 65, 68
McNary, Charles L., 183
McPherson, Aimee Semple, 167
McRae, Thomas Chipman, 44
Meinecke, E.P., 133, 133 *photo*, 280
Mendenhall, William V.,
Merriam, Frank F., 233, 235-236, 263
Merrill, Oscar C., 102
Miller, Donald S., 457-459, 521
Miller, J.M., 134
Montalvo, 3
Moran, Thomas, 22
Morris, Edmund, 65
Morse, C.B., 258
Morse, Grant A., 386, 386 *photo*
Muir, David, 252
Muir, John, 31-33, 37, 39, 44, 58, 106, 285

death of, 113
Hetch Hetchy Valley controversy, 104
National Forest Commission, 45
"Range of Light," 57
relationship with Gifford Pinchot, 45, 112
Shasta Forest Reserve, 73
supports federal forest reserves, 45-46
Mulholland, William, 87
Munns, Edward N., 95, 168, 197

N

Nash-Boulden, Stephen A.,
Nelson, Jesse W., 321
Nelson, M.M., 337, 362
Nixon, Richard M., 429, 452, 455, 461, 465
Noble, John M., 38-39, 42
Norris, Frank, 30-31

O

O'Toole, Randal, 508
Ogden, Peter Skene, 7
Olmsted, Frederick E. "Fritz,"
administration and leadership, 89-90, 110, 112, 119, 132, 137, 159, 171, 221, 249, 330
biography and career, 67, 111-112
California Inspection District 5, 85-86, 89
decentralization of Forest Service, 75, 98
District 5 fire policy, 97
expansion of national forests, 88
Forest Service charter letter, 67
opinion of General Land Office, 75
photograph, 67 *photo*
relationship with Coert DuBois, 118
relationship with Gifford Pinchot, 96, 111
resignation of, 111
Use Book (*Use of the National Forests*), 69
Olmsted, Frederick L., 252
Olson, Culbert L., 236-237
Oppenheimer, J. Robert, 299
Otis, H.G., 35

P

Palmer, A. Mitchell, 161
Pardee, George C., 56, 63, 110, 182
Parkinson, Dan, 338
Peavey, George W., 85
Peterson, Max, 359, 493, 497-499, 508
Pickford, Mary, 165, 175
Pinchot, Gifford
administration and leadership, 74, 243, 294
associates and acquaintances, 74, 79, 170, 203, 322

biography and career, 34
  Civil Service exams, 80
  compared with Stephen T. Mather, 150
  conservation principles and leadership, 61-62, 84, 112, 160, 201, 282, 325-326, 428, 449
  creation of forest reserves, 73-74
  decentralization of Forest Service, 75-76
  dismissal by President Howard Taft, 96-97, 109, 282
  Division of Forestry, 52
  expansion of national forests, 88
  Forest Homestead Act of 1906, 90
  Forest Service charter letter, 66
  Forest Transfer Act, 68
  Hetch Hetchy Valley controversy, 103-104
  joint survey with State of California, 56
  multiple use management, 130, 282
  National Forest Commission, 44-45
  opinion of General Land Office, 42, 57, 75
  Owens aqueduct controversy, 87, 112
  private forests, importance of, 109
  recreation use, 105, 108, 112
  relationship with Frederick E. Olmsted, 67, 74, 98
  relationship with John Muir, 45, 112
  relationship with mining industry, 100
  relationship with Theodore Roosevelt, 68, 74
  resource utilization policies, 47
  Society of American Foresters (SAF), 66
  speeches, 175, 324, 356
  supports federal forest reserves, 45-46
  sustained-yield forest management, 34
  timber industry contracts, 139
  25th anniversary of Forest Service, 220
  water power policy, 101
  work philosophy, 120
Pitchlynn, Paul P.
  biography and career, 171
  Feather River ranger school, 174, 203, 256
  photograph of, 174 *photo*, 187 *photo*
  retires, 321
Plummer, Fred G., 94
Powell, Bradley E., 525-526
Powell, John Wesley, 22
Price, Jay H., 205, 240, 242, 247
Pryce, Diane, 498

R

Rachford, C.E., 142, 145, 151, 153-154, 195-196, 287
Raker, John E., 150
Ralston, William, 27
Reagan, Ronald, 429-430, 448, 489-490, 494-495, 506, 508, 517

Redington, Paul Goodwin,
- administration and leadership, 172, 174, 180, 279, 330
- air patrols, 186, 188
- biography and career, 170-171
- Bureau of Biology, 203, 249
- Clarke-McNary Act of 1924, 183
- cuts in California (State of) budgets, 182
- fire control and protection, 188-189
- light burning controversy, 189-190
- 1924 fire season, 193
- photographs of, 170 *photo*, 171 *photo*
- professionalization of staff, 173
- public relations, 174-175
- re-establishes San Bernardino National Forest, 195
- relationship with Sierra Club, 175
- transfers to Washington, 200, 202
- tribute to Charles Shinn, 80

Reuf, "Boss" Abraham, 63
Reyes, Jacinto Damien, 221-222, 221 *photo*
Richardson, Friend W., 182
Rickel, Lloyd A., 395
Robertson, Dale, 523
Robinson, Bestor, 450
Robinson, Cyril S., 288
Rockefeller, John D., 161
Rockefeller, Laurence, 453
Rockefeller, William G., 110
Rogers, Dave N., 122 *photo*, 322
Rogers, Roy, 332
Rolph, James "Sunny Jim," 167, 230, 243
Roosevelt, Franklin D., 166, 233, 249, 259-260, 278, 283, 285
- Civilian Conservation Corps (CCC), 235-236, 238-240, 289
- World War II, 298, 303

Roosevelt, Theodore, 51, 57, 61, 63, 66, 73, 74, 110, 356
- conservation principles, 65-66, 68, 87, 102
- expansion of national forests, 88
- Hetch Hetchy Valley controversy, 103-104
- relationship with Gifford Pinchot, 68
- relationship with John Muir, 112

Royce, Josiah, 30-31

S

Salant, Richard, 466
Sampson, Arthur W., 358
Sargent, Charles Sprague, 44-45, 75
Schumacher, E.F., 430
Scott, E.L., 186
Scott, Howard, 168
Scoyen, Eivind T., 450

Serra, Junipero, 4
Shinn, Charles Howard, 78-78, 80, 123, 175-176, 322
Shinn, Julia, 79, 123-124
Show, Stuart Bevier (S.B.)
    administration and leadership, 201-203, 206, 207-208, 238, 330, 341, 381, 433
    aerial fire control, 268
    biography and career, 200, 200 *photo*, 201
    California Forest and Range Experiment Station, 204
    Civilian Conservation Corps (CCC), 239-244
    Copeland Resolution and Report, 237-238
    death of, 412
    decentralization of Forest Service, 76
    Feather River Experiment Station, 201
    Feather River ranger school, 174, 202, 294
    fire control and protection, 206, 313, 349
    fire research studies, 171, 192, 201, 205
    general integrating inspections (GIIs), 294, 334
    Great Depression, including labor camps, 230, 237
    ideology and philosophy, 201, 289, 296
    Kings Canyon National Park controversy, 282-285, 325
    light burning controversy, 190, 201
    multiple-use management, 257-258, 282, 284, 296, 373
    National Recovery Administration (NRA) funds, 247
    1924 fire season, 193
    opinion of Stephen T. Mather, 214
    personnel issues and problems, 255-256
    photographs of, 171 *photo*, 323 *photo*, 330 *photo*
    Ponderosa Way, 247
    professionalization of staff, 172-173
    publications, 202, 205, 247, 269, 303
    range management, 273
    ranger districts and workloads, 322-323
    recreation management and values, 211, 213, 215, 231, 279-280
    reforestation problem, 136
    relationship with Edward I. Kotok, 201, 204
    relationship with Ferdinand A. Silcox, 249-250
    relationship with National Park Service (NPS), 214-215, 232, 282-285
    retirement, 326, 330, 380
    San Gorgonio Primitive Area, 365
    watershed management, 210
    wilderness and wilderness area system, 211, 213, 215-219
    wildlife management, 287
    World War II, 296, 302-303
Shuler, Robert P., 167
Silcox, Ferdinand A.,
    administration and leadership, 249-250, 281, 285, 293
    biography and career, 249
    death of, 289, 291
Sinclair, Upton, 161, 233

Smith, C. Stowell, 89, 134
Smith, Herbert A., 126
Smith, Jedediah, 7-8
Smith, Zane G., Jr.
    administration and leadership, 484, 492-493, 496, 506, 523
    biography and career, 451, 492
    consent decree, 496-505
    "herbicide war," 442
    photograph, 492 *photo*
    RARE II, 508-510
    spotted owl controversy, 515
Snyder, Gary, 378
Sparhawk, W.N., 257
Sprague, G. Lynn, 525
Stanfield, Robert N., 197
Starr, Walter, 216
Stegner, Wallace, 1
Steinbeck, John, 234
Stephens, William D., 152
Stewart, Ronald E., 497, 525, 531-532
Stoneman, George, 34
Stuart, Corey, 425
Stuart, Robert Y., 224, 257
    administration and leadership, 228, 238-240, 250
    biography and career, 203, 249
    relationship to Gifford Pinchot, 203
    25th anniversary of Forest Service, 220
Sudworth, George B., 61
Sutter, John, 9
Swain, David C., 213
Sweeley, Frank M., 223
Swift, Lloyd W., 273

T

Taft, Howard, 96, 109-110
Taylor, Frederick W., 119, 337
Thompson, Perry A. "Pat"
    administration and leadership, 331-335, 345, 493
    biography and career, 329-330
    multiple use management, 337, 373
    photographs, 325 *photo*, 330 *photo*
    public involvement and land management planning, 452
    range management, 356
    retires, 337
    San Gorgonio Primitive Area, 366-368
Threlkeld, Richard, 465
Tinker, Earl, 294
Townshed, Francis E., 167
Tracy, Spencer, 165

Trotsky, Leon, 158
Truman, Harry, 348
Tugwell, Rexford Guy, 249
Tully, Gene, 79
Twain, Mark, 35 *photo*, 61

V
Valentino, Rudolph, 165

W
Wallace, Henry C., 197, 238-239, 247, 262, 282, 285
Warren, Earl, 331, 346
Watkins, Carlton, 106
Watts, Lyle F.
    administration and leadership, 320, 324, 337, 340, 348
    San Gorgonio Primitive Area controversy, 366368
Waugh, Frank A., 150
Westenberger, Jane, 435-436, 456, 502
Whalen, Philip, 378
White, Stewart Edward, 79, 106, 189-190
Wiley, C., 89
Wilson, James, 66-68, 74, 296, 324
Wilson, Pete, 510
Wilson, Woodrow, 104, 117, 152
Wood, Adolph, 35
Woodbury, T.D., 95, 130-131, 135, 153-154, 157, 184-186, 206, 221, 259, 259 *photo*, 294
Woods, Samuel E., 380
Work, John, 7
Wothe, Dorothy, 498
Wyckoff, Pete, 481

Z
Zahniser, Howard, 412

Index of Subject
A
Agricultural Adjustment Act (AAA), 234
agricultural and agribusiness interests, 430. *See also* California (state of) and
    Californians:
    agricultural growth.
        Desert Lands Act of 1877, 18
        irrigation interests, 17, 25, 37-38, 50, 65, 104-105, 215, 375
        labor force and, 116, 163, 234-235, 300, 307-308, 328, 375, 430
        World War I and, 118, 152
        Wright Act of 1887, 18
Agricultural Appropriations Act of 1907, 84
Agricultural Appropriations Act of 1908, 111
American Antiquities Act of 1906, 108-109, 439

American Association for the Advancement of Science, 24
American Forest Congress, 66, 324
American Forest Products Industries (AFPI), 345
American Forestry Association (AFA), 36, 44, 65-66, 182, 382, 401, 425
American Indian Religious Freedom Act (AIRFA) of 1978, 459-460, 521-522, 524
American National Livestock Association (ANLA), 356
American Paper and Pulp Association (APPA), 345
Angeles National Forest, 88, 195. *See also* San Gabriel National Forest and/or Reserve, and San Bernardino and Santa Barbara National Forests.
    aerial attack of forest and brush fires, 350 *photo*, 351-352, 352 *photo*
    air patrols, 186-187
    Angeles Crest Highway, 179-180, 254
    anniversary of founding, 324
    California condor survey, 287
    camping and campground development, 148
    Castaic Fire (1947), 352
    Civilian Conservation Corps camps and projects, 243-244
    colonial mining, 5
    Converse Flat Nursery, 95, 136
    Cucamonga Wilderness Study, 478
    cultural resource management, 459
    Devils Canyon-Bear Canyon Primitive Area, 219, 445
    establishment of, 88
    fire prevention, 313
    fire protection and control, 194
    firebreaks, 139
    floods and, 274
    Fork Fire (1970), 472
    GII inspection, 335, 339
    grazing resources, 7
    Henninger Flats Nursery, 94
    history of, 439
    information and education, 126
    Job Corps camps, 433
    Lytle Creek Nursery, 95
    Malibu Fire (1935), 266 *photo*
    maps and recreation information, 147
    Merrick Canyon Nursery, 135
    mining activity, 276
    Monrovia Peak Fire (1953), 354
    municipal camps, 147, 148 *photo*, 274
    Nike missile defense system, 405
    1923 fire season, 192
    1957 fire season, 393
    1970 fire season, 471
    range resources, 270
    ranger work issues, 121
    recreation residences, 40 *photo*, 107-108, 149, 274
    recreational use, 226, 379 *photo*

  San Bernardino Nursery, 135
  San Gabriel Fire (1924), 193-194
  San Gabriel Wilderness, 444-445, 479
  Sheep Mountain Wilderness Study, 478
  St. Francis Dam collapse, 209, 209 *photo*
  supervisor's staff, 79 *photo*
  timber sales and production, 259
  watershed protection and storage, 224, 275, 410
  women and, 318
Angeles Protective Association, 255
Anti-Debris Association, 12
Archaeological and Historical Conservation Act (ACHA) of 1974, 459
Archaeological Resources Protection Act (ARPA) of 1979, 519-520, 522
Associated Farmers of California, 236, 307
Audubon Society, 287-288, 339, 365, 443, 488, 521, 527
Automobile Club of Southern California, 116, 179

B

Bell Aircraft Company, 352
Biltmore Estate and Forestry School, 34, 67, 74
Boy Scouts of America, 367
Bureau of National Parks, 150. *See also* National Park Service.

C

Calaveras Big Tree National Forest, 364
  establishment of, 88
Calaveras County, 23
California (state of) and Californians
  agricultural growth of, 17-18, 27-28, 30, 37-38, 163, 166, 299, 328, 375-376, 429
  aircraft and aerospace industries, 30, 117, 154, 297-298, 327, 375
  American settlement, early 6, 8-9
  Arbor Day, 21
  automotive industry and impacts, 30, 116-117, 121, 146, 328, 377
  Bay Conservation and Development Commission (BCDC), 432
  Board of Agricultural Transactions, 20, 24
  Board of Forestry creation, reconstitution, and activities, 21-22, 25, 34-35, 57, 110, 182, 188
  Board of Trade, 50
  budget and taxes, 429-430, 489-490
  Bureau of Forestry, 79
  Bureau of Mines, 305
  Bureau of Public Roads, 227
  "California lifestyle," 376-377, 429
  colonial history, 3-9
  commerce, 30
  computer industry, 491
  conservation movement, 379
  Council of Defense, 152
  Department of Fish and Game, 151-152, 287, 359, 372, 395, 403, 410, 443, 476, 485, 513
  Department of Forestry and Fire Protection, 211, 472, 485
  Department of Parks and Recreation, 446

Department of Water Resources, 396, 485
Division of Beaches and Parks, 385
Division of Fish and Game, 227
Division of Forestry (CDF), 235, 243, 256, 358, 473
Division of Highways, 377, 448
droughts of 1976-1977, impacts of, 487
ecological movement, 491-492
"Eden," as metaphor, 1, 3, 17, 22-23, 57-62, 487-489
energy crisis, 487
environmental degradation, 379-380, 429-432, 488-492
ethnic and minority groups, 116, 489
exploration of, 3-5, 7-8
forest reserves, support of, 36-37
forests and regeneration of, 28-29, 34-35
fur trapping era, 7-9
Gold Rush era and impacts, 9-11, 13-15, 58-59, 327
Great Depression, 165-168, 227-231
Historic Landmarks Committee, 108-109
legislature and legislative actions, 19, 21, 56, 263, 284
    abolishes Board of Forestry, 34
    Agricultural Labor Relations Act, 430
    Criminal Syndicalist Act of 1919, 161
    Environmental Policy Act (CEQA), 451, 490
    fire control law, 19
    Forest Practices Act, 331, 346
    Forest Protection Act, 57, 110
    Lake Bigler [Tahoe] Forestry Commission established, 20-21
    Lake Tahoe park rejected, 21
    National Forest Acquisition Act of 1934, 263
    reforestation policies, 20
    State Highway Act, 116
    Wilderness Act, 510
manufacturing, 30, 118, 327, 375
metropolitan and urban centers (including suburbanization), 17-18, 25, 29-30, 36-37, 48, 104, 328, 375, 429-432, 446
mining industry, 118, 299
mission era, 4 *photo*, 4-6, 108
motion picture industry and Hollywood, 30, 116, 126, 165-166, 175, 246, 324, 425
Mussel Slough gun battle, 28, 31
New Deal and, 233-237
Northern California, 77, 104, 109, 167
Park Board, 215
park system and development program, 215
Peripheral Canal Project, 487-488, 490
petroleum industry, 29, 117-118, 164, 166, 299, 443
population and growth, 6-7, 10, 19, 27, 29, 327, 359, 376, 402, 417, 427, 431, 481
pre-colonial history, 1-3
progressive movement, 65, 115
racial discrimination, segregation, and conflict, 300-301, 378-379
real estate booms, 29-30, 163-164, 166, 180

    Recreation Commission, 362
    recreation growth and problems, 121, 328, 341, 359, 372
    Red Scare of 1919, 161, 163, 227
    Resource Agency, 485
    San Francisco earthquake, 63
    settlement patterns, 16
    shipbuilding industry, 298
    Southern California, 6, 17, 18, 21, 37, 48, 77, 104, 109, 163-167
    State Emergency Relief Administration (SERA), 235
    tourism, 50, 105-106, 117, 128, 149, 151, 166, 226, 231
    transportation networks, 15-16, 116-117, 128, 146, 164, 178-180, 328
    Unemployment Commission, 167
    water crisis, 487-488
    Water Resources Control Board, 432
    World War II, impact on, 237, 297-301
California Association of Four-Wheel Drive Club, 485
California Cattlemen's Association, 339
California Development Association (CDA), 197-198
California Forest Association (CFA), 408
California Forest Protective Association (CFPA), 139
California Geological Survey, 58
California High Sierra Packers, 485
California Mining Association, 485
California National Forest, 88, 137, 322. *See also* Mendocino National Forest.
    air patrols, 187
    establishment of, 88
    irrigation water supply, 224
    maps and recreation information, 147
    Middle Eel-Yolla Bella Primitive Area, 219
    1923 fire season, 192
    range reconnaissance, 144
    range resources, 225
    wildlife management, 227
California Wildlife Federation, 485
California Ski Association, 366
California Society for Conserving Waters and Protecting Forests, 49
California State Chamber of Commerce, 223, 282-283, 366, 368
California Water and Forest Association, 50
California Wilderness Coalition, 485
Carter (James "Jimmy") administration 487, 494, 506
Caterpillar Tractor Company, 261
Clarke-McNary Act of 1924, 183, 194, 335, 436. *See also* Weeks Act of 1911.
Clean Air Acts of 1967 and 1970, 431
Cleveland (Grover) administration, 44-46
    Washington's Birthday Reserves, 45-47
Cleveland National Forest, 88, 195, 438, 493. *See also* Trabuco Canyon Forest Reserve, and San Bernardino National Forest.
    aerial attack of forest and brush fires and air patrols, 186, 268 *photo*, 438
    Agua Tibe or Tibia Wilderness, 479
    Agua Tibe or Tibia, Primitive Area, 217-218

    California condor, 288
    camping, 218 *photo*
    Civilian Conservation Corps camps and projects, 244
    firefighting, 311 *photo*
    floods and, 274
    GII inspection, 339
    grazing resources, 7
    Hauser Creek Fire (1943), 310
    Inaja Fire (1956), 393, 393 *photo*
    Job Corps camps, 433
    Laguna Fire (1970), 471
    Laguna recreation area, 226, 329 *photo*
    maps and recreation information, 147
    mining activity, 276
    1923 fire season, 192
    1924 fire season, 193
    1950 fire season, 352-353
    1970 fire season, 471
    nurseries, 94
    Oak Grove Nursery, 135
    range resources, 270
    ranger district redrawing, 322
    recreational use, 226
    watershed protection, 224
Cold War, 327, 375-376, 378, 404-405, 489
Columbus Broadcasting Station (CBS), 465
conservation and conservation movement, 19-20, 23-25, 31-36, 89, 334, 495. *See also* environmentalism and environmental movement.
    clear-cutting, or patch cutting practices, 363
    concept and principles of, 54, 68
    fractionalization of, 112-113, 231
    idealistic scientific conservation, 24, 33-34
    Kern Plateau multiple use plan, 426
    measures, 51, 78
    Mineral King winter sports development controversy, 426
    multiple use management and, 372, 426
    Pinchot conservationism demise, 325-326
    pragmatic conservation, 34, 36
    public relations and, 255
    San Gorgonio Primitive Area controversy, 366-368, 374, 413
    Southern California attitude and movement, 48, 224
    utilitarian conservation, 24-25, 33-34, 87, 100, 103-104, 112-113, 160, 231, 326, 428, 464, 523-524
    versus preservationism, 45, 112-113, 285, 296, 389, 464
    World War I, impacts on, 160
Creation Act. *See* Forest Reserve Act of 1891.
cultural resource management (District/Region 5) (CRM). *See also* related individual legislative acts.
    archaeological research, 458, 520
    archaeological sites (prehistoric and historic), 7-8, 151, 458
    cultural resource management, 456-460

    establishment of program, 109
    Executive Order 11593, 457
    historic preservation program, 108-109, 439-440, 456, 457 *photo*
    history program, 458, 522-523
    law enforcement, 519
    national forest archaeological/historical overviews and reports, 521-522
    program objectives, 522-523
    public education, 460 *photo*
    timber sales and, 458
    training programs, 459

D

Del Norte County, 263
Desert Lands Act of 1877, 17-19
Diamond Match Company, 111
Diamond Mountain Forest Reserve, 83. *See also* Plumas National Forest.
    consolidated into Plumas Forest Reserve, 65
    establishment of, 71
Diamond Mountain National Forest, 88. *See also* Plumas and Lassen National Forests.

E

ecology and ecological movement, 427, 486. *See also* Forest Service: District/Region 5 (California), ecosystem management.
Economic Opportunity Act of 1964, 433
    Job Corps centers, 433
    Office Economic Opportunity (OEC), 433-434
Edison Moving Picture Company, 126
Eisenhower (Dwight D.) administration, 373, 376, 394, 399-400
Eldorado National Forest, 46, 136, 192, 205, 316, 451, 452. *See also* Stanislaus and Tahoe National Forests.
    air patrols, 186
    "asbestos" forest, 204
    Bolshevism and, 158
    Camp Sacramento, 147
    Desolation Valley Primitive Area, 218, 445
    Desolation Wilderness, 445-446, 479
    establishment of, 88
    firefighting, 312 *photo*, 353
    GII inspection, 335, 339
    grazing resources, 14
    Heavenly Valley winter sports area, 411
    historic signage, 147
    hydroelectric development, 13, 224, 395-396
    Job Corps camps, 433
    Kirkwood Meadows ski development, 447
    Lake Tahoe Basin, 404, 438, 461
    maps and recreation information, 147
    mining activity, 11-12
    Mokelumne Wild Area, 411,
    Mokelumne Wilderness Study, 478
    NIRA camps, 261
    overgrazing problem, 270

    range reconnaissance, 100
    range resources, 225
    ranger district redrawing, 322
    recreation area created, 212
    recreational conflicts with grazing, 149
    recreational use, 226, 340 *photo*
    supervisor meeting, 138
    timber reconnaissance, 132
    timber sales and production, 131, 225, 316
    watershed protection, 224
    women and, 318
    World War I impacts on grazing, 154
    World War I impacts on mining, 157
Emergency Conservation Work (ECW) Act of 1933, 239. *See also* Forest Service: Civilian Conservation Corps (CCC).
environmentalism and environmental movement, 368, 374, 402, 414, 426, 428, 440, 462, 486, 488, 490, 508-509, 527. *See also* ecology and ecological movement.
    clear-cutting, or patch cutting, practices, 440, 467
    "herbicide war," 442, 442 *photo*
    protests, 441 *photo*
    spotted owl controversy, 515, 527

F

Federal Highway Act of 1921, 178-179, 253
Federal Power Commission (FPC), 209-210, 224, 417
Federal Water Power Act of 1920, 209, 224, 418
fire(s) (California, state of) 34, 48. *See also* individual fires by forest reserve or national forest.
    arson and incendiarism, 97, 99, 141, 192, 228, 237, 461, 516
    attitudes (general) towards, 57, 60-61
    human-caused/accidental, 57, 97, 106, 137, 146, 148, 155, 192, 204, 269, 314, 393, 409, 436
    industry related (mining, timber, livestock, and railroads), 12, 61, 192, 195
    lightning caused, 192
    military, 310
    1910 fire season, 96-97, 110, 136
    1914 fire season, 140
    1917 fire season, 192
    1923 fire season, 192
    1924 fire season, 186, 192-193, 204, 354
    1926 fire season, 203-204
    1932 fire season, 265-266
    1933-1940 fire seasons, 266
    1940 fire season, 268
    1950 fire season, 352-354
    1953 fire season, 354
    1957 fire season, 393
    1960 fire season, 393
    1970 fire season, 471-472
fire control and prevention, 203. *See also* Forest Service: District/Region 5 (California), Riverside Fire Laboratory.
    aerial attack of forest and brush fires, 267-268, 350-353, 350 *photos*, 355, 374, 392-393, 410
    boards of reviews, 194-195

closures of national forests, 314, 314 *photo*
communication planning and systems (heliographs, radios, telephones), 55, 55 *photo*, 137, 139, 187, 264, 340
conservation and, 231
cooperative fire control, 56, 139-140, 186, 310
cooperative fire laws. *See* Clarke-McNary Act of 1924 and Weeks Act of 1911.
Cover Type and Fire Control, 205-206
detection and lookout systems, 96, 99, 137, 139, 155, 186, 189 *photo*, 239, 264
Determination of Hour Control for Adequate Fire Protection, 247
District 5 leadership and tradition, 191, 201
economic fire protection policy, 189, 201
equipment and machinery, 54-55, 137
fire research studies and conferences, 191-192, 202, 204-205, 268-269, 472
firebreaks, including Ponderosa Way, 54-55, 95-96, 139, 230, 247-248, 264
firefighters, fire crews, and fire camps, 56, 99, 140, 167, 193, 229 *photo*, 230, 230, *photo*, 247, 267, 310-311, 350-353, 471-472, 504 *photo*
FIRESCOPE (Firefighting Resources of Southern California), 472-473
FIRESTOP operation, 355, 385
FOCUS (Fire Operation Characteristics Use Simulation), 473
forest reserves (1898-1905), 53
fuel break program, 394, 436, 442
incident command system (ICS), 472-473
large fire organization, 354, 392, 410
leadership, 385
let burn policy, 140-141, 189-190
light burning controversy, 3, 6, 14-15, 99, 138, 189-191, 223
MAFFS (Modular Airborne Fire Fighting System), 474
Mather Field National Fire Control Conference (1921), 191, 191 *photo*, 192
patrol system, including air patrol, 95, 95 *photo*, 96, 99, 136-137, 179 *photo*, 186-188
permits (burning and campfire), 77, 99
road construction, 216-217, 253
safety, 393
SAFETY FIRST program, 473
Southern California Air Attack (SCAA), 392-393
Southern California Firestorm 2003: A Report for the Wildland Fires, 530
suppression, pre-suppression, and costs, 37, 111, 139, 201, 265, 310, 312, 355, 385, 472-473, 486
Systematic Fire Protection in the California Forests, 138-139
urban interface and, 446
women firefighters, 406, 434, 456
women lookouts, 124-125, 155, 155 *photo*, 317
World War I, impacts on, 155
World War II, impacts on, 295, 310-311
fire protection and prevention, 90, 98, 160, 203, 279
chaparral management, 95
economic fire protection policy, 140, 189
fire prevention week, 175
fire protection plans, 99
hazardous fuels problems and reduction, 56, 95, 99, 138-139, 145, 331, 359
law enforcement priorities, procedures, and policies, 141
mapping, 137

planning (forest and ranger district plans), 98, 137-139
research studies, 64
systematic fire protection, 136
World War II, impacts on, 295, 313
Ford (Gerald R.) administration, 480
Forest and Rangeland Renewable Resources Planning Act (RPA) of 1974, 453, 464-465, 485, 505
Forest and Water Society of Southern California, 48
Forest Homestead Act of 1906, 90-91
Forest Reserve Act of 1891, 57, 65, 70, 71, 295-296
Section 24, 36, 41
Forest Transfer Act of 1905, 66, 68
forest reserves, California, 17. *See* individual forest reserves.
concept of, 54
trespass and timber depredations, 42-43, 51
Free Timber Act of 1878, 16, 19
Fresno County, 28, 38, 164, 227
Friends of the Earth, 488
Friends of the River, 485
Fruit Growers Supply Company, 261

G

General Grant National Park, 33, 40, 42, 150, 214, 284
General Land Exchange Act of 1922, 180
General Mining Act of 1866, 10
General Mining Act of 1872, 10, 485, 516-517
General Revision Act. *See* Forest Reserve Act of 1891.
Girl Scouts of America, 454 *photo*
Gold Reserve Act of 1934, 276
Great Northern Paper Company, 110

H

Harrison (Benjamin) administration, 36-41
Hawaii and Pacific Islands, 335-336, 336 *photo*, 407, 436, 492, 505
Healthy Forest Restoration Act of 2003, 531
Historical Society of Southern California, 108
Homestead Act of 1862, 19, 278
Hoover (Herbert C.) administration, 208, 220, 228, 233, 275
Humboldt County, 263, 474
Humboldt Nursery, 409, 470
Humboldt Redwood Park, 182
Humboldt-Toiyabe National Forest, 46. *See also* Toiyabe National Forest.

I

Indian reservations. *See* Native Americans.
Intercontinental Rubber Company, 307
International Workers of the World (IWW), 170
Inyo County, 102
Inyo National Forest, 88, 214, 322, 438, 495. *See also* Mono National Forest.
Ansel Adams Wilderness, 427 *photo*
"asbestos" forest, 204

bighorn sheep sanctuary, 511
  bristlecone pines, 398, 399 *photo*
  campground problems, 362, 397 *photo*
  Civilian Conservation Corps camps and projects, 244, 267 *photo*
  cultural resource management and, 520
  endangered species, 288, 444, 512
  establishment of, 86-87, 105
  fire patrols, 95 *photo*
  GII inspection, 339
  Golden Trout Wilderness, 510-511
  grazing resources, 14, 32 *photo*
  High Sierra Primitive Area, 218, 286, 480
  High Sierra Wilderness, 412, 478-480
  Hoover Primitive Area, 218
  hydroelectric development, 224
  Kern Plateau multiple use plan, 387-389, 402, 425
  Mammoth Mountain Ski Area, 360
  maps and recreation information, 147
  Minarets Wilderness, 411-412
  mining activity, 11, 276, 305-306
  range resources, 225
  recreational survey work, 279
  recreational use, 226, 322, 419 *photo*, 454 *photo*
  timber sales and production, 259
  timber work plans, 93, 130
  transportation and, 253 *photo*
  Upper Kern Wilderness Study, 478
  watershed protection, 224
  White Mountains Wilderness Study, 478
  wildlife management, 321, 475
  World War I impacts on mining, 157
Izaak Walton League, 271, 339, 401, 468

J

John Muir Trail, 129
Johnson (Lyndon Baines) administration, 412-413, 429, 448

K

Kennedy (John F.) administration, 401-403, 411-412, 467
Kern County, 30, 33, 164, 305, 411
Kern National Forest. *See also* Sequoia National Forest.
  establishment of, 88
  recreational conflicts with grazing, 149
Kern Plateau Association, 389
Kings Canyon National Park, 33, 215, 252, 257-258, 282-285, 296, 325, 329, 421, 438
Klamath Forest Reserve, 78. *See also* Klamath National Forest.
  establishment of, 71-72
  livestock, 72
  lumber operations, 72
  mining, 72

Klamath National Forest, 88, 208 *photo*, 416. *See also* Klamath Forest Reserve.
    air patrols, 187
    American settlement, 8
    archaeological/historical overviews and reports, 521-522
    Bald Mountain lookout, 125 *photo*
    Civilian Conservation Corps camps and projects, 244
    Eagle Rock lookout, 98
    Eddy's Gulch lookout, 124
    Etna Primitive Area, 478
    fur trapping era, 7
    GII inspection, 335, 339
    Hog Fire (1977), 474
    insect control work, 134
    Johnson Primitive Area, 478
    land management planning, 506
    law enforcement, 517-518
    Marble Mountain Primitive Area, 219, 220 *photo*
    mining activity, 11, 12 *photo*, 12-13, 276, 278 *photo*, 305-306
    1923 fire season, 192
    1926 fire season, 204
    overgrazing problem, 270, 355 *photo*, 356
    Portuguese Primitive Area, 478
    range reconnaissance, 144
    recreation area created, 212
    recreation use, 445 *photo*
    Red Cap Fire (1938), 266
    Salmon-Trinity Alps Primitive Area, 219, 413, 478-480
    Shackleford Primitive Area, 478
    Snoozer Primitive Area, 478
    spotted owl controversy, 515
    timber sales and production, 225, 259, 316, 348, 441
    watershed protection, 224
Korean War, 327, 347, 373

L

La Moine Timber and Trading Company, 92, 520
Lake Bigler [Tahoe] Forestry Commission, 20-21, 24-25, 49
Lake Tahoe Basin and Lake Tahoe Basin Management Unit (LTBMU), 403-404, 461, 463
    cultural resource management, 459
    Lake Tahoe Visitor Center, 438, 477
    Tahoe Regional Planning Agency (TRPA), 438, 461
Lake Tahoe Forest Reserve, 50, 70. *See also* Tahoe Forest Reserve.
Land and Water Fund Conservation Act, 436, 448, 461
Lassen National Forest, 88, 316. *See also* Lassen Peak National Forest.
    air patrols, 187
    American settlement, 8
    archaeological/historical overviews and reports, 522
    boundary redrawing, 322
    Caribou Peak Primitive Area, 219

    cattle, cattlemen, and ranching, 196
    Civilian Conservation Corps camps and projects, 242 *photo*, 244, 259 *photo*
    cultural resource management and, 520-521
    cut-over lands, 183, 262
    Diamond Mountain National Forest incorporated into, 88
    endangered species, 288
    fur trapping era, 7
    GII inspection, 335, 339
    hydroelectric development, 224
    information and education, 126
    insect control work, 134
    irrigation water supply, 224
    land acquisition program, 272
    1923 fire season, 192
    1926 fire season, 204
    NIRA camps, 261
    overgrazing problem, 270
    range reconnaissance, 144
    range resources, 225
    ranger district redrawing, 322
    recreational use, 226, 411
    reforestation projects, 391
    selective cutting, 261
    Sugarloaf incident, 460-461
    Thousand Lakes Primitive Area, 219
    timber reconnaissance, 132
    timber sales and production, 225, 259, 315-316
    women and, 317
    World War II, impacts on, 314
Lassen Peak Forest Reserve. *See also* Lassen Peak National Forest.
    establishment of, 72-73
    livestock, 73
    logging, 73
Lassen Peak National Forest. *See also* Lassen Peak Forest Reserve.
    renamed Lassen National Forest, 88
Lassen Volcanic National Park, 150, 214, 329, 459
Lava Beds National Monument, 214
livestock industry and interests, 32, 43, 49. *See also* individual national forests and reserves, industry organizations, and range management (District/Region 5).
    attitudes toward burning, 359
    cattle, cattlemen, and ranching, 14, 27, 30, 51, 196, 208, 270, 394
    colonial history, 5-7, 13,
    conservation and natural resources, 19, 23, 38
    controlled or prescribed burning, 358
    Desert Lands Act of 1877, 18
    grazing and grazing conditions, 14, 18, 34, 36, 51-54, 56, 143
    light burning activities and support, 14-15, 39, 358
    sheep and sheepmen, 14-15, 27, 32, 38-39, 43, 51, 142, 149, 196, 208, 394, 394 *photo*

    Sierra forest, and, 59-60
    stock associations, 100
    support Multiple Use Sustained-Yield Act (MUSYA), 01
    trespass, 43, 53-54
    World War I, 146
Los Angeles (City of), 215
Los Angeles Board of Water Commissioners, 87
Los Angeles Chamber of Commerce, 37, 48, 179
Los Angeles County, 15, 28-29, 37-38, 188, 211, 256, 269, 275, 298, 328, 385
Los Padres National Forest, 4, 48, 88. *See alsoi* Pine Mountain and Zaca Lake Forest Reserves, and Santa Barbara and Monterey National Forests.
    air patrols, 187 *photo*
    archaeological/historical overviews and reports, 522
    California condor management, 364-365, 374, 443, 475
    colonial mining, 5
    cultural resource management and, 520
    Earth First protest, 443 *photo*
    floods and, 274
    GII inspection, 335, 339
    grazing resources, 7, 14
    "hippie problem," 446
    Madulce Wilderness Study, 478
    Marble Cone Fire (1977), 474
    mining activity, 276
    named from Santa Barbara National Forest, 263, 287
    1950 fire season, 352
    1970 fire season, 471-474
    ranger district redrawing, 322
    San Marcos Pass Nursery, 94
    San Rafael Primitive Area, 444
    San Rafael Wilderness, 444, 479
    Santa Lucia Wilderness, 510-511
    Sespe Condor Sanctuary, 443, 475, 511
    Sisquoc Falls Sanctuary, 288, 475, 511
    Ventana Wilderness, 445-446, 479
    Watershed management, 364
    Wheeler Springs Fire (1948), 353 *photo*
    World War I, impacts on, 157
    World War II, impacts on, 304
lumber industry, companies and interests, 29, 30, 32, 49, 260, 284. *See also* related individual legislative acts, national forests and reserves and individual lumber companies.
    conservation and natural resources, 19, 21-23
    Desert Lands Act of 1877, 18
    fraudulent timber claims and practices, 16, 41, 48
    relationship with Forest Service, 293
    road construction and, 261
    support state forestry, 293
    watershed, 41
    World War I and, 118

M

Mackinaw Island National Park, 33
Madera County, 362, 412
Mariposa Big Trees Grove, 19
Mariposa County, 305
McKinley (William) administration, 46-50, 65
McRae bill, 44-45
McSweeney-McNary Act of 1928, 205
Mendocino County, 74, 263
Mendocino National Forest, 88, 137, 322-323, 451, 496. *See also* California National Forest.
    Civilian Conservation Corps camps and projects, 244
    cultural resource management, 459
    Eel River Fire (1932), 266
    fur trapping era, 7
    GII inspection, 335, 339
    Job Corps camps, 433
    Log Springs Fire (1940), 266
    Lookout tower, 266 *photo*
    1923 fire season, 192
    range reconnaissance, 144
    Rattlesnake Fire (1953), 354, 374
    timber sales and production, 348, 441
    wildlife management, 321
Mineral Springs Act of 1899, 105
Miners Association, 49
mining industry, 32, 49, 52, 74. *See also* related individual legislative acts, individual national forests and reserves, and minerals management (District/Region 5).
    colonial history, 5
    Comestock Lode, 11
    conservation and natural resources, 19, 21, 23, 65
    fire control, support of,
    Gold Rush era, 9-10
    hydraulic mining, 10, 12, 12 *photo*, 18, 58
    Lake Tahoe region, devastation, 11-12
    light burning, support of,
    locators, 276-277, 279
    mining "homesteaders," 276, 278, 278 *photo*, 279
    quartz mining, 10
    recreational mining, 276-277
    Sierra forest, and, 58-59
    strategic metals development, 155-157, 198
    watershed, 41
    Woodruff vs. North Bloomfield Gravel Mining Company (1884), 13, 18
minerals management (District/Region 5), 90. *See also* related individual legislative acts.
    conflicts with recreation, 198
    conflicts with roads, 198
    cooperation with mining industry, 100, 198, 517
    Forest Ranger's Catechism and, 223
    law abuses and requirements, 101, 396

    mineral and mining claims, 100, 129, 275, 332, 396, 418
    mineral production, 275-276, 494-495, 516-517
    multiple use management, 130, 279, 396, 418
    Multiple Use Sustained-Yield Act (MUSYA) of 1960, 418
    wilderness, conflicts with, 516
    World War I, impacts on, 155-157, 160, 198
    World War II, impacts on, 305-306
Mission Indian Federation (San Jacinto), 194
Modoc Forest Reserve. *See also* Modoc National Forest.
    cattle/sheep conflicts, 51
    establishment of, 50-51, 65
Modoc National Forest, 88, 214, 272, 293, 316, 323. *See also* Modoc Forest Reserve and Warner Mountain National Forest.
    air patrols, 187
    American settlement, 8
    cattle, cattlemen, and ranching, 196
    Civilian Conservation Corps camps and projects, 244
    cultural resource management, 459
    fire control 53 *photo*
    firefighting, 312 *photo*
    fur trapping era, 7
    GII inspection, 335, 339
    hunting, 199, 199 *photo*
    insect control projects, 191
    livestock disease control measures, 142
    1950 fire season, 352
    overgrazing problem, 270
    picnicking on, 222 *photo*
    range reconnaissance, 143-144
    range resources, 225
    range war, 271, 356-358
    Scarface Fire (1977), 474
    sheep and sheepmen, 196
    South Warners Primitive Area, 219
    timber reconnaissance, 132
    timber sales and production, 225
    Warner Mountain National Forest incorporated into, 88
    wild horse and burro roundups, 512
    wildlife management, 321, 331
    World War II impacts, 305
Mono County, 102, 412
Mono National Forest, 46, 88, 215. *See also* Humboldt-Toiyabe and/or Toiyabe National Forest, and Inyo, Sierra, Stanislaus, and Tahoe National Forests.
    "asbestos" forest, 204
    camping, 212 *photo*
    Civilian Conservation Corps camps and projects, 244
    establishment of, 88
    High Sierra ski school, 286
    Hoover Primitive Area, 218
    irrigation water supply, 224

    maps and recreation information, 147
    Mount Dana-Minarets Primitive Area, 218
    recreation area created, 212-213
    timber work plans, 93, 130
    World War I impacts on grazing, 154
Monterey County, 15
Monterey Forest Reserve. *See also* Monterey National Forest.
    establishment of, 73
Monterey National Forest, 88. *See also* Monterey Forest Reserve.
    ranger work issues, 121
    San Benito and Pinnacles National Forests consolidated into, 87
    timber work plans, 93, 130
Mount Shasta Nursery, 391, 409, 470
Muir Woods, 214
Multiple Mineral Development Act of 1954, 396
multiple use management, 68, 113, 130, 150, 257-258, 341, 462
    area planning and, 342, 373
    California condor controversy, 364-365
    concept of, 373, 387, 400
    conservation and, 372
    intensive, 387
    Kern Plateau multiple use plan, 387-389, 425
    Kings Canyon National Park controversy, 285, 326
    Lake Tahoe Basin multiple use plan, 420, 425, 437
    multi-land use plans, 257-258
    multiple purpose plans, 258
    multiple-use principles, plans and policy, 258, 333, 343, 484
    priority of, 404, 425
    ranger districts plans and workloads, 322-323, 403, 414
    recreation management and, 326
    wildlife management and, 326
Multiple Use Mining Act of 1955, 396
Multiple Use Sustained-Yield Act (MUSYA) of 1960, 399-401, 403, 412, 462
    flow chart, 414, 415 *chart*

N

National Academy of Science, 44, 75
    Forest Commission, 44-45
National Aeronautics Administration (NASA), 376, 471, 491
National Environmental Policy Act (NEPA) of 1969, 109, 442, 451-452, 457, 462-464, 482, 516, 524, 529
National Forest Management Act (NFMA) of 1974, 453, 464-465, 469-470, 485, 505, 524, 529
national forests, California. *See* individual forests: Angeles, Cleveland, Eldorado, Humboldt-Toiyabe Inyo, Klamath, Los Padres,
    Modoc, Mono, Plumas, San Bernardino, Sequoia, Shasta, Shasta-Trinity, Sierra, Six Rivers, Stanislaus, Tahoe,
    Toiyabe, and Trinity.
National Historic Preservation Act (NHPA) of 1966, 109, 439-440, 456, 524
National Industrial Recovery Act (NIRA), 234, 260
National Lumber Manufacturers Association (NLMA), 345, 349, 401
National Park Service (NPS), 256
    Civilian Conservation Corps, 243

    compared with Forest Service, 176, 223, 281
    conflicts with Forest Service, 176-177, 203, 213-214, 232, 282-285, 396-397
    cooperation with Forest Service, 176, 214
    establishment of, 113, 150
    High Sierra Primitive Area, 286, 480
    John Muir Wilderness Area, 480
    Mineral King winter sports development controversy, 447-449, 484
    Mission 66, 396
    personnel, 172
    police force, 519
    preservationists and, 150, 160
    Sierra Nevada Framework, 527
National Parks, California,
    See individual parks: General Grant, Kings Canyon, Sequoia, Yosemite.
National Register of Historic Places (NRHP), 457, 520
National Resources Defense Council, 453, 527
National Trails Act of 1968, 444
National Wildlife Federation, 401, 463, 527
National Wool Growers Associations (NWGA), 356
Native Americans, 116. *See also* fire control: firefighters.
    American settlement impacts, 8-9
    burning practices, 2-3, 6, 58, 61
    Chumash Indian pictographs, 444
    cultural resource management and, 520-521
    Ishi, 72
    Klamath Forest Reserve, 72
    Lassen Peak Forest Reserve, 72
    mission system, 4, 6-7
    oppression of, 108
    Plumas Forest Reserve, 70-71
    prehistoric California peoples, 1-2
    protests, 429-430, 459
    Trinity Forest Reserve, 71
Nevada County, 410
Nevada-California Electric Company, 102
Nixon (Richard M.) administration, 429-430, 434, 452, 455, 461, 465
Northern California Electric Company, 102

O

Oakdale Nursery, 391
Occupancy Permits Act of 1915. *See* Term Permits Act of 1915.
Omnibus, or Upstream, Flood Control Act of 1936, 274-275, 369
Orange County, 28, 234, 288, 298, 328
Organic Act of 1897, 47-48, 52, 61, 107, 341, 365, 401, 468-469
    "forest-lieu" clause, 47-48, 50, 70
    forest reserves and rules and regulations, 53

P

Pacific Gas and Electric (PG&E), 63, 102, 164, 210
Pacific Light and Power Company, 102, 146 *photo*

Panama-Pacific California Exposition of 1915, 126
Pasadena Chamber of Commerce, 77
Pine Mountain Forest Reserve. *See also* Santa Barbara Forest Reserve and Los Padres National Forest.
    consolidated into Santa Barbara Forest Reserve, 65
    establishment of, 48
Pinnacles Forest Reserve. *See also* Monterey National Forest
    consolidated into Monterey National Forest, 87
    establishment of, 73
Placer County, 410
Placerville Nursery, 391, 409, 470
Plumas County, 82, 227
Plumas Forest Reserve, 83. *See also* Plumas National Forest.
    Diamond Mountain Forest Reserve consolidated into, 71
    establishment of, 70-71
Plumas National Forest, 88, 316, 322, 453. *See also* Plumas Forest Reserve, and Diamond Mountain National Forest.
    American settlement, 8
    boundary redrawing, 322
    Camping, 128 *photo*
    Civilian Conservation Corps camps and projects, 244
    Claremont Peak lookout, 96
    communication system, 340 *photo*
    cultural resource management, 459
    cut-over lands, 183
    Diamond Mountain National Forest incorporated into, 88
    Feather River Experiment Station, 95, 120, 136, 174, 201
    Feather River Nursery, 185
    Feather River ranger school, 169, 173, 173 *photo*, 174, 202-203, 256, 294, 315, 320-321
    firefighting, 229 *photo*, 309 *photo*, 353
    fishing, 326 *photo*
    GII inspection, 335, 339
    grazing resources, 14
    helicopter logging, 467
    hydroelectric development, 209-210, 224, 396
    management or work plans, 185
    mining activity, 11, 276, 306
    Mount Ingalls lookout station, 96 *photo*
    Murphy Hill Primitive Area, 218
    Nelson Point Fire (1934), 266
    1923 fire season, 192
    1926 fire season, 204
    1950 fire season, 352
    NIRA camps, 261
    office staff, 122 *photo*
    overgrazing problem, 270
    range resources, 225
    recreation area created, 212
    recreational use, 226, 411
    sheep and sheepmen, 196
    timber reconnaissance and marking, 132, 343 *photo*

timber sales and production, 225, 259, 315-316, 345, 408
 timber work plans, 93, 130
 woman lookout, 124
 women staff, 406 *photo*, 456
 World War I, impacts on, 154, 157
 World War II, impacts on, 314
Preservation, preservationists, and preservation movement, 31-36, 44
 aesthetic conservation, 33, 37
 Hetch Hetchy Valley controversy, 102-103, 103 *photo*, 104, 112, 150, 163, 285, 421, 449
 Multiple Use Sustained-Yield Act (MUSYA) of 1960, 401
 National Park Service and, 150
 Owens Aqueduct controversy, 87
 versus utilitarianism, 45, 87, 112-113, 160, 285, 296, 389, 523-524
Progressive movement, 31, 63, 89, 102, 110, 356, 428

R

railroads, 20, 27, 29, 31, 48, 50, 74
 Atchison, Topeka and Santa Fe, 17
 Central Pacific, 16, 50, 59
 Sierra forest, and, 58-59
 Southern Pacific, 16-17, 63, 190
range management (District/Region 5), 90, 98, 160, 203, 495. *See also* individual national forests and reserves,
  and related individual legislative acts.
 administration action program, 320
 administrative and research studies, 273, 358
 advisory grazing boards, 143, 207, 320
 allotments and permits, 141, 148, 207-208, 271, 320, 394-395, 440
 analysis of range conditions and problems, 320-321, 355, 395
 area planning and, 343
 budgets and, 494
 carrying capacity, 141, 208, 270, 272, 356, 395
 cooperation and difficulties with stockmen, 99, 160, 196, 272-273
 Costs and Returns of Controlled Brush Burning for Range Improvement, 358
 disease control measures, 142-143, 196
 District 5 policy, 142-146
 droughts of 1976-1977, impact of, 474
 feed lots and, 418
 fees and fee system, 99, 141-142, 145, 195, 197, 207-208, 228, 238, 249, 270
 fire control and protection and, 154, 197, 269
 five-year range management program, 358
 Forest Problems of California and, 291-292
 Forest Ranger's Catechism and, 223
 land acquisition program, 272
 livestock reductions and controversy, 196-197, 207-208, 271, 273, 332, 355-358
 management plans, 272
 mismanagement, 270
 multiple use management, 130, 257-258, 356
 overgrazing problem, 145, 196, 270-273
 predator control measures, 142, 196, 225

    prescribed burning, 410
    range improvements, 196, 225, 271-273, 356-358, 395, 402, 410, 418
    range reconnaissance, 100, 143-144, 196, 272
    range restoration, 100, 196, 225
    range-transfer program, 356
    recreation, coordination and conflicts with, 269, 271, 273, 356
    relationship to other forest uses, 100, 151
    revegetation measures, 196, 225, 418
    scientific management and Taylorism, 141, 143, 160
    statutory goal, 99
    Supreme Court decision, 141
    timber, conflicts with, 508
    trespass issues, 100, 141
    watershed, conflicts with, 269
    wilderness, conflicts with, 216-217
    wildlife, coordination and conflicts with, 271, 273, 321, 394-395
    work plans, 144-145
    World War I, impacts on, 146, 153-154, 195-196
    World War II, impacts on, 304-305
Reagan (Ronald), administration, 489-491, 494-495, 506, 508, 517
recreation. *See also* recreation management.
    backpacking, 329
    back-to-nature movement, 106
    camping and campers, 105-106, 128, 146, 199, 231, 286, 329, 446
    fishing, 106, 146, 151
    hiking, hikers, and hiking clubs, 106, 146, 286, 419
    horse packing, 148, 445 *photo*
    hunting, 106, 146, 151, 199
    motorists, 146, 199
    negative impacts of, 199
    off-road vehicles (ORVs), 329
    organization camps, 359, 361, 446, 454 *photo*
    picnicking, 105 *photo*, 106, 199
    pre-World War I, 147-152
    private recreation, 149-150, 160, 199
    public recreation, 147-148, 160
    recreational vehicles (RVs), 329
    resorts and hotels, 105, 147, 178, 199
    sightseeing, 106
    winter sports, 106, 286, 359-360, 366-369, 411, 419, 423, 446, 447, 454, 476. *See also* recreational management and individual ski developments on individual national forests.
recreation management (District/Region 5). *See also* individual national forests and reserves, and related individual legislative acts. *See also* wilderness management.
    aesthetic and scenic values, 420
    appropriations, 199-200, 362-363
    area planning and, 342
    automotive rest areas and camps, 147
    budgets and, 494
    camper handbook, 107, 148

campground development, 148-149, 178, 226, 280, 361-362, 372, 398, 411, 454
campground sanitation problems, 362
competition with State of California, 385
conservation and, 231, 385
demands, 359-361, 481
District 5 recreation plan or program, 107
droughts of 1976-1977, impacts of, 476-477
facilities, 150-151, 239, 280-281, 362, 399, 407, 420
fire control and protection and, 195
Forest Problems of California and, 292
Forest Ranger's Catechism and, 223
forests as destination, 117, 146
historic signage, 147
Kern Plateau multiple use plan, 387-389, 420
Kings Canyon National Park controversy, 282-285, 325
Lake Tahoe Basin multiple use plan, 420, 425, 437
landscape architects, 281, 397
maps and recreation information, 147, 199
Mineral King winter sports development controversy, 369, 387, 420-425, 447-450, 453, 481-484
multiple use management and, 130, 257-258, 341, 360-361, 363, 401-402, 411
municipal camps, 147-148, 225-226
Operation Outdoors, 396, 397, 397 *photo*, 398-399
Outdoor Recreation Resource Review Commission (ORRRC), 397, 420, 423-424
planning and policy, 200, 280-281, 362-363
range, conflicts with, 108, 149, 269, 360, 363
ranger attitude toward, 106-107
recreation areas created, 212-213, 216
recreational residences (summer homes), 107, 107 *photo*, 149-150, 199, 211, 213, 226, 279-280, 419 *photo*, 446, 447, 454
relationship with National Park Service (NPS), 176-177, 203, 213-215, 232, 282-285
road construction and, 107, 178-180, 199, 213, 253-254, 280
San Gorgonio Primitive Area controversy, 366-368, 374, 421-422
snow surveys, 286
timber, conflicts with, 108, 360, 363, 419-420, 508
urban interface, 446
user fee system, 411, 446-447
watershed, conflicts with, 105, 360
wilderness, conflicts with, 360, 419, 426
wildlife, conflicts with, 360
Reservoir Salvage Act of 1960, 109
Riverside County, 28, 328
Roads and Trails Act of 1964, 349
Roosevelt (Franklin D.) administration, 166, 233, 235-236, 238-240, 249, 259-260, 278, 280, 283-285, 293, 298, 303, 320
Roosevelt (Theodore) administration, 68-70

S

San Benito County, 227
San Benito National Forest. *See also* Monterey National Forest.
    consolidated into the Monterey National Forest, 85
    establishment of, 87

San Bernardino County, 28, 38, 41, 211, 328
San Bernardino Forest Reserve, 46, 77. *See also* San Bernardino National Forest.
    establishment of, 40-41
    grazing resources, 43
    timber depredations, 43
San Bernardino National Forest, 88, 438, 493, 496. *See also* San Bernardino Forest Reserve and/or Angeles National Forest.
    aerial attack of forest and brush fires, 351 *photo*
    archaeological/historical overviews and reports, 522
    Bear Fire (1972), 467
    camping and picnicking, 146 *photo*, 398 *photo*
    Civilian Conservation Corps camps and projects, 244, 245 *photo*, 247 *photo*
    Cucamonga Primitive Area, 219
    firebreaks, 247 *photo*
    firefighting, 230 *photo*
    floods and, 274, 369
    GII inspection, 335, 339
    grazing resources, 7
    helicopter logging, 467
    hydroelectric development, 224
    irrigation water supply, 224
    Keenbrook Fire (1940), 266
    law enforcement, 229
    mining activity, 276
    motion pictures, 324
    multiple use management, 422 *photo*
    1970 fire season, 471
    nurseries, 94
    range resources, 270
    recreational residences (summer homes), 107, 274
    recreational use, 226
    re-established, 195
    San Gorgonio Primitive and/or Wilderness Area, 218, 359, 365-369, 374
    San Jacinto Primitive Area, 218
    Telegraph Primitive Area, 218
    timber sale and production, 315
    watershed protection and storage, 224, 410
    Whispering Pines Braille Trail, 434
San Bernardino Society of California Pioneers, 41
San Diego County, 28, 288, 298
San Gabriel Forest Reserve, 46, 77, 110. *See also* San Gabriel National Forest and/or Angeles National Forest.
    anniversary of founding, 324
    establishment of, 25, 38, 40-41
    fishing, 84 *photo*
    rangers and job description, 55
    timber depredations, 43
    watershed management, 57
San Gabriel National Forest, 88. *See also* San Gabriel Forest Reserve, and Angeles National Forest.
San Jacinto Forest Reserve, 47
    establishment of, 46

supervisors and rangers, 50 *photo*
San Joaquin Light and Power Company, 102
San Luis Obispo County, 29, 288
San Luis Obispo Forest Reserve
    establishment of, 73
San Luis Obispo National Forest, 88. *See also* Santa Barbara National Forest.
San Mateo County, 21
Sanger Lumber Company, 264
Santa Barbara County, 15, 28-29, 48, 263, 288, 365
Santa Barbara Forest Reserve. *See also* Santa Barbara National Forest.
    establishment of, 65
Santa Barbara National Forest, 48, 88, 221. *See also* Santa Barbara Forest Reserve and/or Angeles, Los Padres, and San Luis Obispo National Forests.
    California condor survey, 287-288
    Civilian Conservation Corps camps and projects, 244
    Cuyama ranger district, 221
    grazing resources and policy, 142
    Indian Canyon Fire (1933), 353 *photo*
    insect control work, 134
    land acquisition program, 263
    Los Prietos Nursery, 135
    Matilija Fire (1932), 265, 269
    Monterey Division, 263
    1923 fire season, 192
    1932 fire season, 265-266
    range resources, 225, 270
    recreational use, 226
    renamed Los Padres National Forest, 262, 287
    San Rafael Primitive Area, 219
    Ventana Primitive Area, 218, 446
    watershed protection, 224
Santa Clara County, 233
Santa Ynez Forest Reserve. *See also* Santa Barbara National Forest.
    consolidated into Santa Barbara Forest Reserve, 65
    establishment of, 48
Save the Redwoods League, 182, 263-264
Selective Service Act of 1940, 318
Sequoia Forest Reserve, 69. *See also* Sequoia National Forest.
Sequoia National Forest, 8, 88, 134, 214, 283, 316, 416, 438, 451. *See also* Kern and Sierra National Forests.
    air patrols, 187
    California condor survey, 287
    cattle, cattlemen, and ranching, 196
    Civilian Conservation Corps camps and projects, 244
    cultural resource management, 459
    cut-over lands, 183
    Dome Lands Wild Area, 411
    endangered species, 288, 444
    establishment of, 88
    firebreaks, 247

    forest officers, 250 *photo*
    GII inspection, 335, 339
    Golden Trout Wilderness, 510-511
    grazing resources, 14
    Great Depression labor camps, 230
    High Sierra Primitive Area, 218, 286, 411, 478, 480
    High Sierra Wilderness, 412, 478-480
    hydroelectric development, 224
    Kern Plateau multiple use plan, 387-389
    Kern River Project, 403, 410
    land acquisition program, 264
    livestock disease control measures, 142
    Mineral King Valley, 421 *photo*
    Mineral King winter sports development controversy, 369, 387, 420-425, 436, 447-450, 453
    Monarch Wilderness Area, 481
    1970 fire season, 471
    range resources, 225
    ranger district redrawing, 322
    recreational conflicts with grazing, 149
    recreational use, 226
    timber sales and production, 225
    women and, 318
    World War I, impacts on, 157
    World War II, impacts on, 305, 314
Sequoia National Game Refuge, 483
Sequoia National Park, 33, 40, 42, 150, 213-214, 282, 329, 421, 423, 438, 448, 482, 484
Shasta Forest Reserve, 77. *See also* Shasta National Forest.
    establishment of, 73-74
Shasta National Forest, 88, 201, 205, 451. *See also* Shasta Forest Reserve.
    1923 fire season, 192
    1926 fire season, 204
    Aerial Fire Control Project, 268
    air patrols, 187
    Civilian Conservation Corps camps and projects, 244, 248 *photo*
    cut-over lands, 183
    Deer Creek Fire (1939), 266
    firefighting, 353
    fur trapping era, 7
    GII inspection, 335, 339
    insect control work, 134
    irrigation water supply, 224
    NIRA camps, 261
    overgrazing problem, 270
    Pilgrim Creek Nursery, 95, 136
    range reconnaissance, 100, 145
    range resources, 225
    recreation area created, 212
    recreational use, 226
    reforestation projects, 136, 391

Salmon-Trinity Alps Primitive Area, 219, 413, 478-480
Shasta Experimental Fire Forest, 206
Sisson Fire (1914), 140
timber reconnaissance, 132
timber sales and production, 92, 225, 259, 316, 348
wildlife management, 151
women and, 317

Shasta-Trinity National Forest. *See also* Stanislaus and Trinity National Forests.
    aerial attack of forest and brush fires, 473 *photo*
    CBS clear-cutting Threlkeld television segment, 466
    cultural resource management and, 520
    Denny community and law enforcement, 517-519
    endangered species, 512
    Job Corps camps, 434
    land management planning, 506
    mining activity, 11
    Mount Shasta Primitive Area, 478
    spotted owl controversy, 476
    timber sales and production, 441
    World War I impacts on mining, 157

Shell Oil Company, 165

Sierra Club, 33, 39, 49-50, 73, 149-150, 175, 188, 214-215, 223, 263-264, 283-284, 332, 363, 450, 453, 463, 466, 467, 485, 506, 521, 527
    Kern Plateau multiple use plan, 389, 413
    Kirkwood Meadows ski development, 447
    Mineral King winter sports development controversy, 420-425, 447-450, 481-484
    Multiple Use Sustained-Yield Act (MUSYA) of 1960, 401, 412
    San Gorgonio Primitive Area controversy, 366, 368, 413, 421-422
    Wilderness Act of 1964, 414

Sierra Forest Reserve, 49, 57, 76, 78-79, 438. *See also* Sequoia and/or Sierra National Forests.
    Civilian Conservation Corps camps and projects, 244
    diaries, letters, and reminiscences, 57-58
    establishment of, 38-41
    grazing resources, 43
    grazing trespass, 141
    rangers and job description, 54-55
    supervisor's meeting, 76, 80
    timber sales, 53

Sierra National Forest, 2 *photo*, 46, 88, 123, 134, 171, 173, 215, 322, 492. *See also* Sierra Forest Reserve and Sequoia National Park.
    air patrols, 187
    anniversary, 438
    campground sanitation problems, 362
    conservation education, 439 *photo*
    cut-over lands, 183
    establishment of, 88
    firebreaks, 247
    firefighting, 126, 353
    GII inspection, 335, 339

GLO timber sales, 92
   grazing resources, 14
   High Sierra Primitive Area, 218, 257, 478, 480
   High Sierra Wilderness, 412, 478-480
   hydroelectric development, 209-210, 224, 284 *photo*
   information and education, 126
   insect control work, 134
   land management planning, 506
   light burning controversy, 189
   Minarets Wilderness, 411
   mining activity, 11
   Monarch Wilderness Area, 481
   Mount Dana-Minarets Primitive Area, 218, 411
   multiple use management, 257-258
   municipal camps, 147
   newsletter, 127
   NIRA camps, 261
   range reconnaissance, 100
   range resources, 225
   ranger district redrawing, 322
   recreational conflicts with grazing, 149
   recreational residences (summer homes), 107 *photo*
   recreational survey work, 279
   recreational use, 226, 383 *photo*, 388 *photo*
   San Joaquin Fire (1939), 266
   San Joaquin insect control project, 190
   San Joaquin Wilderness Study, 478, 511
   Signal Peak lookout, 96
   spotted owl controversy, 515
   timber reconnaissance, 132
   timber sales and production, 181, 225, 259
   timber work plans, 130
   water resources, 304 *photo*
   World War I impacts on grazing, 154
   World War I impacts on mining, 157
Silver Purchase Act of 1933, 275
Six Rivers National Forest, 451. *See also* Klamath and Trinity National Forests.
   CBS clear-cutting Threlkeld television segment, 465-466
   cultural resource management issues, 459, 521
   establishment of, 331
   fur trapping era, 7-8
   GII inspection, 335
   hydroelectric development, 396
   land management planning, 506
   mining activity, 11
   spotted owl controversy, 515
   timber sales and production, 348, 441
Smithsonian Institution, 108
Society of American Foresters (SAF), 66, 111, 190, 197, 382, 400, 485

Sonoma County, 21, 263
Southern California Academy of Science, 48
Southern California Edison and Pacific Light and Power, 102, 209, 283, 403
Southern Pacific Land Company, 413
Standard Lumber Company, 92
Standard Oil Company, 165
Stanislaus Forest Reserve, 47. *See also* Stanislaus National Forest.
    establishment of, 46
Stanislaus National Forest, 88, 140, 293, 316, 438. *See also* Stanislaus Forest Reserve, and/or Eldorado and Mono National Forests.
    air patrols, 186-187
    campground sanitation problems, 362
    Civilian Conservation Corps camps and projects, 244
    cultural resource management, 459
    cut-over lands, 183
    Emigrant Basin Primitive Area, 219, 446
    Emigrant Basin Wilderness, 446, 479
    fire control and protection plans, 98-99, 136, 139, 264
    GII inspection, 339
    grazing resources and policy, 14, 142
    Great Depression labor camps, 230
    Hetch Hetchy Valley controversy, 103
    hydroelectric development, 13, 224
    information and education, 126
    Job Corps camps, 433
    maps and recreation information, 147
    Mark Twain and, 35 *photo*
    mining activity, 11
    Mokelumne Wild Area, 411,
    Mokelumne Wilderness Study, 478
    newsletter, 127
    NIRA camps, 261
    picnicking, 105 *photo*
    Pilot Peak lookout, 96
    Pinecrest campground, 281, 361, 361 *photo*
    range advisory boards, 143
    range reconnaissance, 100, 145
    range resources, 154, 225
    ranger district redrawing, 322
    ranger, 54 *photo*
    recreational conflicts with grazing, 149
    recreational planning and use, 226
    Smith Peak lookout, 498
    timber reconnaissance, 133
    timber sales and production, 92, 225, 259, 316, 408
    water development policy, 101
    women and, 434, 498
    World War I, impacts on, 154
    World War II, impacts on, 316

Stock Raising Homestead Act of 1916, 144-145

Stony Creek Forest Reserve, 77. *See also* Stoney Creek National Forest.
    establishment of, 74

Stony Creek National Forest, 88. *See also* California National Forest.

Students for a Democratic Society (SDS), 378

Sundry Civil Appropriations Act of 1897, 46-47. *See also* Organic Act of 1897.

Sustained-Yield Forest Management Act of 1944, 344-345, 373

T

Tahoe Forest Reserve, 70. *See also* Yuba Forest Reserve and/or Tahoe National Forest.
    establishment of, 49-50
    expansion of, 50
    mining, 70

Tahoe Lumber and Flume Company, 49, 264

Tahoe National Forest, 78, 88, 200, 221, 256, 331, 503. *See also* Tahoe Forest Reserve, and/or Mono and Eldorado National Forests.
    air patrols, 186
    American settlement, 8
    boundary redrawing, 322
    Civilian Conservation Corps camps and projects, 244
    Deadwood Fire (1924), 185 *photo*
    GII inspection, 335, 339
    grazing resources, 14
    hydroelectric development, 13, 209, 224
    Lake Tahoe Basin, 404, 438, 461
    land acquisition program, 264
    mining activity, 11-12, 276
    overgrazing problem, 270
    patrolling, 188 *photo*
    range reconnaissance, 100, 145
    range resources, 225
    recreational use and problems, 226, 341
    sheep and sheepmen, 196
    Sierra Buttes lookout, 405 *photo*
    ski school, 286
    spotted owl controversy, 515
    Squaw Valley Olympics, 404, 411
    supervisor meeting, 138
    timber sales and production, 225, 316
    timber work plans, 130
    women lookouts, 124, 155 *photo*
    World War I impacts on grazing, 154

Tahoe National Park, 49

Taylor Grazing Act of 1934, 286

Term Permits Act of 1915, 149, 280. *See also* recreation management (District/Region 5): recreational residences (summer homes).

Threatened and Endangered Species Act of 1973, 374, 453, 475-476, 512, 516, 524

timber and logging industry, 20, 136. *See also* timber management (District/Region 5).
    colonial history, 5-6

cooperative fire control agreements, 139-140
Desert Lands Act of 1877, 18
hazard fuel reduction,
Lake Tahoe region, devastation, 11-12
logging practices, 18, 34, 36, 38, 43, 52, 181, 428 *photo*
Sierra forest, and, 58-59
timber management and sales, 56, 441
trespass, 54
watershed, 41

Timber and Stone Act of 1878, 16, 19, 36, 41

timber management (District/Region 5), 15, 90, 160, 203, 341. *See also* individual national forests and reserves, and related individual legislative acts.
    administrative costs, 131
    annual allowable cut (AAC), 206, 348-349, 373, 385, 388, 408-409, 416-417, 425, 467
    area planning and, 343-344
    branch of forest management (1920-1935), 183, 261
    brushland reclamation, 93-95, 470, 470 *photo*, 486
    budgets and, 408-409, 485, 494, 508
    Capper Resolution, 181-182, 185
    clear-cutting, or patch, block cutting, practices, 363, 417, 417 *photo*, 440-442, 453, 465-468, 468 *photo*, 469-470, 507
    cruising, marking techniques, and reconnaissance, 93, 130-133, 160, 183-184, 260-261
    cut-over lands, 183
    Deadman Creek timber sale, 363
    depletion, damage to, and abuse, 11, 206, 225
    disease and pathology, 130, 133-134, 206, 391-392, 409
    division of timber management (1935-1974), 261, 315
    Dunning study, 183-185
    Eucalyptus Gold Rush, 135-136
    famine and, 390
    Forest Problems of California and, 291-292
    Forest Ranger's Catechism and, 223
    geologists and, 409
    helicopter logging, 467
    housing demand and crisis of 1950s, 160, 372 *photo*, 373, 407
    insect control, 130, 134-135, 189, 206, 260, 391-392, 442
    intensive management, 440, 467
    Japanese log exportation, 440
    land acquisition programs, 260, 262-264, 332, 345, 385
    landscape management and, 441-442, 507
    long-range planning, 469-470, 506
    marking boards, 131
    Monongahela decision, impact of, 465, 468
    multiple use management, 130, 257-258, 388, 506
    nurseries, 94-95, 391. *See* individual nurseries.
    "old growth" timber inventory, 495
    pest control, 391-392, 409, 442
    priority of Region 5, 388
    private forests and forestry, 109, 239, 260, 291, 332, 345-346, 373, 408
    recreation, conflicts with, 151, 420

    reforestation, regeneration, reproduction, 94-95, 98, 130, 135-136, 183, 185, 206, 223, 238, 260, 262, 385, 390, 390 *photo*, 391, 391 *photo*, 409, 442, 464, 467, 470, 486

    Regional Land and Resource Management Plan, 508

    road construction (timber access) and, 253, 315, 320, 332, 346-349, 372-373, 388-389, 402, 409, 452, 458, 467, 508

    rotation periods, 409, 440

    salvage logging operations, 392, 471

    scientific management, 120, 183-184

    selective cutting practices, 261, 363, 440

    silviculture practices, 91-95, 119-120, 238

    statutory goals, 92

    sustained-yield forest management, 34, 92-93, 181, 184, 206, 231, 257, 260-262, 291, 320, 344, 409, 469. *See also* Sustained-Yield Forest Management Act of 1944, and Multiple Use Sustained-Yield Act (MUSYA) of 1960.

    thinning, release work, and sanitation, 260, 390, 417

    timber crisis, 346

    timber demand, production and utilization, 183, 185, 225, 314-314, 316, 332, 344, 347, 374, 390, 486, 496

    timber policy, 180-186

    Timber Resources for America's Future, 390, 402

    timber sales and harvest levels, 92, 98, 120, 130-131, 134, 154, 159-160, 181, 206, 225, 228, 238, 249, 259-260, 315, 346 *photo*, 347, 349, 372-373, 385, 392, 408-409, 417, 440, 452, 458, 467, 486, 493 *chart*, 496, 508

    timber stand improvements (TSI), 260-261, 390-391, 417

    timber targets, 508

    timber theft, 519

    timber work plans, 93, 98, 130, 185

    viewsheds and visual resource management, 441, 467, 469

    wilderness, conflicts with, 216-217

    World War I, impacts on, 154, 160

    World War II, impacts on, 314-315, 344

Toiyabe National Forest, 88, 404, 438, 461. *See also* Mono National Forest.

Trabuco Canyon Forest Reserve, 46. *See also* Cleveland National Forest.

    establishment of, 40-41

    supervisors and rangers, 50 *photo*

    timber depredations, 43

Transfer Act (1905). *See* Forest Transfer Act.

Trinity County, 74, 518

Trinity Forest Reserve. *See also* Trinity National Forest.

    establishment of, 71

    Ironsides Lookout, 71 *photo*

    livestock, 71

    logging, 71

    settlement, 71

Trinity National Forest, 88, 316. *See also* Trinity Forest Reserve and California National Forest.

    air patrols, 187

    arson, 228

    Bear Wallow Fire (1938), 266

    Civilian Conservation Corps camps and projects, 244

    conscientious objector camp, 319

    fur trapping era, 7

    GII inspection, 335, 339

    hydroelectric development, 224

Middle Eel-Yolla Bella Primitive Area, 219
mining activity, 306, 333 *photo*
overgrazing problem, 270, 356
range resources, 225
recreation area created, 212
recreational use, 226
Salmon-Trinity Alps Primitive Area, 219, 413, 478-480
timber sales and production, 225, 316, 348
women and, 317, 318 *photo*
Truckee River General Electric, 102
Tulare County, 33, 38, 411
Tulare Reserve. *See* Sierra Forest Reserve.
Tuolumne County, 227, 364

U

U.S. Army,
- Air Service and forest patrol, 186
- Army Corps of Engineers, 275, 284, 395
- Civilian Conservation Corps (CCC), 241-242
- Fort McArthur, 241, 243
- Fort McDowell, 241
- Fort Rosecrans, 241
- Fort Scott, 241
- Hunter-Liggett military reservation, 304, 398
- March Aviation Field, 186, 241
- Mather Aviation Field (Sacramento), 187
- Red Bluff Aviation Field, 187
- relationship with Forest Service, 241-242
- Rockfield Aviation Field, 187
- The Presidio, 241

U.S. Bureau of Mines, 516
U.S. Bureau of Sport Fisheries and Wildlife, 403, 443
U.S. Civil Service Commission, 83
U.S. Coast Survey, 44
U.S. Congress, 10, 33-34, 36, 38, 42, 44-47, 70, 108, 128-129, 144, 149, 188, 195, 197, 373-374, 397, 400, 444, 450, 453, 459, 481, 485, 494, 508, 511. *See also* related individual legislative acts and bills.
U.S. Defense Department, 376, 491
U.S. Department of Agriculture,
- Bureau of Entomology, 110, 134
- Bureau of Forestry, 42, 66-67
    - cooperative agreements, 110
    - joint survey with State of California, 56-57
    - Klamath Forest Reserve, 72
    - named changed to Forest Service, 68
    - private forests, importance of, 110
- Bureau of Plant Industry, 280
- Division of Biological Survey, 249
- Division of Forestry, 67, 74
    - cooperation with General Land Office, 42, 52-53, 57

creation of, 24
elevated to bureau, 42
map of forest reserves (1904), 26
map of forest reserves (1907), 64

Forest Service. *See also* Forest Transfer Act of 1905.
    Agricultural Adjustment Administration (AAA), 234, 272
    anniversaries, 221-223, 295-296, 324
    California Inspection District 5, 85-86, 118
    centralization of power, 293-294, 325, 338
    changes name of "districts" to "regions,' 220
    changes name of "forest reserves" to "national forests," 84
    charter letter, 66-68, 74, 296, 324
    Civil Service exams, 76, 80, 82-83
    Civil Works Administration (CWA), 235-236, 246, 248, 254
    Civilian Conservation Corps (CCC), 229, 235-236, 248, 254, 319
        decline of, 281, 289
        fire control manpower, 240, 265, 265 *photo*
        origins, organization, and projects, 239-240, 250, 252-253, 260, 264-265, 271-272, 280-281, 362
    Civilian Public Service (CPS), 318-319
    Copeland Report (*A National Plan for American Forestry*), 237-239, 257, 260, 285
    commodity development, 402, 495
    custodial management, 159
    decentralization, 75-76, 85, 87-88, 294
    District/Region 4 (Intermountain Region), 381
    District/Region 5 (California). *See also* mining, range, recreation, timber,
        watershed, wilderness, and wildlife management.
        "active" custodial management, 159-160, 180, 197, 231-232, 372-373
        Accelerated Public Works (APW), Program, 404-406, 411
        accomplishments, 222-223
        administration staff and clerk positions, 121-125, 453-454
        administrative sites, facilities, and improvements, 125, 127-129, 180, 239, 243-244, 252, 289, 384, 454
        advisory councils, 332, 338, 347-348, 382, 448
        Agricultural Conservation Program (ACP), 408
        "analysis paralysis," 529-531
        annual work plans, 336
        area planning or long range sub-regional planning, 341-344, 373, 384, 402. *See also* Forest Service: District/
            Region 5 (California), management plans and guides.
        building locations, 85, 86 *photo*, 89-90, 125, 177, 177 *photo*, 251, 251 *photo*, 252, 323-324, 324 *photo*
        budgets and, 228, 454, 494-495
        California Forest and Range Experiment Station, 204-206, 210, 223, 264, 269, 291, 407
        Civilian Conservation Corps (CCC) accomplishments, 290
        conscientious objectors and, 318-319
        consent decree,496-505, 525-526
        conservation education program, 435, 435 *photo*, 436, 456
        consolidations, renaming, additions, and eliminations of, 84-88
        controlled or prescribed burning program and policies, 140-141
        cooperation with State of California, 110, 135, 182, 486
        creation of, 61-62, 88-89

discrimination (sex and minority). *See* Forest Service: District/Region 5 (California), consent decree and/or equal employment opportunity (EEO) issues.
ecosystem management, 507, 523-524
Environmental Impact Statement (DEIS) for the Pacific Southwest Region, 507-508, 514
equal employment opportunity (EEO) issues, 434, 455-456, 496, 498-499. *See also* Forest Service: District/Region 5 (California), consent decree.
fire mission and policy, 97
Forest and Rangeland Renewable Resources Planning Act (RPA) and, 505, 507
forest/agricultural land question, 90-91
general integrating inspections (GIIs), 294, 325, 334-335, 338-339, 345, 349, 356, 358, 372, 384-387, 395, 406-407, 436, 455
Great Depression labor camps, 229-230, 237
headquarters, 172-173, 177-178, 180, 323-324, 455
information and education activities, 82, 126, 174-175, 193, 222-223, 245-246, 254-255, 291, 292, 301-303, 331-334, 338-339, 385-386, 434-435, 437, 452, 466
inspections, "unscheduled," 339-341
land-use planning, consolidation, exchanges and extensions, 180, 183, 344
Lassie television program, 425, 426 *photo*, 434, 450
law enforcement issues, 228-229, 461, 486, 516-520
lookouts and lookout system. *See* fire control: detection and lookout systems.
map of California's national forests (1911), 114
map of California's national forests (1919), 162
minorities and, 434
motion picture and television production, 255, 386, 425-426
multiple use management plans and guides, 239, 344, 414
multiple use management plans and guides, Lake Tahoe Basin, 420, 425, 437
multiple use management plans and guides, Northern California (NORCAL), 414, 416-420, 425, 437, 462-465
multiple use management plans and guides, Northwest Subregion, 402
multiple use management plans and guides, Southern California (SOCAL), 344, 403, 425, 437
National Forest Management Act (NFMA) of 1974 and, 505-506
newsletters, 126-127, 153, 168, 175, 220, 228, 302
organization and formation of, 88-91
organization and staffing, 250-251, 337-338
personnel and Bolshevism, 158-159, 168-169
personnel and moral issues, 69 *photo*, 120
personnel and part-time staffing, 494
personnel and unionism, 159
personnel and World War I, 157
personnel division, 256
personnel reductions, 495-496
professionalization of staff, 120, 172-173, 202-203, 256, 384, 485
public involvement and land management planning, 437, 452, 462, 464, 469, 530
public relations. *See* Forest Service: District/Region 5, information and education activities.
publications, 223-224, 262, 262 *photo*, 333, 339, 348, 352, 386, 400, 406, 446, 492, 501, 505
rangers, ranger districts and workloads, 322-323, 372, 384, 450
receipts (25% fund), 227, 228, 238, 248-249, 259, 316, 440, 485
Redwood Experimental Forest, 264
Regional Land and Resource Management Plan, 507-508
regional meeting (1912), 119, 121

    research branch, 185
    Riverside Forest Fire Laboratory, 355, 393, 472
    salaries and working conditions, 168-170, 172-173
    San Dimas Experimental Forest and Station, 269, 319, 358
    Sierra Nevada Framework, 526-529
    standardization and, 120, 127
    state and private forestry, 331, 407-408, 436
    supervisor autonomy, 85
    supervisor meetings, 76, 97-100, 143, 190
    supervisors and ranger qualities, responsibilities, and workloads, 76-84, 120-121
    Sustaining Ecosystems: A Conceptual Framework, 523-524
    Today and Tomorrow: Report of Citizens Committee on U.S. Forest Service Management Practices in California, 485-486
    transportation planning (roads and trails construction), 55, 81 *photo*, 127-129, 178-180, 239, 246-247, 252-253, 253 *photo*, 254, 258
    unionism, 153
    World War I, impacts on, 152-159, 169, 172, 174. *See also* World War I.
    World War II, impacts on, 295, 295, 301-319. *See also* World War II.
    women and, 123-125, 155, 172, 175, 317-318, 318 *photo*, 406, 406 *photo*, 434, 455-456. *See also* Forest Service: District/Region 5 (California), consent decree.
District/Region 6 (Pacific Northwest), 331, 426
ecosystem management, 492
education program, 291-293
Environmental Program for the Future (EPFF), 452-453, 459, 462, 467
Federal Emergency Relief Administration (FERA), 235-236, 246
Framework for Change, 455
Framework for the Future, 452, 462
highway system, 127-128
historic preservation and, 109
inspection districts, 76, 85
inspection system, 294, 334
Joint Congressional Committee on Forestry (JCCF), 289-291, 302
Monongahela decision, 465, 468
National Recovery Administration (NRA) camps, codes, funding, 236, 246-249, 254, 259-261, 271-272
Operation Multiple Use, 401-402
organization and staffing, 250-251
origins of, 112
Pacific Southwest Region (PSW), 336, 492. *See* Forest Service: District/Region 5 (California).
Pacific Southwest Region (PSW) Station, 336, 407, 497-498, 502. *See also* Forest Service: District/Region 5, California Forest and Range Experiment Station.
prohibition of additional forests, 84-85
Public Works Administration (PWA), 234, 236
ranger's ten commandments, 82
Reconstruction Finance Corporation (RFC), 233, 275
recreation policy, 148, 150
recreation residences, 108
reorganization and restructuring of, 281-283, 455
scientific management, and Taylorism, 118-119, 337, 344
Smokey Bear campaign, 269, 313

        statutory role and wages, 120, 153, 169-170, 172
        timber reconnaissance, 133
        timber sale contracts and regulations, 93, 133, 195, 260
        timber work plans, 130
        25th anniversary, 220-221
        Use Book (*Use of the National Forests*), 69, 81-82, 100, 130, 143
        "wise use" philosophy, 231, 486
        Woodsy Owl campaign, 452
        Works Progress Administration (WPA), 235-236, 252
    U.S. Department of Interior
        Bureau of Land Management (BLM), 357, 364, 527
        Bureau of Reclamation (BOR), 110, 234
        General Land Office (GLO), 321-322
            administration and organization, 47, 52-57
            conservation and, 51
            cooperation with USDA Bureau of Forestry and Division of Forestry, 52-53, 110
            Division R, 52, 66, 69
            establishment of California forest reserves, 37-41, 50, 65
            Forest Reserve Manual, 54, 69, 106
            legal actions, 43
            management of forest reserves, 41-42, 61, 75, 76-77
            patrolling forest reserves, 42, 54, 221
            rangers and job description, 54-55 , 80
            telephone lines, 55
            timber sales, 53-54, 92
            transfer of forest reserves to USDA, 249. *See also* Forest Transfer Act of 1905.
        Geological Survey (USGS), 44, 47, 61, 66-67, 110
        Grazing Service, 282
        Indian Service, 243, 256
        Reclamation Service, 129
U.S. Federal Bureau of Investigation (FBI), 283, 317
U.S. Fish and Wildlife Service (FWS), 319, 392, 395, 511, 527
U.S. Office of Civil Defense (OCD), 311-312
U.S. Supreme Court, 141, 260-261, 481-482, 499, 521
U.S. Vanadium Corporation, 306
Union Oil Company, 165
University of California, Berkeley, 243, 256 *photo*, 257, 376, 378, 451
utilitarianism. *See* conservation and conservation movement: utilitarian conservation.

V

Ventura County, 15, 28, 48, 288, 328, 364
Vietnam War, 429, 442, 489

W

Walt Disney Productions and Studios, 313, 423-425, 447, 450, 481-483. *See also* recreation management,
        Mineral King winter sports development controversy.
War Resisters League, 319
Warner Mountains Forest Reserve. *See also* Modoc National Forest.
    cattle/sheep conflicts, 51
    establishment of, 50-51, 65

incorporated into Modoc National Forest, 88
watershed management (District/Region 5), 15, 90, 203. *See also* individual national forests and reserves, and related individual legislative acts.
    All-American Canal, 163
    area planning and, 342
    brushfields (chaparral), protection of, 211
    California Water Plan (CWP), 375, 395, 403, 417
    Central Valley Project (CVP), 234, 274
    Colorado Aqueduct, 164
    Colorado River Compact of 1922, 211
    conservation, 284
    cooperative management and research, 210-211, 410
    damage to and abuse, 11, 15, 34, 41, 56
    drought of 1919, 196
    droughts of 1862-1864, 14, 27
    droughts of 1923-1924, 196
    droughts of 1976-1977, 471, 474-477
    education and, 333
    erosion control and soils, 36, 273, 275, 342, 370, 486
    fire control and, 341
    floods and flood control, 12, 18, 34, 36, 48, 197, 274, 369-370, 418
    Forest Problems of California and, 291-292
    Forest Ranger's Catechism and, 223
    Great Flood of 1862, 13
    Great Flood of 1938, 274, 276
    Hetch Hetchy Valley controversy, 102, 103, 103 *photo*, 104, 112, 150, 163, 285, 421, 449
    hydroelectric development, 13, 36, 50, 101-103, 116, 129, 164, 203, 209, 224, 249, 273, 284, 369, 417-418
    intensive management of, 369-370
    inventory and survey of watersheds, 275, 370, 410
    La Crescenta-Montrose Flood of 1933, 274
    Lake Tahoe area, 264
    leadership, 385
    Los Angeles River Flood Control Project, 275
    multiple use management, 257-258, 273, 370
    municipal water needs, 410
    Munns flood study, 197
    Owens or Los Angeles Aqueduct, 86-87, 104-105, 112-113, 163, 274
    protection of, 34-35, 37, 39, 48-49, 51, 65, 77, 85, 94, 224, 343, 370
    range, conflicts with, 269
    recreation and, 151, 418
    restoration of, 372
    Sacramento Flood of 1940, 274
    Sacramento Municipal Utility District (SMUD), 395
    San Gorgonio Primitive Area controversy, 366-369, 374
    St. Francis Dam collapse, 209
    statutory goal, 101
    supply and shortages, 34
    timber, conflicts with, 508
    urban watershed protection, 105

water districts, 18
water policies, 197
water storage, 369
wilderness, conflicts with, 216-217
Weeks Act of 1911, 110-111, 139, 183, 262. *See also* Clarke-McNary Act of 1924.
    National Forest Reservation Commission, 110, 263-264
    25 percent fund, 111
Western Forestry and Conservation Association (WFCA), 382, 387, 424
Western Lumber Manufacturers, Inc., 339
Western Mining Association, 339
Western Pine Association (WPA), 292, 347
Western Timber Association (WTA), 408, 463, 485, 506
Western Wood Products Association (WWPA), 382, 441
Weyerhaeuser Timber Company, 110
Whiskeytown-Shasta-Trinity National Recreation Area, 436, 463
Wild and Scenic Rivers Act of 1968, 444, 516
    Middle Fork Feather Wild and Scenic River, 463
Wild Horses and Burros Protection Act of 1971, 512, 512 *photo*
Wilderness Act of 1964, 374, 412-414, 444-445, 477, 479, 516
wilderness and wilderness movement, 33-34, 211, 413. *See also* wilderness management.
    definition of, 478
    loss of, 200
    preservationist concept and protection of, 200, 231, 363
wilderness management. *See also* individual national forests and reserves, and related individual legislative acts. *See also* Wilderness Act of 1964.
    establishment of wilderness or primitive areas, 217-219
    handicap access, 509
    Kern Plateau multiple use plan, 389
    Mineral King winter sports development controversy, 423
    multiple use management, 401, 477
    National Wilderness Preservation System, 413, 446, 478, 480, 510
    permit system, 481
    policy, 366
    primitive areas, 203, 217-219, 225, 231, 286, 454
    reclassification of primitive areas to wilderness areas, 411-412, 444, 479
    Regulation L-20 (1929) and amendments, 217-219, 286, 365
    road construction and, 107, 453
    Roadless Area Review and Evaluation (RARE I and II), 413, 477-479, 492, 508-510
    San Gorgonio Primitive Area controversy, 366-368, 374, 389, 413
    "U" Regulations, 285-286
    wild areas, 286
    wilderness area system and criteria, 215-216
    wilderness areas, 285, 454
Wilderness Society, 368, 483, 527
    Kern Plateau multiple use plan, 389, 413
    San Gorgonio Primitive Area controversy, 413
    Wilderness Act of 1964, 412, 414
wildlife, 15, 160, 374
    anadromous fish, 12-13, 58, 511

    big game (antelope, bear, bighorn sheep, deer, elk), 13, 58, 226-227, 395, 454, 511
    depletion, damage to, and abuse, 11, 13, 58-61
    Gold Rush, and, 13, 58
wildlife interest groups, 400
wildlife management (District/Region 5), 98, 151-152, 286. *See also* individual national forests and reserves, and related individual legislative acts. *See also* Threatened Endangered Species Act of 1973.
    area planning and, 343
    bighorn sheep, 287, 372, 475, 511, 514
    California condor, 232, 287-289, 364-365, 374, 443, 475, 513
    deer management programs, 151-152, 331, 372, 394, 410
    droughts of 1976-1977, impact of, 475-476
    endangered species, 227, 232, 288, 374, 443, 476, 511-514
    Forest Ranger's Catechism and, 223
    habitat protection and improvement, 286, 370-371, 374, 395, 410, 475-476, 486, 511
    multiple use management and, 257-258, 341, 395, 419
    protection of, 232
    public interest in, 332
    recognition of wildlife values, 371
    spotted owl controversy, 476, 514-515, 526, 527-528
    state game refuges and hatcheries, 151-152, 160, 226, 238, 370
    timber, conflicts with, 508
World War I, 124, 142, 145-146, 186, 259
    armistice, 157
    declaration, 116-117
    insect control work, 135
World War II, 160, 264, 267, 281, 286, 294-295, 297, 325, 442, 489
    Aircraft Warning Service (AWS), 308, 317-318
    air-ground rescue work, 317
    assistance to farm families, 303-304
    Civil Air Patrol (CAP),
    Forest Firefighters Service (FFFS), 312
    guayule rubber project, 295, 306, 306 *photo*, 307-308, 319
    Japanese balloon fire bombs, 308-310
    sabotage fears, 317
    War Advertising Council, 313
    War Production Board (WPB), 303
    War Relocation Authority (WRA), 301, 319
Wright Act of 1887, 17-18

Y

Yale University Forestry School, 56, 67, 96, 170-171, 201, 203, 381
Yellowstone National Park, 33, 44-45
Yellowstone Park Timberland Reserve, 36-37, 322
Yosemite National Park, 33, 37, 40, 42, 103, 105, 112, 150, 213, 215, 219, 329. *See also* Yosemite Valley State Park.
Yosemite Valley State Park, 32. *See also* Yosemite National Park.
    creation 19
    mismanagement and return to federal status, 21, 33
Yuba Forest Reserve. *See also* Tahoe Forest Reserve.
    consolidated into Tahoe Forest Reserve, 73
    establishment of, 73

Z

Zaca Lake Forest Reserve. *See also* Santa Barbara Forest Reserve, and Los Padres National Forest.
    consolidated into Santa Barbara Forest Reserve, 65
    establishment of, 48